DIENES IN THE DIELS–ALDER REACTION

to Graziella and our children
—F. F.

to Brunella, Angela, and Paolo
—A. T.

***Library of Congress Cataloging in Publication Data*:**

Fringuelli, Francesco.
  Dienes in the Diels–Alder reaction/Francesco Fringuelli and Aldo Taticchi.

  p. cm.
  "A Wiley-Interscience publication."
  Includes bibliographical references.
  ISBN 0-471-85549-9
  1. Diolefins. 2. Diels–Alder reaction. I. Taticchi, Aldo. II. Title.

QD305.H7F69 1990 89-16738
547′.412—dc20 CIP

Printed in the United States of America

10 9 8 7 6 5 4 3 2 1

DIENES IN THE DIELS–ALDER REACTION

FRANCESCO FRINGUELLI
ALDO TATICCHI

University of Perugia
Perugia, Italy

A WILEY-INTERSCIENCE PUBLICATION
John Wiley & Sons, Inc.
NEW YORK / CHICHESTER / BRISBANE / TORONTO / SINGAPORE

PREFACE

In light of increasing interest in the theoretical, mechanistic, and synthetic aspects of the Diels–Alder reaction, we felt, based on our personal experiences, that a book dealing with principles, recent developments, and synthetic applications of this reaction could be useful.

In consideration of the books and reviews already published on the subject, we decided to focus our attention on one of the reactants of the intermolecular Diels–Alder reaction, the diene, with the purpose of giving a picture of the potential use of different types of dienes in synthetic organic chemistry.

The first chapter describes pedagogically the general principles of the reaction and reports some recent methodologies not generally described elsewhere. Chapters 2 to 6 deal with various classes of dienes. Each chapter provides comments on the salient features (mechanism, reactivity, catalysis, regiochemistry, stereochemistry, etc.) of the cycloadditions, tables of Diels–Alder reactions for many kinds of dienes, and literature references. The comments are not exhaustive and emphasize some particularly interesting properties of the class of dienes under examination. A large part of the literature from 1978–1987 is covered; some papers published in 1988 are also noted. The introduction provides details on how the reader can use the tables to obtain the various types of information that they contain.

The book is directed to undergraduate and graduate students and to academic and industrial researchers concerned especially with organic synthesis.

FRANCESCO FRINGUELLI
ALDO TATICCHI

Perugia, Italy
February 1990

CONTENTS

INTRODUCTION

The intermolecular Diels–Alder reactions of most of the dienes that appeared in the literature from 1978 to 1987 have been tabulated. In addition to the reactants, the tables report the yield of the reaction, the literature reference(s), and indicate the reactants generated in situ, those optically active, the optically active catalysts, if kinetic data are quoted in the original paper, and if the work is theoretical in character. To this end suitable symbols have been used.

The dienes have been ordered in five main classes based on both the structural features of the molecule and the diene moiety involved in the cycloaddition reaction. The dienophiles have been ordered in 10 principal groups (A through J) and subgroups (A1, A2, . . . , J1, J2) on the basis of the atoms and their substituents which constitute the dienophilic center. This "classification" has been used to collect dienophiles that are formally different but have the same dienophilic center. Readers can therefore find in the tables of Diels–Alder reactions of various kinds of dienes the desired dienophile and the cycloaddition reactions of interest to them. The names of the dienophiles in the tables are those reported in the original papers; some generalizations have been made. Only the yields of the reactions expressly quoted in the literature are reported in the tables. The usual abbreviations have been employed.

CLASSIFICATIONS

DIENES

Class	Example
Open Chain	
a. 1,3-Butadienes	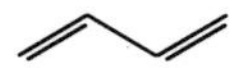
b. Hetero-1,3-butadienes	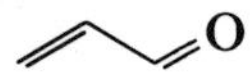
Outer Ring	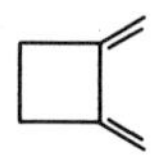
Inner–Outer Ring	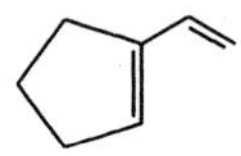
Across Rings	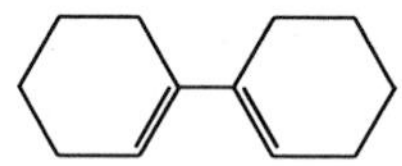
Inner Ring	
a. Carbocyclic dienes	
b. Aromatic hydrocarbons	
c. Heterocyclic dienes	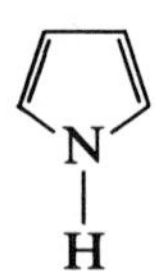

ABBREVIATIONS

DIENOPHILES

AC	Acrolein
ACN	Acrylonitrile
BQ	1,4-Benzoquinone

DMAD	Dimethylacetylenedicarboxylate
DMF	Dimethylfumarate
DMM	Dimethylmaleate
MA	Maleic anhydride
MAC	Methyl acrylate
MP	Methylpropiolate
MTA	Methacrolein
MVK	Methyl vinyl ketone
NEM	*N*-Ethylmaleimide
NMM	*N*-Methylmaleimide
NMTAD	*N*-Methyl-1,2,4-triazoline-3,5-dione
NPM	*N*-Phenylmaleimide
NPTAD	*N*-Phenyl-1,2,4-triazoline-3,5-dione
NQ	1,4-Naphthoquinone
TCNE	Tetracyanoethene

General

Ac	Acetyl
Ar	Aryl
Bn	Benzyl
Bs	Brosyl, *p*—$BrC_6H_4SO_2$—
Bu	*n*-Butyl
Bu^i	Isobutyl
Bu^t	*t*-Butyl
Et	Ethyl
Fu	Furyl
Glu	D(+)-Glucose
Me	Methyl
(P)	Polymer
Ph	Phenyl
Pr	*n*-Propyl
Pr^i	Isopropyl
Py	Pyridyl
R	Alkyl if not otherwise specified
t	*trans*
Th	Thienyl
THF	Tetrahydrofuran
THP	Tetrahydropyran
Ts	Tosyl, *p*-$MeC_6H_4SO_2$—
W	H, OH, OR, R, Ar, NR_2, Cl
X, Y	Heteroatoms

DIENOPHILIC CENTER

A $>C{=}C<$

A1	$CH_2{=}CH{-}COW$	A4	$-CH{=}\overset{\mid}{C}{-}COW$
A2	$CH_2{=}\overset{\mid}{C}{-}COW$	A5	$>C{=}CH{-}COW$
A3	$-CH{=}CH{-}COW$	A6	$>C{=}\overset{\mid}{C}{-}COW$

A 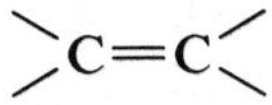

A7 WOC—CH=CH—COW

A8 $-CH{=}CH_2$

A9 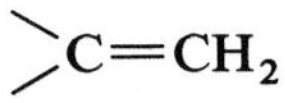

A10 —CH=CH—

A11 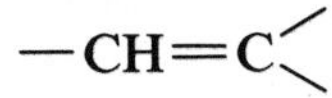

A12 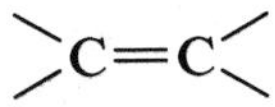

A13 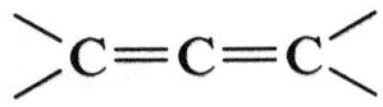

A14 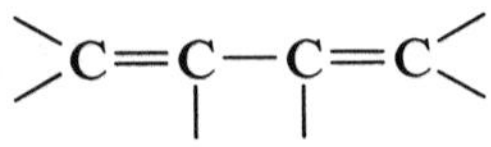

B 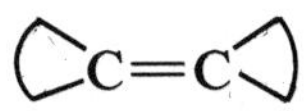

B1 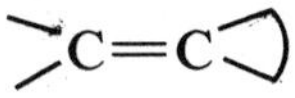

B2 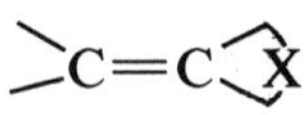

B3 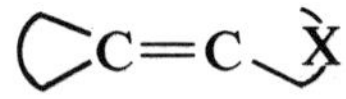

C 

C1 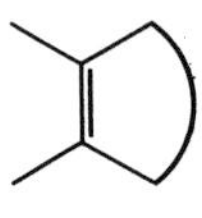

C2 

C3 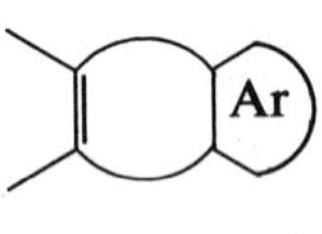

C4 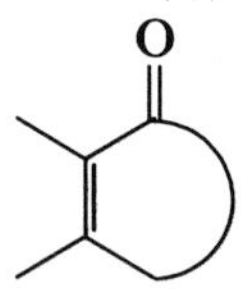

C5 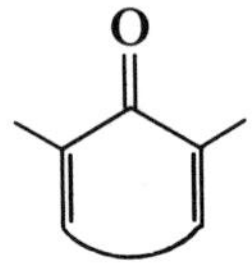

C6 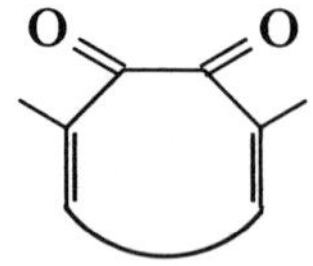

C 

C7 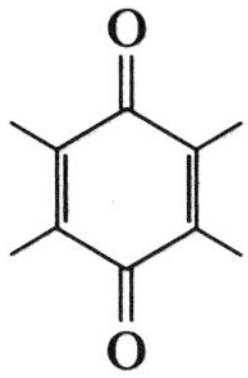

C9 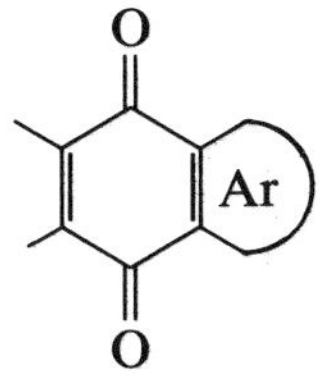 or

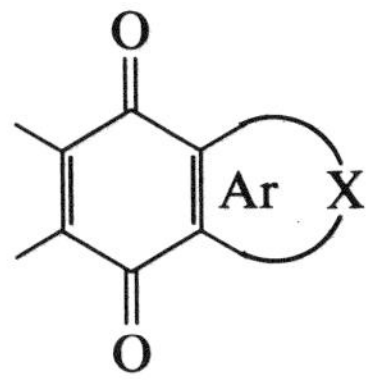

C8 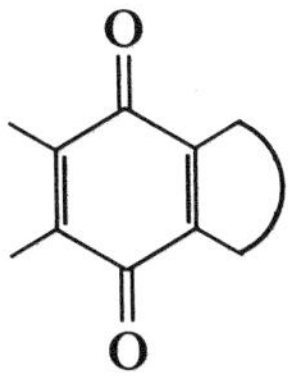 or

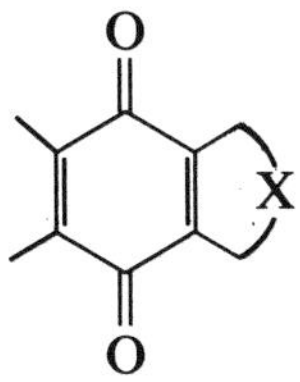

C10 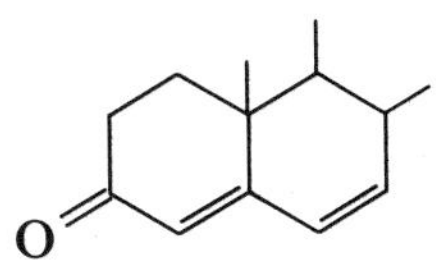

D   

D1 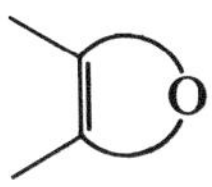

D5 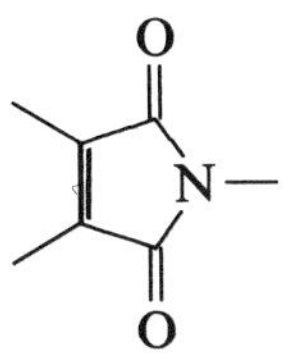

D2 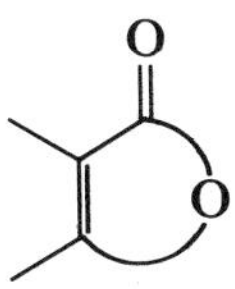

D6 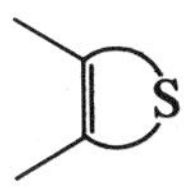

D3 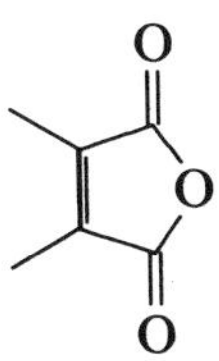

D7 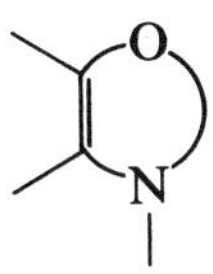

D4 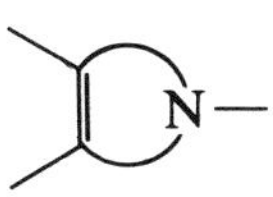

D8 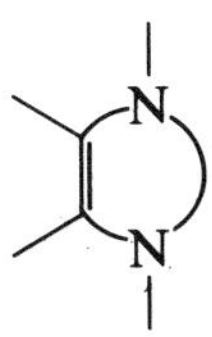

D  (*Continued*)

D9 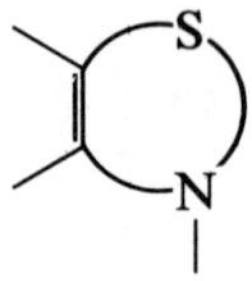

D10 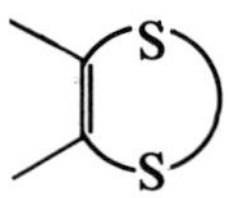

D11 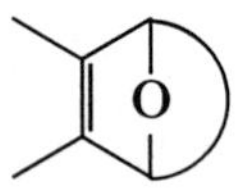

D12 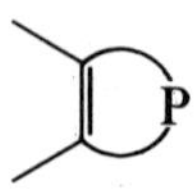

D13 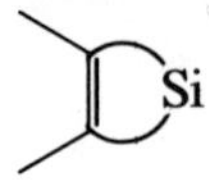

D14 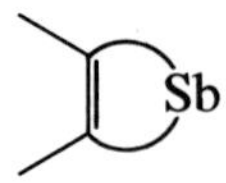

D15 

E —C≡C—

E1	CH≡C—COW	E5	—C≡C—
E2	—C≡C—COW	E6	
E3	WOC—C≡C—COW	E7	
E4	CH≡C—		

F 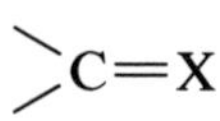

F1 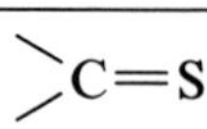

F2 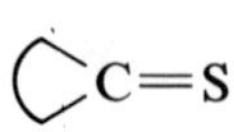

F3 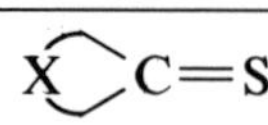

F4 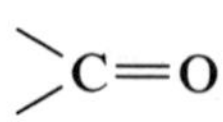

F >C=X (*Continued*)

F5	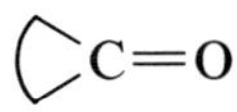	F8	>C=P—
F6	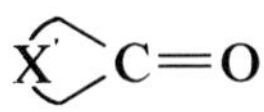	F9	>C=Si<
F7	>C=N—	F10	>C=Se

G 

G1	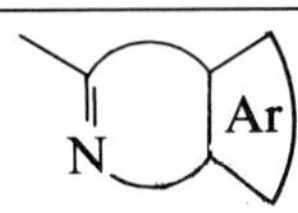	G3	
G2		G4	

H —C≡X X≡X

H1	—C≡N	H3	—C≡P
H2	$-\overset{+}{N}\equiv N$		

I X=X

I1	—N=N—	I4	
I2	S=S	I5	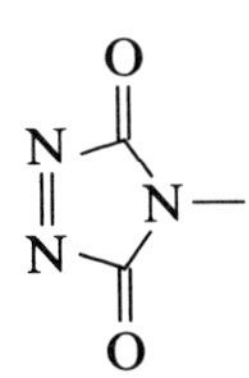
I3	O=O	I6	—Ge=Ge—

J	X=Y		
J1	—N=O	J2	—N=S

SYMBOLS

[ ]	Compound generated in situ
( )	Theoretical paper
(*)	Optically active compound
(k)	Kinetic data are reported
Cat.(*)	Optically active catalyst

1

DIELS–ALDER REACTION

1.1 INTRODUCTION

The Diels–Alder reaction is one of the most fundamental and useful reactions in synthetic organic chemistry. It is a widely used method for forming carbon–carbon, carbon–heteroatom, and heteroatom–heteroatom bonds. It is a $[_{\pi}4_s + _{\pi}2_s]$ cycloaddition in which a conjugated diene component undergoes a stereospecific addition reaction with another component, called a dienophile; the reaction product is called an adduct (Eq. 1.1).

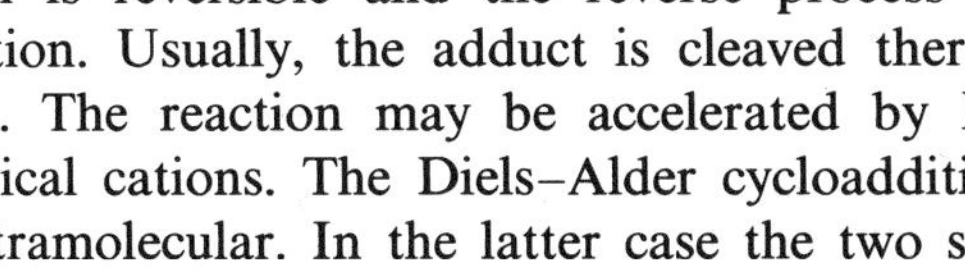

(1.1)

The cycloaddition is reversible and the reverse process is called a retro Diels–Alder reaction. Usually, the adduct is cleaved thermally to give the starting materials. The reaction may be accelerated by Lewis acids, high pressure, and radical cations. The Diels–Alder cycloaddition may be intermolecular and intramolecular. In the latter case the two structural features, diene and dienophile, are incorporated in the same compound (Eq. 1.2). Otto Diels and Kurt Alder shared the 1950 Nobel Prize in Chemistry for this discovery. In the following sections we deal with the various aspects of the reaction in some detail.

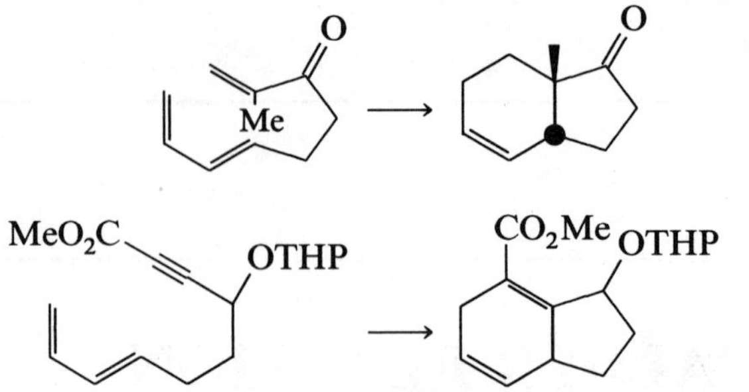

(1.2)

1.2 DIENE

The diene is the 4π component of the Diels–Alder reaction. Although the number and variety of dienes is very high, they may be classified as follows[1]:

(a) Open Chain

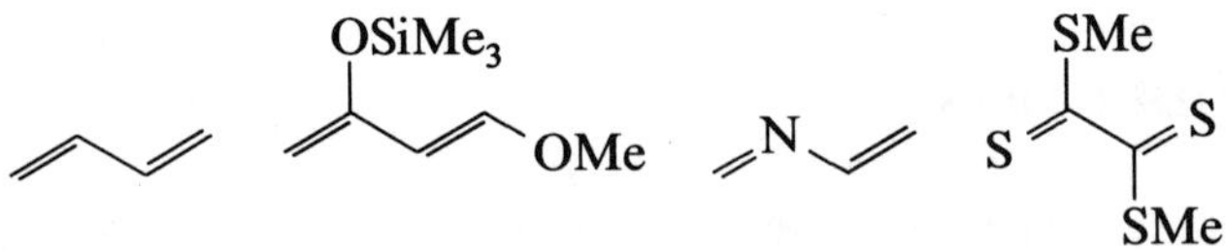

(b) Outer Ring

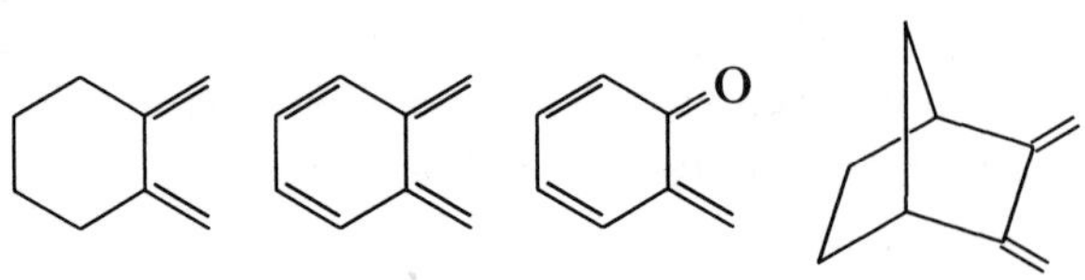

(c) Inner–Outer Ring

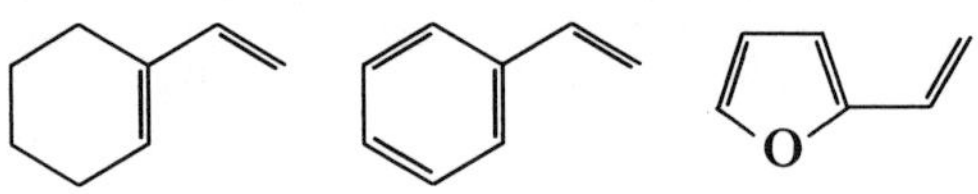

(d) Across Rings

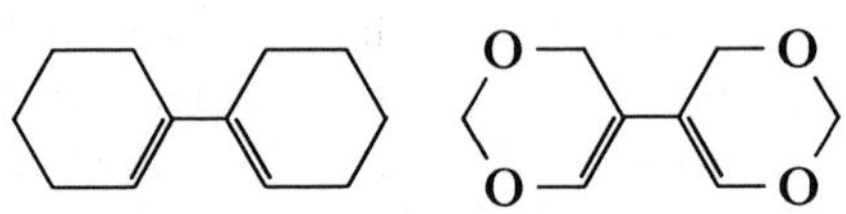

(e) Inner Ring

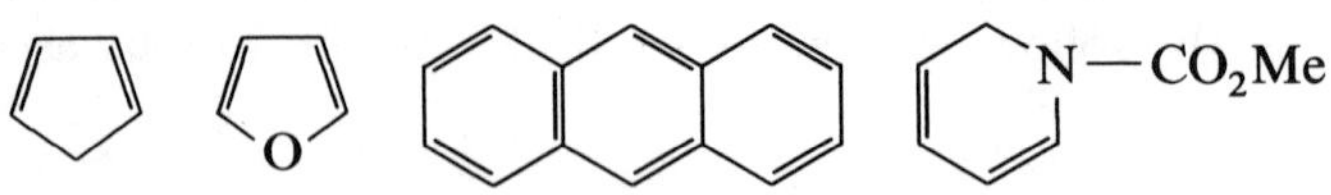

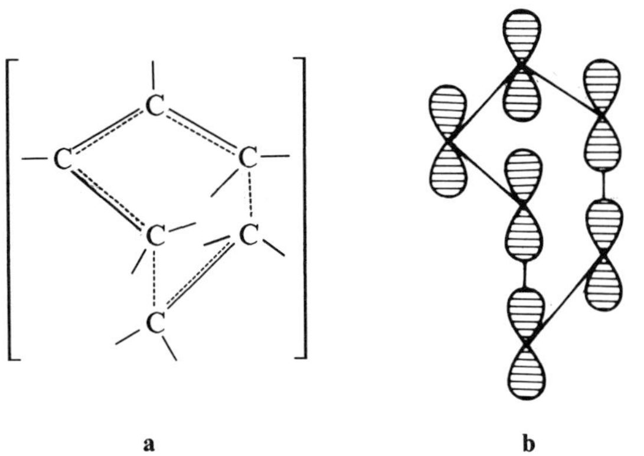

Figure 1.1 Transition state of Diels–Alder reaction.

As diene and dienophile interact in the transition state as depicted in Fig. 1.1*a*, the diene can react only in the *cisoid* conformation.[2] This condition is essential in order to obtain overlapping of the *p*-orbitals in the diene and dienophile (Fig. 1.1*b*). Furthermore, only this diene conformation has the geometry required for the formation of a double bond in a six-membered ring. This view is strongly supported by experimental evidence. Whereas cyclic dienes, constrained by the ring to a *transoid* conformation (Eq. 1.3), do not give the Diels–Alder reaction, *cisoid* cyclic dienes (Eq. 1.4) are reactive.

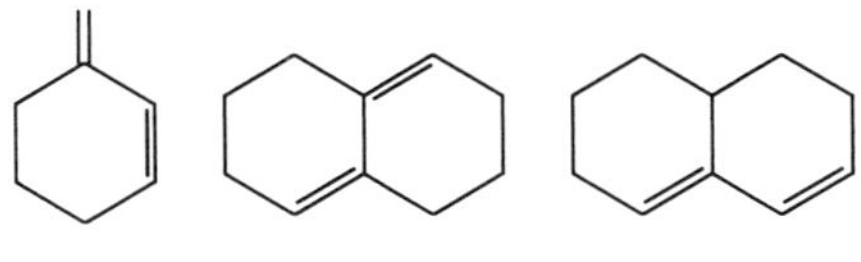

(1.3)

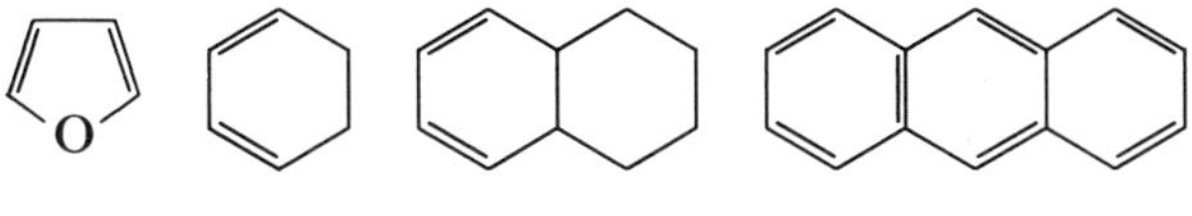

(1.4)

Open-chain dienes exist in two conformations,[3] *cisoid* and *transoid*, and the position of the equilibrium depends on the ground-state energy differences between the two conformers. Usually, open-chain dienes assume the more stable *transoid* conformation. For example, the *transoid* form of 1,3-butadiene is at least 2 kcal/mol more stable than the *cisoid* because of the absence of steric interference of hydrogen atoms in the former conformation. Open-chain dienes that have a strained *cisoid* conformation are unreactive or show low reactivity. For example, whereas (E)-piperylene reacts with MA in benzene, at reflux temperature, its (Z)-isomer (Eq. 1.5) requires a higher reaction temperature, a longer reaction time, and a catalyst.[4,5] The methyl substituent in the (Z)-isomer destabilizes both the *cisoid* conformation of the diene (Eq. 1.6) and the transition state by steric effects, therefore causing the decrease in reactivity.

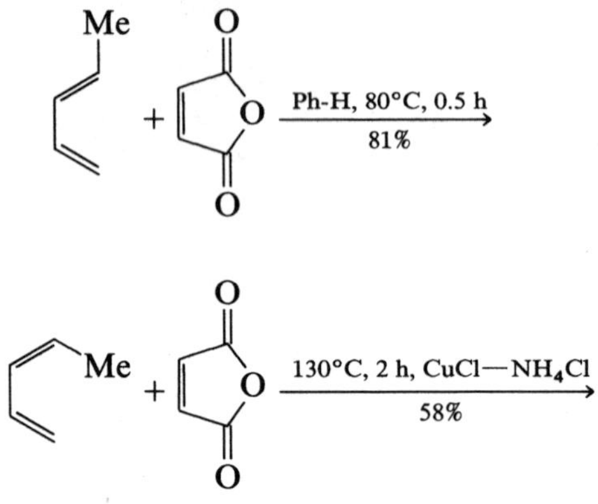

(1.5)

The substituent also influences the diene reactivity by electronic effects. Whereas electron-donating substituents determine an increase of reactivity in a "normal" demand Diels–Alder reaction (see Sec. 1.5), electron-attracting groups cause the opposite effect.

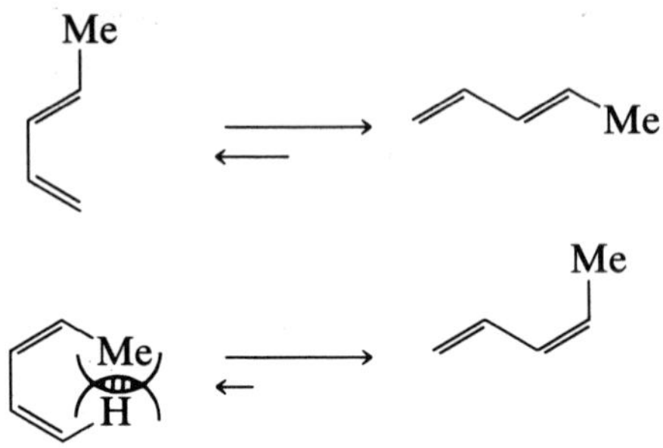

(1.6)

The reactivity order of some butadienes in the reactions with MA and TCNE is the following[2b,6]: (*E*)-1-methoxy-1,3-butadiene > 2-methoxy-1,3-butadiene > 2,3-dimethyl-1,3-butadiene > (*E*)-1-methyl-1,3-butadiene > 2-methyl-1,3-butadiene > 1,3-butadiene.

1.3 DIENOPHILE

The dienophile, the 2π reacting partner in a Diels–Alder reaction, is a molecule containing a double or triple bond. The reaction of ethene with 1,3-butadiene (Eq. 1.7) requires strong experimental conditions and affords poor yield of cyclohexene because of the low reactivity of dienophile.

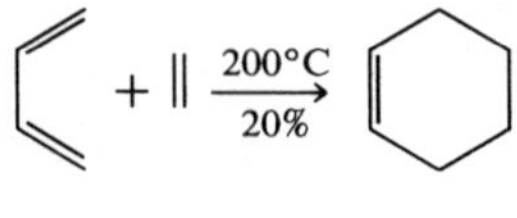

(1.7)

TABLE 1.1 Relative Rate Constants of the Reactions of Cyclopentadiene with Cyanoethenes

Dienophile	k_{rel} (20°C; Dioxane)
TCNE	43×10^6
Tricyanoethene	48×10^4
1,1-Dicyanoethene	45.5×10^3
(*E*)-1,2-Dicyanoethene	91
(*Z*)-1,2-Dicyanoethene	81
Acrylonitrile (ACN)	1

Electron-attracting groups attached to ethene increase its reactivity markedly (Eq. 1.8) toward "normal" electron demand reactions.

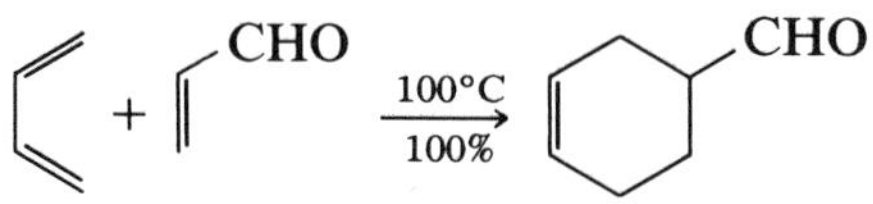

(1.8)

Carbon–carbon double or triple bonds are activated in "normal" demand reactions (Sec. 1.5) by electron-withdrawing substituents. A strong increase in reactivity is observed by introducing more than one electron-attracting group in an ethylenic double bond as shown by the values of the relative rate constants of the reactions of cyclopentadiene with cyanoethenes[7] (Table 1.1).

Classification of the dienophiles is difficult because of their large number and variety. Nevertheless, it may be useful to report some examples of the most common types of dienophiles used in synthetic organic chemistry.

Acrolein (AC)

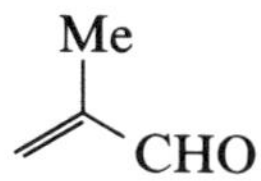

Methacrolein (MTA)

Methyl acrylate (MAC)

Methyl vinyl ketone (MVK)

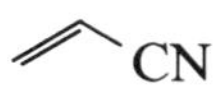

Acrylonitrile (ACN)

$(CN)_2C{=}C(CN)_2$

Tetracyanoethene (TCNE)

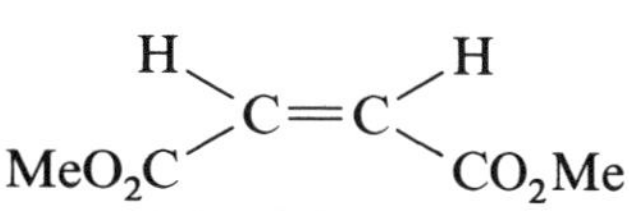

Dimethyl maleate (DMM)

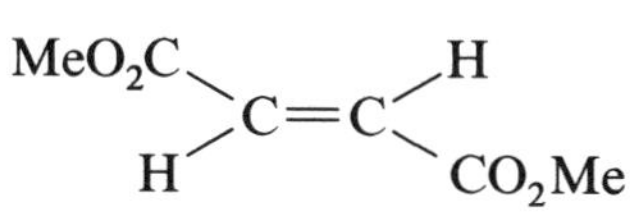

Dimethyl fumarate (DMF)

$CH_2{=}C{=}CHMe$

1,2-Butadiene

1,3-Butadiene

$CH{\equiv}CO_2Me$

Methyl propiolate

(MP)

$MeO_2C{-}C{\equiv}C{-}CO_2Me$

Dimethylacetylene dicarboxylate (DMAD)

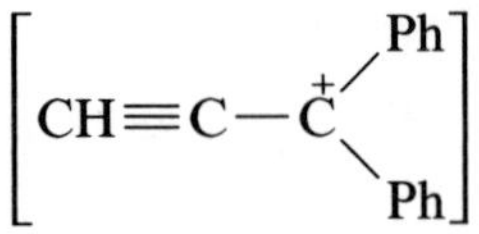

Diphenylpropargyl cation

$Me_2C{=}S$

Tioaceton

$Ph{-}N{=}O$

Nitrosobenzene

$Ar{-}N{=}N{-}CN$

Arylazocyanide

$^1O_2$

Singlet oxygen

$^1S_2$

Singlet sulfur

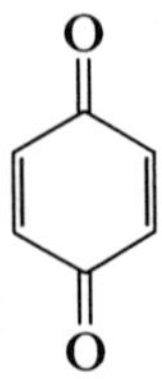

1,4-Benzoquinone (BQ)

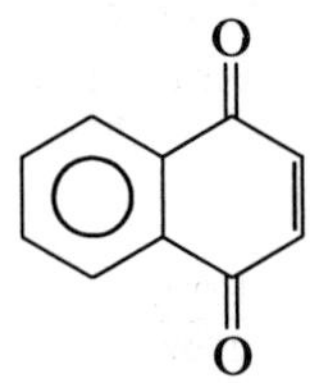

1,4-Naphthoquinone (NQ)

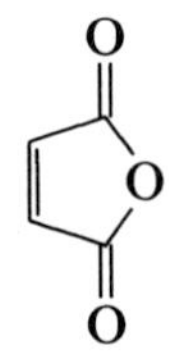

Maleic anhydride (MA)

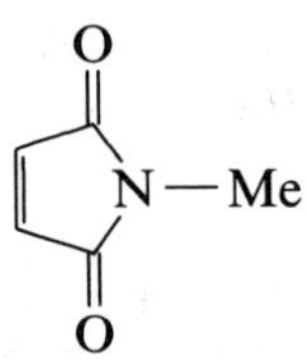

N-Methylmaleimide (NMM)

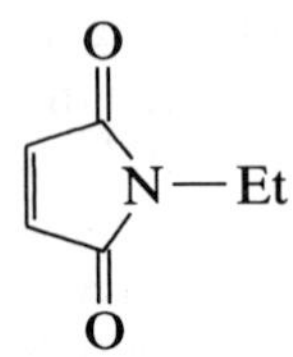

N-Ethylmaleimide (NEM)

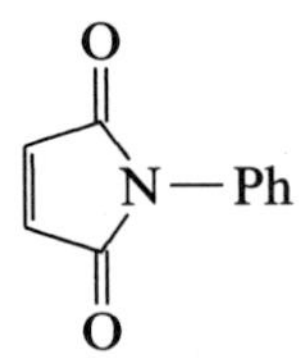

N-Phenylmaleimide (NPM)

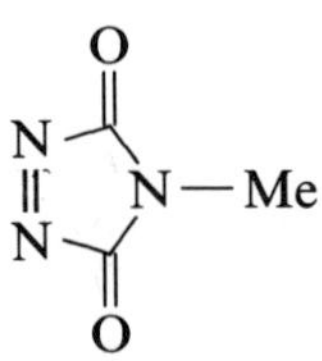

N-Methyl-1,2,4-triazoline-3,5-dione (NMTAD)

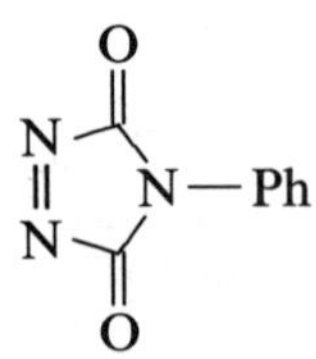

N-Phenyl-1,2,4-triazoline-3,5-dione (NPTAD)

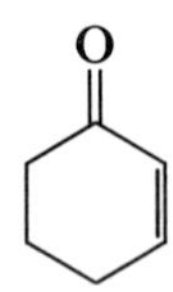

Cyclohexen-2-one

Cyclopentadiene

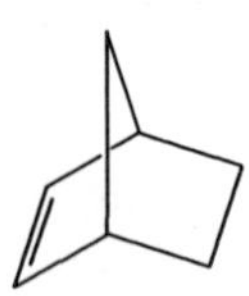

Norbornene

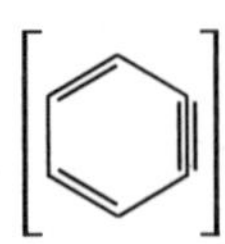

Benzyne

Allyl and propargyl cations, as well as arynes, are highly reactive dienophiles and because of their unstability are trapped in situ (Eq. 1.9).

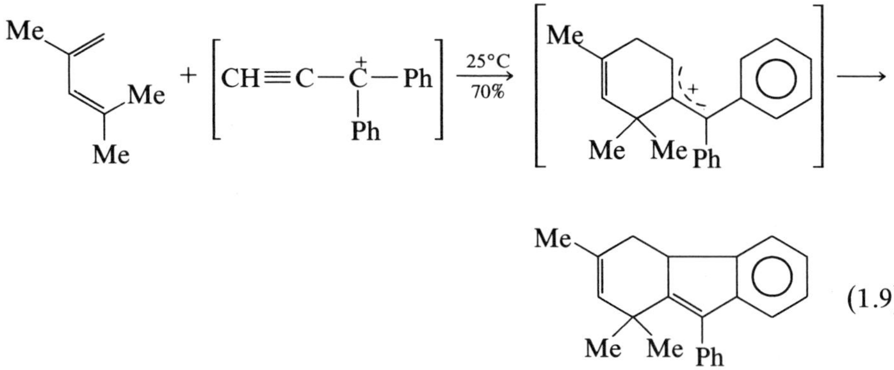

(1.9)

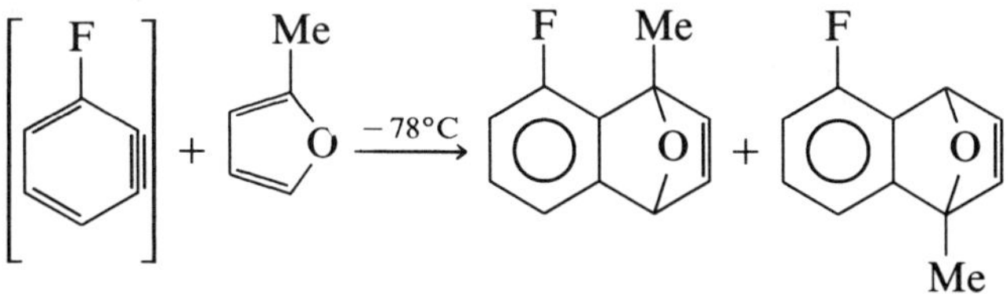

1.4 MECHANISM

The Diels–Alder reaction is bimolecular, the transition state involving one molecule of diene and one of dienophile. Extensive investigations[8–11] on the reaction mechanism have emphasized that all the Diels–Alder cycloadditions cannot be interpreted in terms of only one mechanism. However, most of the thermal cycloadditions can be described by a symmetry-allowed one-step mechanism (Eq. 1.10).

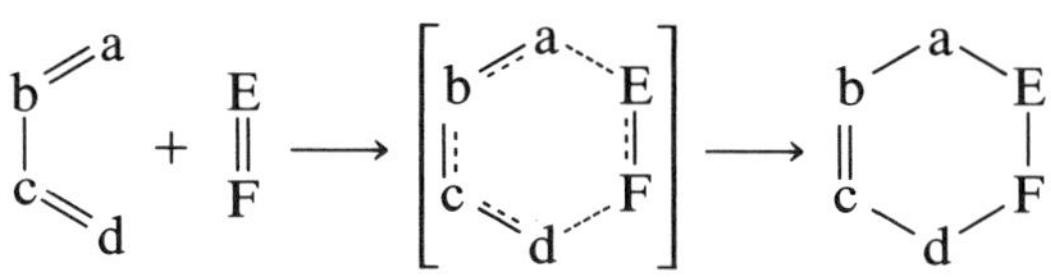

(1.10)

In some cases the reaction occurs by a two-step mechanism (Eq. 1.11) involving a biradical or zwitterion intermediate.

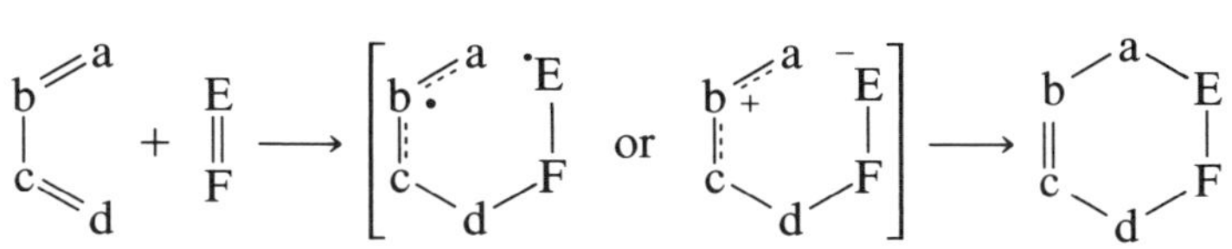

(1.11)

The biradical intermediate has two unpaired electrons with opposite spins. According to the concerted mechanism, diene and dienophile approach each other in parallel planes and interact in the transition state in the way depicted in Fig. 1.1.

Chemical evidence that supports the concerted mechanism is:

1. The *syn* stereospecificity of the reaction. A two-step mechanism seems to be inadequate to rationalize this property, although a priori it does not exclude a stereospecific reaction. Indeed, if the radical or zwitterion intermediate changes into the products faster than it rotates about single bonds, the two-step mechanism can account for the *syn* stereospecificity.
2. The low solvent effects on the reaction rate. Indeed, the rate of the Diels–Alder reaction is essentially independent of solvent polarity. This ought to rule out a zwitterion intermediate.
3. Kinetic isotope effect investigations[12] on the secondary α-deuterium isotope effects in a retro Diels–Alder reaction of 9,10-dihydro-9,10-ethanoanthracene and its bridge deuteriated analogs. These allow one to conclude in favor of a concerted mechanism. In view of the principle of microscopic reversibility, the mechanism of the forward and reverse reaction has the same rate-determining transition state.
4. The large negative values of both activation entropy and activation volume. These give further support to a concerted mechanism with a geometrically highly oriented transition state.

The synchronous or nonsynchronous nature of the transition state has been the source of great debate.[8] A synchronous mechanism is a concerted mechanism in which bond-breaking and bond-forming processes are simultaneous and have proceeded to almost the same extent in the transition state.[13] A transition state in which the formation of one σ-bond proceeds entirely or partially in advance of the second one is called unsymmetrical.[10e] Experimental as well as theoretical evidence in favor of a concerted synchronous mechanism for the prototype Diels–Alder reaction of 1,3-butadiene with ethene has been reported.[10b]

A nonconcerted mechanism has been proposed by Jensen and Foote[14] to explain the results of the reactions of 4-phenyl-1,2,4-triazoline-3,5-dione with some substituted butadienes [(E, E)-, (Z, E)-, and (Z, Z)-2,4-hexadiene]. They hypothesized that for dienes with at least one methyl group in the Z-configuration, the reaction proceeds through one or more intermediates, which can be trapped when the reactions are performed in methanol. Parallel results that give further support to this mechanistic hypothesis were obtained by Clennan[15] in the reactions of NPTAD and *tert*-butoxylated butadienes.

According to frontier molecular orbital theory (Sec. 1.5), the Diels–Alder reaction is controlled by the "in-phase" interaction of the two HOMO–LUMO molecular orbitals closest in energy. This condition is achieved only when the

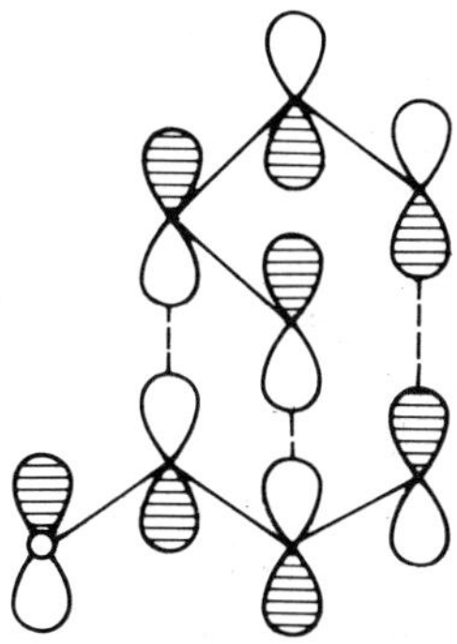

Figure 1.2 HOMO–LUMO orbital arrangement in the *endo* transition state for the butadiene–AC interaction.

two MOs interact suprafacially on each component. For instance, in the case of cycloaddition of AC with 1,3-butadiene, the "in-phase" dominating interaction occurs between HOMO diene–LUMO dienophile suprafacially (Fig. 1.2).

1.5 REACTIVITY AND FMO THEORY

Frontier molecular orbital theory has been used successfully to explain reactivity and selectivity phenomena in cycloaddition reactions.[8,16–21] Diels–Alder reactions have been classified by Sustmann[22] as three general types, depending on the three possible arrangements of HOMO and LUMO MOs (Fig. 1.3).

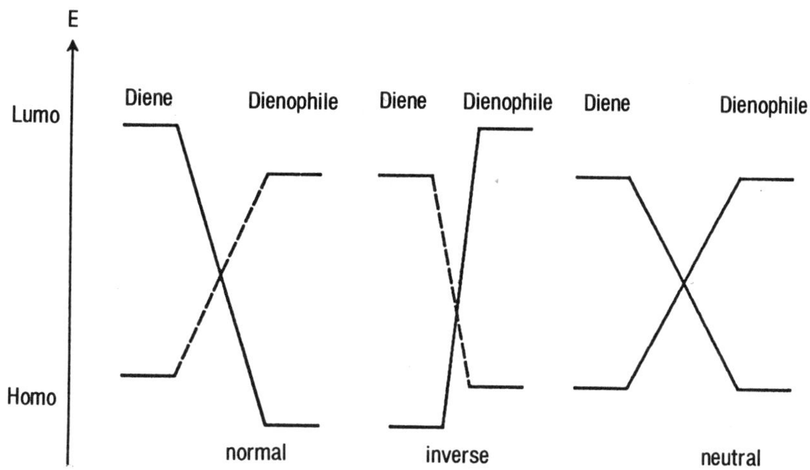

Figure 1.3 HOMO–LUMO orbital arrangements for the Diels–Alder reaction.

These three situations are termed "normal" electron demand, "inverse" electron demand, and "neutral" electron demand Diels–Alder reactions.

In the normal electron demand reaction, the dominating orbital interaction (corresponding to the lowest HOMO–LUMO energy separation) is the HOMO diene–LUMO dienophile. The inverse electron demand reaction is controlled by LUMO diene–HOMO dienophile interaction. In the neutral electron demand reaction, neither frontier orbital interaction dominates.

The reactivity of a Diels–Alder reaction depends on the lowest HOMO–LUMO energy separation that can be achieved by the reacting partners. All factors that lower the HOMO–LUMO distance increase the reaction rate because of a smaller energy difference between interacting FMOs, making the contribution to the stabilization of the transition state greater. Reactivity is explained on the basis of the effects that substituents exercise on the MOs of diene and dienophile. Electron-withdrawing substituents lower the energy of both HOMO and LUMO molecular orbitals, whereas electron-donating substituents raise their energies (Fig. 1.4). In the case of the normal electron demand reaction (HOMO diene controlled), electron-donating substituents in the diene accelerate the reaction, whereas electron-attracting groups retard it (Fig. 1.5). As reactivity increases when the energy difference between interacting FMOs decreases, the typical normal Diels–Alder reaction is one between an electron-rich diene with an electron-deficient dienophile. Sustmann has found a correlation between HOMO diene–LUMO dienophile energy separations and the logarithm of the rate constants of the cycloadditions of MA and TCNE with substituted butadienes.[23] This correlation gives quantitative support to the theoretical reactivity model. Most Diels–Alder reactions are of normal electron demand type.

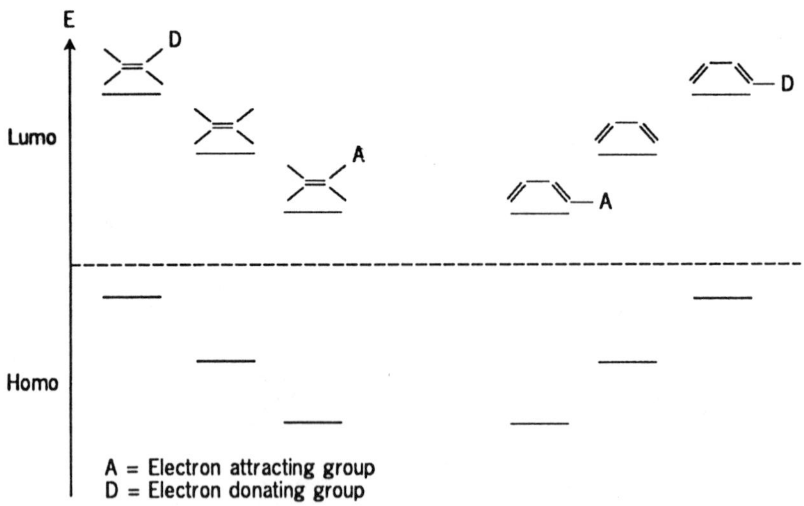

Figure 1.4 Electronic substituent effects on orbital energies.

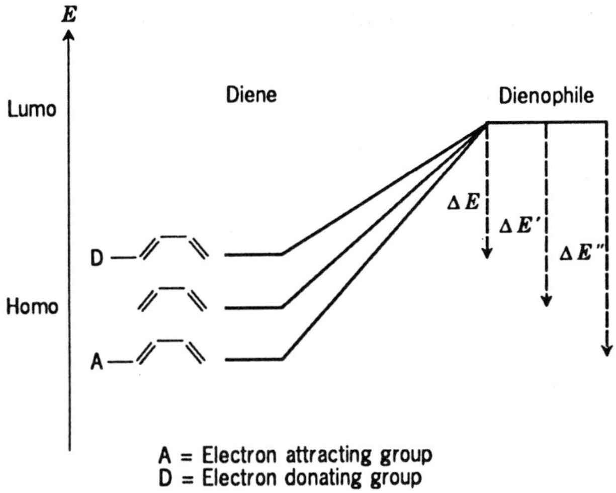

Figure 1.5 HOMO–LUMO arrangements for a normal Diels–Alder reaction.

In the case of inverse cycloaddition (LUMO diene controlled), the electronic effect of the substituents influences the reactivity in an opposite way. Electron-withdrawing groups in the dienophile will retard the reaction, while electron-donating groups accelerate it. Conversely, electron-withdrawing groups in the diene will accelerate the reaction while electron-donating ones retard it. An example is the reaction of tetrazine-3,6-dimethyldicarboxylate and *para*-substituted styrenes[24] (Eq. 1.12).

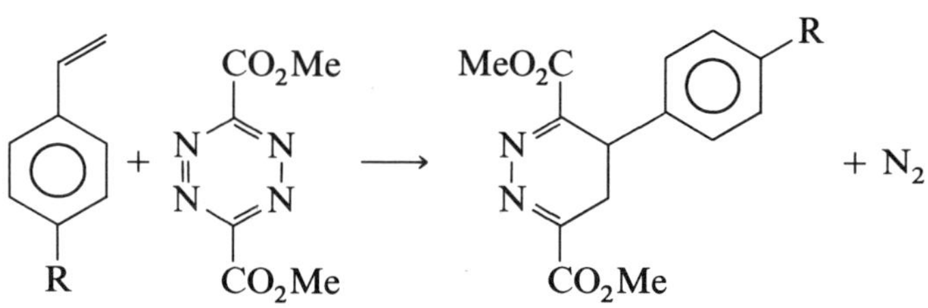

(1.12)

R	H	NO_2	OMe
$k_2 \times 10^5\ \mathrm{mol^{-1} \cdot s^{-1}}$	6500	872	25,400

The inverse electron demand Diels–Alder reaction customarily employs an electron-deficient diene and an electron-rich dienophile. This type of reaction is widely used in the synthesis of heterocyclic compounds[25] and in the chemistry of carbohydrates and related highly functionalized natural products.[9,26–28] For instance, the cycloaddition of α-methoxymethylene-sub-

TABLE 1.2 Relative Rate Constants of the Diels–Alder Reactions of Substituted Tetraphenylcyclopentadienones with Substituted Styrenes in Toluene at 60°C

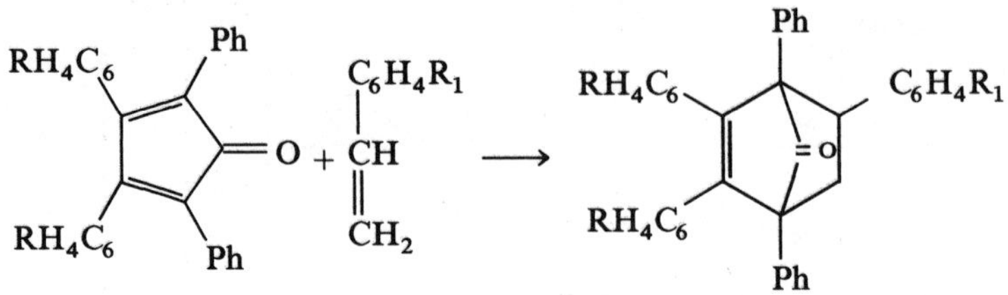

Diene (R)	Dienophile (R_1) 4-NMe_2	4-OMe	H	4-Cl	3-NO_2	4-NO_2
H	4.63	1.40	1	1.07	1.08	1.20
p-OMe	3.95	1.18	1	1	1.39	1.86
p-NMe_2	2.39	1.29	1	2	3.1	5.55

stituted 1,3-dicarbonyl compounds with (*E*)-dimethoxyethene (Eq. 1.13) allows the synthesis of 3,4-dihydro-2*H*-pyrans to be achieved.[26]

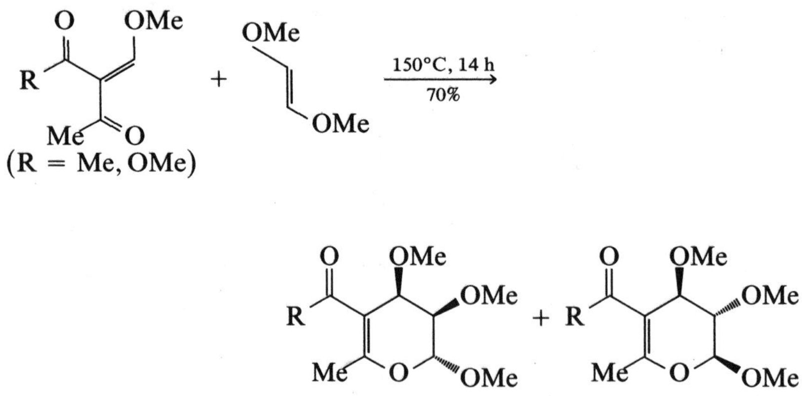

(1.13)

In neutral Diels–Alder reactions, both frontier orbital interactions influence the reactivity (HOMO–LUMO controlled). The first example of neutral cycloadditions was reported in 1973 by Konovalov and Solomonov[29] and concerns the reactions of substituted tetraphenylcyclopentadienones with substituted styrenes (Table 1.2). The data reported in the table show that rate constants are almost insensitive to the substituents in the diene and/or in the dienophile.

1.6 REGIOSELECTIVITY

In principle, a cycloaddition involving two unsymmetrical reactants can give two regioisomer adducts, resulting from two different orientations of the reactants (Eq. 1.14). Usually, the relative position of the substituents in the adduct is indicated, such as in disubstituted benzene by using the classical words *ortho*, *meta*, and *para*. Although used, in the current literature, however, this system is rough and inadequate and has encountered difficulties for adducts from disubstituted dienes and dienophiles.[30] The cycloadditions of MAC to 2-ethoxy-1,3-butadiene and 2-cyano-1,3-butadiene show different regioselectivities (Eq. 1.15). Indeed, whereas the reaction with the first diene is totally *para* regioselective, the reaction with 2-cyano-1,3-butadiene is less regioselective, affording a 5.2 : 1 mixture of *para/meta* isomers.[31] Also, the cycloadditions of isoprene and (*E*)-piperylene with 1,4-benzoquinones (Eq. 1.16) show a different regiochemical outcome.[32] Whereas the reactions with isoprene are regiounselective, those with (*E*)-piperylene are highly regioselective. The regioselectivity is influenced by catalyst. Table 1.3 summarizes *para/meta* ratios for the cycloadditions of some 2-substituted butadienes with MAC performed under thermal and catalyzed conditions.[33,34] Independently of the electronic nature of the substituent, 2-substituted butadienes preferentially give the *para* isomer. This preference is increased in catalyzed reactions.

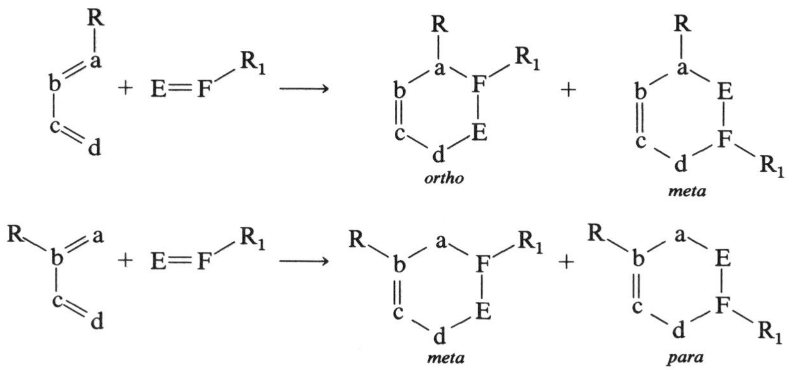

(1.14)

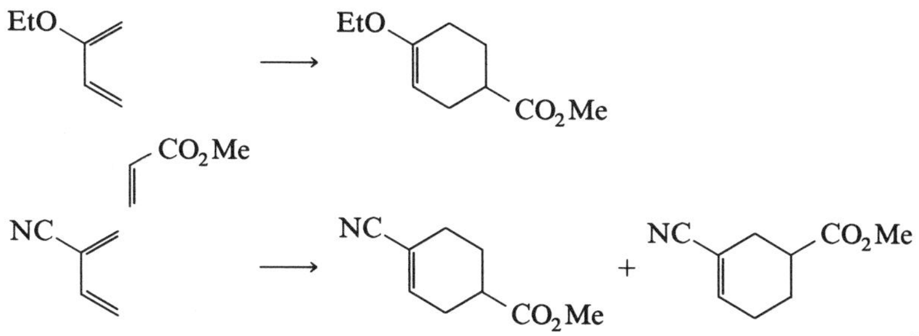

(1.15)

TABLE 1.3 Para / Meta Ratios of Diels–Alder Reactions of 2-Substituted Butadienes and MAC

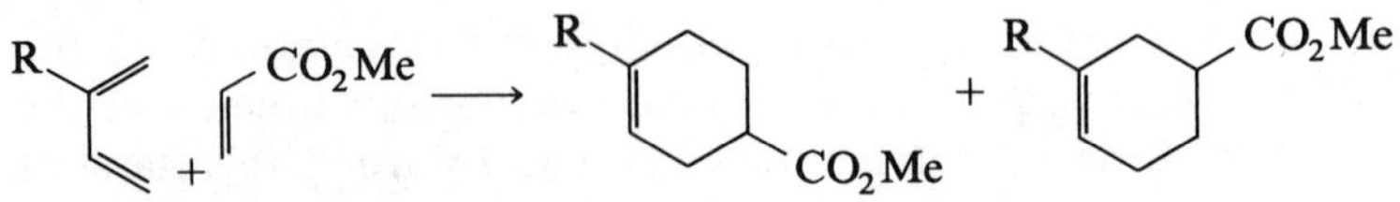

	Diene (R)					
Reaction	Me	Ph	CN	CF_3	Cl	CH_2SiMe_3
Thermal	2.3	4	5.2	1.2	6.7	5.2
Catalyzed	19	32.3	2.7	1	49	199

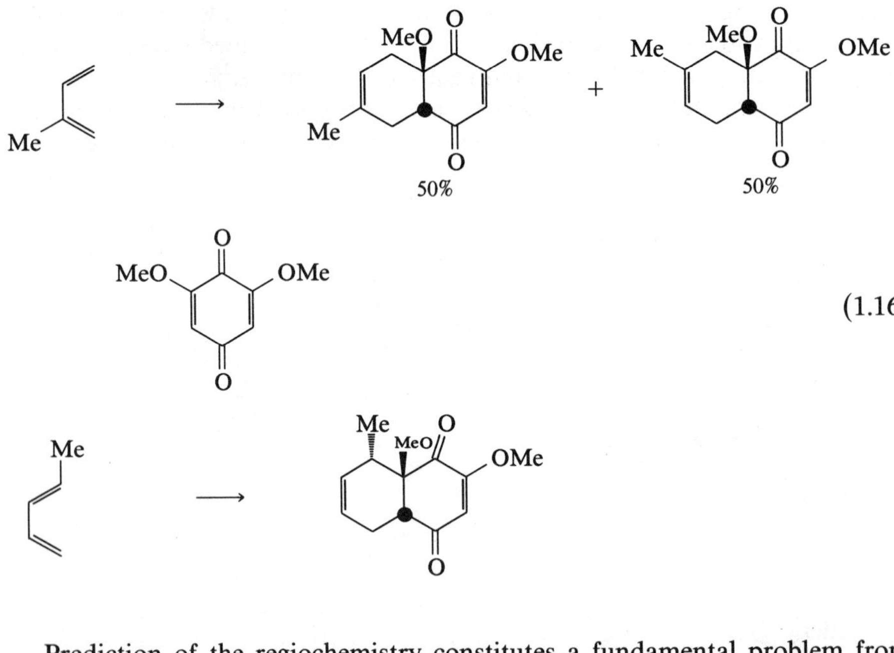

(1.16)

Prediction of the regiochemistry constitutes a fundamental problem from the theoretical aspect as well as for synthetic uses of the Diels–Alder reaction. This problem, claimed by Huisgen[35] to be "the biggest unsolved problem in the field," is a stimulating challenge for those working in this field today. FMO theory has been used to explain and predict the regioselectivity of many Diels–Alder reactions.[8,16,19,31,33,36] Recently, another approach to the discussion of Diels–Alder regiochemistry has been proposed[37] which is based on the matching of complementary reactivity surfaces for diene and dienophile.

The FMO theory–based approach developed by Houk provides two generalizations to predict regioselectivity[36b]:

1. The principal stabilization of the transition state will arise from interaction of HOMO–LUMO pairs of addend frontier orbitals which are closest in energy.
2. The larger terminal coefficient on each addend will become bonded preferentially in the transition state.

Considering these generalizations, let us explain the regioselectivities for the cycloadditions discussed previously (Eq. 1.15) between MAC and 2-ethoxy- and 2-cyano-1,3-butadiene. The calculated coefficients of carbon atoms are[31]

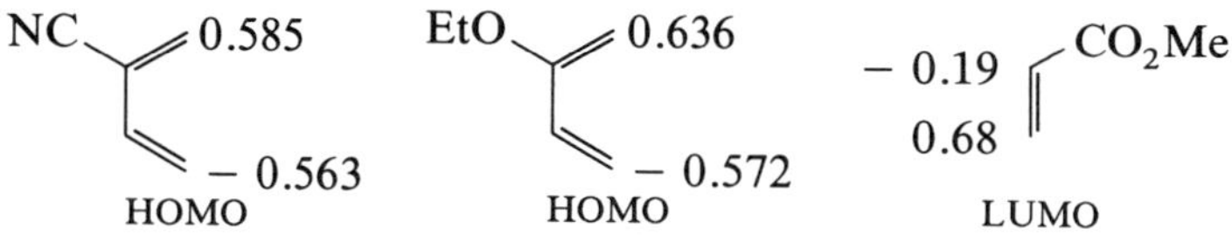

Since these two reactions are typical normal electron demand Diels–Alder cycloadditions, they will be controlled by the HOMO diene–LUMO dienophile interaction. Let us first consider the cycloaddition between MAC and 2-ethoxy-1,3-butadiene. The general rules allow one to predict a strong preference in favor of the *para* isomer because the HOMO diene terminal coefficients have different magnitudes.

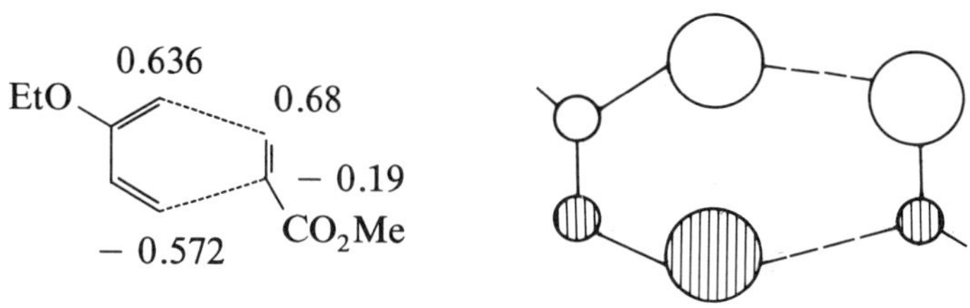

In the reaction of MAC with 2-cyano-1,3-butadiene, a preference for the *para* regioisomer can be predicted again, but because the difference between the HOMO diene terminal coefficients is smaller, a lower regioselectivity must be expected.

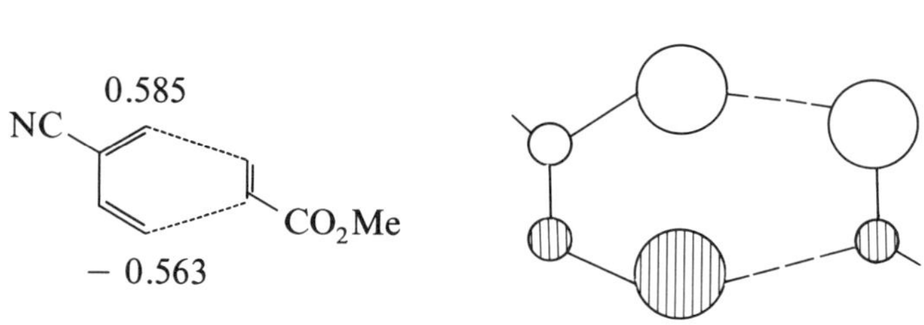

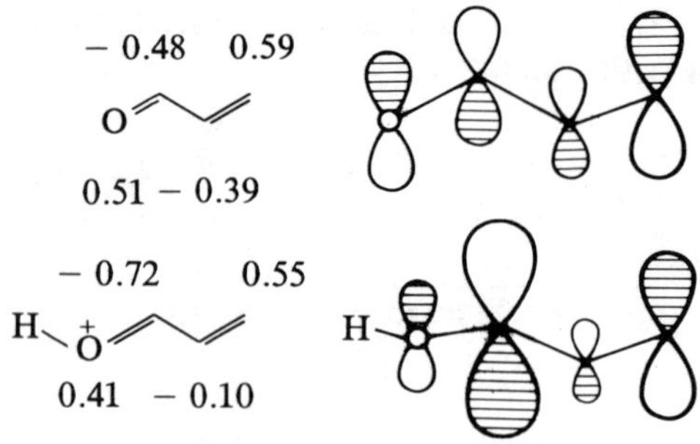

Figure 1.6 Effect of protonation in the LUMO of the acrolein.

Sometimes there are discrepancies between the prediction and the experimental results. Alston and co-workers[38] have found that regioselectivity can still be predicted by considering the secondary orbital interactions.

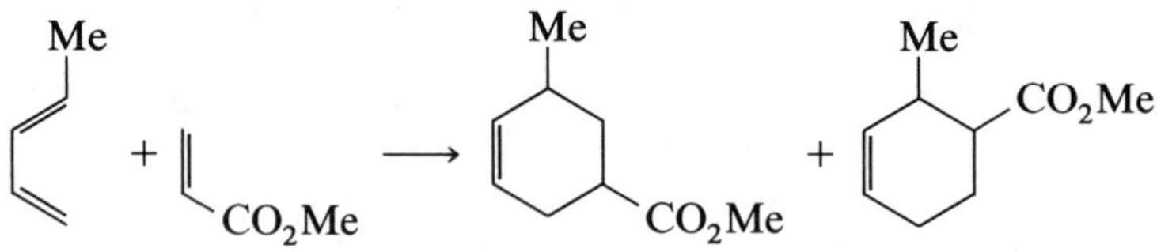

(1.17)

Regioselectivity may be affected by Lewis acid catalyst. For example, in the reaction of (E)-piperylene and MAC (Eq. 1.17), the *ortho/para* ratio increases from 9 : 1 (thermal reaction) to 49 : 1 ($AlCl_3$-catalyzed reaction).[39] Houk explains the role of the catalyst[40] by considering the significant changes in the frontier orbitals of the dienophile acrolein caused by complexation (Fig. 1.6) with Lewis acid. The LUMO amplitude at the carbonyl carbon as well as the difference in the magnitudes of the LUMO coefficients in the alkene moiety increases due to the influence of the protonated carbonyl oxygen. The greater difference in magnitudes of the LUMO coefficients makes the cycloaddition more regioselective. According to Houk, this model allows predictions on the regioselectivity of normal and inverse electron demand reactions.

The important role of the secondary orbital interactions in the Lewis acid–catalyzed Diels–Alder reaction of AC has been emphasized in a recent theoretical study by Ottenbrite,[41] based on *ab initio* calculations of frontier molecular orbital energies and coefficients.

1.7 STEREOSELECTIVITY

Alder and Stein[42] first observed that the relative configuration of groups of the reactants is preserved in the adduct(s), and translated this observation into a rule, known as the *cis* principle. For example, the reactions of 1,3-butadiene with DMM and DMF give *cis*- and *trans*-cyclohexene-4,5-dimethyldicarboxylate, respectively, and the reaction of (E,E)-hexadiene with MA gives an adduct with the methyl groups in *cis* configuration (Eq. 1.18).

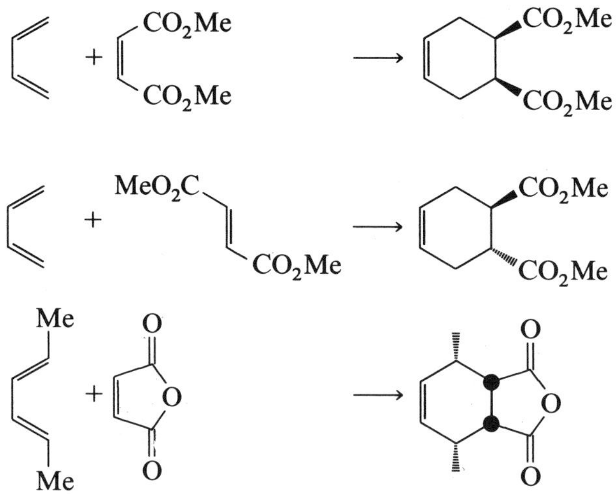

(1.18)

The retention of the relative configuration of groups of the reactants in the adduct can be explained by the fact that the cycloaddition occurs suprafacially on each component. This means that as a suprafacial addition is a *syn* addition, the Diels–Alder reaction is a stereospecific *syn* addition of the diene with respect to dienophile and of the dienophile with respect to the diene.

Moreover, the two components of a Diels–Alder reaction, approaching each other in parallel planes, a priori may interact in two different orientations affording *endo* and *exo* adducts. The two methods of interaction (*endo* addition, *exo* addition), as well as the two adducts, are illustrated in Fig. 1.7 for the reaction between cyclopentadiene and MA. The terms *endo* and *exo* were initially used to designate only the stereochemistry of the adducts, not the stereochemical mode of interaction of the reactants in the transition state. This has been the source of some confusion.

Tentatively, *endo* addition can be defined as that particular spatial arrangement of reactants in which the more bulky side of the diene is under the more bulky side of the dienophile. Conversely, in the *exo* addition the more bulky side of one component will be under the smaller side of the other component.

A priori, considering regioisomerism, as well as stereoisomerism of the reaction, the interaction of two unsymmetrical reactants can give several isomeric adducts. For example, the cycloaddition of a 1-substituted 1,3-butadiene with a monosubstituted alkene can give four adducts (Eq. 1.19). It has been found that most of the Diels–Alder reactions show a preference for *endo* addition. Whereas the reaction of MA with (E)-1-(trichloroacetamido)-1,3-butadiene is totally *endo* diastereoselective,[43] the cycloaddition of MAC and cyclopentadiene[44] gives a 3.2 : 1 mixture of *endo*/*exo* adducts (Eq. 1.20). A further example of different *endo-exo* diastereoselectivity is observed in the

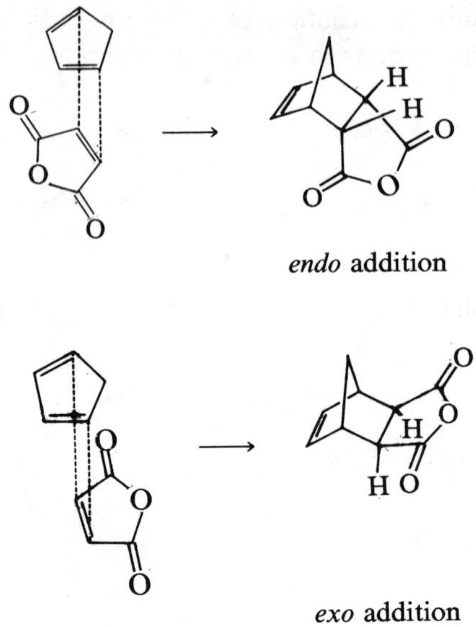

Figure 1.7 Endo-exo addition for the reaction of MA and cyclopentadiene.

cycloadditions of 2-methyl-2-cyclohexenone with (E)-piperylene[45] and (E)-1-methoxy-1,3-butadiene[46] (Eq. 1.21).

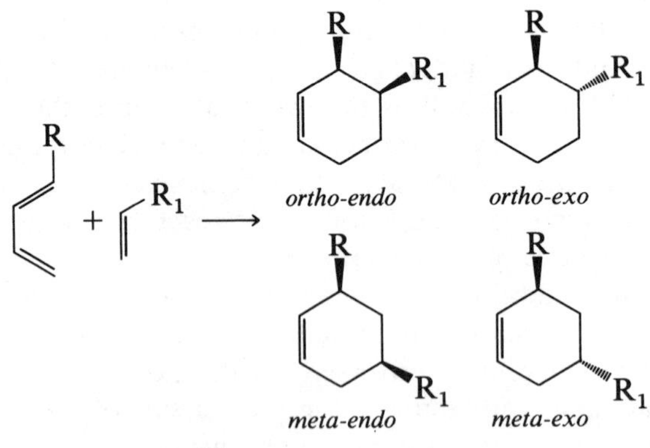

(1.19)

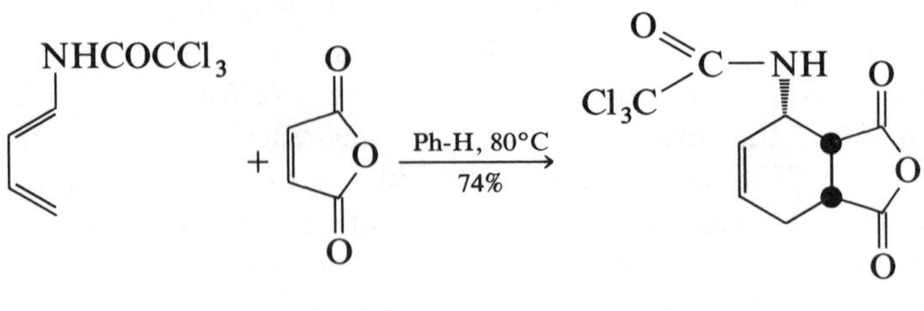

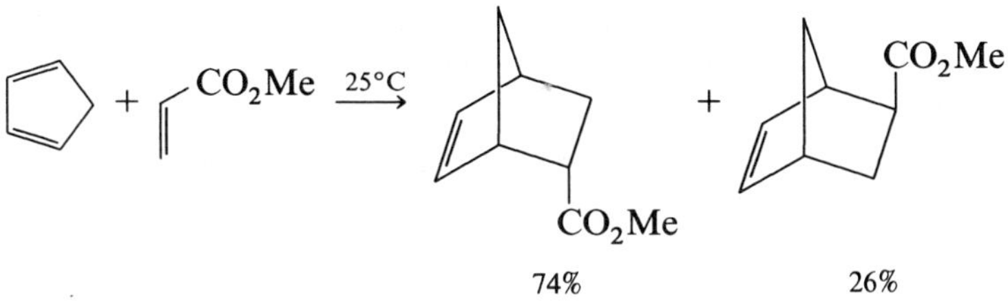

(1.20)

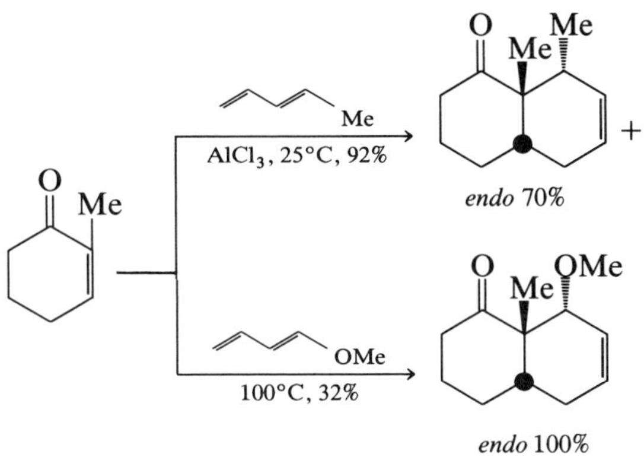

(1.21)

The preference for the *endo* addition was first justified by the Alder–Stein rule[47] of maximum accumulation of double bonds, and later, by stabilizing secondary orbital interactions,[8,19,48] inductive[49] or charge-transfer interactions,[50] and by the geometrical overlap relationship of the π-orbitals at the primary centers.[51] Secondary orbital interaction between atoms that will not be bonded in the adduct (Fig. 1.2) accounts satisfactorily for the *endo* preference in most of the cycloadditions. Sometimes, more than one effect must be considered to understand the *endo-exo* stereoselectivity. Stephenson and co-workers,[52] studying the cycloaddition between *cis,cis*-1,4-dideuterio-1,3-butadiene and MA (Eq. 1.22), determined the minimum intrinsic energy advantage [ΔG (80°C) = 1.2 kcal/mol] that can be attributed to the secondary orbital interaction. These results are in general agreement with those obtained previously by Houk,[53] who studied the stereoselectivity of other cycloadditions.

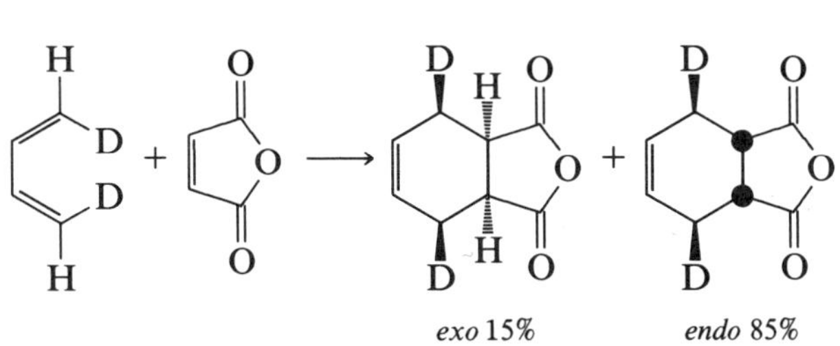

(1.22)

The *endo-exo* diastereoselectivity is affected by Lewis acid catalyst. Thus $AlCl_3$-catalyzed cycloadditions of cyclopentadiene with allenic esters[54] are more *endo* diastereoselective than are the thermal ones (Eq. 1.23).

The influence of the catalyst has been explained[8,40] by FMO theory by considering that in the Lewis acid complexed dienophile, the LUMO amplitude of the carbonyl increases (Fig. 1.6) and, as a consequence, the stabilizing effect of the secondary orbital interaction in the *endo* transition state increases. The influence of the catalyst, moreover, can also depend on the nature of the catalyst. The *endo-exo* diastereoselectivity can also be affected by solvent.[55]

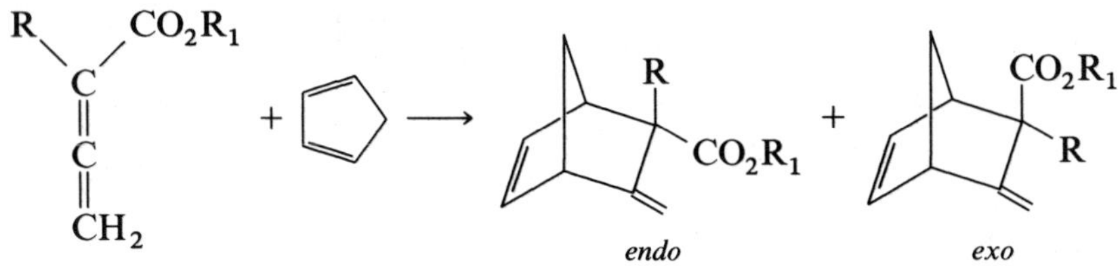

(1.23)

R	R_1	*endo : exo* Thermal	*endo : exo* Catalyzed
H	Et	64 : 36	86 : 14
Me	Me	40 : 60	76 : 24

Another aspect of the stereochemistry of the Diels–Alder reaction becomes important when the two faces of the π-bond system of the interacting diene and/or dienophile are not equivalent. This phenomenon manifests itself whenever the plane through the multiple-bond system of neither one nor both of the reactants represents a symmetry plane.[56] In this case the cycloaddition gives adducts that are diastereoisomers. The two ways of addition are called *syn* and *anti* with respect to the group, or structural moiety, that makes the two faces different. The *syn* and *anti* additions of a general dienophile to a monosubstituted cyclopentadiene are illustrated in Fig. 1.8. Some examples of nonsymmetrical reactants among the most common described in literature are

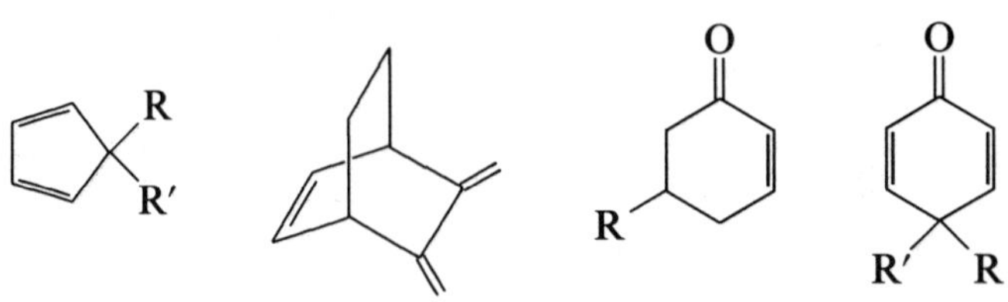

The cycloaddition of α-chloroacryloyl chloride to 5-methoxymethylcyclopentadiene (Eq. 1.24), which is a key initial stage of the "Corey bicycloheptene

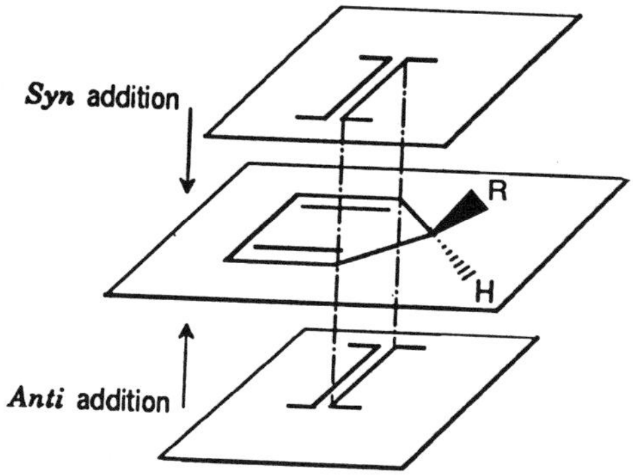

Figure 1.8 *Syn* and *anti* addition to a monosubstituted cyclopentadiene.

route"[57] to prostaglandines, is totally *anti* diastereofacial selective. The reaction of 4-methyl-4-hydroxy-2,5-cyclohexadienone with (E)-piperylene is completely *syn* diastereoselective with respect to hydroxy groups[58] (Eq. 1.25). Opposite diastereofacial selectivity is observed in the thermal and $SnCl_4$-catalyzed reactions of (E)-piperylene with 2,4,5-trimethyl-4-hydroxycyclohex-2,5-dien-1-one[58] (Eq. 1.26).

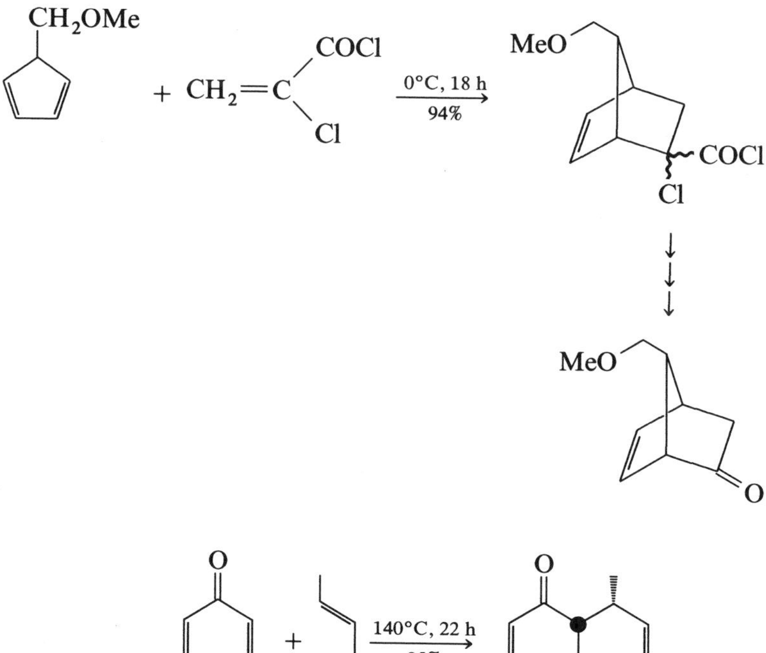

(1.24)

(1.25)

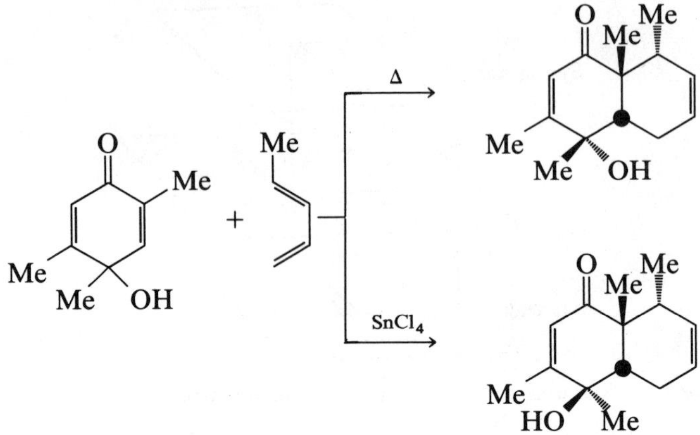

(1.26)

For the catalyzed process, the observed diastereoselectivity presumably results from complexation of the hydroxyl group by $SnCl_4$, thereby transforming it into the larger of the two geminal substituents. The cycloadditions of 1,3-butadiene to 4-alkyl-substituted 2-cyclohexenones (Eq. 1.27) show different diastereofacial selectivity.[56d]

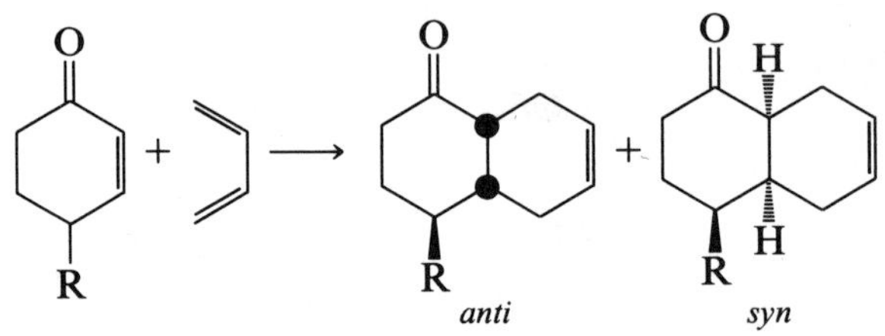

(1.27)

R	*anti*	*syn*
Me	55%	45%
Pr^i	67%	33%
Bu^t	100%	0%

The *anti* facial selectivity increases going from the methyl to the *t*-butyl group. The diastereofacial selectivity of substituted 2-cyclohexenones has been investigated by the authors[56d] and rationalized in terms of a unifying stereoelectronic pathway and conformational reactivity.

The reaction between methyl-*N*-benzoyl-3-aza-2-oxabicyclo[2.2.2]oct-5-ene-6-carboxylate and 1,3-butadiene is *syn* diastereoselective to the heterocyclic bridge[56c] (Eq. 1.28).

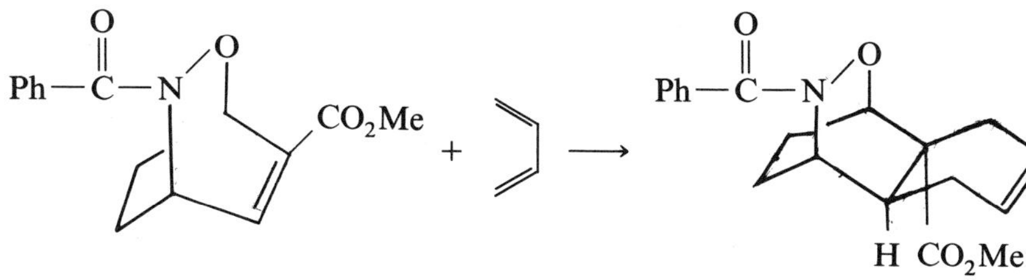

(1.28)

Extensive investigations on diastereofacial selectivity of Diels–Alder reactions of rigid exocyclic dienes grafted onto norbornene or bicyclo[2.2.2]octane skeleton with several dienophiles have been carried out by Vogel and coworkers.[56a, 59]

1.8 CATALYST

The accelerating effect of the Diels–Alder reaction in acid media has been observed by Wassermann[60] since 1942, but the influence of the catalyst on the reaction rate was estimated to be small. Yates and Eaton,[61] in 1960, first reported the remarkable acceleration of the reactions of anthracene with MA, BQ, and DMF catalyzed by aluminum trichloride. Since 1960, Lewis acid catalyzed Diels–Alder reactions have received increased attention because the catalyst allows the reactions to be carried out under milder conditions and makes cycloadditions with low-reactive dienophiles attractive to synthetic organic chemistry once again. For example, whereas thermal cycloaddition of isoprene to 2-methyl-2-cyclohexenone occurs at 150°C in poor yield, the $AlCl_3$-catalyzed reaction proceeds at room temperature, for a shorter time, and in higher yield[62] (Eq. 1.29).

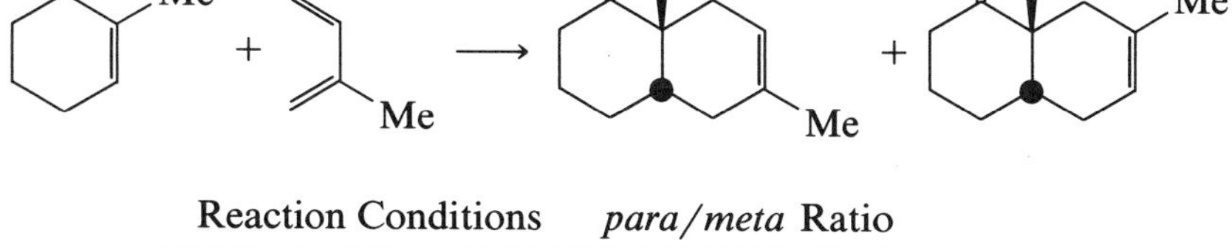

Reaction Conditions	*para/meta* Ratio
150°C, 142h, 20%	1.9
25°C, $AlCl_3$, 17h, 97%	36

(1.29)

The catalyst gives a complex with the basic center of the electron-withdrawing group of the dienophile causing a greater polarization of the dienophile center.

There are, however, some limitations in using Lewis acid catalyst, as in the case of the cycloadditions involving acid-sensitive polyfunctional dienes and/or dienophiles. Furthermore, diene polymerization can occur even to a large extent, especially when electron-rich dienes are used. To preserve valuable functionalities in reactants and adducts, milder catalysts, such as lanthanide complexes, have been used.[63]

In contrast to the increasing interest in the Lewis acid catalysis, protic-catalyzed Diels–Alder cycloadditions have not received much attention. In 1979, Olah[64] reported a highly efficient catalysis by Nafion-H, a perfluorinated resin sulfonic acid catalyst. For example, BQ reacts with isoprene in the presence of Nafion-H at room temperature and in high yield (Eq. 1.30).

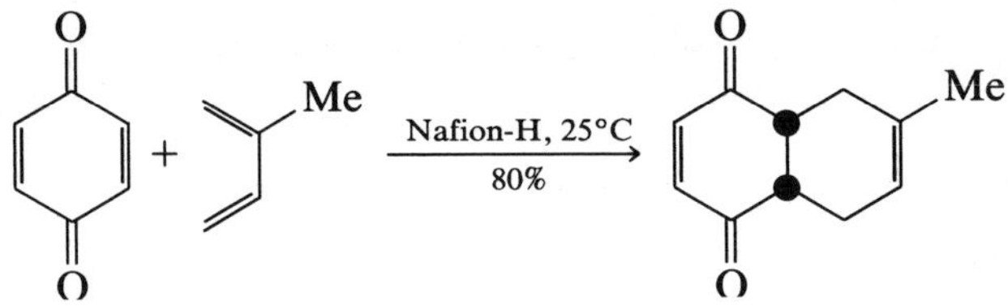

(1.30)

Lewis acid catalyzed cycloadditions not only proceed faster than their thermal counterparts, but are also usually more regioselective and stereoselective. These properties also make these reactions of interest from a theoretical point of view, as they seem to constitute an exception to the reactivity–selectivity principle. A theoretical interpretation of the apparently anomalous behavior of the Lewis acid catalyzed reactions has been elaborated by Epiotis[65] and Branchadell.[66]

FMO theory (Sec. 1.5) has been used successfully to explain the role of the Lewis acid catalyst.[19,40,67] Donor–acceptor interaction between dienophile

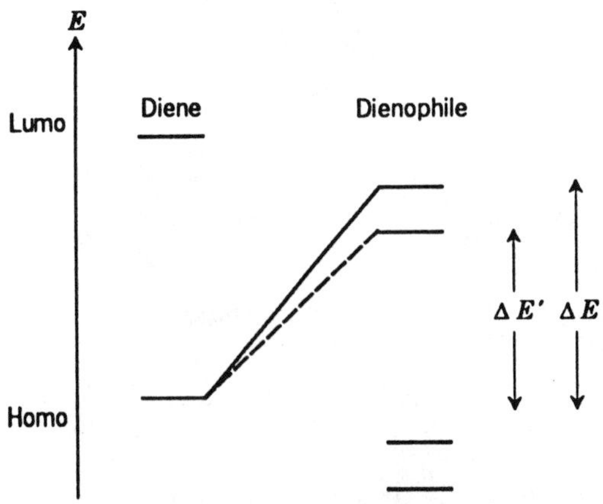

Figure 1.9 HOMO–LUMO arrangements for catalyzed (dashed line) and uncatalyzed (solid line) normal electron demand Diels–Alder reaction.

and catalyst lowers the energy of the highest-occupied molecular orbital and of the lowest-unoccupied molecular orbital of the dienophile. As a consequence, in the case of a normal electron demand cycloaddition, the separation between MOs will decrease and the stabilization of the transition state will increase (Fig. 1.9).

In 1981, Bauld and co-workers discovered the cation radical-catalyzed Diels–Alder reaction.[98,99] The conversion of neutral or electron-rich dienophiles to the corresponding cation radicals increases their reactivity remarkably. Thus the dimerization of 1,3-cyclohexadiene under cation radical catalysis occurs under milder conditions (0°C, 15 min) than those of its thermal counterpart (200°C, 30 h). Cation radical cycloaddition is a symmetry-allowed stereospecific reaction.

The adduct (A) is formed via either the cation radical of the dienophile (DP) or the cation radical of the diene (D) according to the following mechanism:

$$DP + Ar_3N^{+\cdot} \rightleftarrows Ar_3N: + DP^{+\cdot}$$

$$DP^{+\cdot} + D \rightleftarrows A^{+\cdot}$$

$$A^{+\cdot} + DP \rightarrow A + DP^{+\cdot}$$

1.9 PRESSURE

Intermolecular Diels–Alder reactions have large negative activation volumes and thus can be markedly accelerated by high pressure. In the intramolecular processes, the activation volumes are somewhat less negative than those of intermolecular reactions. High-pressure kinetic studies have been widely used for investigating both structure and properties of the transition state of the Diels–Alder reaction. Eckert and co-workers[68] have made a valuable contribution in this field. High-pressure cycloadditions are also used in synthetic organic chemistry. Indeed, this methodology allows reactions of heat-sensitive and/or low-reactive materials to be carried out.[69] Thus, whereas the cycloaddition of the methyl-2,4-pentadienoate with BQ at 15 kbar proceeds at room temperature (r.t.) in good yield[70] (Eq. 1.31), at atmospheric pressure the yield is lower (28%).

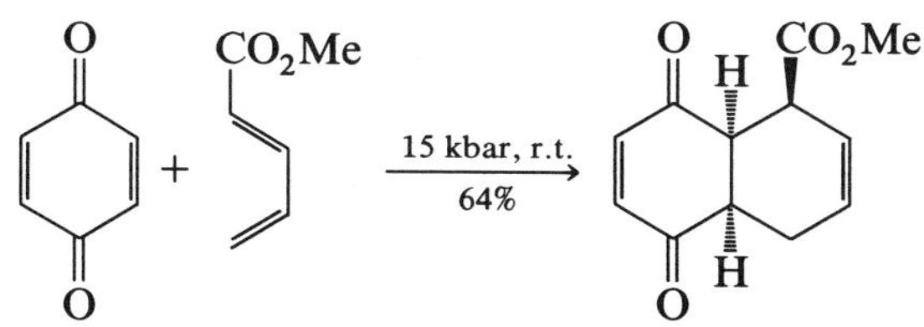

(1.31)

Application of high pressure to the cycloaddition of (*E*)-1-methoxy-1,3-butadiene with ethyl pyruvate[71] (Eq. 1.32) increases the yield from 27% to 85%.

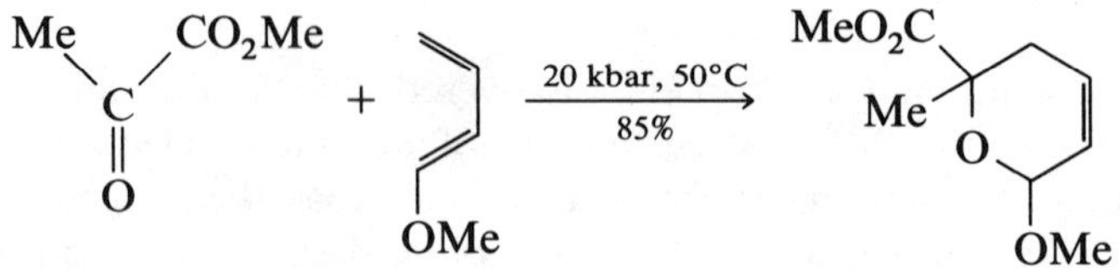

(1.32)

1.10 SOLVENT

Solvent effects on the Diels–Alder reaction have been investigated extensively by several authors.[8] These studies have shown that the reaction rate varies a little with the polarity of the solvent. In going from a polar solvent to a nonpolar one, the reaction rate can vary by more or less than a factor of 10. This low solvent dependence is generally considered to be an indication of the lack of significant differences in polarity between reactants and transition state. Table 1.4 summarizes the values of the rate constants of the isoprene–MA reaction in a wide range of solvents.[72]

The data indicate that the rate of the reaction varies little with the solvent and there is little correlation with the dielectric constant. Therefore, although the reaction involves a rather polar dienophile, the transition state must be only a little more polar than the reactants.

Investigations of the mechanism by which the solvent acts when reactants, products, and the transition state are weakly polar have been carried out by Desimoni and co-workers.[73] A correlation between kinetic data and the acceptor numbers of the solvent have been observed and interpreted. In the dimerization reaction of cyclopentadiene performed in a wide range of sol-

TABLE 1.4 Rate Constants for the Reactions of Isoprene and MA in Several Solvents at 30.3°C

Solvent	Dielectric Constant	Rate Constant ($10^5 k_1$, $1 \cdot mol^{-1} \cdot s^{-1}$)
Benzene	2.3	17.5
Anisole	4.3	25.2
Veratrole	4.5	31.0
Chlorobenzene	5.6	25.2
o-Dichlorobenzene	7.5	66.7
Benzonitrile	26.5	34.3
Nitrobenzene	36.1	52.9
Nitromethane	39	33.2

vents, from nonpolar *n*-hexane to polar water, the reaction rate changes only by a factor of 4.[74]

Rate enhancements have been observed in several intermolecular cycloadditions carried out in ethylene glycol.[75] For example, the reaction of 2,6-dimethyl-1,4-benzoquinone with methyl-1-butadienylcarbonate in ethylene glycol is 26 and 14 times as great as those in benzene and in acetonitrile, respectively. In view of the fact that the value of the dielectric constant of acetonitrile (36.2) is almost equal to that of ethylene glycol (37.7), the polarity of the solvent cannot be an important parameter in determining the increase in the reaction rate. The observed rate enhancement has been ascribed to the molecular aggregation.

1.11 REACTIONS IN WATER

Most Diels–Alder reactions are carried out in organic solvents. Because the solvent is not a very important parameter in the optimization of synthetic applications, the choice of solvent in a cycloaddition does not constitute a crucial problem. The nature of the solvent may be selected in terms of its influence on the *endo-exo* diastereoselectivity of certain cycloadditions.[55]

In recent years, several reports on cycloadditions carried out in an aqueous medium have been published. In 1980, Breslow reported a remarkable acceleration of reactions performed in water.[76] The reaction of 9-hydroxymethylene anthracene with NEM in water (Eq. 1.33) is 200 times faster than that in acetonitrile.

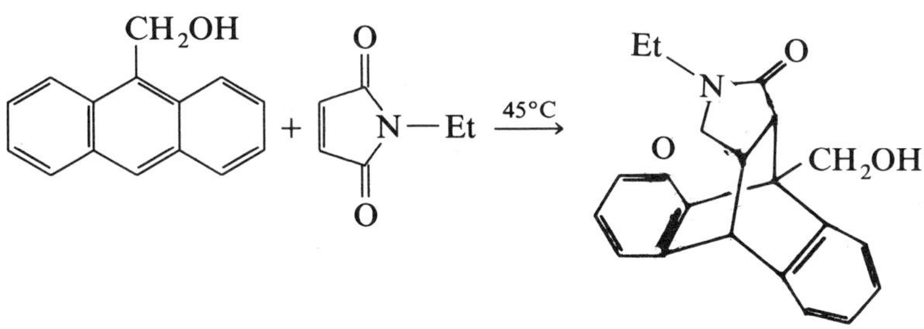

(1.33)

The main effect of water has been ascribed to hydrophobic packing of the diene with the dienophile in the transition state.

Cycloadditions in water show an increased preference for *endo* attack.[77] Grieco and co-workers have also extensively investigated the aqueous intermolecular Diels–Alder reactions.[78] Some examples which illustrate that the reaction rate, as well as the *endo-exo* diastereoselectivity, are significantly affected by water are reported in Eq. 1.34.

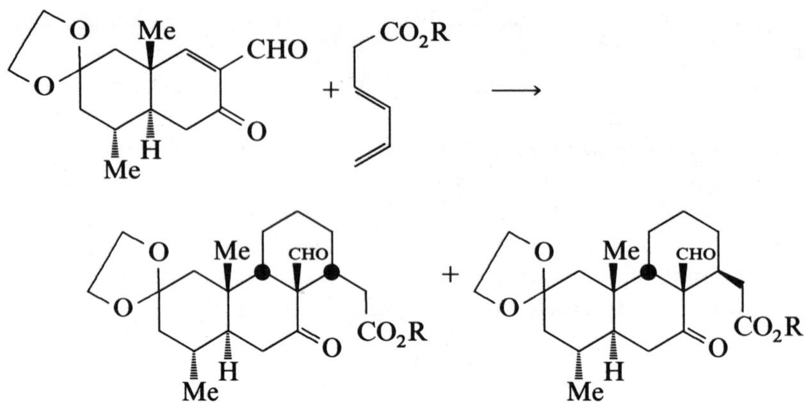

R	Reaction Conditions	*endo/exo* Ratio
Et	PhMe, 100°C, 36 h, 97%	0.9
Na	Water, r.t., 4.5 h, 90%	3

(1.34)

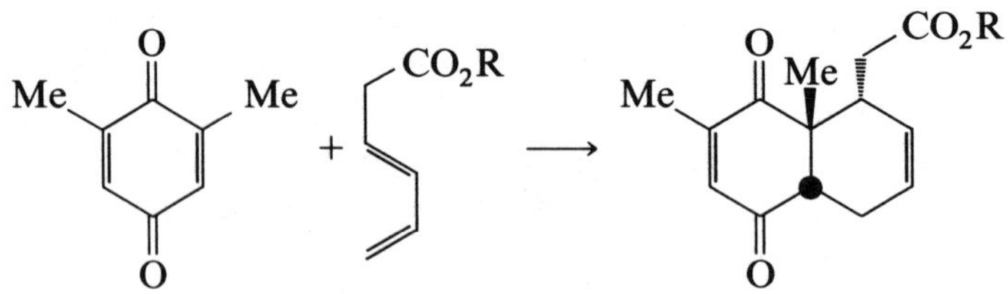

R	Reaction Conditions	Yield
Me	PhMe, r.t., 1 week	Trace
Na	Water, r.t., 1 h	77%

The water effect has been explained by invoking some type of micellar catalysis. Aqueous detergent solutions do not affect the reaction selectivity.[77a]

1.12 REACTIONS IN HETEROGENEOUS PHASE

Endo-exo diastereoselectivity as well as the reaction yield can be influenced when the cycloadditions are performed in solid phases. *Endo/exo* ratios for the reaction of cyclopentadiene with AC, ACN, and MAC,[79] carried out in solid phases and in homogeneous phases (HP), are reported in Table 1.5. The *endo-exo* ratios are almost independent of temperature (40 to 70°C).

Insoluble polymer supports have also been used for accomplishing cycloadditions. In principle this methodology is attractive because it could allow better control of selectivity together with an easier workup of the reaction mixture and isolation of the adducts from excess reactants and by-products.

TABLE 1.5 ***Endo/Exo*** **Ratios for the Reactions of Cyclopentadiene and AC, ACN, and MAC on Activated Neutral Alumina, Cellulose, Montigel, and in Homogeneous Phases (HP)**

Dienophile	Al_2O_3	Cellulose	Montigel	HP
AC	1.07	2.94	2.5	3
ACN	1.63	1.56	2.6	1.45
MAC	32.3	2.94	13.7	3

Thus polymer-bound benzyl acrylate, prepared from polymer-bound benzyl alcohol and acryloyl chloride, interacts with (*E*)-1-phenyl-1,3-butadiene (Eq. 1.35) to give the corresponding polymer-bound adduct, which by treatment with tetra-*n*-butyl ammonium hydroxide affords *ortho*- and *meta*-free adducts in 94 : 6 ratio.[80]

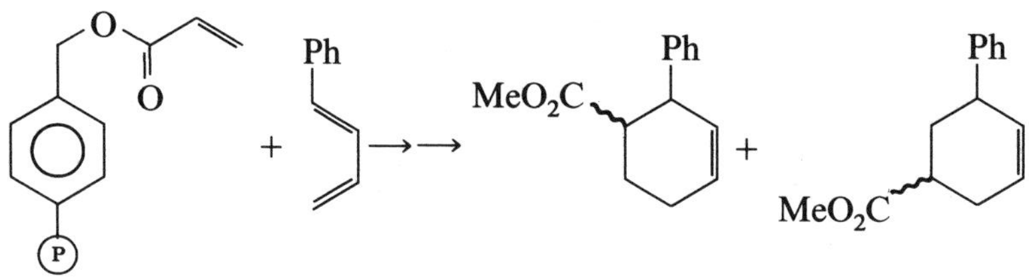

(1.35)

1.13 INTRAMOLECULAR DIELS–ALDER REACTION

Despite its potentiality in the construction of complex polycyclic molecules,[81,82] the intramolecular version of the Diels–Alder reaction, in which both diene and dienophile are incorporated in the same compound, has become of interest to chemists only in the last 15 years. An extensive and exhaustive review of this topic was written recently by Ciganek.[11]

In the intramolecular Diels–Alder reaction, two rings are formed in one step (Eq. 1.36), of which only one is formed by the [4 + 2] cycloaddition. The second ring is usually five- or six-membered. An intramolecular cycloaddition can occur in two different regiochemical modes, giving either fused and/or bridged products if the chain is attacked at C-1 of the diene, or two different bridged adducts when the chain is attacked at C-2 of the diene (Eq. 1.37).

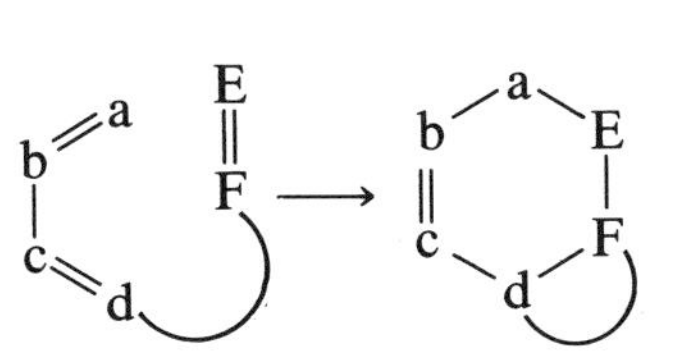

(1.36)

When the chain is attacked at C-1, the regiochemistry depends on the diene configuration (E, Z) and on the chain length. Normally, however, most of the

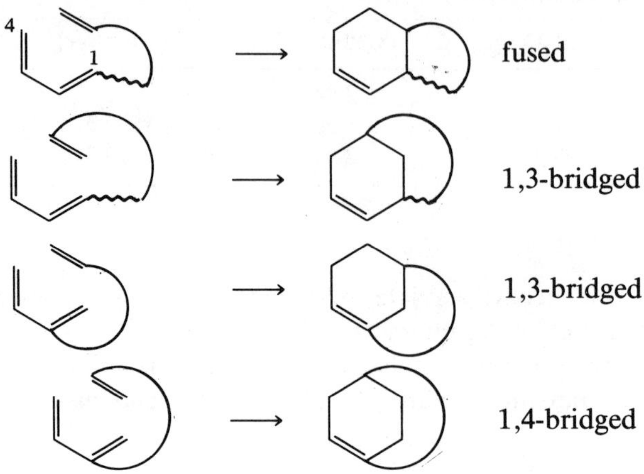

(1.37)

intramolecular Diels–Alder reactions give the fused isomers exclusively. When the chain is at C-2 of the diene, the preference for the formation of 1,3- or 1,4-bridged adducts is strongly dependent on chain length.

The *endo–exo* diastereoselectivity in the intramolecular Diels–Alder reaction depends on the configuration of the diene moiety and the length of the chain between diene and dienophile. A suitable choice of the chain length and diene configuration (E and Z) can offer a method of controlling the selectivity of the reaction.

Intramolecular Diels–Alder reactions have been used widely in the total synthesis of natural products.[9,83] For example, the spontaneous cyclization of an intermediate trienol in the course of the oxidation reaction is the key step in the stereoselective total synthesis of (±)-torreyol[84] (Eq. 1.38).

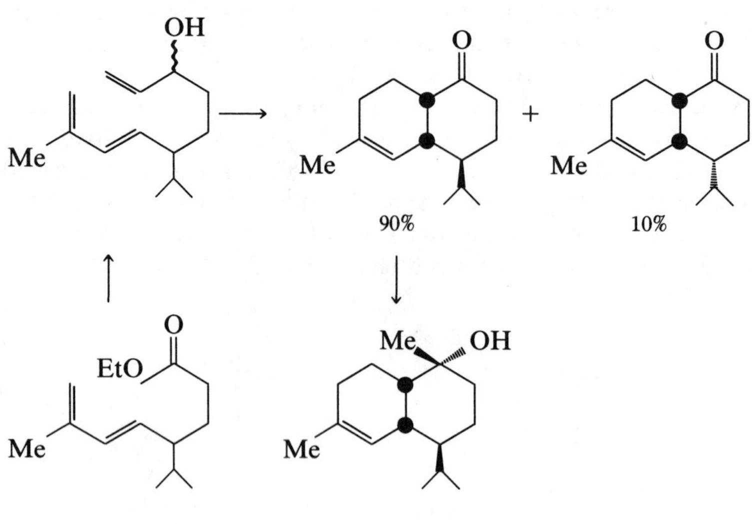

(1.38)

Allene intramolecular cycloaddition has been used to develop a new approach for the synthesis of N-substituted indoles, which are useful intermediates in the synthesis of indol alkaloids[85] (Eq. 1.39).

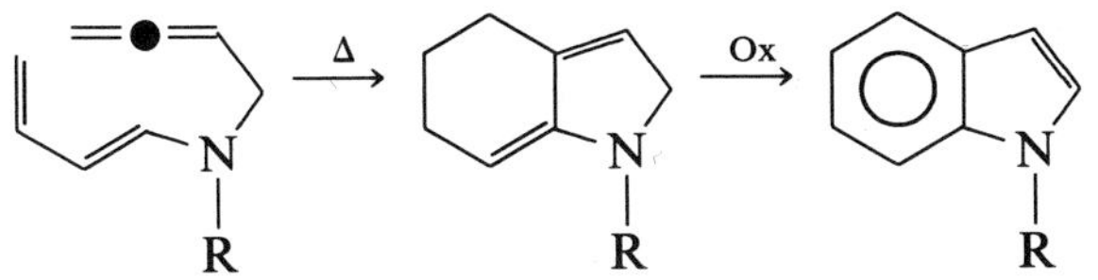

(1.39)

The synthesis of δ-coniceine, the simplest indolizidine alkaloid, is based on the intramolecular imino Diels–Alder cycloaddition of an unstable *N*-acylimine intermediate[86] (Eq. 1.40).

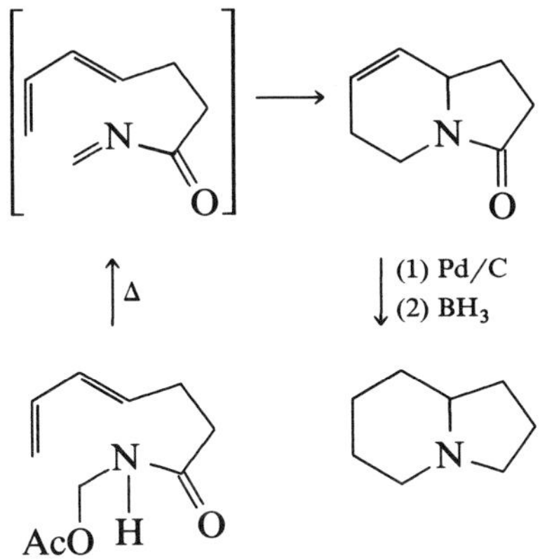

(1.40)

1.14 MULTIPLE DIELS–ALDER REACTION

Multiple Diels–Alder reaction is a process characterized by the sequence of two [4 + 2] cycloadditions. Various types of multiple cycloadditions have been described in the literature. A multiple cycloaddition entails a sequence in which the initial monoadduct is involved, directly or after some other transformation, in a second cycloaddition that affords the final polycyclic bisadduct. Let us describe the various types of multiple cycloaddition processes and illustrate them with some examples.

Domino and Timed Diels–Alder Reactions

The terms "domino" and "timed" both indicate a multiple process characterized by a sequence of two cycloadditions: The first is always intermolecular, and the second is inter- or intramolecular in the timed reaction and intramolecular in the domino reaction. The latter cycloaddition[87] is the first example of a multiple process. The conversion of 9,10-dihydrofulvalene into polyfused cyclopentanoid systems (Eq. 1.41) illustrates the domino reaction. When a tetrahydrofuran solution of 9,10-dihydrofulvalene and DMAD is allowed to warm slowly from −78°C to room temperature, a mixture of two polycyclic diadducts is obtained in 40% overall yield. The intermolecular cycloaddition gives a mixture of diastereofacial monoadducts, which then undergo the intramolecular cycloaddition. An example of a timed Diels–Alder reaction in which the starting bisdiene undergoes intermolecular and intramolecular cycloaddition is illustrated in Eq. 1.42.[88,89]

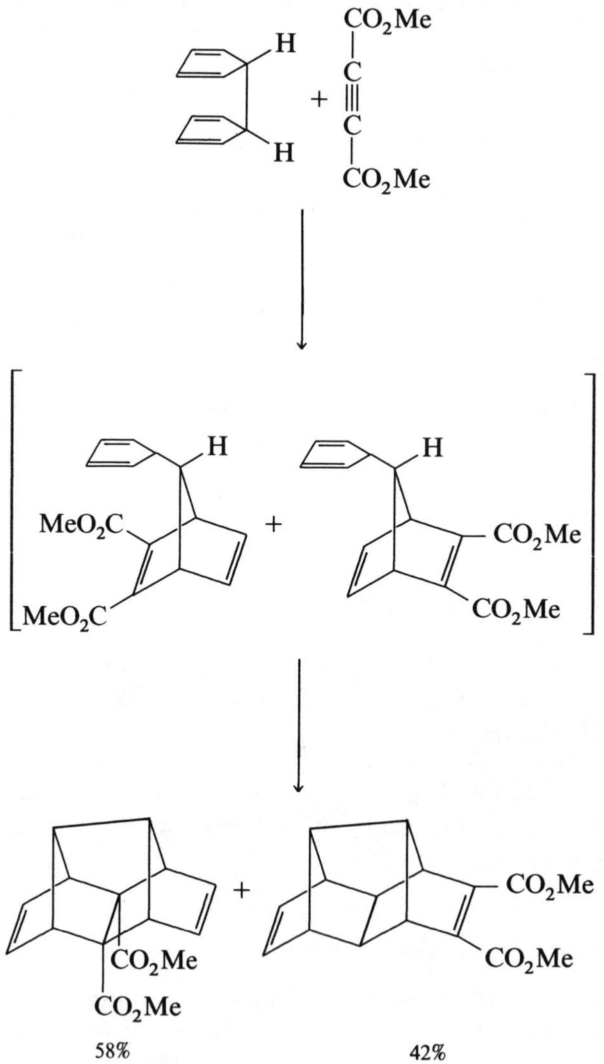

(1.41)

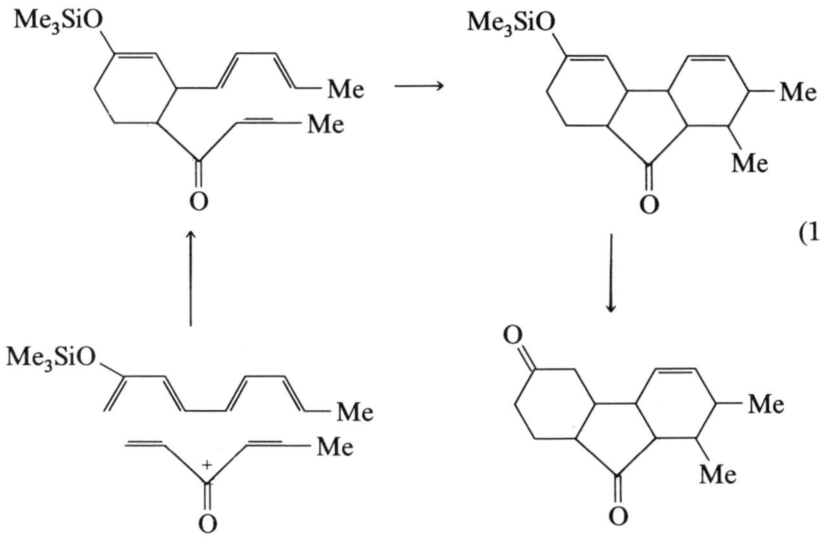

(1.42)

Tandem Diels–Alder Reaction

This process is based on the sequence of two cycloadditions in which the monoadduct resulting from the first is converted to a diene or dienophile component which then undergoes a second cycloaddition. Tandem Diels–Alder reactions involving the conversion of the initial monoadduct to a new diene are the most usual and the generalized sequence may be depicted as follows:

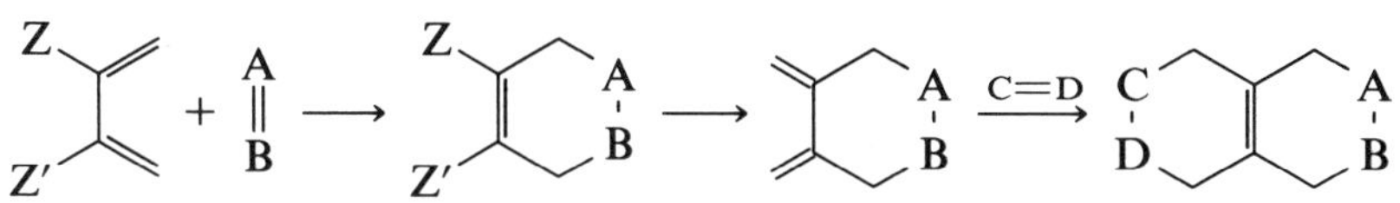

According to this scheme, the choice of the starting diene constitutes a crucial problem since the conversion of the initial monoadduct to a new diene component must occur smoothly. 2,3-Bis(trimethylsilyl)methyl-1,3-butadiene has been used by Trost as the starting diene.[90] Oxidative bisdesilylation by *N*-bromosuccinimide in tetrahydrofuran converts the initial monoadduct to a 1,2-dimethylenecyclohexane diene type that subsequently undergoes the second cycloaddition (Eq. 1.43). 2-Dimethylaminomethyl-3-trimethylsilyl-1,3-butadiene has also been used as a starting diene.[91] In this case the monoadduct is converted to the 1,2-dimethylenecyclohexane derivative "via conjugate 1,4-elimination" of the corresponding tetralkylammonium salt (Eq. 1.44).

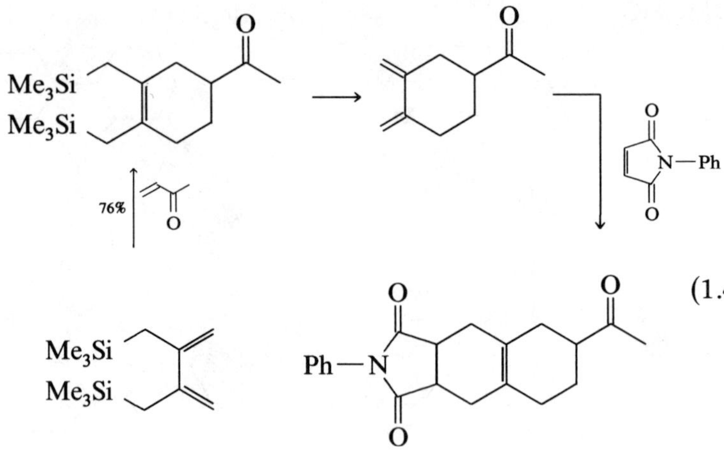

(1.43)

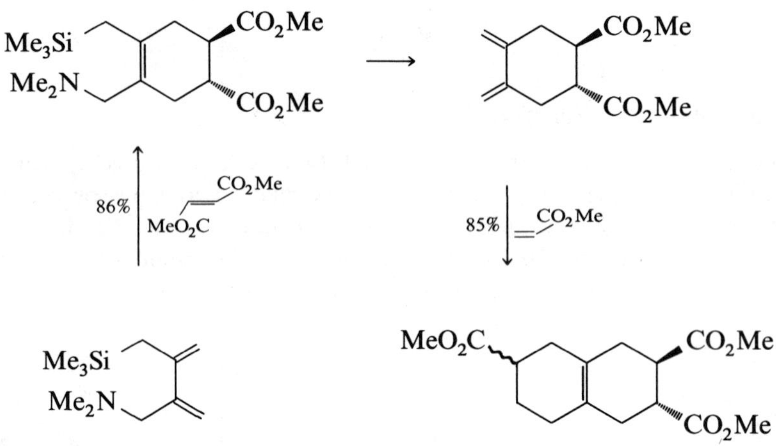

(1.44)

Diene Transmissive Diels–Alder Reaction

Conceptually, this multiple process involves an initial condensation of a cross-conjugated triene with a dienophile, affording a monoadduct containing a new diene unit which then undergoes a second intermolecular cycloaddition. The generalized cycloaddition sequence may be depicted as in Eq. 1.45.

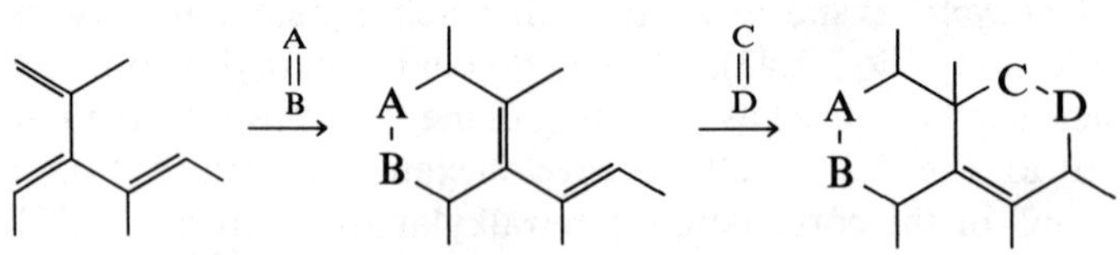

(1.45)

Typical examples[92] are illustrated in Eqs. 1.46 and 1.47.

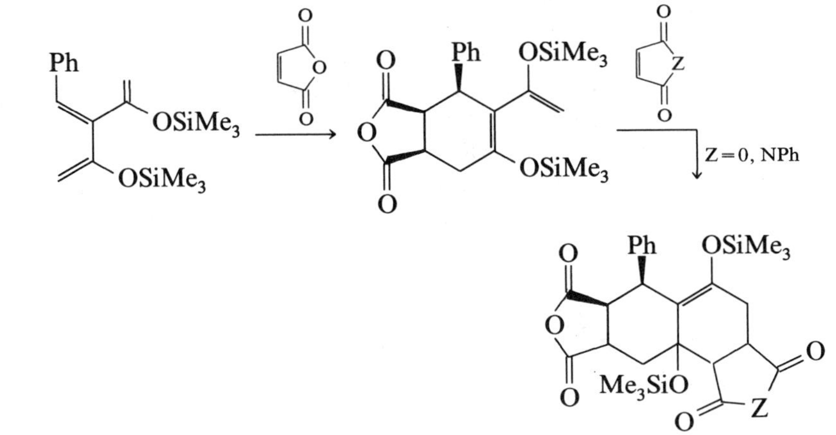

(1.46)

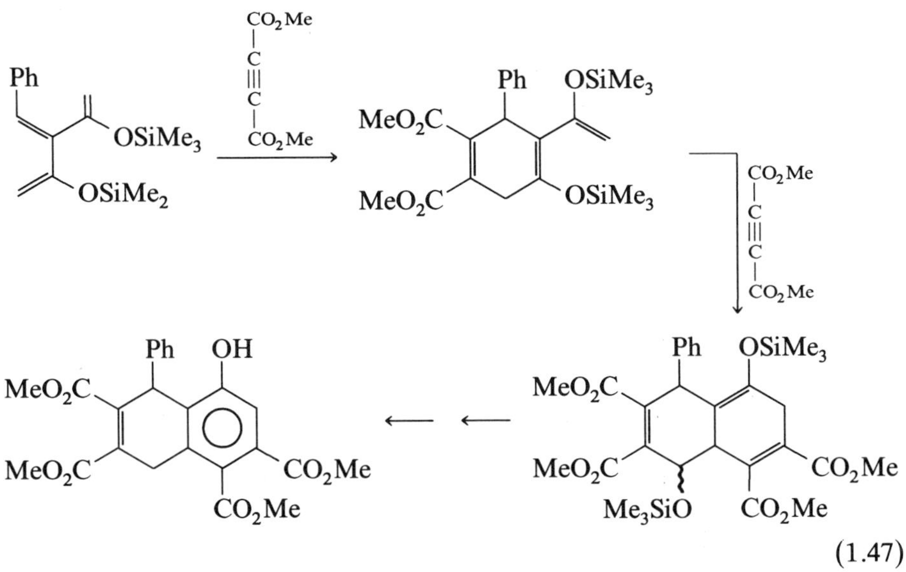

(1.47)

1.15 RETRO DIELS–ALDER REACTION

The Diels–Alder reaction is reversible; the reverse reaction, called retro Diels–Alder reaction, converts the adduct(s) to diene(s) and dienophile(s). The cracking of dicyclopentadiene to give cyclopentadiene, and the kinetic investigation of the secondary α-deuterium isotope effects on 9,10-ethanoanthracenes

(Sec. 1.4), are examples of retro Diels–Alder reaction. These cycloreversions have been studied with respect to their mechanistic aspects and especially for their use in preparative organic chemistry.[8,93,94] Molecules that are difficult to obtain by conventional procedures have often been prepared successfully using this reaction. Let us examine some examples that illustrate these points.

The synthesis of *cis*-crotonaldehyde has been performed by a three-step route[95] from easily accessible starting materials (Eq. 1.48). Intermolecular cycloaddition of (*E*)-crotonitrile with cyclopentadiene gives a mixture of *endo-exo* adducts that, after reduction and thermolysis, yield *cis*-crotonaldehyde stereospecifically.

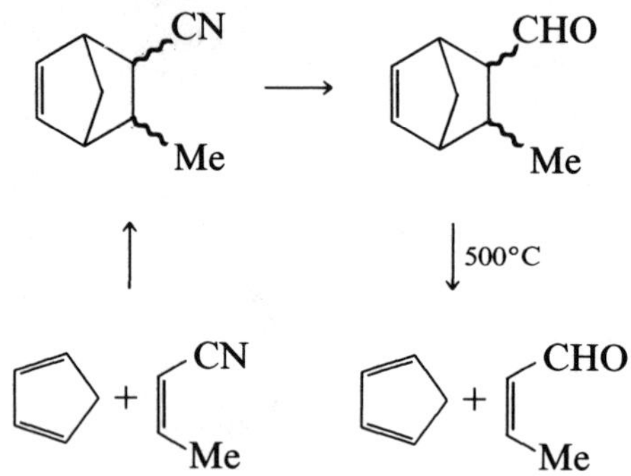

(1.48)

A valuable cycloreversion-based synthesis of β-substituted furanes, a class of compounds somewhat difficult to prepare, has been accomplished according the sequence illustrate in Eq. 1.49.[96]

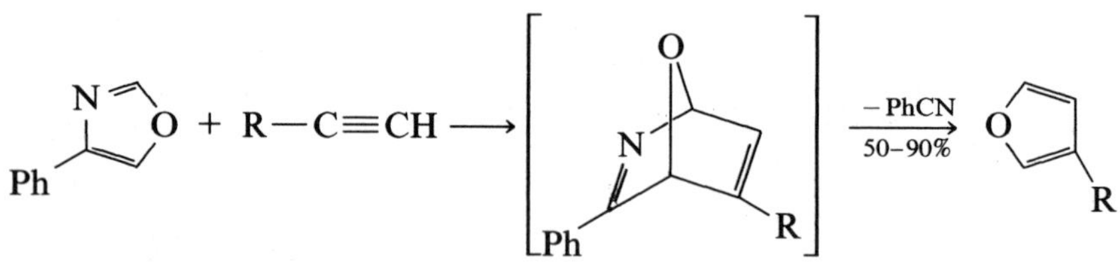

(1.49)

The combination of Diels–Alder cycloaddition–cycloreversion reactions represents a useful tool to protect and then regenerate ethylenic double bonds. An efficient procedure for protecting the diene moiety of ergosterol esters by cycloaddition with *N*-(*p*-nitrophenyl)-1,2,4-triazolinedione has been reported by Barton[97] (Eq. 1.50). The diene moiety of the ergosterol esters is regenerated by cycloreversion under mild conditions. These few examples show the flexibility, simplicity, and wide variety of applications of the cycloreversion reactions.

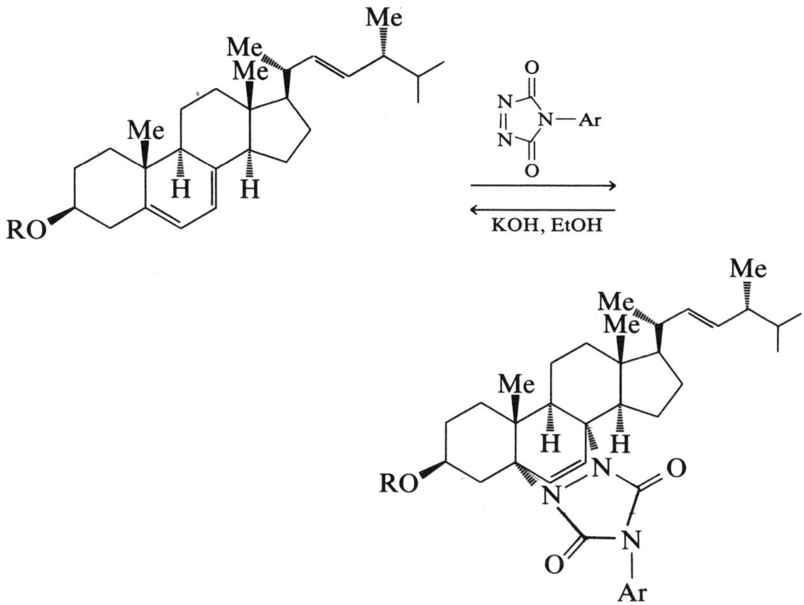

(1.50)

1.16 ASYMMETRIC INDUCTION

Among the asymmetric organic reactions used to produce optically active compounds, a Diels–Alder reaction in both the intermolecular and intramolecular versions merits particular attention. Indeed, two σ-bonds are formed in this process and therefore up to four chiral centers may be produced at binding sites. Asymmetric induction in the intermolecular process may be achieved by starting with chiral dienophiles, chiral dienes, and using chiral catalysts.[100,101]

Chiral Dienophiles

Chiral-conjugated carboxylic esters such as acrylates and fumarates are the most used dienophiles. Thus the cycloaddition of (+)phenylmenthyl acrylate to 5-benzyloxymethylcyclopentadiene catalyzed by aluminum chloride[102] gives the *endo* adduct in high yield and with high diastereotopic face differentiation (d.e. 89%) (Eq. 1.51).

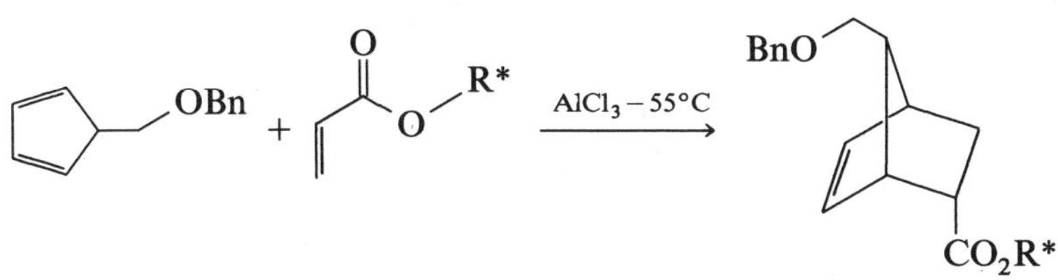

R* = 8-phenylmenthyl

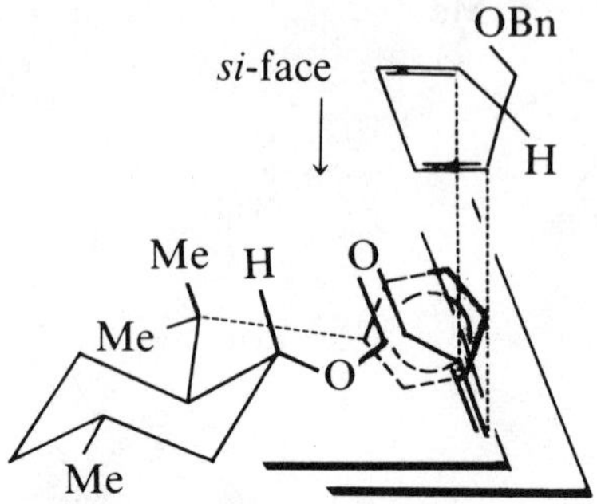

(1.51)

The *si*-face directing effect has been explained[100] considering a transition state where the ester carbonyl group is antiplanar with the olefinic bond and synplanar with the alkoxy-C,H bonds, which causes the phenyl group to shield the *re*-face and makes easier the dienophile to the *si*-face.

Better results in the Diels–Alder cycloadditions of cyclopentadiene with chiral acrylates have been achieved by Oppolzer using isoborneol derivatives as chiral auxiliaries.[103] In the cycloaddition of *cis*-2-neopentyloxyisobornyl acrylate with cyclopentadiene (Eq. 1.52), induction is virtually quantitative. The result has been explained by considering the staggered conformation of the neopentyloxy chain, in which the *re*-face is shielded by the *t*-butyl group.

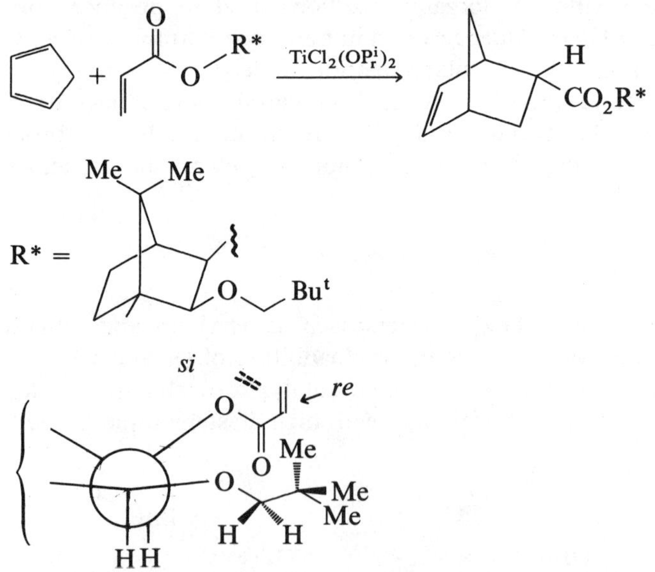

(1.52)

Dependence of optical yield on the pressure has bee observed in the cycloaddition of (−)-di-(*R*)-menthylfumarate with 1,3-butadiene and isoprene[104] (Eq. 1.53). In all cases the *S*-configuration is induced at C-1 and C-2 and the optical yield varies in a range from 6.2 to 12.8% and increases with increased pressure. Historically, this is the system in which Korolev and Mür first demonstrated the possibility of asymmetric induction in the Diels–Alder reaction.[105]

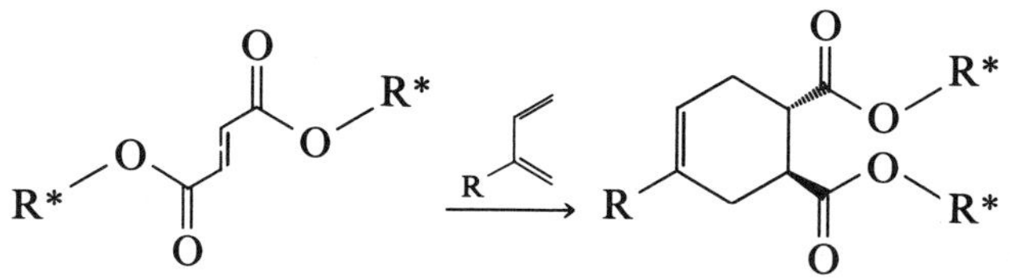

(1.53)

R* = menthyl; R = H, Me

The cycloaddition of optically active 2-methylene-imidazolidines with (E,E)- and (E,Z)-methyl-2,4-hexadienoates (Eq. 1.54) leads to diastereomeric N,N-acetals, which can be cleaved by dilute acids to produce optically active cyclohexenones.[106] The optical yields are not high. In this sequence the 2-methyleneimidazolidines are used as chiral ketene equivalents. This reaction is a rare example of an enantioselective inverse electron demand Diels–Alder cycloaddition.

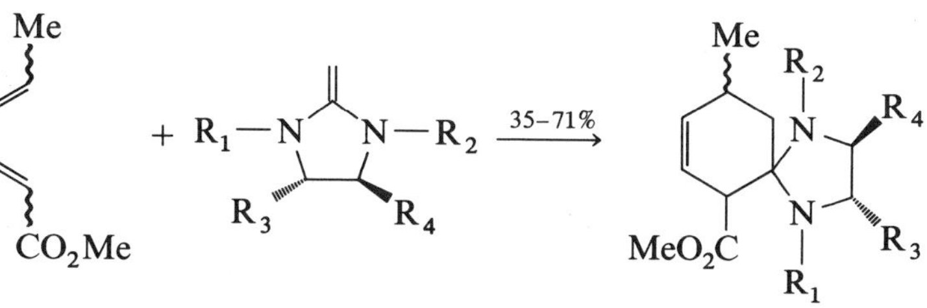

(1.54)

Chiral Dienes

Asymmetric Diels–Alder reactions of chiral dienes have been investigated less extensively. Common chiral dienes are those bearing chiral allylic substituents or sugar residue linked at C-1. Let us examine one example to show the use of chiral dienes.

Trost and co-workers[107] have developed a useful working model for enantioselectivity in Diels–Alder reaction using an (S)-O-methylmandeloxy group as a chiral inducing agent. The cycloaddition of 1-(S)-O-methylmandeloxy-1,3-butadiene with AC in the presence of $BF_3 \cdot Et_2O$ gives a 82 : 18 mixture of the 3R,4S and 3S,4R isomers (Eq. 1.55).

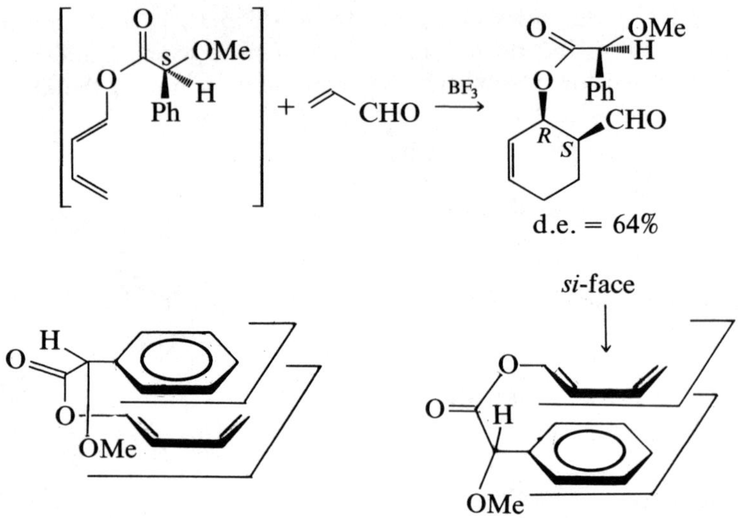

(1.55)

A B

The analogous reaction with juglone occurs with virtually complete asymmetric induction. The enantiotopic face differentiation has been explained by considering that on the basis of a π-stacking model, two conformations **A** and **B** can be envisioned for the mandeloxybutadiene, of which one, **A**, shows severe nonbonded interaction, and hence the conformer **B** undergoes preferential dienophile addition opposite the phenyl-shielded diene face.

Chiral Catalyst

Although the use of chiral catalyst should be attractive, little work has been done in this field to date. The first significant results have been achieved in the cycloadditions of cyclopentadiene with AC, MAC, and MTA catalyzed by chiral alcoxyaluminum dichlorides (Eq. 1.56).[108]

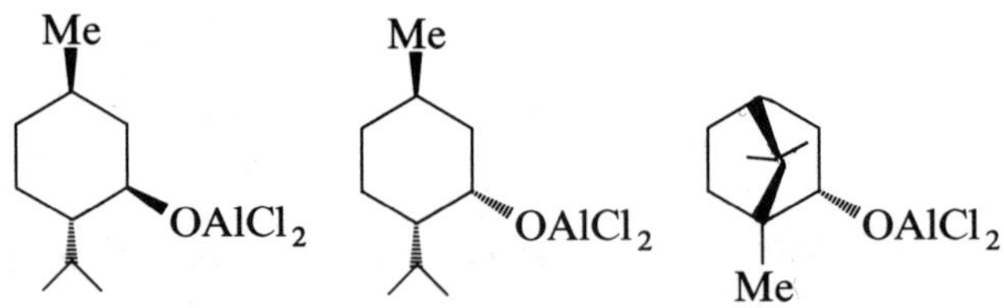

(1.56)

REFERENCES

1. J. March, "Advanced Organic Chemistry," 3rd Edition, Wiley-Interscience, New York (1985).
2. a. A. S. Onishenko, "Diene Synthesis", Israel Program for Scientific Translations, Jerusalem (1964).
   b. J. Sauer, *Angew. Chem.* (*Int. Ed. Engl.*), **6**, 16 (1967).

3. E. L. Eliel, N. L. Allinger, S. J. Angjal, and G. A. Morrison, "Conformational Analysis", Wiley, New York (1965).
4. C. A. Stewart, Jr., *J. Org. Chem.*, **28**, 3320 (1963).
5. T. J. Brocksom and M. G. Constantino, *J. Org. Chem.*, **47**, 3450 (1982).
6. A. Wassermann, "Diels–Alder Reactions", Elsevier, New York (1965).
7. J. Sauer, H. Wiest, and A. Mielert, *Chem. Ber.*, **97**, 3183 (1964).
8. J. Sauer and R. Sustmann, *Angew. Chem. (Int. Ed. Engl.)*, **19**, 779 (1980).
9. G. Desimoni, G. Tacconi, A. Barco, and G. P. Pollini, "Natural Product Synthesis through Pericyclic Reactions", *ACS Monograph*, **180**, American Chemical Society, Washington, D.C. (1983).
10. Inter alia:
    a. R. B. Woodward and T. J. Katz, *Tetrahedron*, **5**, 70 (1959).
    b. K. N. Houk, Y.-T. Lin, and F. K. Brown, *J. Am. Chem. Soc.*, **108**, 554 (1986) and references cited therein.
    c. M. J. S. Dewar and A. B. Pierini, *J. Am. Chem. Soc.*, **106**, 203 (1984).
    d. F. Bernardi, A. Bottoni, M. A. Robb, M. J. Field, I. H. Hillier, and M. F. Guest, *J. Chem. Soc. Chem. Commun.*, 1051 (1985).
    e. M. T. H. Liu and C. Schmidt, *Tetrahedron*, **27**, 5289 (1971).
11. E. Ciganek, *Org. React.*, **32**, 1 (1984).
12. M. Taagepera and E. R. Thorton, *J. Am. Chem. Soc.*, **94**, 1168 (1972).
13. M. J. S. Dewar, S. Olivella, and J. J. P. Stewart, *J. Am. Chem. Soc.*, **108**, 5771 (1986).
14. F. Jensen, and C. S. Foote, *J. Am. Chem. Soc.*, **109**, 6376 (1987).
15. E. L. Clennan and A. D. Earlywine, *J. Am. Chem. Soc.*, **109**, 7104 (1987).
16. K. N. Houk, *Acc. Chem. Res.*, **8**, 361 (1975).
17. K. Fukui, *Fortschr. Chem. Forsch.*, **15**, 1 (1970).
18. M. J. S. Dewar, "The Molecular Orbital Theory of Organic Chemistry", McGraw-Hill, New York (1969).
19. I. Fleming, "Frontier Orbitals and Organic Chemical Reactions", Wiley, New York (1976).
20. R. Sustmann, *Pure Appl. Chem.*, **40**, 569 (1974).
21. W. C. Herndon, *Chem. Rev.*, **72**, 157 (1972).
22. R. Sustmann, *Tetrahedron Lett.*, 2721 (1971).
23. R. Sustmann and R. Shubert, *Angew. Chem. (Int. Ed. Engl.)*, **11**, 840 (1972).
24. J. Saüer and D. Lang, *Angew. Chem.*, **76**, 603 (1964).
25. D. E. Davies, T. L. Gilchrist, and T. G. Roberts, *J. Chem. Soc. Perkin Trans. 1*, 1275 (1983).
26. R. R. Schmidt and M. Maier, *Tetrahedron Lett.*, **23**, 1789 (1982).
27. G. Desimoni and G. Tacconi, *Chem. Rev.*, **75**, 651 (1975).
28. D. L. Boger and S. N. Weinreb, "Hetero Diels–Alder Methodology in Organic Synthesis", *Organic Chemistry, A Series of Monographs*, **47**, Academic Press, New York (1987).
29. A. I. Konovalov and B. N. Solomonov, *Dokl. Akad. Nauk SSSR, Ser. Khim.*, **211**, 1115 (1973).

30. E. C. Angell, F. Fringuelli, L. Minuti, F. Pizzo, A. Taticchi, and E. Wenkert, *J. Org. Chem.*, **51**, 5177 (1986).
31. O. Eisenstein, J. M. Lefour, N. T. Anh, and R. F. Hudson, *Tetrahedron*, **33**, 523 (1977).
32. I.-M. T. Larsoon, M. D. Rozeboom, and K. N. Houk, *Tetrahedron Lett.*, **22**, 2043 (1981).
33. P. V. Alston and R. M. Ottenbrite, *J. Org. Chem.*, **40**, 1111 (1975).
34. A. Hosomi, H. Iguchi, J. Sasaki, and H. Sakurai, *Tetrahedron Lett.*, **23**, 551 (1982).
35. R. Huisgen, *J. Org. Chem.*, **33**, 2291 (1968).
36. a. P. V. Alston and D. D. Shillady, *J. Org. Chem.*, **39**, 3402 (1974).
 b. K. N. Houk, *J. Am. Chem. Soc.*, **95**, 4092 (1973).
 c. N. D. Epiotis, *J. Am. Chem. Soc.*, **95**, 5624 (1973).
 d. O. Eisenstein, J.-M. Lefour, and N. T. Anh, *J. Chem. Soc. Chem. Commun.*, 969 (1971).
37. S. D. Kahn, C. F. Pau, L. E. Overman, and W. J. Hehre, *J. Am. Chem. Soc.*, **108**, 7381 (1986).
38. P. V. Alston, R. M. Ottenbrite, and D. D. Shillady, *J. Org. Chem.*, **38**, 4075 (1973).
39. T. Inukai and T. Kojima, *J. Org. Chem.*, **36**, 924 (1971).
40. K. N. Houk and R. W. Strozier, *J. Am. Chem. Soc.*, **95**, 4094 (1973).
41. O. F. Guner, R. M. Ottenbrite, D. D. Shillady, and P. V. Alston, *J. Org. Chem.*, **52**, 391 (1987).
42. K. Alder and G. Stein, *Angew. Chem.*, **50**, 510 (1937).
43. L. E. Overman, R. L Freerks, C. B. Petty, L. A. Clizbe, R. K. Ono, G. F. Taylor, and P. J. Jessup, *J. Am. Chem. Soc.*, **103**, 2816 (1981).
44. Y. Kobuke, T. Fueno, and J. Furukawa, *J. Am. Chem. Soc.*, **92**, 6548 (1970).
45. E. C. Angell, F. Fringuelli, L. Minuti, F. Pizzo, B. Porter, A Taticchi, and E. Wenkert, *J. Org. Chem.*, **51**, 2649 (1986) and references cited therein.
46. F. Fringuelli, L. Minuti, L. Radics, A. Taticchi, and E. Wenkert, *J. Org. Chem.*, **53**, 4607 (1988).
47. J. G. Martin and R. K. Hill, *Chem. Rev.*, **61**, 537 (1961).
48. Inter alia: D. Ginsburg, *Tetrahedron*, **39**, 2095 (1983) and references cited therein.
49. A. Wassermann, *J. Chem. Soc.*, **825**, 1511 (1935).
50. R. B. Woodward and H. Baer, *J. Am. Chem. Soc.*, **66**, 645 (1944).
51. W. C. Herndon and L. H. Hall, *Tetrahedron Lett.*, 3095 (1967).
52. L. M. Stephenson, D. E. Smith, and S. P. Current, *J. Org. Chem.*, **47**, 4171 (1982).
53. K. N. Houk, *Tetrahedron Lett.*, 2621 (1970).
54. Z. M. Ismail and H. M. R. Hoffmann, *J. Org. Chem.*, **46**, 3549 (1981).
55. J. A. Berson, Z. Hamlet, and W. A. Mueller, *J. Am. Chem. Soc.*, **84**, 297 (1962).
56. Inter alia:
 a. M. Avenati and P. Vogel, *Helv. Chim. Acta*, **66**, 1279 (1983).
 b. F. K. Brown, K. N. Houk, D. J. Burnell, and Z. Valenta, *J. Org. Chem.*, **52**, 3050 (1987).
 c. D. L. Boger and M. Patel, *Tetrahedron Lett.*, **27**, 683 (1986).

d. E. C. Angell, F. Fringuelli, F. Pizzo, B. Porter, A. Taticchi, and E. Wenkert, *J. Org. Chem.*, **51**, 2642 (1986).
e. S. D. Kahn and W. J. Hehre, *Tetrahedron Lett.*, **27**, 6041 (1986).

57. E. Corey, T. Ravindranathan, and S. Tereshima, *J. Am. Chem. Soc.*, **93**, 4326 (1971).
58. D. Liotta, M. Saindane, and C. Barnum, *J. Am. Chem. Soc.*, **103**, 3224 (1981).
59. P. Vogel, in "Stereochemistry and Reactivity of Systems Containing π Electrons", W. H. Watson, Ed., of the series *Methods in Stereochemical Analysis*, Verlag Chemie, Weinheim, West Germany (1983); Proceedings of the Symposium "Stereochemistry and Reactivity of Pi Systems".
60. A. Wassermann, *J. Chem. Soc.*, 618 (1942).
61. P. Yates and P. Eaton, *J. Am. Chem. Soc.*, **82**, 4436 (1960).
62. F. Fringuelli, F. Pizzo, A. Taticchi, T. D. J. Halls, and E. Wenkert, *J. Org. Chem.*, **47**, 5056 (1982).
63. a. M. Bednarski and S. Danishefsky, *J. Am. Chem. Soc.*, **105**, 3716 (1983).
    b. S. Danishefsky and M. Bednarski, *Tetrahedron Lett.*, **24**, 721 (1984).
64. G. A. Olah, D. Meidar, and A. P. Fung, *Synthesis*, 270 (1979).
65. N. D. Epiotis and S. Shaik, *J. Am. Chem. Soc.*, **100**, 1 (1978).
66. a. V. Branchadell, A. Oliva, and J. Bertran, *Chem. Phys. Lett.*, **97**, 378 (1983).
    b. V. Branchadell, A. Oliva, and J. Bertran, *J. Mol. Struct.* (*Theochem.*), **138**, 117 (1986).
67. D. J. Nelson, *J. Org. Chem.*, **51**, 3185 (1986).
68. J. R. McCabe and C. A. Eckert, *Acc. Chem. Res.*, **7**, 251 (1974).
69. a. W. J. Le Noble and H. Kelm, *Angew. Chem.* (*Int. Ed. Engl.*), **19**, 841 (1980).
    b. W. J. Le Noble and T. Asano, *Chem. Rev.*, **78**, 407 (1978).
70. W. G. Dauben and W. R. Baker, *Tetrahedron Lett.*, **23**, 2611 (1982).
71. J. Jurczak, M. Chimielewski, and S. Filipek, *Synthesis*, 41 (1979).
72. M. J. S. Dewar and R. S. Pyron, *J. Am. Chem. Soc.*, **92**, 3098 (1970).
73. a. A. Corsico Coda, G. Desimoni, E. Ferrari, P. P. Righetti, and G. Tacconi, *Tetrahedron*, **40**, 1611 (1984).
    b. G. Desimoni, G. Faita, P. P. Righetti, N. Tornaletti, and M. Visigalli, *J. Chem. Soc. Perkin Trans. 2*, 437 (1989).
74. A. Wassermann, *Monatsh. Chem.*, **83**, 543 (1952).
75. T. Dumaus, W. Hoekstra, M. Pentaleri, and D. Liotta, *Tetrahedron Lett.*, **29**, 3745 (1988).
76. R. Breslow and D. Rideout, *J. Am. Chem. Soc.*, **102**, 7816 (1980).
77. a. R. Breslow, U. Maitra, and D. Rideout, *Tetrahedron Lett.*, **24**, 1901 (1983).
    b. R. Breslow and U. Maitra, *Tetrahedron Lett.*, **25**, 1239 (1984).
78. a. P. A. Grieco, P. Garner, and Z. He, *Tetrahedron Lett.*, **24**, 1897 (1983).
    b. P. A. Grieco, K. Yoshida, and P. Garner, *J. Org. Chem.*, **48**, 3137 (1983).
79. H. Parlar and R. Baumann, *Angew. Chem.* (*Int. Ed. Engl.*), **20**, 1014 (1981).
80. V. Yedidia and C. C. Leznoff, *Can. J. Chem.*, **58**, 1144 (1980) and references cited therein.
81. W. Oppolzer, *Angew. Chem.* (*Int. Ed. Engl.*), **16**, 10 (1977).

82. D. F. Taber, "Intramolecular Diels–Alder and Alder Ene Reactions", Springer-Verlag, New York (1984).
83. S. M. Weinreb, *Acc. Chem. Res.*, **18**, 16 (1985).
84. D. F. Taber and B. P. Gunn, *J. Am. Chem. Soc.*, **101**, 3992 (1979).
85. K. Hayakawa, T. Yasukouchi, and K. Kanematsu, *Tetrahedron Lett.*, **27**, 1837 (1986).
86. S. M. Weinreb, N. A. Khatri, and J. Shringapure, *J. Am. Chem. Soc.*, **101**, 5073 (1979).
87. L. A. Paquette and M. J. Wyvratt, *J. Am. Chem. Soc.*, **96**, 4671 (1974).
88. G. A. Krauss and M. J. Tashner, *J. Am. Chem. Soc.*, **102**, 1974 (1980).
89. J. M. Tornare and P. Vogel, *Helv. Chim. Acta*, **68**, 1069 (1985).
90. B. M. Trost and M. Shimizu, *J. Am. Chem. Soc.*, **104**, 4299 (1982).
91. A. Hosomi, K. Otaka, and H. Sakurai, *Tetrahedron Lett.*, **27**, 2881 (1986).
92. a. O. Tsuge, E. Wada, and S. Kanemasa, *Chem. Lett.*, 239 (1983).
    b. O. Tsuge, E. Wada, S. Kanemasa, and H. Sakoh, *Bull. Chem. Soc. Jpn.*, **57**, 3221 (1984).
    c. O. Tsuge, S. Kanemasa, H. Sakoh, and E. Wada, *Bull. Chem. Soc. Jpn.*, **57**, 3234 (1984).
    d. S. Kanemasa, H. Sakoh, E. Wada, and O. Tsuge, *Bull. Chem. Soc. Jpn.*, **59**, 1869 (1986).
93. J.-L. Ripoll, A. Rouessac, and F. Rouessac, *Tetrahedron*, **34**, 19 (1978).
94. M.-C. Lasne and J.-L. Ripoll, *Synthesis*, 121 (1985).
95. M. Perrier and F. Rouessac, *Nouv. J. Chim.*, **1**, 367 (1977).
96. D. Liotta, M. Saindane, and W. Ott, *Tetrahedron Lett.*, **24**, 2473 (1983).
97. D. H. R. Barton, X. Lusinschi, and J. S. Ramirez, *Tetrahedron Lett.*, **24**, 2995 (1983).
98. D. J. Bellville, D. D. Wirth, and N. L. Bauld, *J. Am. Chem. Soc.*, **103**, 718 (1981).
99. a. D. J. Bellville and N. L. Bauld, *J. Am. Chem. Soc.*, **104**, 2665 (1982).
    b. N. L. Bauld, D. J. Bellville, R. Pabon, R. Chelsky, and G. Green, *J. Am. Chem. Soc.*, **105**, 2378 (1983).
100. W. Oppolzer, *Angew. Chem.* (*Int. Ed. Engl.*), **23**, 876 (1984).
101. J. D. Morrison, "Asymmetric Synthesis", Academic Press, New York, vol. 3 (1984).
102. E. J. Corey and H. E. Ensley, *J. Am. Chem. Soc.*, **97**, 6908 (1975).
103. W. Oppolzer, C. Chapuis, G. M. Dao, D. Reichlin, and T. Godel, *Tetrahedron Lett.*, **23**, 4781 (1982).
104. J. Jurczak, *Bull. Chem. Soc. Jpn.*, **52**, 3438 (1979).
105. A. Korolev and V. Mur, *Dokl. Akad. Nauk SSSR*, **59**, 251 (1948).
106. U. Gruseck and M. Heuschmann, *Tetrahedron Lett.*, **28**, 2681 (1987).
107. B. M. Trost, D. O'Krongly, and J. L. Belletire, *J. Am. Chem. Soc.*, **102**, 7595 (1980).
108. S. Hashimoto, N. Komeshima, and K. Koga, *J. Chem. Soc. Chem. Commun.*, 437 (1979).

2

OPEN-CHAIN DIENES

2.1 1,3-BUTADIENES

Reactivity, Selectivity, Synthesis The 1,3-butadienes react with a large number of dienophiles and the six-membered ring of the adduct is carbocyclic or heterocyclic. General pictures are illustrated in Schemes 2.1 and 2.2, where 1,3-butadiene is used as a representative structure of this class of dienes.

The majority of these cycloadditions are HOMO diene–LUMO dienophile controlled (Sec. 1.5), and therefore the reactivity of the diene is increased on the electronic ground by electron-donating substituents and decreased by electron-withdrawing groups. Tables 2.1 and 2.2 illustrate the substituent effect on the reactivity of typical 1,3-butadienes in the cycloaddition with TCNE. Hammett reaction constants have been calculated for the Diels–Alder reactions of aryl-1,3-butadienes with several dienophiles,[456] but their mechanistic significance has been questioned.[457]

Electron-donating groups at C-1 in *trans* configuration increase the reactivity of the diene more than when they are bonded at C-2. A substituent at C-1 in *cis* configuration strongly reduces the reactivity of the diene because of steric interactions that destabilize the *cisoid* conformation. Sometimes [2 + 2] cycloadditions are competitive[342] or unexpected Diels–Alder adducts are obtained[87] (Sec. 1.4).

The study of substituent effect on regio- and stereoselectivity requires careful quantitative analyses, which have been possible in recent years due to the greatly improved chromatographic techniques. 1-Substituted-1,3-butadienes react with electron-deficient dienophiles such as acrylates and their analogs to give predominantly *ortho-endo* isomers. The electronic nature and size of the

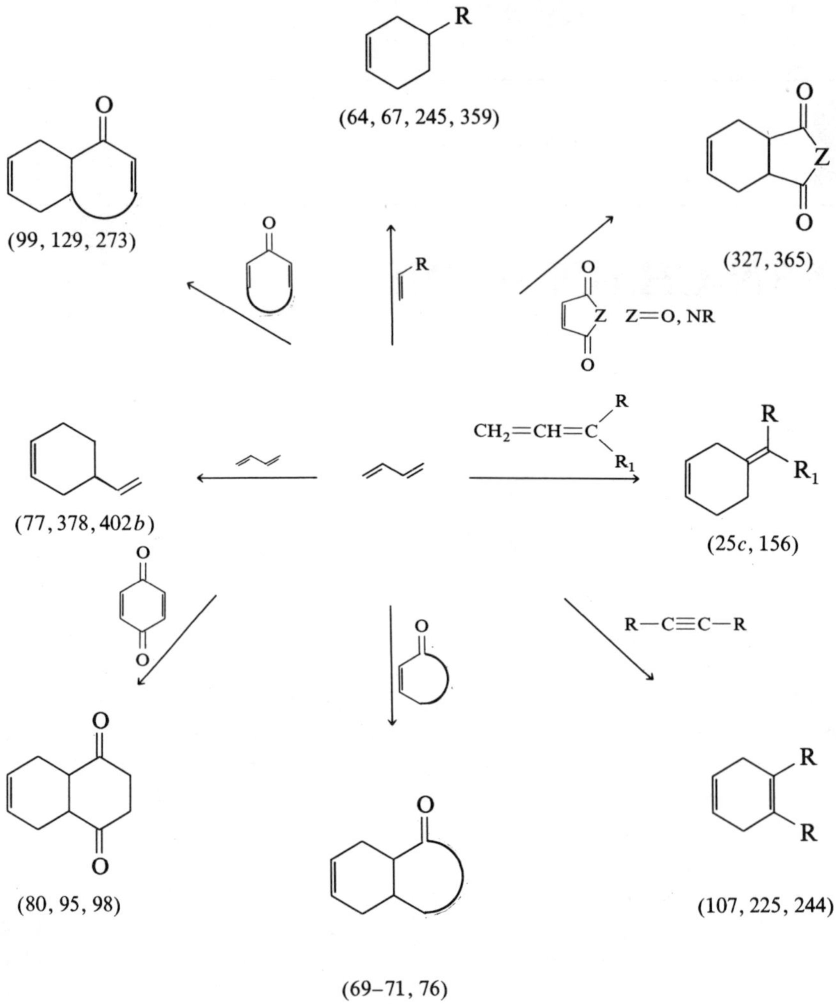

Scheme 2.1 (Reference numbers in parentheses.)

substituent influence the stereoselectivity more than the regioselectivity of the reaction (Scheme 2.3 and Table 2.3). A substituent at C-2 gives the *para* adduct as the major regioisomer (Table 2.3). When two different substituents are present, one of them works as regiodirector and controls the regiochemical outcome of the reaction. The regiocontrol of the cycloaddition of 2-methoxy-3-phenylthiobutadiene (**1**) with MVK, for example,[359,365] is due to the sulfur substituent and adducts **2** and **3** are obtained in a 4 : 1 ratio. Other examples of regiodirector groups are reported in Table 2.4.

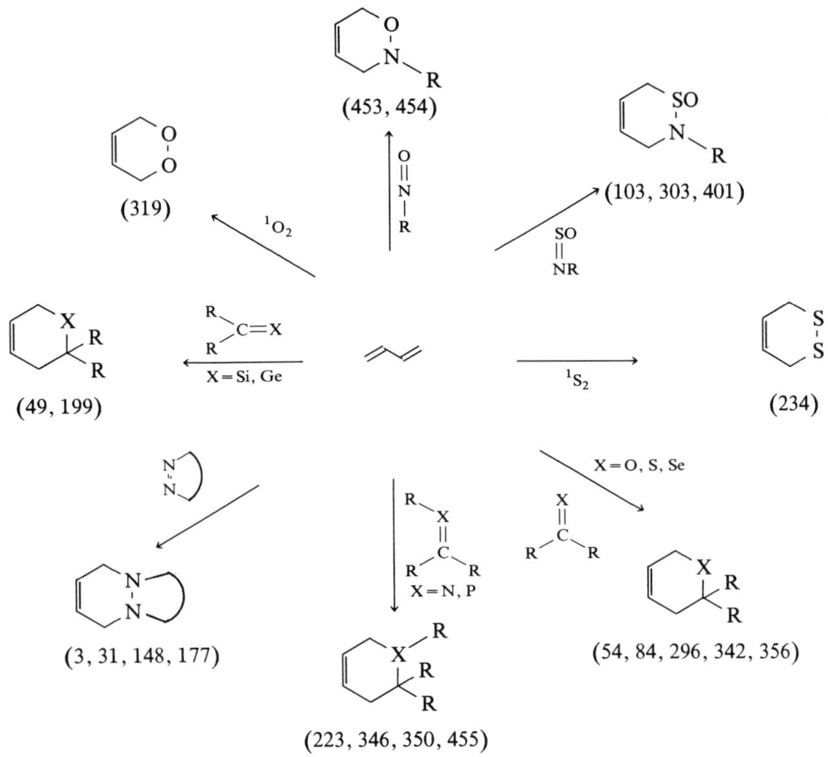

Scheme 2.2 (Reference numbers in parentheses.)

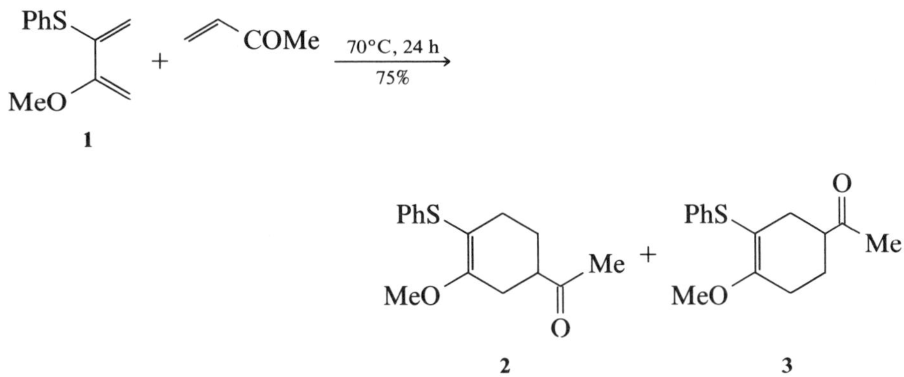

The regio- and stereoselectivity of the cycloadditions can be strongly influenced by catalyst and pressure (Secs. 1.8 and 1.9). The amount of *ortho-endo* adduct of the reaction of 1-*t*-butyl-1,3-butadiene with MAC (Scheme

TABLE 2.1 Relative Rate Constants of the Reactions of Substituted 1,3-Butadienes with TCNE in CH_2Cl_2 at 20°C

	R			
Diene	H	Me	Ph	OMe
	1	103	385	50,934
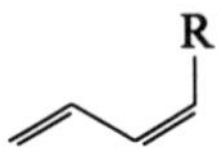	1	0.002		1,125
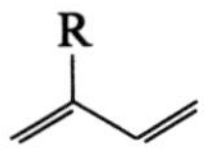	1	45	191	1,750
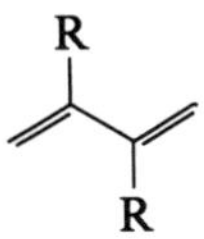	1	1,439		
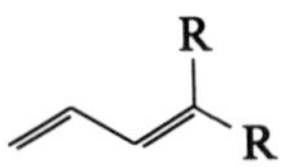	1	1.1		
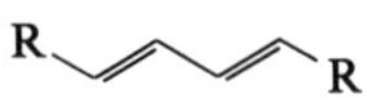	1	1,663	43	49,775
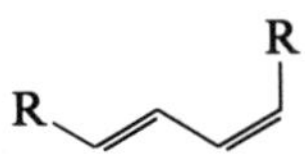	1	0.12		

Source: Ref. 21.

TABLE 2.2 Relative Rate Constants of the Reactions of *trans*-1-Aryl-1,3-butadienes with TCNE in CH_3CN at 20°C

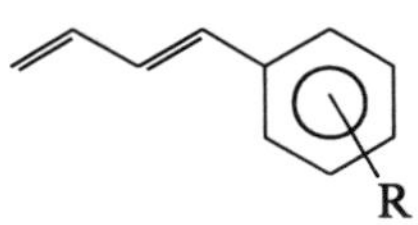

R	k_{rel}
p-NMe_2	976
p-OMe	40
p-Me	3.4
H	1
m-F	0.29
m-Cl	0.26
m-CF_3	0.17
p-NO_2	0.03

Source: Ref. 173.

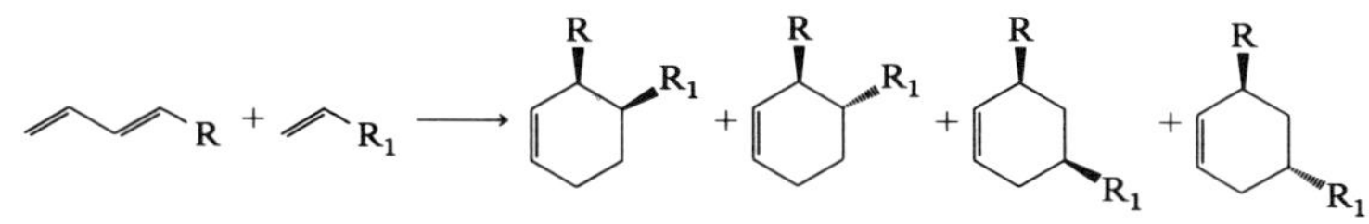

R	R_1	*ortho-endo*	*ortho-exo*	*meta-endo*	*meta-exo*
Me	CO_2Pr	42.7	44.5	6.3	6.5
Pr^i	CO_2Me	48.3	41.0	5.7	5.0
Bu^t	CO_2Me	53.8	34.9	6.7	4.6
OMe	CO_2Me	75.4	24.6	—	—

Scheme 2.3 Distribution (%) of regio- and stereoisomers in the cycloadditions of 1-substituted 1,3-Butadienes with acrylates. (From Ref. 458.)

TABLE 2.3 Regioselectivity in the Cycloadditions of Monosubstituted 1,3-Butadienes with MAC

R	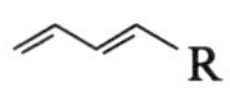 (% *ortho*)	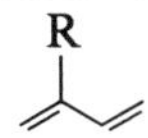 (% *para*)
Me	88	70
Ph	> 98	> 96
COOH	83	
$NHCO_2Et$	98	
OMe	88	70
SPh	> 98	80
SiR_1	75 (R_1 = Me)	77 (R_1 = Et)

Source: Refs. 90 and 459.

TABLE 2.4 Regioselectivity in the Cycloadditions of Disubstituted 1,3-Butadienes with Various Dienophiles

1,2[a] (*ortho* director)	1,4 (*ortho* director)	2,3 (*para* director)
Me > Ph	Me > $SiEt_3$	Ph > Me
Me > OEt	Ph > Me ≃ COOH	SPh > Me
Me > OAc	$NHCO_2Et$ > Me	SPh > OMe
Me > Cl	$NHCO_2Bn$ > SPh	SPh > OAc
Bu > $NHCOCl_3$	OAc > Et	Cl > Me
SPh > OMe	SPh > OAc	

Source: Refs. 90, 359, 365, 459, and 460.

[a] The first-named substituent is bonded at C-1.

2.3) is increased to 62.7% when the reaction is carried out at 6 kbar,[458] and the ratio of adducts **2** and **3** becomes 19 : 1 if the reaction is executed[359] in the presence of $ZnCl_2$. The FMO theory is capable to predict the regiochemistry of cycloadditions of monosubstituted dienes with electron-deficient dienophiles (Sec. 1.5) but sometimes fails in the cycloadditions of disubstituted dienes. An alternative and more recent approach, based on the matching of complementary reactivity surfaces for diene and dienophile[459] (modeling chemical reactivity), appears to be successful where the FMO model misassigns.

The reactivity and selectivity of Diels–Alder reactions of 1,3-butadienes are poorly influenced by organic solvents, but sometimes they are greatly improved by water (Sec. 1.11). Table 2.5 compares the data of reactions carried out in water and toluene or benzene of dienophiles **4–7** with 1-carboxylate-1,3-butadienes (**8–13**) and 1-glyco-1,3-butadienes, which have the butadienyl part linked in the anomeric carbon of α and β-D-glucose (**14**, **16**), and their peracetylated derivatives **15** and **17**. The sodium carboxylate group and the glucose unit make the dienes water soluble. In water, the yields of reactions are high and the rate and stereoselectivity are greatly enhanced.

Similarly, water-soluble dienophiles such as benzyliminium ions generated in situ react in aqueous medium[213] selectively with 1,3-butadiene and its methyl derivatives, which are poorly soluble in water. The cycloaddition can therefore be achieved even if one of the reactants is only sparingly soluble. An exception is the water-insoluble diene **15**, which does not give reactions even with soluble dienophile.[191] Use of alcoholic or hydroalcoholic solvent led to a decrease in the reaction rate.[182,191,213,259]

The water effect has been much debated.[461a] The polarity of the water,[461] hydrophobic effects,[461] micellar aggregation,[259] and entropy-driven association of the reactants[191] have been cited to explain rate and selectivity enhancement.

TABLE 2.5 Reactivity and Stereoselectivity of Diels–Alder Reactions of 1,3-Butadienes in Water and Toluene or Benzene

Reactants		Solvent	Temperature (°C)	Time (h)	*Endo/Exo*	Yield (%)	Ref.
8	**6**	H_2O	25	1	> 100	77	182
9	**6**	Ph-Me	25	168	> 100	Traces	182
8	7	H_2O	60	24	2	67	182
9	7	Ph-Me	110	96	0.5	84	182
10	5	H_2O	25	4	3	99	259
11	5	Ph-H	25	228	0.9	52	259
12	**6**	H_2O	25	0.2	> 100	93	306
13	**6**	Ph-H	80	12	> 100	69	306
14	4	H_2O	20	3.5	> 100	90	191
15	4	Ph-Me	80	168	6.7	80	191
16	4	H_2O	20	6	32	78	191
17	4	Ph-Me	80	168	5.2	94	191

Alkyl- and aryl-1,3-butadienes have been used extensively to study the reactivity, the mechanism, and stereochemistry of the cycloaddition. The alkyl and aryl groups can hardly be manipulated, however, and therefore on the point of view of the synthesis, all the chemical transformations are executed on the dienophilic part of the adduct. Some target structures that can be obtained from cycloadditions and subsequent chemical transformations are reported in Table 2.6.

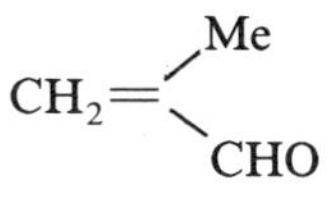

4

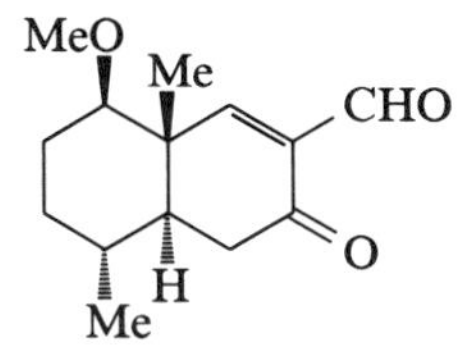

5

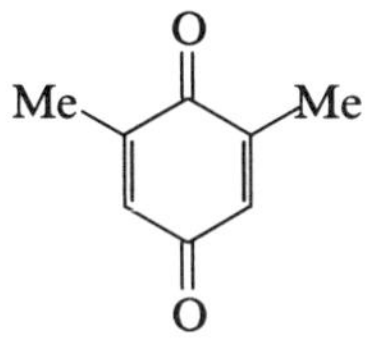

6

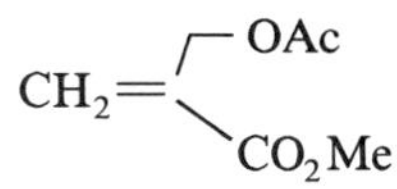

7

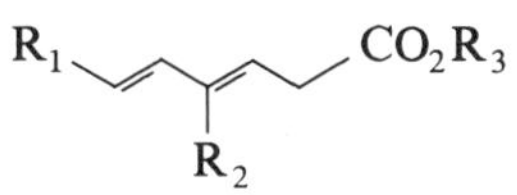

8 R_1 = H, R_2 = H, R_3 = Na
9 R_1 = H, R_2 = H, R_3 = Me

10 R_1 = H, R_2 = Me, R_3 = Na
11 R_1 = H, R_2 = Me, R_3 = Et

12 R_1 = OMe, R_2 = H, R_3 = Na
13 R_1 = OMe, R_2 = H, R_3 = Me

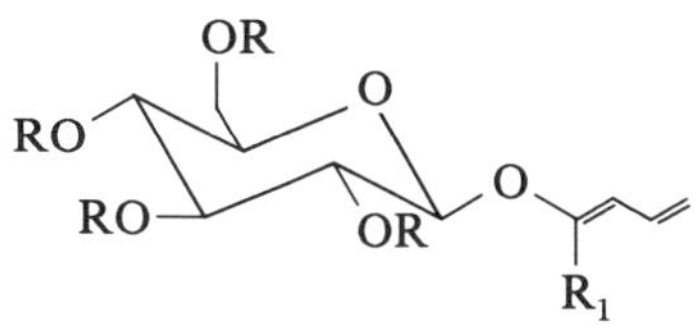

14 R = R_1 = H
15 R = Ac, R_1 = H
18 R = Ac, R_1 = $OSiMe_3$

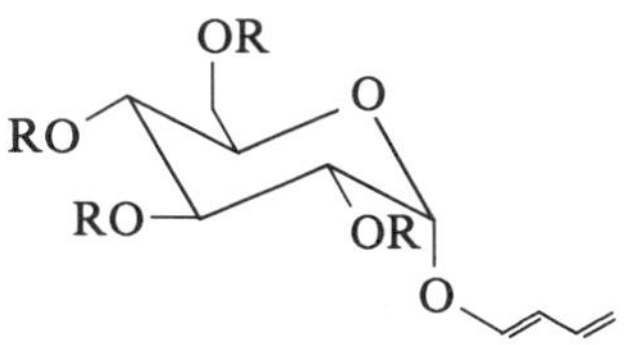

16 R = H
17 R = Ac

The nitro adducts (entry 1a) can be reduced to amines or converted to ketones and the N=O dienophilic functionality allows the synthesis of δ-amino alcohols (entry 1b). Pyridines are formed by spontaneous dehydrogenation of the adducts coming from the thermal cycloaddition with cyanogen (entry 1c), and *N*-sulfinyl dienophile cycloadditions lead to dihydrothiazine oxides, which are stereospecifically converted into acyclic unsaturated vicinal amino alcohols (entry 1d) by a two-step procedure.

The presence of heterosubstituents on the diene affects the regiochemistry and stereochemistry of the cycloaddition and permits chemical transforma-

TABLE 2.6 Synthetic Applications of 1,3-Butadienes

Diene	Dienophile	Target Structure	Ref.
1. 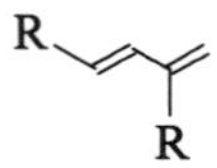 (R = H, Me)	a. $R_1CH{=}CHNO_2$	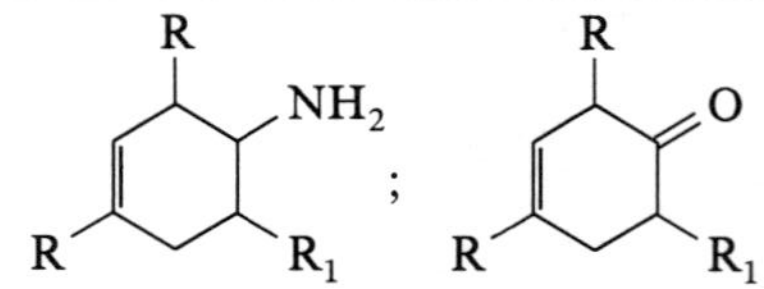	473
	b. 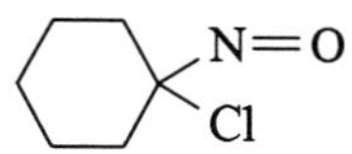	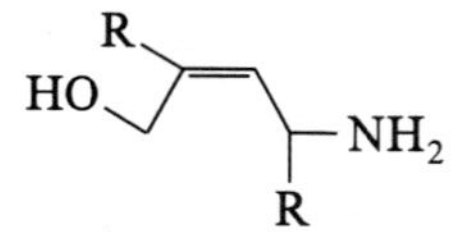	471
	c. $(C{\equiv}N)_2$	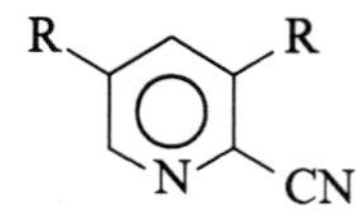	470
	d. $OS{=}NCO_2Bn$	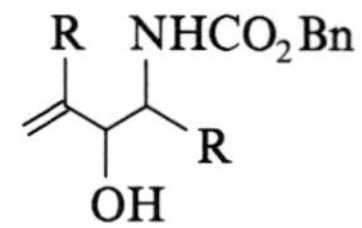	472
2. 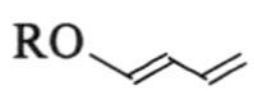 (R = Me, Ac)	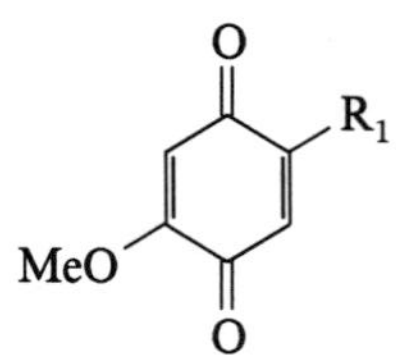	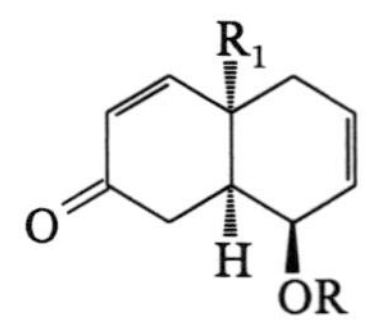	465

3.	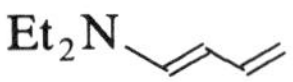		$CH_2{=}CH{-}CHO$	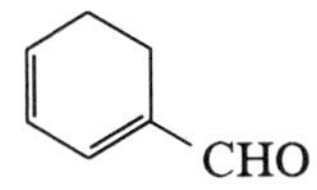	465
4.	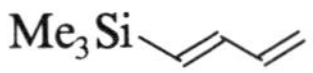	a. 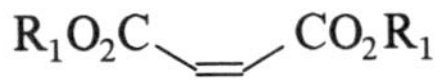	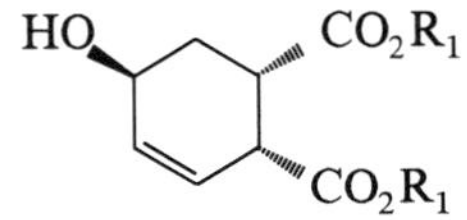	465
		b. 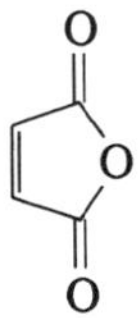	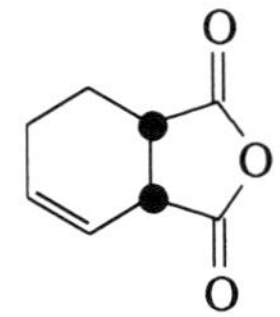	465
5.	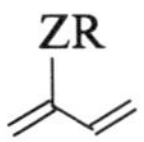 ($ZR = OMe, SPh$)	$CH_2{=}CH{-}CO_2Me$	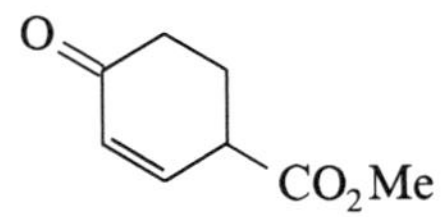	465
6.	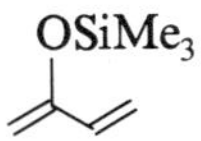	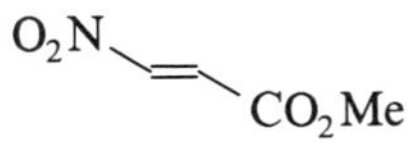	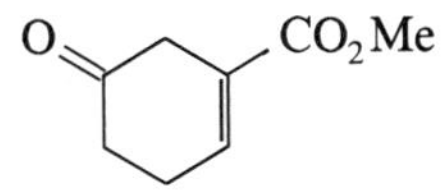	463
7.	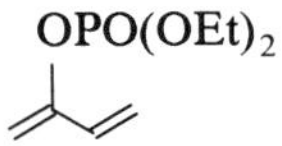	$CH_2{=}CH{-}COMe$	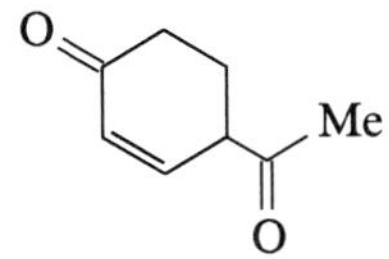	462

TABLE 2.6 (*Continued*)

Diene	Dienophile	Target Structure	Ref.
8. 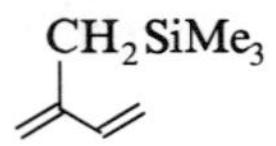	$CH_2{=}CH{-}COMe$	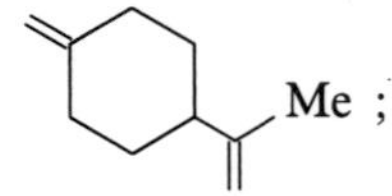 ; 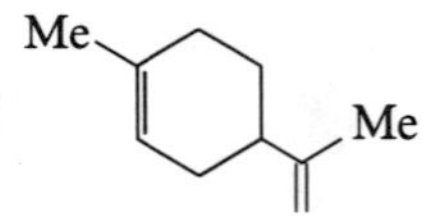	469
9. 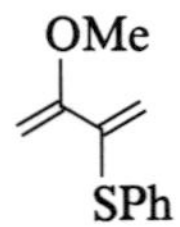	a. 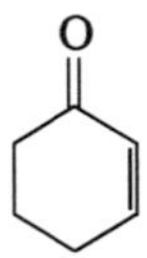	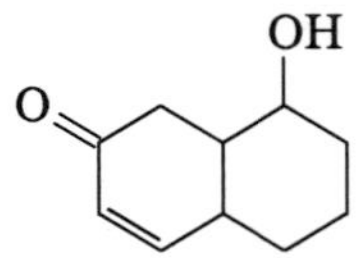 ; 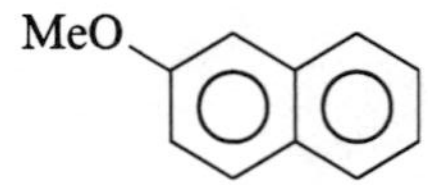	365
	b. $CH_2{=}CH{-}COMe$	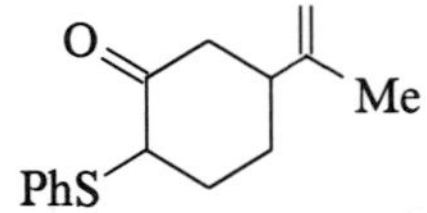 ; 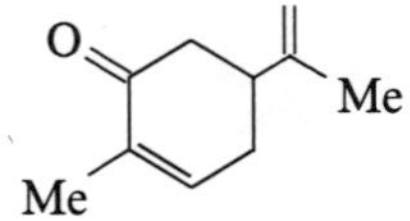	365
10. 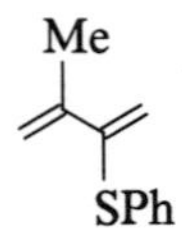	$CH_2{=}CH{-}COMe$	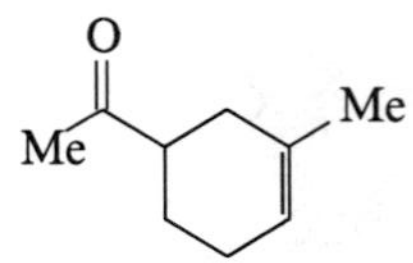	359
11. 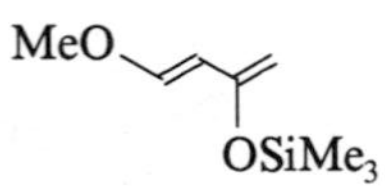	a. $[CH_2{=}NMe_2]^+\ I^-$	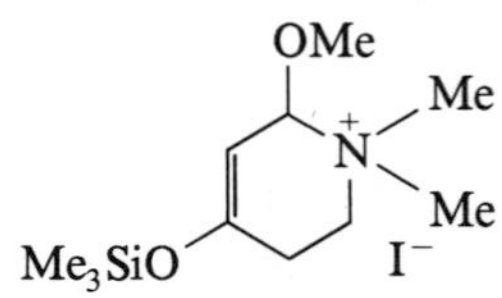 ; 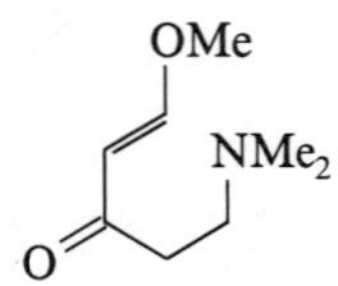	464

	b. 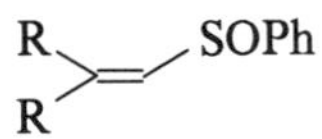		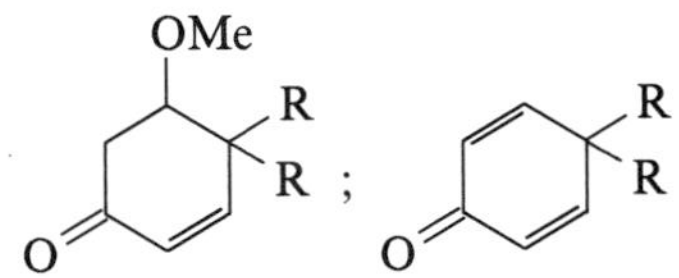	467
	c. $CH_2=CH-C(H)=O$		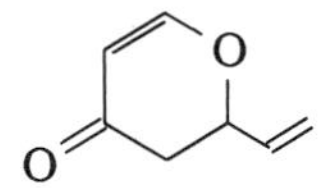	461b
	d. $CH_2=CH-COR$		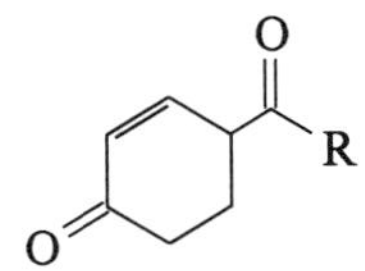	461b
12.	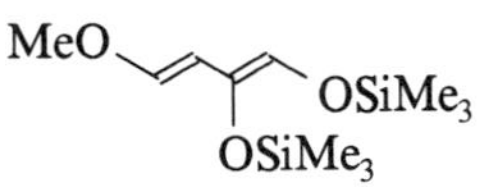	$BnOCH_2-C(H)=O$	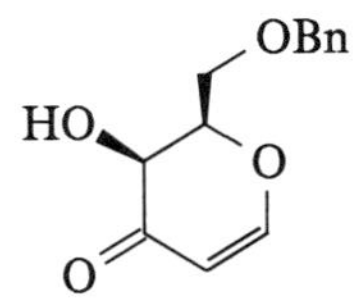	468
13.	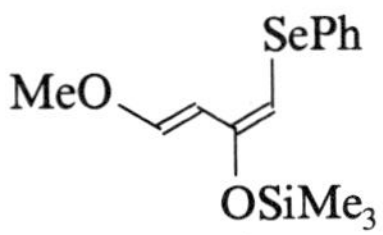	$CH_2=C(R_1)CO_2R_2$	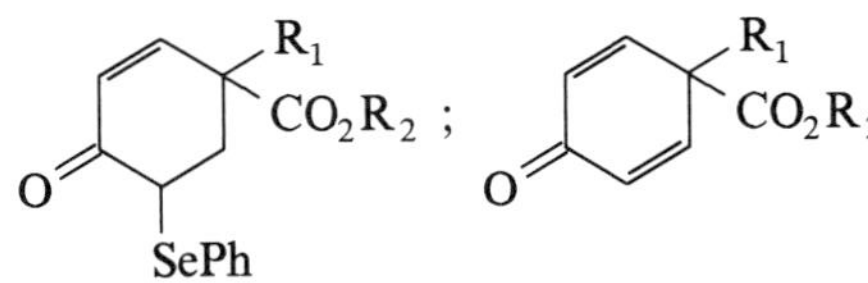	466

tions which improve considerably the synthetic utility of the reaction. Entries 2–9 of Table 2.6 illustrate the synthetic potentiality of some heterosubstituted 1,3-butadienes.

Acyloxy and alkoxy groups bonded at C-1 of the 1,3-butadienylic unit (entry 2) make it possible to prepare angularly functionalized hexalones, which are useful intermediates in the synthesis of natural products. Diethylamino and trimethylsilyl groups at C-1 are also versatile functionalities (entries 3 and 4). The initial *ortho* adducts can be changed into cyclohexadienes by the elimination of the amino group under acidic conditions, or they can undergo allylic displacement of the allylsilane moiety and migration of the double bond with or without introducing an hydroxy group.

The $OSiMe_3$, $OPO(OEt)_2$, SPh, OMe, and OAc groups at C-2 (entries 5–8) are potential ketonic functionalities, and the CH_2SiMe_3 group allows the formation of methylene and propenyl groups. Cycloaddition with β-nitroacrylates is highly *para*-regioselective with respect to the nitro group. This peculiarity allows the synthesis of compounds with ketonic and ester groups in the *meta* position after elimination of the nitro group by 1,8-diazobicyclo[5.4.0]undec-7-ene (DBU) (entry 6).

Sulfur substituents are regiocontrol elements (entries 9 and 10) and useful building blocks for further functionalizations, and after desulfurization, they allow the formation of cycloadducts with the substituents in reverse Diels–Alder regiochemical preference.

Among the heterosubstituted-1,3-butadienes, the best known is the (*E*)-1-methoxy-3-trimethylsilyloxy-1,3-butadiene (Danishefsky's diene; Table 2.6, entry 11), which is readily available, highly reactive, and synthetically versatile. The two heterogroups and their 1,3-relationship allow the synthesis of α,β-unsaturated ketones (entries 11b and d) and dienones (entry 11b). When the dienophiles are aldehydes and the cycloaddition is catalyzed by Lewis acids, dihydropyranones (entry 11c) are prepared. The OMe group is the *ortho*-directing group. In cycloaddition with iminium salts (entry 11a) the adducts are cleaved with NaOH to give α-aminoenones.

The synthetic potentiality and reactivity of Danishefsky's diene is improved by $OSiMe_3$ and SePh groups at C-4 (entries 12 and 13). Thus a precursor of a β-talose derivative can be synthetized (entry 12) by using a suitable dienophile and proper reaction conditions (i.e., benzyloxyacetaldehyde; BF_3; $-78°C$). A SePh group in *E*-configuration (entry 13) does not decrease the facility of the Diels–Alder reaction as one would expect, and thus functionalized α,β-unsaturated ketones and dienones can be prepared.

Facial Selectivity Study of the diastereofacial selectivity of butadienes has been developed during the past 15 years, and most of the dienes studied have a butadienyl part linked to a chiral fragment. Typical chiral auxiliaries are the D-carbohydrates (i.e., **14–22**), which also allow asymmetric induction. With dienophiles of the general formula $RR_1C=Z$ ($R = H$, Me; $R_1 = CHO$, CO_2Me, CO_2Bu; $Z = CH_2, O$), the cycloaddition reaction creates two new

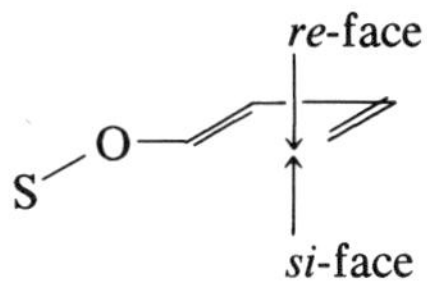

The *re* and *si* faces are defined with reference to the prochiral carbon atom C-1 linked to the oxygen atom[191]; S = sugar residue

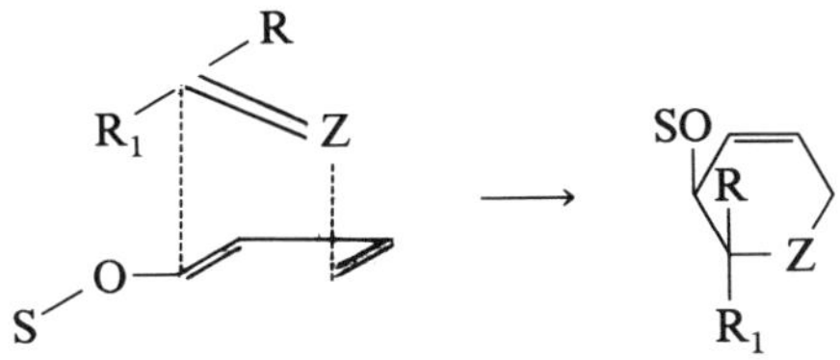

exo-re-ortho

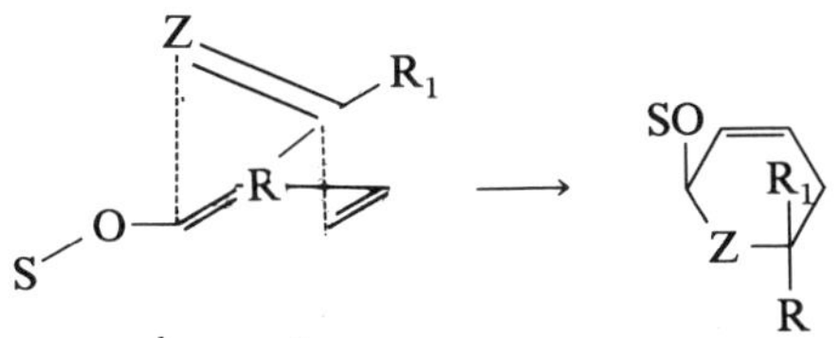

endo-re-meta

Scheme 2.4

chiral centers and leads, in principle, to four pairs of diastereoisomers as a consequence of *endo* and *exo* additions of *meta*- and *ortho*-oriented diene on the *re* and *si* faces of the prochiral diene (Scheme 2.4).

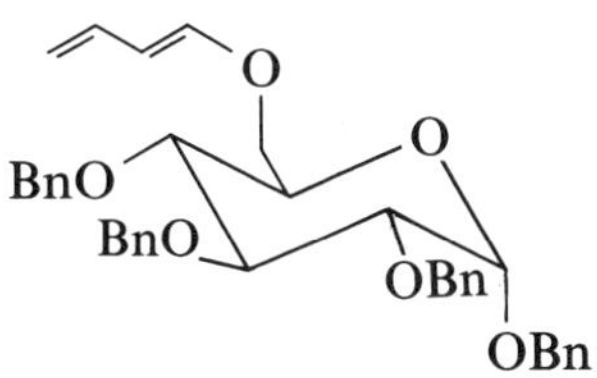

19

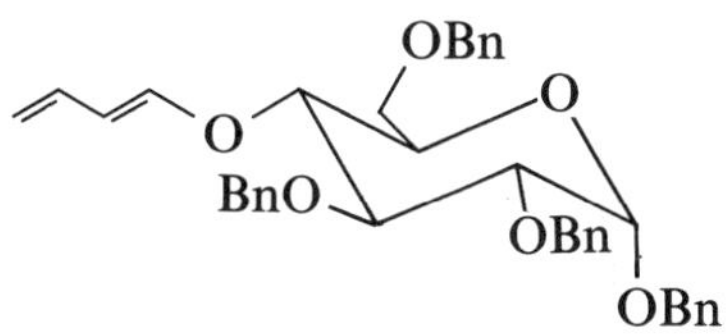

20

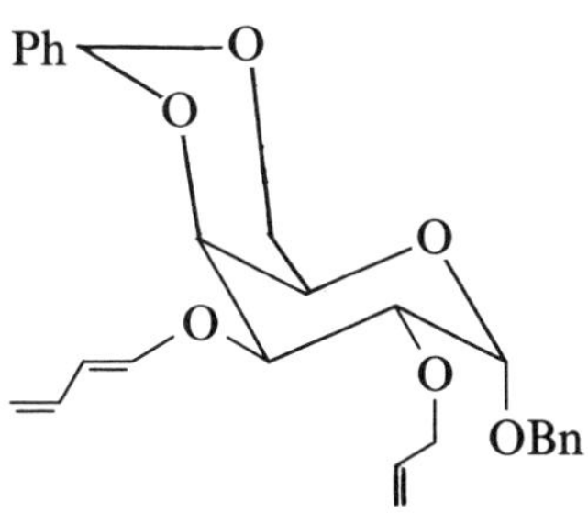

21

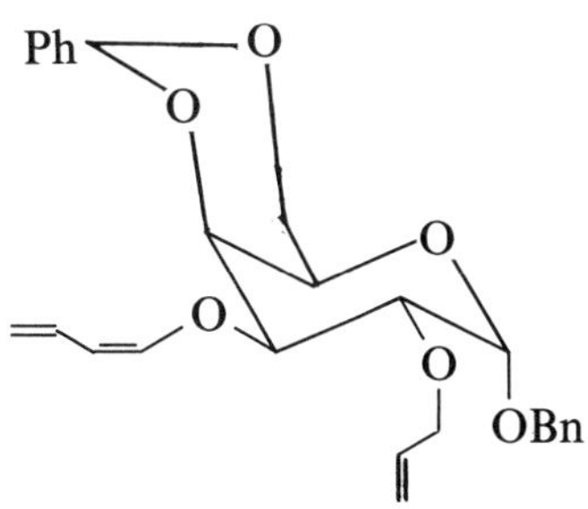

22

The addition of *n*-butylglyoxylate in benzene to glycodienes **19–21** is completely *meta*-regioselective, as expected, but there is no preference for *exo* or *endo* addition.[191,474,475] The facial selectivity is dependent on the position of butadienyl moiety (**19**: *re/si* = 18/82; **20**: *re/si* = 4/96; **21**: *re/si* = 97/3). Both the *trans* diene **21** and the *cis* diene **22** react with diethylmesoxalate to give more adduct from the *re*-face (*re/si* = 80/20).

When D-glucose is the chiral fragment (**14** and **16**) the diene has a hydrophilic part and the cycloaddition can be carried out in aqueous solvent.[191] The cycloadditions of **14** and **16** in water show rate and stereoselectivity enhancement compared to the analogous reactions of peracetylated derivatives **15** and **17** in organic solvents. Both **14** and **16** show high or exclusive *endo* selectivity, a complete *ortho* regioselectivity but a low and opposite facial-selectivity.

Diene **18**, a key reagent in the synthesis of (+)4-demethoxydaunomycinone, reacts with a variety of dienophiles, showing a total *endo* selectivity and a 70 to 90% *re*-facial selectivity. The glycocycloadducts can be converted[191,476] into functionalized chiral cyclohexanes by removing the sugar part or into disaccharide analogs after functional group manipulations.

Antipodal menthyloxy dienes **23** have been synthesized by Bednarski and Danishefsky[296] and cyclocondensed with benzaldehyde in the presence of lanthanide complexes as catalyst. The reactions are highly *endo*-specific but show a low facial selectivity when the achiral $Eu(fod)_3$ is used. Using the chiral catalyst $Eu(hfc)_3$ and permuting each of the chiral auxiliaries, the cycloadditions of (−)menthyloxydienes give 75 to 93% of adducts **24**. The observed high *si*-facial selectivity is a consequence of a strong interactivity of the two chiral elements.

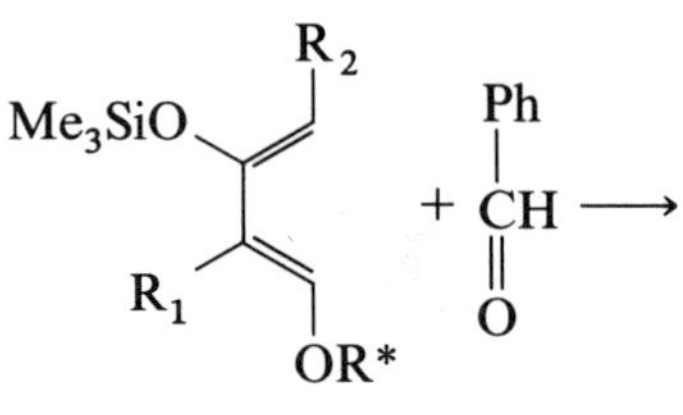

23 $R^* = (+)$ or $(-)$ menthyl
$R_1 = H, Me, OAc; R_2 = H, Me$

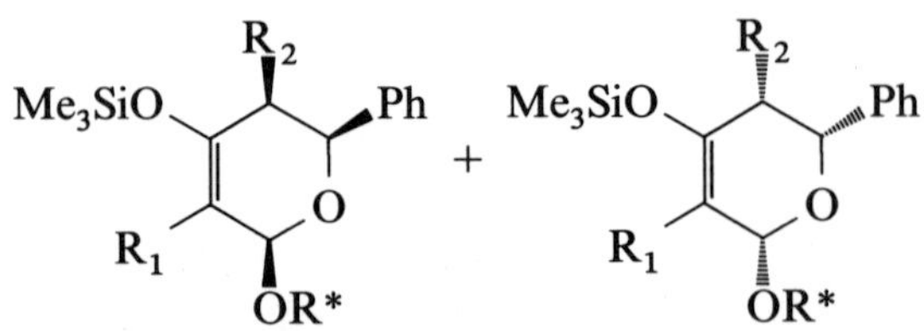

re-face addition *si*-face addition

24

A total facial selectivity has been observed in the reaction of optically active diene **25** with juglone **26** as a consequence of steric steering of the phenyl group, which directs the dienophile to the *si*-enantiotopic face of the diene.

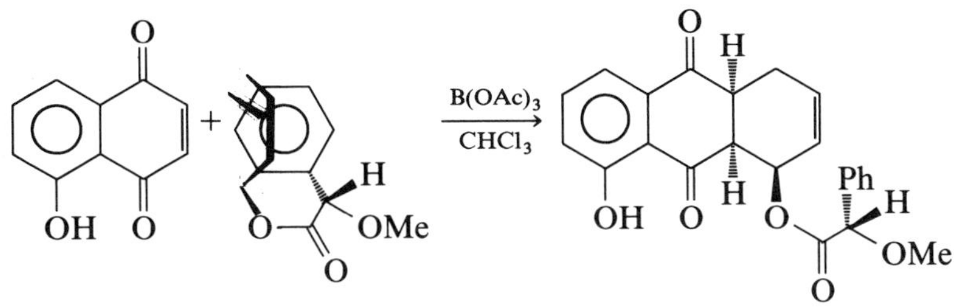

26 **25**

The effect of an adjacent chiral center on the diastereoselectivity of the reaction has been evaluated by using dienes **30**–**32** (Scheme 2.5). These dienes react[329,373] *endo* selectively and with increased *re*-facial selectivity with NPM; the results have been explained in terms of conformational reactivity. The (*Z*)-2-methoxymethyl substituent of **32** disfavors the conformer **29** and forces the reaction to proceed exclusively from conformer **27**. An empirical rule

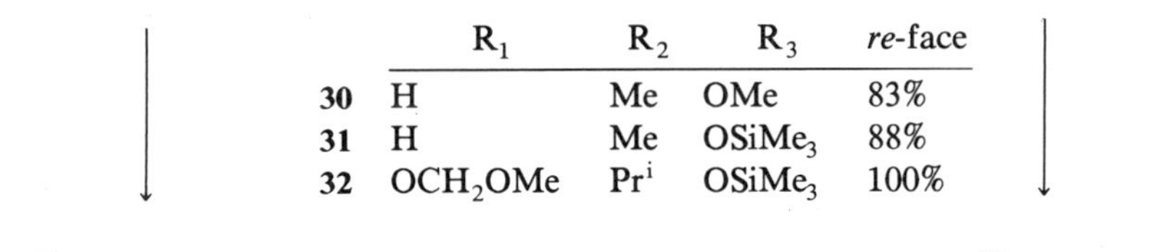

27 **28** **29**

	R_1	R_2	R_3	*re*-face
30	H	Me	OMe	83%
31	H	Me	$OSiMe_3$	88%
32	OCH_2OMe	Pr^i	$OSiMe_3$	100%

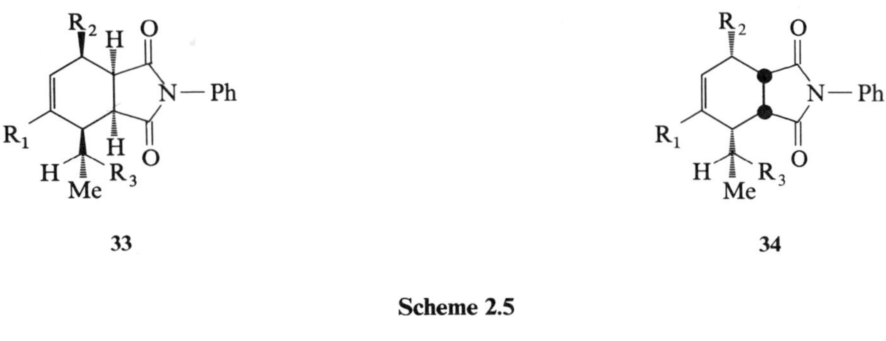

33 **34**

Scheme 2.5

states[329] that a chiral R-carbon bearing an alkoxy group and bonded to C-1 of a butadienyl moiety induces *re*-facial selectivity and a S-carbon induces *si*-facial selectivity.

Recent theoretical studies indicate[477b] that the diastereofacial selectivity involving chiral dienes is influenced by electrostatic interactions and the reaction between an electron-rich diene and an electron-deficient dienophile occurs preferentially on the more nucleophic face of the diene. This means that the approach of a dienophile (an "electrophile") to dienes **30**–**32** must occur *syn* to the alkoxy group of the energetically favored conformer **28**.

Trienes and Tetraenes The linearly conjugated trienes with the central double bond having a *cis* configuration are very poor reactive molecules and under drastic conditions prefer an intramolecular cyclization to a Diels–Alder reaction because steric hindrances obstruct the planar cisoid conformations.[478–480] Examples are the *cis*-1,3,5-hexatriene (**35**: $R_1 = R_2 = R_3 =$ H), the *cis-trans*-1,3,5-heptatriene (**35**: $R_1 = R_2 =$ H, $R_3 =$ Me), and the *cis*-2,5-dimethyl-1,3,5-hexatriene (**35**: $R_1 = R_2 =$ Me, $R_3 =$ H), which are unreactive toward MA but cyclize to corresponding cyclohexadienes **36**.

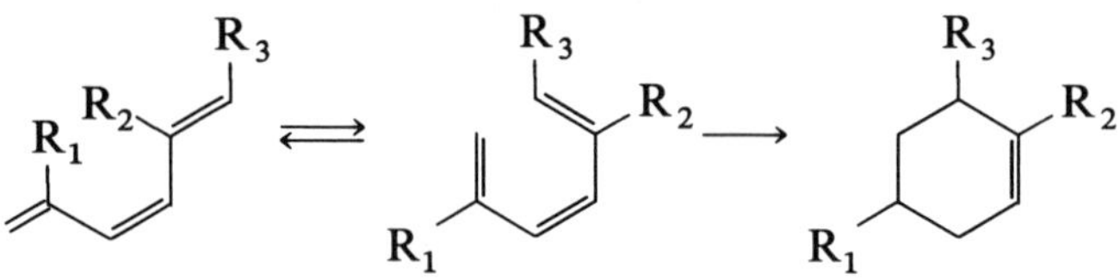

35 **36**

When the configuration of the central double bond of the triene is *trans*, a planar double-cisoid conformation can be energetically favored. The cycloaddition reaction occurs under moderate conditions, and unsymmetrical molecules can give two adducts as the result of two possible 1,4-additions to *a* and *b* dienic functionalities (**37** and **38**). *Trans*-1,3,5-hexatriene (**37**: $R_1 = R_2 =$ H) and *trans-trans-trans*-1,6-diphenyl-1,3,5-hexatriene (**37**: $R_1 = R_2 =$ Ph) readily react with MA to give single adducts.[481]

The *trans*-triene **37** ($R_1 = R_2 =$ H) reacts with TCNE at 20°C about 3500 times faster than does its *cis* isomer.[21] In the *trans-trans*-1,3,5-heptatriene (**37**: $R_1 =$ Me, $R_2 =$ H) the methyl group differentiates the 1,4-positions of dienic functionalities, and two adducts are obtained in the reaction with MA. The 1,4-dienic system far from the methyl group (*b* in compound **37**: $R_1 =$ Me, $R_2 =$ H) is the most reactive.[481] Also, the *cis-trans*-1,3,5-heptatriene (**38**: $R_1 = R_3 = R_4 =$ H; $R_2 =$ Me) and B-allocymene (**38**: $R_1 = R_2 = R_3 = R_4$ = Me) have two nonequivalent 1,4-positions, but the dienic system *a* is much less reactive than *b* because the steric hindrance of the methyl group prohibits the cisoid conformation. As a consequence, these compounds add a dienophile molecule only at the *b* dienic functionality.[246,478]

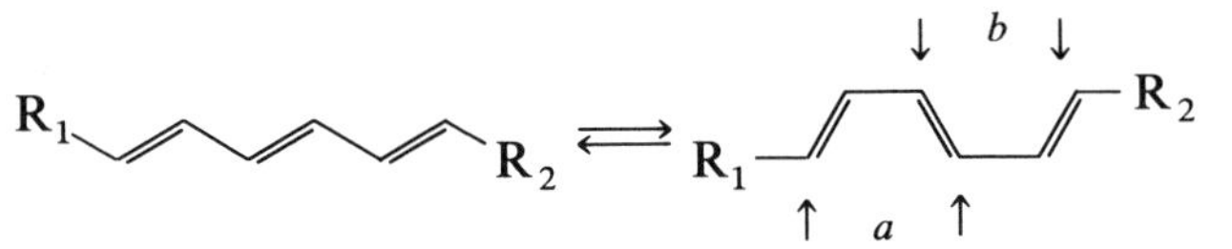

37

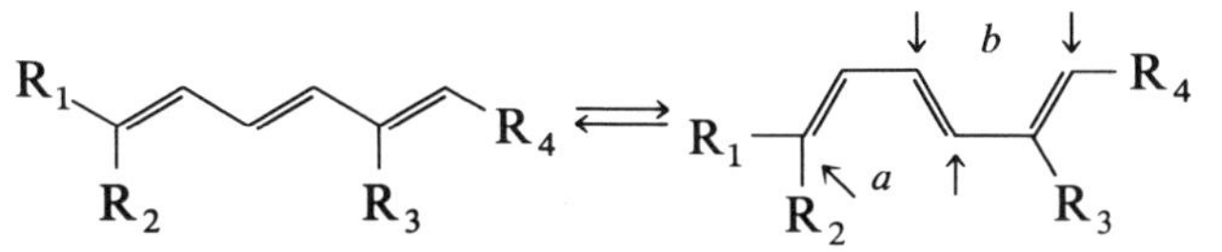

38

In cross-conjugated trienes (i.e., **39**) Diels–Alder reaction produces another dienic functionality for a second cycloaddition (Sec. 1.14). The synthetic use of these trienes depends on the reactivity of the monocycloadduct, which must be much less than that of the starting triene to permit the orderly incorporation of two different dienophiles. Several substituted cross-conjugated trienes have been synthesized, but the chemoselective control of the cycloaddition has been rather limited.

The triene **39** (R = Ph) gives predominantly bis adducts **41** and **42** with cyclic dienophiles via *endo* mono cycloadducts **40** and a mixture of *exo* **43** and *endo* monoadducts **44** with acyclic dienophiles[377c] (Scheme 2.6).

The methoxytriene **39** (R = OMe) similarly affords only monoadducts with acyclic dienophiles because the elimination of methanol from monoadducts **45** is faster than the second cycloaddition, but with cyclic dienophiles it provides both mono **40** and bis cycloadducts **41** selectively.[403b] A double Diels–Alder reaction has therefore been achieved[403a] by using maleimides or MA as the first dienophile (**39** → **40**) and acetylenes as the second (**40** → **46**). The hydrolytic desilylation (**46** → **47**) inverts the configuration of the methoxy group.

The use of a cross-conjugated triene system coupled with selective control of cycloadditions is offered by triene synthetic equivalents such as 1-methyl-3-methylene-4-pentenyltosylate[239] (**48**). The first cycloaddition with NPM (**48** → **49**) occurs in 99% yield, and the second cycloaddition with DMAD gives a mixture of **50** and **52** in 87% yield after elimination of *p*-toluenesulfonic acid from **49**. The prevalent bisadduct is **50** formed by the addition of dienophile from the less hindered side of diene **51**.

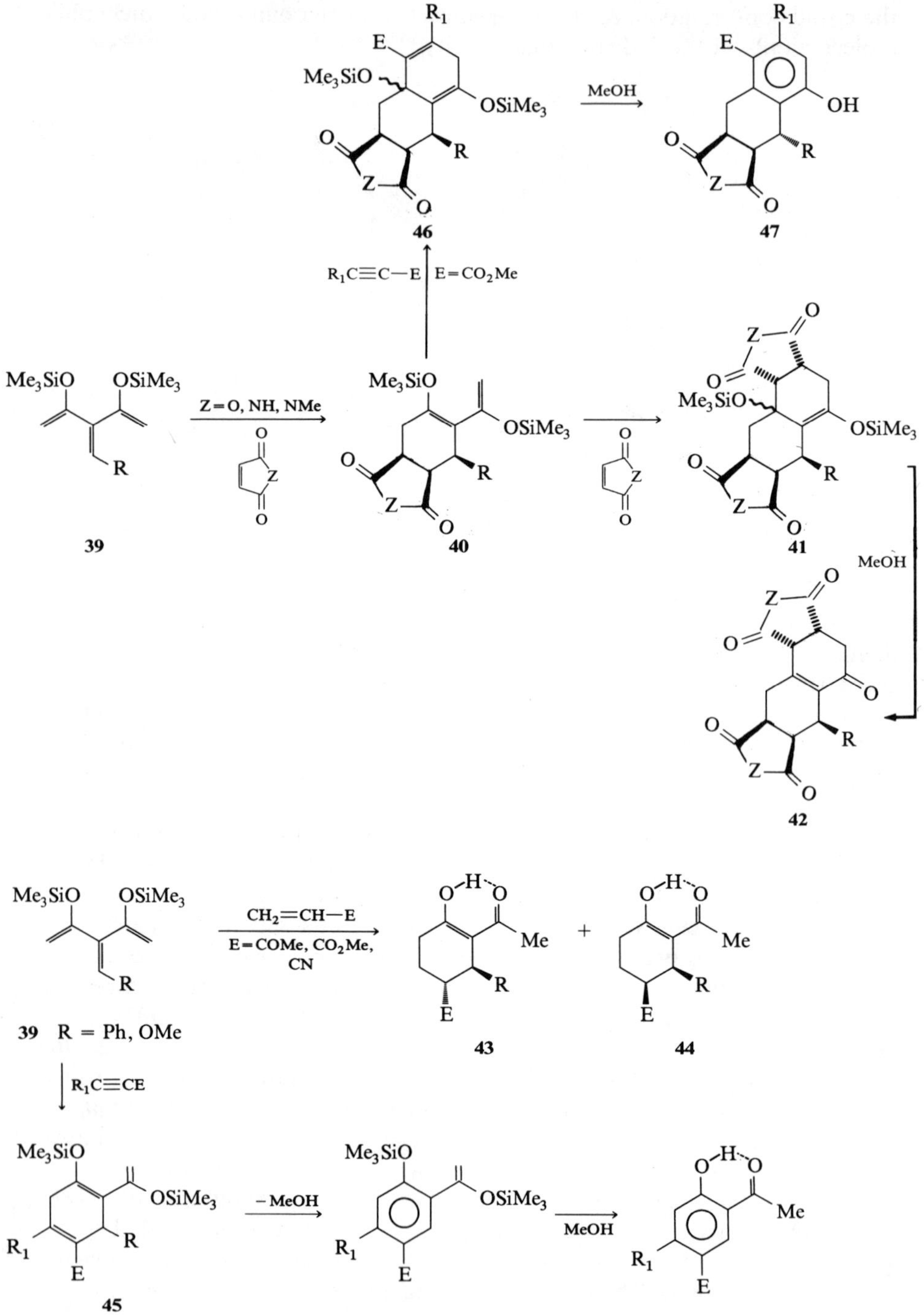

MeOH
46
47
$R_1C{\equiv}C{-}E$
$E{=}CO_2Me$
$Z{=}O, NH, NMe$
39
40
41
MeOH
42
$CH_2{=}CH{-}E$
$E{=}COMe, CO_2Me, CN$
43
44
39 R = Ph, OMe
$R_1C{\equiv}CE$
45
– MeOH
MeOH

Scheme 2.6

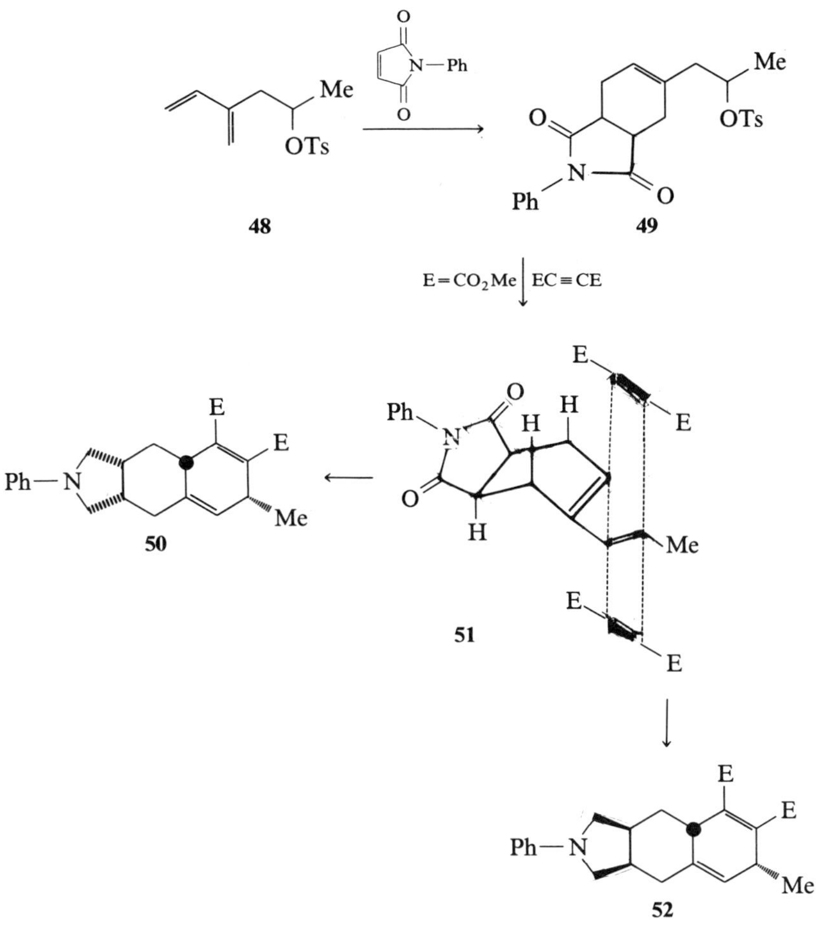

The *trans-trans* conjugated tetraenes have two diene units, and if one is more reactive than the other, timed Diels–Alder reactions with different dienophiles can be carried out. This strategy has been followed in the synthesis of compounds having the fluorenone skeleton[130] (Eq. 1.42).

Dienes as Dienophiles The cyclodimerization of 1,3-butadienes is a Diels–Alder reaction where one molecule of the reactants behaves as diene and another as dienophile. Unactivated dienes require drastic reaction conditions, and the yield and selectivity of the reaction are generally low in the absence of catalyst.

At 250°C the conversion of 1,3-butadiene (**53**) to 4-vinylcyclohexene (**54**) is less than 0.1% in the absence of catalyst but more than 20 to 30% in the presence of large-pore zeolites or carbon molecular sieves.[77] This catalytic effect has been attributed to the porosity of zeolites and carbon, which permits concentration of the diene within the cavities. The cyclodimerization of 2,4-dimethyl-1,3-pentadiene (**55**) is achieved easily, selectively, and with good yield in the presence of systems able to generate radical cations such as 4-bromophenylaminiumhexachloroantimonate and Fe^{III}-doped K10 montmorillonite combined with 4-*t*-butylphenol.[402] The same catalysts make the Diels–Alder reaction possible between the 2,5-dimethyl-2,4-hexadiene (**57**) and the unactivated 1,3-cyclohexadiene (**56**), where the aliphatic diene behaves exclusively as dienophile.[402] With the highly reactive pentachloro-1-azacyclopentadiene (**58**) and hexachlorocyclopentadiene (**59**) the cycloadditions of a variety of 1,3-butadienes occur selectively and without catalyst.[482,483]

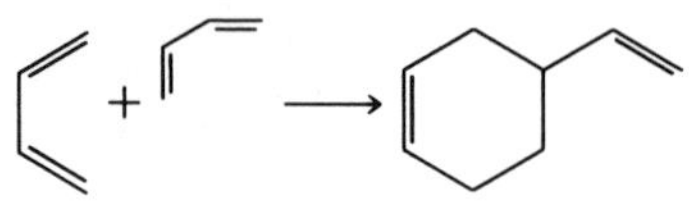

53 **54**

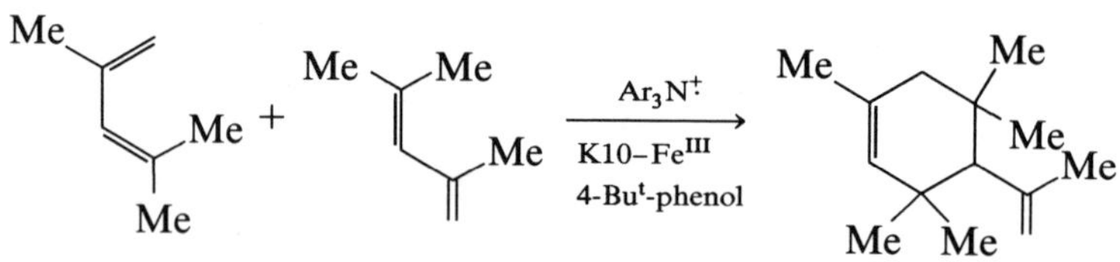

55

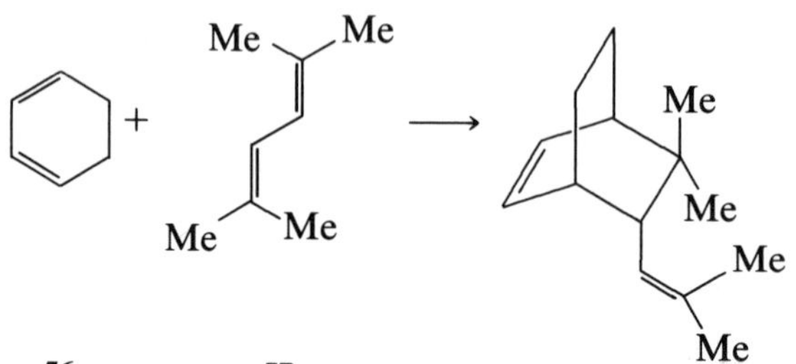

56 **57**

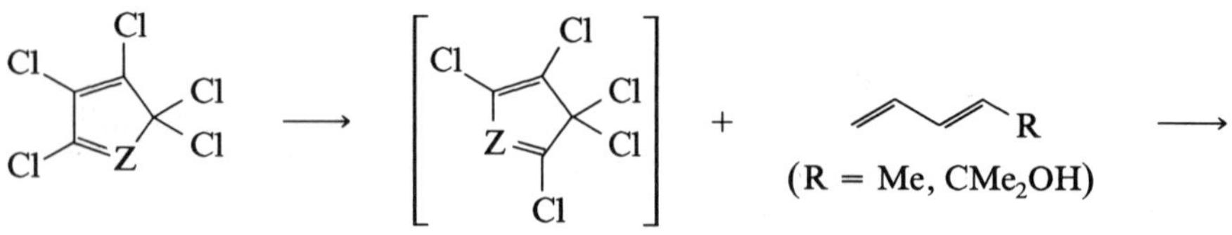

58 Z = N
59 Z = CCl

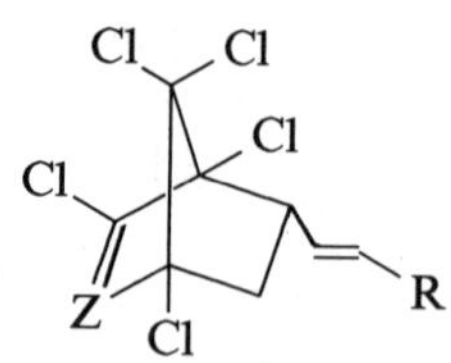

Tables of Diels–Alder Reactions See Tables 2.7 to 2.33; see also the general remarks about classifications, abbreviations, and symbols in the Introduction.

TABLE 2.7 Diels–Alder Reactions of 1,3-Butadiene

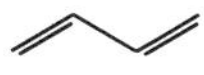

	Dienophile	Yield (%)	Ref.
A1	Acrylates	78–100	85
A1	AC	58	(1a), 67
A1	Acrylic acid	68	19
A1	Camphor-10-sulfonamide acrylate (*)	98	52
A1	D-Pantolactone acrylate (*)	73	68
A1	*N*-Acryloylsultam (*)	93	59
A2	2-Methylenediisobutylglutarate	47	5
A3	Crotonaldehyde	76	38, 72
A3	(*E*)-Ethyl-3-nitroacrylate	81	136
A4	2-(4-Chlorophenylseleno)-2-butenal	80	48
A4	3-Chloro-2-methylpropenal (*Z* and *E*)	12–41	64
A5	(*E*)-Ethyl-3-nitrocrotonate	67	136
A7	Dimenthyl fumarate (*)	56–100	15, 88
A8	Acrylic monomers, Cat.(*)		29
A8	Ethene		(1)
A8	Propene (k)		2
A8	*S*-*p*-Tolyl-*N*-tosyl-*S*-vinylsulfoximine	90	51
A8	Vinyl ethers		10
A8	Vinylgermane		35
A9	1,1-Bis(benzenesulfonyl)ethene	75	60
A12	TCNE (k)		21
A13	Allenes		(50)
A13	Fluoroallene	57	25
A13	α-Ethenylidenecyclanones	13–52	66
A14	Butadiene	20	77
B1	2,2-Difluoromethylenecyclopropane	63	81
B2	5-Methylene-2,2-dimethyl-1,3-dioxane-4,6-dione	50	65
B3	Isopropylidenecyclohexylidene malonates	16	78
C1	Cyclopropenone ketal	79	79
C1	1,2,3-Trichloro-3-fluorocyclopropene	65	6
C1	5,6-Epoxycyclooctene	87	12
C1	Cylobuteno[16α, 17α]pregnene		28
C1	Tris(trifluoromethyl)trifluoroacetylcyclopropene		27
C2	Hexacyclotetradecenes	90	39
C2	Hexakis(trifluoromethyl)bicyclo[2.2.0]hexa-2,5-diene	64	14
C2	Hexakis(trifluoromethyl)benzvalene	78	13
C2	Norbornadiene	92	22
C3	Acephenantrylene	23	82
C4	[Bridgehead enones]	41	83
C4	(−)Carvone(*)	86	71
C4	2-(Phenylthio)-cyclopent-2-en-1-one	82	24
C4	2-Cycloalkenones	62–95	9, 42

TABLE 2.7 (*Continued*)

	Dienophile	Yield (%)	Ref.
C4	4-Alkyl-2-cyclohexenones	70–95	69, 76a
C4	5-Alkyl-2-cyclohexenones	29–97	8, 20, 70
C4	Bicyclo[4.3.1]dec-6-en-8-one	39	33
C5	2,5-Cyclohexadien-1-ones	90–95	7, 32, 45
C7	2-Acetyl-1,4-benzoquinone	74	61
C9	1,2-Dihydro-4,7-dihydroxycyclobuta[*b*]-naphthalene-3,8-dione	96	63
C9	[1,4-Naphthoquinones]	80	80
C10	6-Dehydrotestosterone acetate	41	17
D2	2-Methoxy-5-oxo-5,6-dihydro-2*H*-pyran	73	4
D2	6-*O*-Acetyl-2,3-dideoxyhex-2-enopyranosid-4-ulose (*)	81	47
D2	Butenolides (*)	32–80	75
D3	Carbomethoxymaleic anhydrides (k)	75	37, 40
D3	MA (k)		21, 86
D4	2-Aryl-Δ^2-pyrroline-4,5-diones	6	30
D6	2-Thiolene-1,1-dioxides		55
D6	3-Chlorothiete-1,1 dioxide	59	53
D6	3-Methoxycarbonyl-2,5-dihydrothiophenes	38–55	57
D7	Nitrosocarbonyl compounds	64–81	58
D8	6-Acetylpyrimidone	3	46
E2	Tributylstannioacetylenes	71–74	62
E3	Di-*t*-butylacetylene dicarboxylate	99	23
E3	DMAD	98	44
E5	Acetylenedicarbaldehyde monoacetal	60	74
E5	Bis(trimethylsilyl)acetylene	78	73
F1	Monosaccharide-*O*-thioformates (*)		3
F1	Methylalkane dithiocarboxylates	20–60	26
F1	*N*,*N*-Dialkylcyanothioformamide	56–81	11
F2	Adamanthane thione	56	84
F3	*N*-Methyldithiophtalimide	87	54
F4	Formaldehyde		(56)
F7	Hexafluoro-2-acyliminopropane	63	200
F9	Silaethene		49
H8	1,2,3,6-Tetrahydrobenzocyclobutene-3,6-dione	94	36
H9	1,4-Anthraquinones	75–94	34, 41
H11	1-Ethyl-4,5,8-trioxo-1,4,5,8-tetrahydro-3-quinolinecarboxylic acid	73	43
I1	Arylazocyanides	85	18
I5	NPTAD (k)		87
I5	4-(α-Phenylethyl)-1,2,4-triazoline-3,5-dione (*)		31
J1	1-Chloro-1-nitrosocyclohexane	42	16

TABLE 2.8 Diels–Alder Reactions of (*E*)-1,3-Pentadiene (Piperylene)

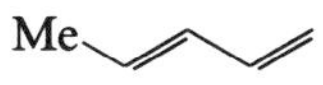

	Dienophile	Yield (%)	Ref.
A1	AC		38, (116)
A1	MAC	71	72
A3	Acrylates		(90), 96, (102), 115
A3	α, β-Unsaturated carboxymides (*)	77–84	114
A4	2-(4-Chlorophenylseleno)-2-butenal	92	48
A7	Dimenthyl fumarate (*)	39	88
A7	(*E*)-Methyl-4-oxobutenoate	> 90	93
A8	ACN		111
A10	α, β-Unsaturated acyliron complex	72	119
A10	(*E*)-1-Nitro-2-(phenylsulfinyl)ethene	54	120
A12	TCNE (*k*)	71	21
A13	Allenic ketones	59–75	66, 106
C2	Hexakis(trifluoromethyl)benzvalene	85	13
C4	(−)Carvone (*)	88	71
C4	2-Cycloalkenones	28–92	42, 97, 105, 109
C4	4-Alkyl-2-cyclohexenones	83–93	69, 118
C4	5-Alkyl-2-cyclohexenones	80–91	70, 118
C4	*trans*-$\Delta^{1,2}$-3-Octalone	90	42c
C4	6-Alkyl-2-cyclohexenones	70–95	76a
C5	2,5-Cyclohexadienones	88–95	45, 99
C6	*o*-Quinones	19–87	112
C8	1,2,3,6-Tetrahydrobenzocyclobutene-3,6-dione	87	36
C9	1,4-Naphthoquinones	46–95	89, 98
C9	1,4-Benzoquinones	61–85	61, 95, 98, 110
D2	2-Methoxy-5-oxo-5,6-dihydro-2*H*-pyran	66	4
D2	3-Carbethoxy-5,5-dimethyl-2(5*H*)-furanone	75	104
D3	Maleic anhydrides (k)	81–93	21, 86, 101
D5	*N*-Phenylmaleimides	69–71	91, 101
D10	1,4-Benzodithiin-1,1,4,4-tetroxide	96	121
E2	Ethyl-β-phenylsulfonylpropiolate	75	100
E2	Methyl(tributylstannio)propiolate	66	62
E5	Bis(trimethylsilyl)acetylene	76	73
E5	Diethylacetylene	49–64	107
F1	*S*-Methyl-*O*-ethyldithiooxalate	79	94
F2	Adamanthane thione	80	84
F2	Thione-*S*-imide	62	92
F3	*N*-Methyldithiophtalimide	100	54
F7	Hexafluoro-2-acyliminopropanes	69	200
H2	4-Nitrobenzendiazonium fluoroborate	68	113
H3	β-Methoxydihydronaphthalene	69	108
I1	Arylazocyanides	47–100	18
J1	1-Chloro-1-nitrosocyclohexane	47	16
J1	2,3-5,6-Di-*O*-isopropylidene-1-*C*-nitroso-α-D-mannofuranosyl chloride (*)	68	117
J2	*N*-Sulfinyl-*p*-toluamide	95	103

TABLE 2.9 Diels–Alder Reactions of (*E*)-1-Methoxy-1,3-butadiene

	Dienophile	Yield (%)	Ref.
A1	AC	60	67
A1	MAC (k)		115
A2	1-Acetylvinyl-1-*p*-nitrobenzoate	66	135
A7	DMF	96	90
A7	(*E*)-Methyl-4-oxobutenoate	> 90	93
A11	Cyanoethenes	84–98	126
A12	TCNE (k)	85	21
C1	1,2-Bromochlorocyclopropene		134
C1	Cyclopropene		134
C1	Cyclopropenone ketal	72	79
C1	Tetrachlorocyclopropene		134
C4	Cycloalkenones	32–55	42d
C5	Cyclohexa-2,5-dienones	66–97	125, 129
C9	1,2,3,4-Tetrahydro-1,4-anthraquinones	94	41
C9	9-Chloro-10-hydroxy-1,4-anthraquinone	93	131
D2	Ethyl-6-*O*(*t*-butyldimethylsilyl)-2,3-dideoxy-α-D-glicero-hex-2-enopyranosid-4-ulose (*)	56	128
D3	MA (k)		21, 86
D4	Isoquinolinopyrrolinediones	36–80	30
F4	1,2,3,4-Di-*O*-isopropylidene-α-D-galactopyranos-6-ulose (*)	56–72	132
F4	2,3-*O*-Isopropylidene-D-glyceraldehyde (*)	11–86	133
F4	2-Furaldehyde	73	124
F4	Acetaldehyde	62	124
F4	Benzaldehyde	50–80	124, 127
F4	Hexanal	28	124, 127
F4	Menthyl glioxylate (*)	95	123
F4	Phenyl trifluoromethyl ketone	81	124

TABLE 2.10 Diels–Alder Reactions of (*E*)-1-Acetoxy-1,3-butadiene

	Dienophile	Yield (%)	Ref.
A1	Acrylates	69–87	141
A2	1-Acetoxy-1-carboxytrichloroethylethene	81	141
A2	1-Acetylvinyl-1-*p*-nitrobenzoate	66	135
A2	2-Methylprop-2-enal	46	142
A3	4-(Diphenylphosphinyl)-3-buten-2-one	55	143
A3	(*E*)-1-Carboxytrichloroethyl-2-phenylethene	77	141
A3	(*E*)-Ethyl-3-nitroacrylate	57	136
A3	Ethyl-β-trichloroethyl fumarate	71	141
A5	Ethyl-3-nitrocrotonate (*E* and *Z*)	16–26	136
A7	DMF	73	90, 140
A7	(*E*)-Dibenzoylethene	93	152
A9	1,1-Bis(benzenesulfonyl)ethene	57	60
A9	2-Nitro-1-butene	40	144
A9	α-Chloroacrylonitrile		147
A10	β-Substituted styrenes	37–73	139
C1	Tetrachlorocyclopropene		134
C3	Acephenantrylene	> 20	82
C4	[Bridgehead enones]	55–63	83
C5	Tricyclo-2,5-cyclohexadien-1-one	90	32
C6	*p*-Tropoquinone	69	137
C8	1,2,3,6-Tetrahydrobenzocyclobutene-3,6-dione	84	36
C9	1,4-Anthraquinone	58	34
C9	2-Acyl-1,4-naphthoquinones	43–74	61
C9	5-Hydroxy-1,4-naphthoquinone	87	138a, 146
C9	5-Hydroxy-1,4-naphthoquinone, Cat.(*)	> 68	150
C9	[1,4-Naphthoquinones]	75	80
C10	6-Dehydrotestosterone acetate (*)	14	17
D2	3-Carbethoxy-5,5-dimethyl-2(5*H*)-furanone	60	104
D3	MA	100	139
D11	2-Carbomethoxy-7-oxabicyclo[2.2.1]hept-2-ene	83	149
E2	Ethyl(diethoxyphosphinyl)propynoate		145
I4	[5-Phenylpyrazol-3-one]	67	148

TABLE 2.11 Diels–Alder Reactions of (*E*)-1-Trimethylsilyloxy-1,3-butadiene

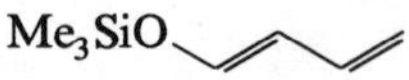

	Dienophile	Yield (%)	Ref.
A1	AC	53	154
A3	(*E*)-Crotonaldehyde	69	154
A7	(*E*)-4-Oxobutenoate	> 90	93
A7	(*E*)-Dibenzoylethene	93	152
A12	TCNE	50	155
A13	(Phenylsulfonyl)propadiene	75	156
C1	Halocyclopropenes		134
C8	2,3,4*a*,9*a*-Tetrachloro-4*a*,9*a*-dihydro-anthracene-1,4,9,10-tetrone	56	158
C8	2-Chloroquinizarindiquinone	97	158
C9	1,4-Naphthoquinones	71–100	159a, 160
C9	5-Hydroxy-1,4-naphthoquinone, Cat.(*)	73–96	157
C9	9-Chloro-10-acetoxy-1,4-anthraquinone	40	131
C9	9-Chloro-10-hydroxy-1,4-anthraquinone	100	131
C9	[2-Butanoyl-1,4-benzoquinone]	77	151
D2	Alkyl-2,3-dideoxy-α-D-glicero-hex-2-enopyranosid-4-uloses (*)	> 51	128
D3	MA	98	155
F4	Ethyl mesoxalate	55	153

TABLE 2.12 Diels–Alder Reactions of (*E*)-2,4-Pentadienoic Acid and Its Esters

R		Dienophile	Yield (%)	Ref.
H	A1	Acrylates		90
	C9	1,4-Naphthoquinones	18–48	164
Me	A1	Polymer-bound benzylacrylate	52	161
	A2	1-Acetylvinyl-1-*p*-nitrobenzoate	85	135
	C1	Cyclopropenone ketal	65	79
	C7	1,4-Naphthoquinones	27–86	163a, 164
	F5	Indantrione	77	165
	F6	5-Ketobarbituric acid	50	165
	F6	Mesoxalic acid ester	32	165
	G1	4,9-Dihydro-3*H*-pyrido[3,4*b*]indoles	71–87	166, 167
	I4	[5-Phenylpyrazol-3-one]	67	148
Et	A1	AC	76	162
$CHMeCH_2Ph$ (*)	C7	BQ	98	163b
CHMePh (*)	C7	BQ	98	163b
$CH_2-CH(OMe)Ph$ (*)	C7	BQ	88	163b
(−)Menthyl (*)	C7	BQ	51	163
(−)8-Phenylmenthyl (*)	C7	BQ	62	163b

TABLE 2.13 Diels–Alder Reactions of (*E*)-1-Amino- and 1-Amido-1,3-butadienes

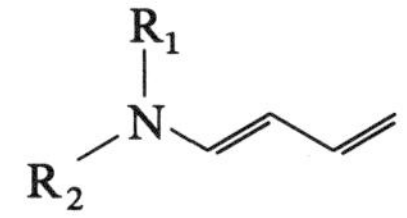

R_1	R_2		Dienophile	Yield (%)	Ref.
Et	Et	A7	(*E*)-Diaroylethenes	56–100	152
		A10	β-Substituted styrenes		170
		E6	[Arynes]	8–35	169
$COCCl_3$	H	A1	MAC		90
CO_2Bn	H	C9	1,4-Naphthoquinones	56–64	171
CO_2Bu^t	H	A3	Methyl-β-nitroacrylate	68	168
COBn	Bn	E3	DMAD	80	192
COPh	Bn	A1	Acrylates	71–86	192
		A8	ACN	98	192
COPh	Ph	A8	ACN	97	192
		E3	DMAD	97	192
CO_2Bn	Bn	D3	MA	94	192
$-CH_2-CH_2-O-CH_2-CH_2-$		A7	(*E*)-Dibenzoylethene	92	152
$-CH_2-CH_2-CH_2-CH_2-CH_2-$		A7	(*E*)-Dibenzoylethene	99	152

TABLE 2.14 Diels–Alder Reactions of (*E*)-1-Aryl-1,3-butadienes

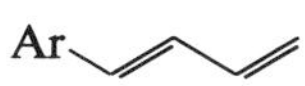

Ar		Dienophile	Yield (%)	Ref.
Ph	A1	MAC		90
	A1	Copolymer-bound benzylacrylate	80	161
	A12	1,1-Dicyano-2,2-ditrifluoromethylethene	80	173
	A12	TCNE (k)	100	21, 173
	D3	MA (k)	100	21, 86
	E2	Ethyl (diethoxyphosphinyl)propynoate	100	145
	J1	1-Chloro-1-nitrosocyclohexane	53	16
	J2	*N*-Sulfinyl-*p*-toluamide	95	103a
$R-C_6H_4$	A12	1,1-Dicyano-2,2-ditrifluoromethylethene	51–85	173
(R = *p*-NMe_2, *p*-Me, *m*-F, *m*-Cl, *m*-CF_3, *m*-NO_2)	A12	TCNE	75–86	173
p-$OMeC_6H_4$	A1	MAC		90
	A12	1,1-Dicyano-2,2-ditrifluoromethylethene	82	173
	A12	TCNE	78	173
	J2	*N*-Sulfinyl-*p*-toluamide	80	103a
p-$NO_2C_6H_4$	A1	MAC		90
	J2	*N*-Sulfinyl-*p*-toluamide	80	103a
(2-Me, 3-NO_2)C_6H_3	D3	MA	89	174
	F7	Carbobenzoxymethylenimine	52	174
[3,4-(OCH_2O)C_6H_3]	A7	Fumaric acid		172

TABLE 2.15 Diels–Alder Reactions of (*E*)-1-Substituted 1,3-Butadienes

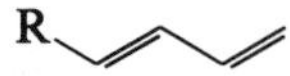

R		Dienophile	Yield (%)	Ref.
Pr^i	A2	Acrylates		115
Bu^t	A1	MAC		115
Bn	C7	1,4-Quinones		181
$CH{=}CH_2$	A12	TCNE (k)		21
CH_2CO_2H	A2	2-Methylene-4-(phenylmethoxy)butanal	78	186
CH_2CO_2Na	A2	Methyl-α-(acetoxymethyl)acrylate	67	182
	A2	2-Methylene-4-(phenylmethoxy)butanal	98	186
	C7	Dialkyl-1,4-benzoquinones	73–77	182
CH_2CO_2Me	A2	Methyl-α-(acetoxymethyl)acrylate	84	182
	A2	2-Methylene-4(phenylmethoxy)butanal	59	186
	C7	2,6-Dimethyl-1,4-benzoquinone	Traces	182
CH_2CO_2Et	C7	[1,4-Benzoquinones]	94–98	151
$C(OH)Me_2$	A1	AC	24	175
	A8	Vinylcarbinol	16	175
	E4	Ethynylcarbinol	21	175
$(CH_2)_2CH{=}CH_2$	C8	5,6,7,8-Tetrahydro-1,4-naphthoquinone		176
$(CH_2)_2CH{=}CHMe$	A1	AC	39	178
$(CH_2)_2CO_2^-$	A2	Acroleins	67–70	182
	A9	2-Nitropropene	51	182
	C7	1,4-Benzoquinones	53–86	182, 184
	D11	2-Carbomethoxy-7-oxabicyclo[2.2.1]-hept-2-ene	94	182
	E3	DMAD	60	182
$(CH_2)_2CO_2Et$	C7	2,6-Dimethyl-1,4-benzoquinone	Traces	182
$(CH_2)_2NH_3^+Cl^-$	C7	1,4-Benzoquinones	77–99	190
$(CH_2)_2OBn$	C7	[1,4-Benzoquinones]	95–100	151
$(CH_2)_3NHCO$	A1	MVK	72	190
	A2	Acroleins	79–81	190
$NaO_2C{-}(CH_2)_2$	C7	1,4-Benzoquinones	56–81	190
	D11	2-Carbomethoxy-7-oxabicyclo[2.2.1]-hept-2-ene	89	190
$(CH_2)_4NO_2$	A9	α-Chloroacrylonitile	73	185
	C4	2-Cyclohexenone	52	185
	C7	1,4-Benzoquinones	90–100	185
	D3	MA	91	185
	E3	DMAD	60	185
	F4	Diethyl mesoxalate	60	185
$CH_2NH_3^+Cl^-$	C7	1,4-Benzoquinones	86–98	190
CN	A8	Ethene		(1a)
	G1	4,9-Dihydro-3*H*-pyrido[3,4-*b*]indoles	86	166b
CONHCHMePh (*)	C7	BQ	92	163b
OH	A8	Ethene, AC, ACN		(1a), (183)
OEt	A7	(*E*)-Dibenzoylethene	91	152
OCOPh	A7	(*E*)-Dibenzoylethene	88	152
OCOCH(OMe)Ph (*)	C9	5-Hydroxy-1,4-naphthoquinone	98	179
$OCOCH_2$—Ⓟ	I1	[Diimine]		177
O-Glu-α-(D) (*)	A2	MTA	78–94	191

TABLE 2.15 (*Continued*)

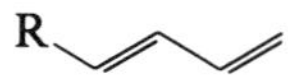

R		Dienophile	Yield (%)	Ref.
O-Glu-β-(D) (*)	A1	MAC	66	191
	A2	MTA	80–92	191
$OSiMe_2Bu^t$	C9	5-Hydroxy-1,4-naphthoquinone, Cat.(*)	82	157
	F6	Mesoxalic acid ester	59	165
$OSiEt_3$	C9	5-Hydroxy-1,4-naphthoquinone, Cat.(*)	73–82	157
SCN	A7	DMF	15	189
	A7	DMM	15	189
	A8	Acrylates	60–77	189
	A12	TCNE	70	187
	D5	NPM	70	187, 189
	D11	2-Carbomethoxy-7-oxabicyclo[2.2.1]hept-2-ene	78	190
	I5	NPTAD		187
SPh	A2	MAC		90
	C1	Cyclopropenone ketal	33	79
	D11	2-Carbomethoxy-7 oxabicyclo[2.2.1]hept-2-ene	90	149
$SiMe_3$	A1	MAC	42	180
	C7	2,6-Dimethyl-1,4-benzoquinone	32	180
	D3	Maleic anhydrides	34–74	180
	E1	MP	40	180
	F4	Diethyl mesoxalate	48	180
$SiMe_2SiMe_3$	D3	MA	82	180
$Si(OMe)_3$	C12	TCNE	50	155
	D3	MA	98	155
$SnCl_4$	A2	3-Acryloyl-5-methoxy-1,4-naphthoquinone	100	188

TABLE 2.16 Diels–Alder Reactions of (*Z*)-1-Substituted 1,3-Butadienes

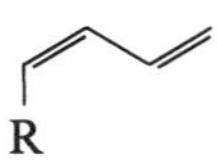

R		Dienophile	Yield (%)	Ref.
Me	A12	TCNE (k)	98	21
	D3	Maleic anhydrides (k)	27–58	21, 101, 122
	D5	NPM	35	101, 122
	E2	Ethyl-β-phenylsulfonyl propiolate	75	100
$CH{=}CH_2$	A12	TCNE (k)	99	21
$CH_2SiMe_3(E, Z)$	A7	DMM	42	198
Ph	A12	TCNE (k)	85	21
OMe	A12	TCNE (k)	90	21
SCN	D5	NPM	45	187

TABLE 2.17 Diels–Alder Reactions of 2-Methyl-1,3-butadiene (Isoprene)

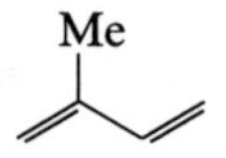

	Dienophile	Yield (%)	Ref.
A1	MVK	39	198, 207
A1	Methyl acrylates	84	198, 201, 207
A1	D-Pantolactone acrylate (*)	76	68
A2	Acroleins	34–71	64, 204
A2	Methyl α-(diethylphosphonate)acrylate	95	57
A2	1-Acetylvinyl-1-arene carboxylate	70–98	135
A3	4-(Diphenylphosphinyl)-3-buten-2-one	80	143
A3	α, β-Unsaturated carboxyimides (*)	83–85	114
A3	(*E*)-3-Trimethylsilylacrilic acid	73	211
A3	(*E*)-Carbomethoxyvinyltrimethyl ammonium salt	80	214
A3	Tributylstannylalkenones	91–95	218
A4	2-(4-Chlorophenylseleno)-2-butenal	87	48
A7	Dimenthyl fumarate (*)	100	15
A8	Phenylvinylsulfone	93	197, 208
A8	Cr-, Mo-, W$(CO)_5$-vinyl complexes (k)	61–87	206
A8	*S*-*p*-Tolyl-*N*-tosyl-*S*-vinylsulfoximine	95	51
A8	Acetals of acrolein	27–62	219
A9	Acrylonitriles	62–81	201, 215, 216
A10	Nitrovinylsilane	94	210
A10	α, β-Unsaturated acyliron complexes	38–95	119
A12	TCNE (k)	70	21
A13	Allenic ketones	34–58	66, 106
B2	2-Oxoindolin-3-ylidenes	38–74	205
B2	5-Methylene-2,2-dimethyl-1,3-dioxane-4,6-dione	80	65
C1	Cyclopropenoneketal	96	79
C2	Hexakis(trifluoromethyl)bicyclo[2.2.0]hexa-2,5-diene	80	14
C4	2-(Phenylthio)cyclopent-2-en-1-one	93	24
C4	3-Nitro-2-cyclohexen-1-one	99	97
C4	Alkyl-2-cycloalkenones	73–90	42, (212)
C4	2-Carbomethoxy-2-cyclohexen-1-one	60	105, 209
C4	(−)Carvone (*)	80	71
C4	4-Alkyl-2-cyclohexenones	40–60	69
C4	5-Alkyl-2-cyclohexenones	72–92	70
C4	Methyl-2-cyclohexenones	73–96	76

TABLE 2.17 (*Continued*)

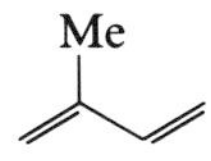

	Dienophile	Yield (%)	Ref.
C5	2,5-Cyclohexadienones	32–85	45, 129
C7	[2-Carbomethoxy-1,4-benzoquinone]	100	151
C7	2-Methoxy-5-methyl-1,4-benzoquinone	> 80	95
C7	1,4-Benzoquinones	40–80	43, 98, (110), 196
C7	1,2,3,6-Tetrahydrobenzocyclobutene-3,6-dione	77	36
C8	4*a*,9*a*-Epoxy-4*a*,9*a*-dihydroanthracene-1,4,9,10-tetrone	95	194
C9	1,4-Naphthoquinones	40–90	89, 98, 193a
C9	1,4-Anthraquinones	88	34, 203
C10	3-Keto-4,6-dienic steroid	67	17
D2	2-Methoxy-5-oxo-5,6-dihydro-2*H*-pyran	79	4
D2	3-Carbethoxy-5,5-dimethyl-2-(5*H*)-furanone	86	104
D2	Butenolides (*)	22–84	217
D3	Maleic anhydrides (k)	44–90	21, 37, 40, 195
D6	2-Thiolene-4-one-1,1-dioxides		55
D6	3-Methoxycarbonyl-2,5-dihydrothiophene	16	57
D8	6-Acetylpyrimidone		46
D10	1,4-Benzodithiin-1,1,4,4-tetroxide	98	121
E2	Ethyl(diethoxyphosphinyl)propynoate		145
E2	3-Trimethylsilylpropiolic acid	66	211
E3	DMAD	100	23
E5	Acethylenes	13–72	73, 107
F1	*N*, *N*-Dialkylcyanothioformamides	42–99	11
F2	Thione-*S*-imide	94	92
F2	Adamanthanethione	82	84
F3	*N*-Methyl dithiophtalimide	60	54
F7	Polifluoro-2-acyliminopropanes	30	200
F7	[Methylenebenzyliminium chloride]	59	213
G2	1,4-Di-*t*-butyl-1-germabenzene		199
H1	Methyl coumalate	32	202
I1	4-Chlorophenylazocyanide	55	18
I4	[5-Phenylpyrazol-3-one]	78	148
J1	1-Chloro-1-nitrosocyclohexane	32	16
J2	*N*-Sulfinyl-*p*-toluamide	85	103a

TABLE 2.18 Diels–Alder Reactions of 2-Trimethylsilyloxy-1,3-butadiene

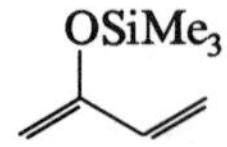

	Dienophile	Yield (%)	Ref.
A1	MVK	60	222
A1	MAC	35	222
A1	AC	87	226
A7	Diethyl fumarate	71	220, 222
A8	Trifluoromethylethene	17	224
A13	(Phenylsulfonyl)propadiene	60	156
B3	Isopropylidene-1,4-dioxaspiro[4.5]decan-8-ylidene malonate	97	78
C4	3-Nitro-2-cyclohexen-1-one	70	97
C4	Bicyclic enedione	79	221
C4	2-Carbomethoxy-2-cyclohexenone	26	105
C8	4*a*,9*a*-Epoxy-4*a*,9*a*-dihydroanthracene-1,4,9,10-tetrone	90	227
C9	9-Chloro-10-hydroxy-1,4-anthraquinone	40	131
D4	2-Aryl-Δ^2-pyrroline-4,5-diones	11–33	30
E5	(Trimethylsilyl)ethynylmethoxychromiumcarbene complex	> 58	225
F1	Monosaccharide *O*-thioformates (*)		3
F4	Diethyl mesoxalate	30	153
F7	[Ethyl-*t*-butylazomethinedicarboxylate]	30	223
G1	Dihydroarylpyridines	30–55	228

TABLE 2.19 Diels–Alder Reactions of 2-Methoxy-1,3-butadiene

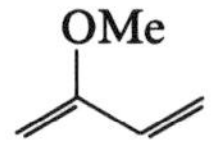

	Dienophile	Yield (%)	Ref.
A11	Cyanoarylethenes	84–98	126
A12	TCNE (k)	90	21
C5	4,4-Dialkoxycyclohexa-2,5-dien-1-ones	30–72	129
C7	BQ	75	230
C9	1,4-Naphthoquinones	100	193a, b
C9	9-Chloro-10-hydroxy-1,4-anthraquinone	70	131

TABLE 2.20 Diels–Alder Reactions of 2-Substituted 1,3-Butadienes

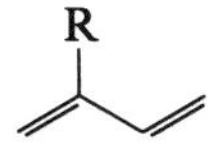

R		Dienophile	Yield (%)	Ref.
CH_2SiMe_3	A1	Acrylates	58–83	198, 207
	A1	AC	69	207
	A1	MVK	64	207
	A7	DMF	79	207
	A7	DMM	79	207
	A8	Trifluoromethylethene	38	224
	A10	(*E*)-β-Trifluoro-methylstyrenes	77–90	224
	C4	Cycloalkenones	56–70	207
	C7	BQ	88	207
	C9	NQ	88	198
	D3	MA	100	207
	E3	DMAD	96	198, 207
CH_2SnMe_3	A1	MAC	58	198, 207
	A1	MVK	94	198, 207
	E3	DMAD	94	198, 207
CH=CHMe	A1	MVK	98	239
	D5	NPM	92	239
	E3	DMAD	77	239
$CH_2CHMeOTs$	A1	MVK	74	239
	C9	NQ	72	239
	D5	NPM	99	239
$(CH_2)_2CH=CMe$ | $Me_2C=CH(CH_2)_2$	A1	MVK	66	246
$(CH_2)_2CHMe(CH_2)_3$ | $Me_2CH(CH_2)_3CHMe$	C7	2,3-Dimethyl-1,4-benzoquinone	> 35	231
$(CH_2)_2CH=CMe_2$	A1	Alkyl vinyl ketones	57–73	246
	A8	Phenylvinylsulfone	89	208
	I2	[Singlet sulfur]	75	234
1-Adamantyl	D3	MA	70	449
	D5	NPM	87	449
	E3	DMAD	65	449
$(CH_2)_2SOPh$	A1	MVK	93	243
	A1	MAC	69	243
	A1	AC	86	243
	A7	DMM	75	243
	D5	NPM	83	243
	E3	DMAD	75	243
CHO	A14	[2-Formyl-1,3-butadiene]	50	235
$C(OH)Me_2$	A7	DMM	95	237
	D5	NPM	> 70	237
C(OH)MeEt	A7	DMM	> 50	237

TABLE 2.20 (*Continued*)

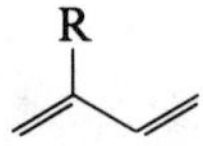

R		Dienophile	Yield (%)	Ref.
$C(OH)R'$— p-$R''C_6H_4$	A7	DMM	80–82	298
(R' = Ph, Me; R'' = H, OMe)	D5	NPM	86–94	298
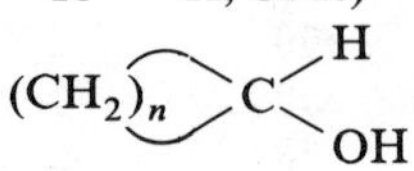 (n = 2, 3, 4)	A7	DMF	92–97	240
	D5	NPM	100	240
	E3	DMAD	89–100	240
CO_2Et	A14	[2-Carbethoxy-1,3-butadiene]	90	235
Ph	A3	(E)-3-Trimethylsilylacryloly-chloride	89	211
	A3	(E)-4-(Tributylstannyl)-3-buten-2-one	87	218
	A12	TCNE (k)	98	21
	D3	MA (k)		21
	J1	1-Chloro-1-nitroso-cyclohexane	58	16
	J2	N-Sulfinyl-p-toluamide	90	103a
OEt	F1	[Thioaldehydes]	38–76	233a
OAc	C8	Anthracene-1,4,9,10-tetrone	16	236
$OSiMe_2Bu^t$	F1	[Thioaldehydes]	0–74	233
[$SiMe_3$]	A1	Acrylates	63–70	229
	A1	Alkyl vinyl ketones	68–72	229
	A7	DMM	62	229
	E3	DMAD	76	244
$SiEt_3$	A1	Ethyl acrylate	77	232
	A2	MTA	69	232
	A7	DMF	81	232
	C7	BQ	85	232
	D3	MA	75	232
	E3	DMAD	87	232
[$POPh_2$]	A2	Acrylates	85	242
	A3	Benzalacetone	38	242
	A3	Benzalacetophenone	33	242
	A3	Crotonaldehyde	55	242
	A14	Butadiene	28	242
	I1	Diethylazodicarboxylate	82	242
[SPh]	A1	MVK	97	238
	A1	Ethyl acrylate	95	238
	A8	ACN	90	238
	E2	Ethyl propynoate	93	238
[SO-p-MeC_6H_4]	A1	MAC	97	393
	A2	Methyl methacrylate	74	393
	A7	DMF	89	393

TABLE 2.20 (*Continued*)

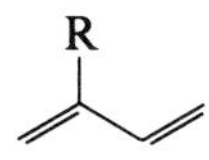

R		Dienophile	Yield (%)	Ref.
SO_2Ph	A1	MAC	74	245
[Ts]	A2	Methyl methacrylate	93	393
	A7	DMF	98	393
	D3	MA	99	393
Cl	C8	Anthracene-1,4,9,10-tetrone	79	236
	D3	Carbomethoxymaleic anhydride (k)		37
	E2	Ethyl(diethoxyphosphinyl)propynoate		145
[SePh]	A1	MVK	96	238
	A1	Ethyl acrylate	94	238
	A8	ACN	81	238
	E2	Ethyl propynoate	79	238
$SnBu_3$	A1	MVK	69	241
	A2	Acrylates	60–72	241
	A7	DMF	69	241
	C7	BQ	67	241
	D2	Butenolide	33	241
	D3	MA	75	241
	D5	NPM	80	241

TABLE 2.21 Diels–Alder Reactions of 1,1-Disubstituted 1,3-Butadienes

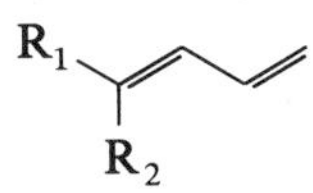

R_1	R_2		Dienophile	Yield (%)	Ref.
Me	Me	A12	TCNE (k)	99	21, 250
$—CH_2-CH_2—$		A12	TCNE	77	249
		C7	BQ	56	249
		D3	MA	77	249
		E3	DMAD	54	249
OMe	OMe	A11	Cyanoethenes	84–98	126
		F4	Diethyl mesoxalate	95	247
		F4	Hexachloroacetone	92	247
		F4	Benzoyl cyanide	90	247
		F4	Trichloroacetaldehyde	87	247
$—O-CH_2-CH_2-O—$		C9	2,6-Dichloronaphthazarin	94	159a
		C9	2,7-Dichloronaphthazarin	> 22	159a
		C9	NQ	81	159b
		C9	5-Hydroxy-1,4-naphthoquinone	66	159b
OCOPh	NMe_2 (*E*, *Z*)	I5	NPTAD		248

TABLE 2.22 Diels–Alder Reactions of 1,2-Disubstituted 1,3-Butadienes

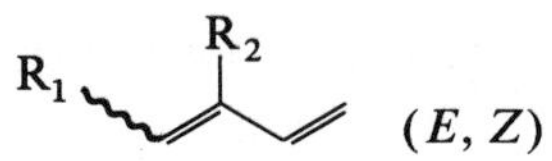

R_1	R_2			Dienophile	Yield (%)	Ref.
Me	Me	(*E*)	A10	α,β-Unsaturated acyliron complex	73	119
			F3	*N*-Methyldithiophtalimide	100	54
Me	Ph	(*E*, *Z*)	F3	*N*-Methyldithiophtalimide	100	54
		(*Z*)	J1	1-Chloro-1-nitrosocyclo-hexane	33	16
Me	OAc	(*Z*)	C4	Bicyclic enedione	> 65	221
Me	$OSiMe_3$	(*Z*)	B3	Isopropylidenecyclo hexylidene malonates	73	78
			F4	Diethyl mesoxalate	24	153
$CH_2CH{=}CMe_2$	Me	(*E*)	A1	MVK	80	246
Bu	$NHCOCCl_3$	(*Z*)	A1	MAC	87	260
Ph	Me	(*E*, *Z*)	F3	*N*-Methyldithiophtalimide	100	54
Ph	Ph	(*E*, *Z*)	F3	*N*-Methyldithiophtalimide	100	54
CH_2CO_2Et	Me	(*E*)	C4	*trans*-Octalone	40	251
			G2	Isopropylidene malonates	60–84	255
CO_2R	Me	(*E*, *Z*)	A2	AC	76	162
(R = H, Na, Et)		(*E*)	C4	*trans*-Octalones	82–100	259
NHAr	CO_2Me	(*Z*)	A8	Acrylates	24–83	252
			D3	MA	> 30	252
			D5	*N*-Arylmaleimides	85–98	252
OEt	Me	(*E*)	H1	Ethyl cyanoformate	53	256
OEt	Et	(*E*)	H1	Ethyl cyanoformate	56	256
OAc	Et	(*E*)	C9	Naphthazarin	99	254
[$SiMe_3$	Me]	(*E*)	E3	DMAD	90	244
SMe	SMe	(*Z*)	E3	DMAD	96	257
SPh	OMe	(*Z*)	A1	MVK	85	(253), 258
			F7	[Ethyl-*t*-butylazomethine dicarboxylate]	56	223

TABLE 2.23 Diels–Alder Reactions of (*E*)-1-Methoxy-3-trimethylsilyloxy-1,3-butadiene

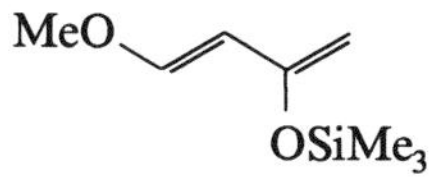

	Dienophile	Yield (%)	Ref.
A1	MVK	86	261
A1	MAC	85	261
A2	MTA	71	261
A3	(*E*)-Methyl crotonate	30	261
A4	Methyl-2-phenylsulfinylmethylenepropyonate	83	262
A5	4-Phenylsulfinyl-pent-3-en-2-one	75	262
A8	Phenylvinylsulfones	85–93	264
A8	Trifluoromethylethene	25	224
A11	2-Nitro-2-pentenes (*Z* and *E*)	63–71	144
A13	(Phenylsulfonyl)propadiene	82	156
B2	Heterocyclic methylene sulfoxide	57	263
C4	2-Methyl-2-cyclohexenone	53	261
C4	2-Cycloalkenones	68–98	83, 262
C4	Bicyclic enedione	91	221
C4	3-(Benzyloxycarbonyl)cyclopentenone	34	273
C5	[3-Carbobenzoxycyclopentadienone]	89	273
C7	2,5-Dimethylbenzoquinone	90	98
C8	Dihydroanthracenetetrones	70	158, 227
C9	1,4-Naphthoquinones	62–100	193
C9	9-Chloro-10-hydroxy-1,4-anthraquinone	93	131
D3	MA	90	261, 272
D4	Benzazepinopyrrolinedione	56	265
E3	DMAD	79	261
E5	Propynylchromiumcarbene complex	100	225
F1	Thiobenzophenone	68	84
F2	Adamanthane thione	64	84
F3	Bn-*N*-CBn-Pyroglutamate sufoxides (*)	57	269
F4	*n*-Butyl glyoxalate	80	267
F4	Diethyl mesoxalate	30	153
F4	Aldehydes	40–98	268, 270
F6	2-Phenyl-Δ^2-pyrroline-4,5-diones	57–96	30, 271
F7	[Acylimines]	60–84	223
F8	Phosphaalkene		266
G1	3,4-Dihydroarylpyridines	39–73	228
I5	NPTAD	84	148

TABLE 2.24 Diels–Alder Reactions of (*E*)-1-Substituted 3-Methyl-1,3-butadienes

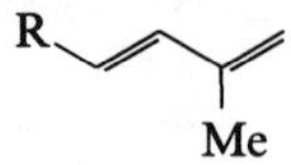

R		Dienophile	Yield (%)	Ref.
$B(OCH_2)_2$	D3	MA	100	282
	D5	NPM	100	282
$B(OCMe_2)_2$	A1	MAC	69	282
	A8	ACN	50	282
$BO_2C_6H_4$	D3	MA	100	282
	D5	NPM	100	282
Me	A8	3,3-Diethoxypropene	48	219
	A8	(Ferra-β-diketonato)BF_2 acrylates	26–34	279
	A13	α-Ethenylidenecyclanones	28–50	66
	C4	2-Carbomethoxy-2-cyclohexenone	34	105
	C4	Alkylcyclohexen-2-ones	62–94	76c
	C4	(−)Carvone (*)	86	71
	D2	3-Carbethoxy-5,5-dimethyl-2(5*H*)-furanone	93	104
	F1	*S*-Methyl-*O*-methyl dithiooxalate	82	94
	F3	*N*-Methyldithiophtalimide	100	54
Ph	J2	*N*-Sulfinyl-*p*-toluamide	78	103b
$NPrCO_2Me$	E3	DMAD	79	192
$O(CH_2)_2OCH_2OMe$	J2	Benzylsulfinylcarbamate	91	289
$O(CH_2)_2OSiMe_2Bu^t$	J2	Benzylsulfinylcarbamate	97	289
$O(CH_2)_2R'$ (R′ = OMe, SPh, SePh)	A1	MVK	76–80	286, 287
OAc	A2	MTA	46	142
	A3	Cinnamic acid	77	141
	A3	(*E*)-β-Nitromethyl acrylate	63	283
	C9	[NQ]	85	80
	D8	6-Acetylpyrimidone	31	46
$OSiMe_3$	A1	AC		154
	A2	Methyl-α-(1-methyl-2,5-dioxa-cyclopent-1-yl)acrylate	88	277
	C7	BQ	> 50	193c, f
	C9	1,4-Naphthoquinones Cat (*)	51	150, 193c
$SiMe_3$	A1	MAC	51	180, 274
	E1	MP	42–54	180, 274
	D3	MA	60	180
[$SiMe_3$]	E3	DMAD	85	244

TABLE 2.25 Diels–Alder Reactions of (*E*)-1-Methoxy-3-Substituted 1,3-Butadienes

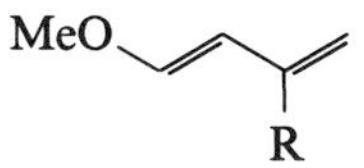

R		Dienophile	Yield (%)	Ref.
Me	A11	Cyanoethenes	84–98	126
	C9	5-Hydroxy-7-methoxy-1,4-naphthoquinone	70–90	138
	C9	5,7-Dihydroxy-1,4-naphthoquinone	80	138
$CH(CH_2)_5$	A8	Phenylvinylsulfone	53	287
$CMe(OCH_2)_2$	C8	Anthracene 1,4,9,10-tetrone	> 80	275
	C8	[5-Methoxy-10-hydroxy-1,4-dioxo-5,8-ethanoanthracen-9-yl-2,2-dimethyl propanoate]	28	278
CH_2SEt	A1	MVK	70	285
	A1	MAC	78	285
CH_2SPh	A1	MVK	73	285, 287
	A1	MAC	71	285
	A1	AC	59	285
OMe	D3	MA	98	284
	E3	DMAD	44–63	284
	I5	NPTAD	82	284
$OSiMe_2Bu^t$	A11	Acetoxydimethylmaleate	78	281
	A11	(*Z*)-Benzyl-4-[(*t*-butyl)diphenyl-silyloxy]-2-pivaloyl-2-butenoate	83	281
	D2	Methyl-2,3-dideoxy-α-D-glicero-6β-methylhex-2-enopyranosid-4-ulose (*)	93	276
	F1	[Thioaldehydes]	90	233b
	F4	*n*-Butylglioxylate	85	267
Cl	C7	2,3-Dimethoxy-1,4-benzoquinone	93	280

TABLE 2.26 Diels–Alder Reactions of (*E*)-1,3-Disubstituted 1,3-Butadienes

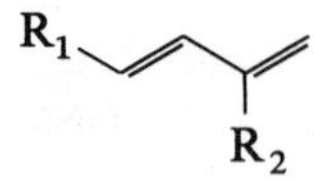

R_1	R_2		Dienophile	Yield (%)	Ref.
Me	OAc	C6	*o*-Quinones	17–85	112
Me	$OPO(OEt)_2$	C4	2-Carbomethoxy-2-cyclohexenone	66	295
		C7	BQ	60	295
Me	SO_2Ph	A1	Vinyl ketones	68–85	245
$(CH{=}CH)_2Me$	$OSiMe_3$	A3	1,4-Hexadien-3-one	35	130
$(CH{=}CH)_2Me$	$OSiMe_2Bu^t$	A8	1-Phenyl-4-penten-1-yn-3-one	55	130
$CHMeOSiMe_2Bu^t$ | $CHNHCO_2Bn$	$OSiMe_2Bu^t$	E1	MP		293
[Ph	OAc]	D3	Citraconic anhydride	41	291
[*p*-$OMeC_6H_4$	OAc]	D3	Citraconic anhydride	55	291
α-Fu	Me	A8	Ethyl-4-oxo-5-hexen-2-ynoate	37	130
$O(CH_2)_2OMe$	CH_2SPh	A1	MVK	57	285
		E1	MP	61	180
OBu^t	$OSiMe_3$	C9	9-Chloro-10-hydroxy-1,4-anthraquinone	75	297
OAc	$OSiMe_3$	C9	9-Chloro-10-hydroxy-1,4-anthraquinone	80	297
O-Menthyl	$OSiMe_3$ (*)	F4	Benzaldehyde Cat.(*)		296
$OSiMe_3$	$OSiMe_3$	C9	1,4-Naphthoquinones		193c
		C9	9-Chloro-10-hydroxy-1,4-anthraquinone	84	297
		F6	Benzazepinopyrrolinedione	44	265
β-D-Glu$(OAc)_4$	$OSiMe_3$ (*)	C7	1,4-Benzoquinones	32–59	294
		C8	4*a*,9*a*-Epoxy-4*a*,9*a*-dihydro-anthracene-1,4,9,10-tetrone	> 57	290
$SiMe_3$	$OSiMe_3$	A1	MAC	63	180
Cl	$CMe(OCH_2)_2$	C8	1,4,9,10-Anthracenetetrone	88	275

TABLE 2.27 Diels–Alder Reactions of 1-Methyl-4-Substituted 1,3-Butadienes

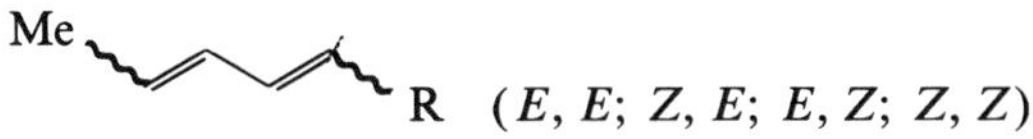

R			Dienophile	Yield (%)	Ref.
Me	(*E*, *E*)	D4	Δ^3-Pyrrolin-2-ones	19–69	305
	(*E*, *E*)	D5	[3-Acyl-3-pyrroline-2,5-diones]	12–71	326
	(*E*, *E*)	E5	Bis(trimethylsilyl)acetylene	15	73
	(*E*, *E*)	F7	[Methylenebenzyliminium chloride]	62	213
	(*E*, *E*)	H2	4-Nitrobenzenediazonium fluoborate	58	113
	(*E*, *E*)	I5	NPTAD (k)	99	87
	(*Z*, *E*)	I5	NPTAD (k)	99	87
	(*Z*, *Z*)	I5	NPTAD (k)		87
	(*E*, *E*)	J2	*N*,*N'*-Bis-tosylsulfurdiimide	48	303
	(*E*, *Z*)	J2	*N*,*N'*-Bis-tosylsulfurdiimide	70	303
	(*E*, *E*)	J2	*N*-Sulfinylbenzylcarbamate	89	301
	(*E*, *Z*)	J2	*N*-Sulfinylbenzylcarbamate	57	301
	(*E*, *E*)	J2	*N*-Sulfinyl-*p*-toluamide	75	103b
Ph	(*E*, *E*)	H2	4-Nitrobenzenediazonium fluoborate	39	113
	(*E*, *E*)	J2	*N*-Sulfinyl-*p*-toluamide	68	103b
p-$NO_2C_6H_4$	(*E*, *E*)	J2	*N*-Sulfynyl-*p*-toluamide	70	103b
CH(Me)OR' (*)		A12	TCNE	100	304
(R' = Me, $SiMe_3$)	(*E*, *E*)	D5	NPM		329
CH_2OAc	(*E*, *E*)	A1	Ethyl acrylate		302
	(*E*, *E*)	C9	1,4-Naphthoquinone-2-methyl sulfoxide	74	80
	(*E*, *E*)	E1	Ethyl propiolate		302
CH_2OCOBu^t	(*E*, *E*)	C7	Carbomethoxy-1,4-benzoquinone		302
$(CH_2)_7X^+Bu_3$ $^-OSO_2Me$ X = N, P	(*E*, *E*)	I5	*N*-Aryltriazoline-3,5-diones	94–99	307
$(CH_2)_7OAc$	(*E*, *E*)	I3	[Singlet oxygen]		319
$(CH_2)_7O$-β-D-Malto-pyranosyl	(*E*, *E*)	D5	*N*-Arylmaleimides		323
	(*E*, *E*)	I5	*N*-Aryltriazoline-3,5-diones		323
$(CH_2)_7OSO_3Na$	(*E*, *E*)	D5	*N*-Arylmaleimides		323
	(*E*, *E*)	I5	*N*-Arytriazoline-3,5-diones	99	323
CO_2Me	(*E*, *E*)	B2	2-Methyleneimidazolidines (*)	35–71	299
	(*E*, *Z*)	B2	2-Methyleneimidazolidines (*)	50–57	299
	(*Z*, *E*)	B2	2-Methyleneimidazolidines (*)	39	299
	(*E*, *E*)	C7	1,4-Benzoquinones	72	315
	(*E*, *Z*)	D3	MA	40	314
	(*E*, *E*)	F8	Phosphaalkene		266
CO_2Et	(*E*, *E*)	C1	Cyclopropenone ketal	70	79
$NHCO_2Et$	(*E*, *E*)	A9	α-Methylthioacrylonitrile	88	216
OMe	(*E*, *E*)	A11	Cyanoethenes	84–98	126
$SiMe_3$	(*E*, *E*)	A1	MAC	66	180
	(*E*, *E*)	D3	MA	69	180
	(*E*, *E*)	E1	MP	68	180

TABLE 2.28 Diels–Alder Reactions of 1,4-Disubstituted 1,3-Butadienes

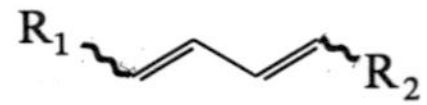

$(E, E;\ E, Z)$

R_1	R_2			Dienophile	Yield (%)	Ref.
Et	Ph	(E, E)	H2	4-Nitrobenzenediazonium fluoborate	30	113
Et	OAc	(E, E)	A1	β-Trichloroethyl acrylate	69	141
Et	$(CH_2)_6OAc$	(E, E)	I3	[Singlet oxygen]		319
Et	OCOCH(OMe)Ph (*)	(E, E)	A1	AC		179
Pr	SPh	(E, E)	D11	Methyl(1*R*, 4*S*)-7-oxabicyclo-[2.2.1]hept-2-ene-2-carbo-xylate (*)	77	149
Ph	Ph	(E, E)	C1	Chlorofluorocyclopropene	11–58	6
		(E, E)	D5	Malehymides (k)		320
		(E, E)	E4	Ethynyl-*p*-tolylsulfone	58	317
		(E, E)	F3	1,1′-Thiocarbonyl-bis(1,2,4-triazole)	67	316
		(E, E)	H2	4-Nitrobenzenediazonium fluo-borate	69	113
Ph	$R'C_6H_4$ (R′ = Me, CN, NO_2, OMe, Cl)	(E, E)	D3	MA (k)		86, 320
Ph	*p*-$OMeC_6H_4$	(E, E)	H2	4-Nitrobenzenediazonium fluoborate		113
Ph	CO_2Et	(E, Z)	D3	MA	60	314
o-$NO_2C_6H_4$	*o*-NO_2, *p*-$OMeC_6H_3$	(E, E)	D5	Maleimides	55–95	321
CHMeOMe	CHMeOMe	(E, E)	A12	TCNE		304
CH_2CO_2Me	OMe	(E, E)	C7	2-Methyl-6-methoxy-1,4-benzoquinone	69	306
CH_2CO_2Na	OMe	(E, E)	C7	1,4-Benzoquinones	50–93	306
$(CH_2)_nOR'$ (n = 1, 2, 7; R′ = H, Me, $SiMe_3$, OAc)	OMe	(E, Z)	F4	Diethyl mesoxalate	32–45	313

CO_2Me	CO_2Me	(*E*, *E*)	F5	Indantrione	71	165
		(*E*, *E*)	F6	5-Ketobarbituric acid	87	165
		(*E*, *E*)	F6	Mesoxalic acid ester	60	165
$NHCO_2Et$	$NHCO_2Et$	(*E*, *E*)	C9	Quinolinediones	56–62	310
$NHCO_2Et$	$OSiMe_2Bu^t$	(*E*, *E*)	F4	*p*-Nitrobenzyl glyoxylate	38	309
$NHCO_2Bn$	SPh	(*E*, *E*)	A1	Phenyl vinyl ketone	89	322
$NHCO_2Bn$	SOPh	(*E*, *E*)	A1	AC	45	322
$NHCO_2Bn$	SO_2Ph	(*E*, *E*)	A1	AC	78	322
$NHCO_2Et$	OCOPh	(*E*, *E*)	C12	TCNE	88	311
		(*E*, *E*)	D3	MA	73	311
CO_2Me	OMe	(*E*, *E*)	C1	Cyclopropenone ketal	80	79
		(*E*, *E*)	C4	2-Cyclopentenone	40	325
OMe	$OSiMe_3$	(*E*, *E*)	C9	2-Acetyl-1,4-naphthoquinone	43	61
OMe	SePh	(*E*, *E*)	D3	MA	> 80	327
OBu^t	OBu^t	(*E*, *E*)	A10	Fumaronitrile	18	324
		(*E*, *Z*)	D3	MA	33	324
OAc	OAc	(*E*, *E*)	A3	(*E*)-Ethyl-3-nitroacrylate	91	136
		(*E*, *E*)	C7	Acyl-1,4-benzoquinones	40	61
		(*E*, *E*)	C9	2-Chloronaphthazarindiacetate		159a
		(*E*, *E*)	D6	3-Thiolen-2-one	33	318
		(*E*, *E*)	E1	Propiolaldehyde	89	311
OAc	$SiMe_3$	(*E*, *E*)	A1	MAC	80	312
		(*E*, *E*)	D3	Maleic anhydrides	74–76	312
OAc	SPh	(*E*, *E*)	A2	MTA		(253)
OCH_2OMe	CO_2Me	(*E*, *E*)	A1	MAC	70	325
		(*E*, *E*)	D3	MA	80	325
		(*E*, *E*)	D8	5,6-Dihydro-2-pyrone	75	328
OCH_2OMe	CO_2Bn	(*E*, *E*)	A3	(*E*)-Methyl crotonate	40	328
OCH_2OEt	CO_2Et	(*E*, *E*)	E5	Diethylpropinylamina	40	325
$OSiMe_2Bu^t$	$OSiMe_2Bu^t$	(*E*, *E*)	C1	Spirodienone	48	300
$SiMe_3$	$SiMe_3$	(*E*, *E*)	D3	MA	63	180
SPh	CH_2CH_2—$THP(OMe)_2$		D11	Methyl(1*R*, 4*S*)-7-oxabicyclo-[2.2.1]hept-2-ene-2-carboxylate (*)	86	149

TABLE 2.29 Diels–Alder Reactions of 2,3-Dimethyl-1,3-butadiene

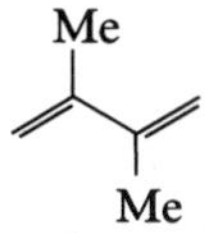

	Dienophile	Yield (%)	Ref.
A1	AC	60	67, 352
A3	[*n*-Butylhept-2-en-1-oate]	92	339
A7	Dimenthyl fumarate (*)	70	88
A7	Menthylphenethyl fumarate (*)	70	357
A8	Vinylnonafluorobutylsulfone	100	332
A8	Vinyl-substituted complexes (Cr, Mo, W)	75–80	206
A8	Acrolein acetals	52–65	219
A8	Phenylvinylsulfone	94	208
A8	Trimethyl silylvinyl ketones	84–94	308
A9	Morpholino-2-acrylonitrile	81	215
A11	(Ferra-β-diketonato)BF_2 complex		279
A13	1,1-Difluoroallene		25b
A13	α-Ethenylidenecyclanones	77–88	66
A13	Allenic ketones	55–74	347
B1	Dimethylene-2,4-cyclobutanedione		333
B2	(*E*)-4-Benzal-1-phenyl-5-pyrazolone	90	343
B2	5-Methylene-2-dimethyl-1,3-dioxane-4,6 dione	72	65
C1	Cyclopropenone ketal	57	79
C1	Cyclobuteno[16α, 17α]pregnene		345
C1	Tris(trifluoromethyl)trifluoroacetylcyclopropene		27
C2	Hexakis(trifluoromethyl)bicyclo[2.2.0]hexa-2,5-diene	76	14
C4	3-Nitrocycloalkenones	83	97
C4	2-Methyl-2-cyclohexenone	80	42a
C4	2-Carbomethoxy-2-cyclohexenone	61	105
C5	[3-Carbobenzoxycyclopentadienone]	84	273
C6	*o*-Quinones	27–77	112
C7	1,2,3,6-Tetrahydrobenzocyclobutene-3,6-dione	77	36
C7	1,4-Benzoquinones	48–78	61
C8	1-Ethyl-4,5,8-trioxo-1,4,5,8-tetrahydro-3-quinolin carboxylic acid	83	43
C8	Chloroquinizarin diquinones	90–96	158
C9	1,4-Naphthoquinones	21–93	139, 196, 355
C9	1,4-Anthraquinones	96	34, 41
D2	2-Methoxy-5-oxo-5,6-dihydro-2*H*-pyran	91	4
D2	3-Carbethoxy-5,5-dimethyl-2(5*H*)furanone	91	104
D2	Hexenopyranosiduloses (*)	43–56	338
D4	1-Methyl-4-cyano-2(1*H*)pyridone	72	334
D4	4-(Methoxycarbonyl)-1-methyl-2-(1*H*)pyridone	84	344
D6	3-Thiolen-2-one	81	318

TABLE 2.29 (*Continued*)

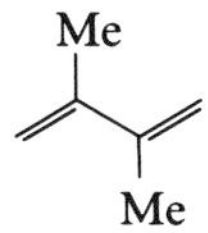

	Dienophile	Yield (%)	Ref.
D7	3-Acetyl-2(3*H*)oxazolone	26	351
D9	4-Isothiazolin-3-one-1-oxide	90	331
E2	3-Trimethylsilylpropiolic acid	70	211
E2	Acetylenedicarbaldehydemonoacetal	68	74
E3	Di-*t*-butylacetylenedicarboxylate	99	23
E4	Ethynyl-*p*-tolylsulfone	100	317
E4	1-(*t*-Butyldimethylsilyl)-2-propyn-1-one	80	308
E5	Acetylenedicarbaldehyde monoacetal	95	335
E5	Diethylacetylene	93	107
E5	Trimethylsilylethinylmethoxychromiumcarbene complex	67	225
E5	Bis(trimethylsilyl)acetylene	72	73
E5	(Tributylstannio)propiolonitrile	69	62
F1	Thiooxalates	100	330
F1	Cyanothioformamides	46–97	11
F1	*S*-Methyl-*O*-ethyldithiooxalate	75	94
F1	1,1′-Thiocarbonylbis(1,2,4-triazole)	99	316
F1	10-Chloro-10-sulfinylcamphor (*)	80	336
F1	Sulfoximino sulfines (*)	23–90	336, 349
F1	[Diethylthioxomalonate-*S*-oxide]	42	340
F1	Bis(trifluoromethyl)sulfene	41	342
F1	Dithioacetate	70–73	26
F1	[Thioaldehydes]	50–61	233a
F2	Thione-*S*-imide	61	92
F2	Adamanthanethione	82	84
F3	α-Oxosulfines	66–89	353
F4	Menthyl glioxylate (*)	19–46	123b
F7	[Methylenebenzyliminium chloride]	64	213
F8	Phosphalkenes	50–84	350
F10	Pentacarbonyl(selenobenzaldehyde)tungsten	53–85	356
G2	1,4-Di-*t*-butyl-1-germabenzene	20	199
G3	3,4-Dimethyl-2*H*-phosphole	35	346
I1	Arylazocyanides	70–94	18
I2	[Diatomic sulfur]	35–60	234
J1	1-Chloro-1-nitrosocyclohexane	40	16
J1	Nitrosobenzene		354
J2	*N*-Sulfinyl-*p*-toluamide		103b
J2	[Substituted cation sulfurdiimides]	92–100	341

TABLE 2.30 Diels–Alder Reactions of 2,3-Disubstituted 1,3-Butadienes

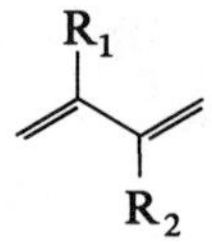

R_1	R_2		Dienophile	Yield (%)	Ref.
Me	SPh	A1	MVK	91	359
		A8	ACN	77	359
Ph	Ph	F1	*N*-Benzoyl-*N*-(4-nitrophenyl)cyanothioformamide	66	11
		I2	[Singlet sulfur]	85	234
*m*Cl—C_6H_4	*m*Cl—C_6H_4	E3	DMAD	77	364
CN	CN	A13	1,1-Difluoroallene	1	25b
[CN	CN]	C1	Indene		366
CH_2OH	CH_2OH	C7	BQ	84	358
CH_2OAc	CH_2OAc	C7	BQ	68	358
CO_2Me	CO_2Me	C7	BQ	72	358
CH_2SiMe_3	CH_2NMe_2	A1	MAC	82	361
		A7	DMF	86	361
		F4	*n*-Butyl glioxylate	90	361
CH_2SiMe_3	CH_2SiMe_3	D5	NPM	100	367
		E3	DMAD	82	367
CH_2Cl	CH_2Cl	D3	MA	74	369
		E3	DMAD	84	369
CH_2I	CH_2I	I5	NPTAD	60	148
OMe	OMe	A10	Nitrostyrene	57	139
		F7	[Ethyl-*t*-butylazomethine dicarboxylate]	16	223
OMe (R′ = H, OH, OMe)	S-*p*-R′C_6H_4	A1	MVK		(253)
OME	S—$C_4H_3N_2$	A1	MVK	50	(253), 365
OEt	OEt	I5	4-Alkyl-1,2,4-triazole-3,5-diones (k)		362
OAc (R′ = H, OH, OMe)	*S*-*p*-R′C_6H_4	A1	MVK		(253)
		A8	Acrylates	55–94	365
		C4	Bicyclic enedione	46	221
		C4	Cycloalkenones	33–83	146, 365
$OSiMe_3$	$OSiMe_3$	E3	DMAD	66	363
$SiMe_3$	$SiMe_3$	D3	MA	80	360
SPh	Cl	A1	MVK	68	368
SPh	Br	A1	MVK	40	368
Cl	SePh	A1	MVK	72	368
SePh	Br	A1	MVK	70	368

TABLE 2.31 Diels–Alder Reactions of Trisubstituted 1,3-Butadienes

R_1	R_2	R_3		Dienophile	Yield (%)	Ref.
			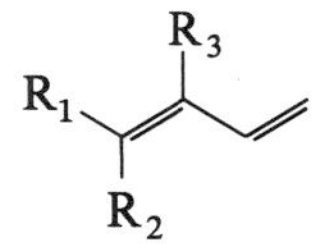			
Me	Me	$(CH_2)_2Ph$	I5	NPTAD	33	371
	$=CHPO(OR)_2$	Me	E3	DMAD		407
NMe_2	NMe_2	Me	A8	ACN	90	390
NMe_2	OMe	Me	E3	DMAD	70	390
NMe_2	F	Me	A8	ACN	69	390
NMe_2	Cl	Me	A8	ACN	> 60	390
			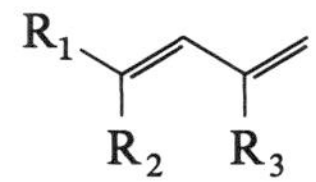			
Me	Me	Me	A8	3,3-Diethoxypropene	61	219
			A14	1,1,3-Trimethyl-1,3-butadiene	27–90	378, 402
			E4	[1,1-Diphenylpropargyl cation]	70	372
OMe	OMe	$OSiMe_3$	A13	Methyl-2,3-alkadienoates (*)	52–59	382
			B2	7-Chloro-4,6-dimethoxy-2-(1-phenylsulfinylethylidene)-3(2*H*)benzofuranone	54	387
			B2	(*Z*)-2-Ethylidene-7-chloro-4,6-dimethoxycumaran-3-one	55	391
			C9	Naphthoquinonesulfoxide	90	80
			F4	Diethyl mesoxalate	66	247
			F4	Aldehydes	69–85	405

TABLE 2.31 (*Continued*)

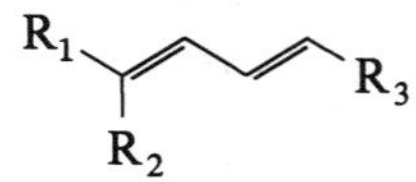

R_1	R_2	R_3		Dienophile	Yield (%)	Ref.
Me	Me	$CH{=}CMe_2$	A12	TCNE		250
Me	Me	$N(CH_2)_4$	A7	DMF	82	404
OMe	OMe	OMe	A11	Cyanoethenes	94	126
			C9	1,4-Naphthoquinones	76–100	380
			F4	Methylethyl mesoxalate	70	247

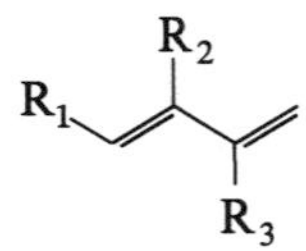

R_1	R_2	R_3		Dienophile	Yield (%)	Ref.
Me	Me	Me	A13	Allenes		(50)
Ph	$C(OSiMe_3){=}CH_2$	$OSiMe_3$	E3	DMAD	> 51	377
$(CH_2)_4CH(OCH_2)_2$	$OSiMe_3$	Me	C7	2,3-Dichloro-1,4-benzoquinone	78	408
$[NO_2$	Me	$CH_2NO_2]$	A8	Styrene	26	394
OMe	Me	$OSiMe_3$	A4	Ethyl-2-(4-chlorophenylthio)-4-oxopent-2-enoate	64	395
			A8	Vinylsulfones	69	381
			C7	1,4-Benzoquinones	89–100	386
			E3	DMAD	87	388

OMe	Me	$SiMe_3$	C8	Morpholino-1,4-benzoquinone	92	397
OMe	$C(OSiMe_3){=}CH_2$	$OSiMe_3$	D3	MA	> 39	403
			D5	NMM	63	377
OMe	OAc	$OSiMe_3$	C7	BQ	81	400
			E1	Ethyl propiolate	87	400
			E3	DMAD	84	400
OMe	OAc	$OSiMe_2Bu^t$	A5	Acetoxydimethylfumarate	83	281
O-Menthyl	Me	$OSiMe_3$ (*)	F4	Benzaldehyde, Cat. (*)		296
O-Menthyl	OAc	$OSiMe_3$ (*)	F4	Benzaldehyde, Cat. (*)		296
$OC_6H_9(OBn)_2$	OAc	$OSiMe_2Bu^t$	F4	Acetaldehyde	51	375
$OSiMe_3$	Me	$OSiMe_3$	Cl	1-Carbethoxy-3-acetoxycyclo-hexen-1-ene	72	383
[$SiMe_3$	Me	Me]	E3	DMAD	85	244
$PO(OEt)_2$	$CH_2CH{=}CH_2$	Me	D5	NPM	66	292
SPh	OMe	Me	C8	Morpholino-1,4-benzoquinone		397
SPh	$OCH_2CH{=}CH_2$	Me	C8	Morpholino-1,4-benzoquinone	> 75	384
SCN	Me	Me	D5	NPM	70	187
Cl	Cl	SPh	A1	AC	57	368
Cl	Cl	SePh	A1	MVK	47	368
Cl	Br	SPh	A1	AC	17	368
Cl	Br	SePh	A1	MVK	62	368
Cl	Br	Br	A1	MVK	51	368
Cl	I	I	A1	MVK	13	368

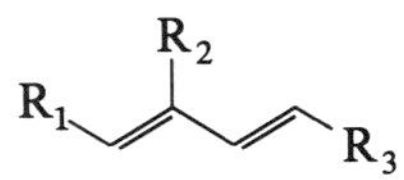

Me ($R' = SiMe_2Bu^t$, COEt, Me, THP)	Me	CH_2OR'	A4	Dimethyl-2-[2′-(benzyloxycarbonyl-amino)-3′-phenyl-propylidene]malonate	9–90	379

TABLE 2.31 (*Continued*)

R_1	R_2	R_3		Dienophile	Yield (%)	Ref.
Me	Me	$CH(OCH_2)_2$	A4	Dimethyl-2-[2′,-(benzyloxycarbonyl-amino)-3′-phenyl-propylidene]malonate	51	379
Me	Me	CH_2OBn	J2	*N*-Sulfinyltoluenesulfonamide	90	401
Me	Me	OBn	J2	*N*-Phenylsulfinyl-*p*-toluenesulfonamide	91	301
Me	$OSiMe_3$	OMe	A2	MTA	56	388
$CH(OH)CH_2OBn$	OBn	CN (*)	D5	NPM	51	409
$CHMeOSiMe_3$	OCH_2OMe	Pr^i (*)	A1	AC	72	373
OMe	Me	OMe	C9	1,4-Naphthoquinones		193a
OAc	Me	OAc	C9	1,4-Naphthoquinones		193a
$OSiMe_3$	Me	Me	A1	MAC	89	376
SePh	$OSiMe_3$	OMe	A2	Ethyl-α-phenyl acrylate	50	388
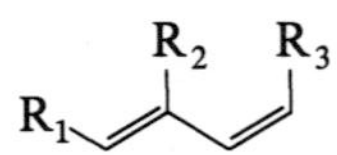						
CO_2Et	OEt	OEt	F4	Diethyl mesoxalate	54	313
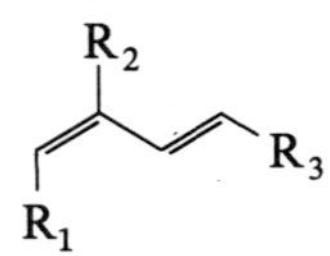						
nC_6H_{11}	$C_6H_5SO_2$	Et	Al	MVK	48	374

TABLE 2.32 Diels–Alder Reactions of Tetra- and Hexasubstituted 1,3-Butadienes

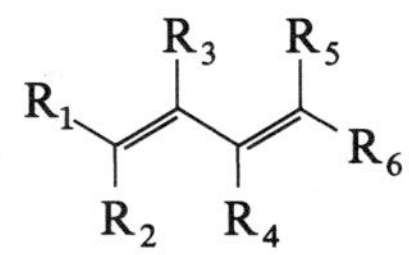

R_1	R_2	R_3	R_4	R_5	R_6		Dienophile	Yield (%)	Ref.
						1, 1, 2, 3			
Me	Me	CH_2Bn	Me	H	H	I5	NPTAD	85	371
CO_2Me	$OSiMe_3$	CO_2Me	$OSiMe_3$	H	H	C7	2,6-Dichloro-1,4-benzoquinone	35	385
$OSiMe_3$	Me	CO_2Et	OMe	H	H	C7	2,6-Dichloro-1,4-benzoquinone	80	389
						1, 1, 2, 4			
OMe	OMe	NMe_2	H	H	NMe_2	D3	MA	26	410
						D5	NPM	96	410
OMe	OMe	OMe	H	H	NMe_2	C9	1,4-Naphthoquinones	70–95	399
						D5	NPM	69	399
OMe	OMe	OEt	H	H	NMe_2	D5	NPM	40	399
						1, 1, 3, 4			
$OSiMe_3$	Me	H	Me	H	Me	C7	[2-Carbomethoxy-1,4-benzoquinone]	96	151
						1, 1, 4, 4			
Me	Me	H	H	Me	Me	I5	NPTAD		87
$—CH_2—CH_2—$		H	H	$—CH_2—CH_2—$		E3	DMAD	12	406
$=CH_2$		H	H	$=CH_2$		C7	BQ	90	396
						D3	MA	80	396
						D5	NPM	60	396

TABLE 2.32 (*Continued*)

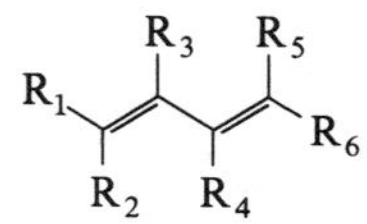

R_1	R_2	R_3	R_4	R_5	R_6		DDienophile	Yield (%)	Ref.
						E3	DMAD	40–50	370
$=CMe_2$		H	H	$=CMe_2$		I5	NPTAD	87	392
						J2	Sulfinylalanine	64	392
$=CMePh$		H	H	$=CMePh$		I5	NPTAD	33	392
$=CPh_2$		H	H	$=CPh_2$		I5	NPTAD	45	392
					1, 2, 3, 4				
Me	H	Me	Me	Me	H	J2	*N*-Sulfinyltoluenesulfonamide	> 83	401
Me	H	Me	Me	H	Me	J2	*N*-Sulfinyltoluenesulfonamide	> 85	401
Me	H	$OSiMe_3$	Me	H	OMe	F4	Aldehydes	49–66	270
Ph	H	Me	Me	H	Ph	H1	4-Nitrobenzenediazonium fluoborate	74	113
OMe	H	Me	$OSiMe_3$	H	Me	E3	DMAD	58	388
O-Menthyl	H	Me	$OSiMe_3$	H	Me (*)	F4	Benzaldehyde, Cat. (*)		296
[NO_2	H	Me	Me	H	NO_2]	A8	Styrene	56	394
					1, 2, 3, 4, 5, 6				
F	F	F	F	F	F	A8	*p*-Methoxystyrene	80	398
Me	Me	—S(O)—		Me	Me	I3	[Singlet oxygen]	72	451
						I5	4-Alkyl-1,2,4-triazoline-3,5-diones		450

TABLE 2.33 Diels–Alder Reactions of Deuterio-1,3-butadienes

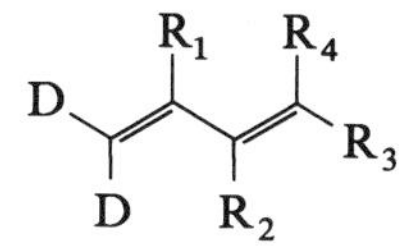

Diene		Dienophile	Yield (%)	Ref.
R_2 = Me, R_1 = R_3 = R_4 = H	A3	(*E*)-Methyl-β-cyanoacrylate		438
	A8	ACN		438
	A10	Fumaronitrile		438
R_1 = CD_3, R_2 = CH_2Bn, R_3 = R_4 = H	A1	AC	60	371
[R_1 = R_2 = H, R_3 = R_4 = D]	A7	Fumaroyl chloride	81	439
	D3	MA	> 80	439
R_1 = Me, R_2 = H, R_3 = R_4 = D	A3	(*E*)-Methyl-β-cyanoacrylate		438
	A8	ACN		438
	A9	Vinylidene cyanide		438
	A10	Fumaronitrile		438
R_1 = R_2 = Ph, R_3 = R_4 = D	A7	Fumaroyl chloride	91	439
	D3	MA	> 80	439
R_1 = R_2 = R_3 = R_4 = D	A7	Fumaroyl chloride	> 80	439

2.2 HETERO-1,3-BUTADIENES

Reactivity, Selectivity, Synthesis In the realm of heterobutadienes the α,β-unsaturated aldehydes and ketones have been studied in detail. They react preferentially with electron-rich dienophiles; the cycloadditions are highly regioselective and are HOMO dienophile–LUMO diene controlled (Diels–Alder reactions with inverse electron demand; Sec. 1.5). A survey of reactions is reported in Scheme 2.7, where the acrolein is used as a representative diene. Some brief comments are made about these cycloadditions.

The cycloaddition with enol ethers is a powerful synthetic route to dihydropyrans, which are intermediates in the synthesis of several natural products, such as carbohydrates. Dienes that do not have an electron-withdrawing substituent are unreactive or require severe reaction conditions in the absence of catalyst.[497] The crotonaldehyde (**60**) gives a mixture[419,496,498] of **61** and **62** with ethoxyvinyl ether under thermal conditions (3 h, 225°C), but in the presence of Yb(fod)$_3$, only the adduct **61** is obtained. The activated dienes **63** and **64** react *endo* selectively to give the adduct **65** (70°C; 90% yield) and two diastereoisomers, **67** and **66** (25°C; 60% yield; ratio 2 : 1), respectively. The

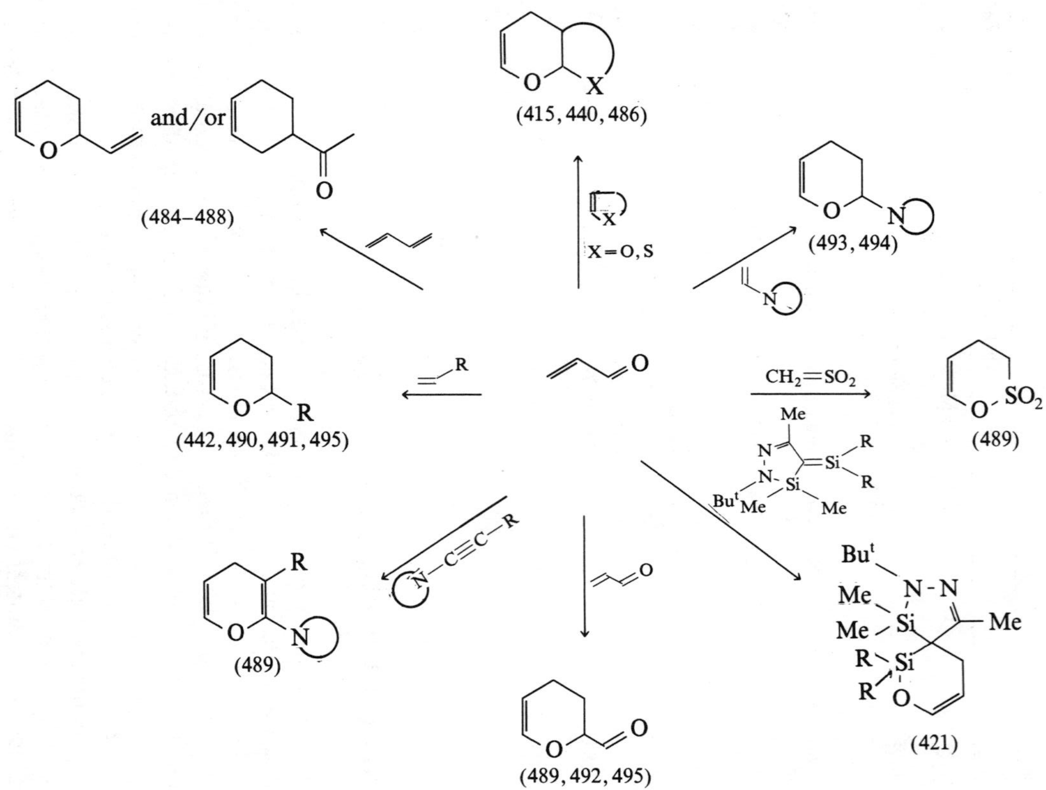

Scheme 2.7 (Reference numbers in parentheses)

adducts **65** and **67** have been converted into desoxy sugars.[423]

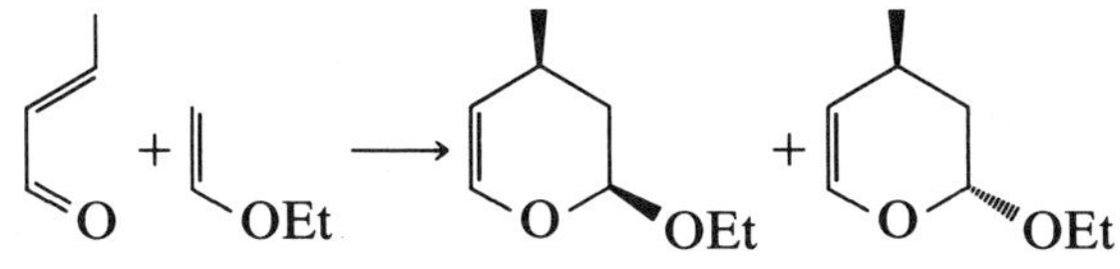

60 **61** **62**

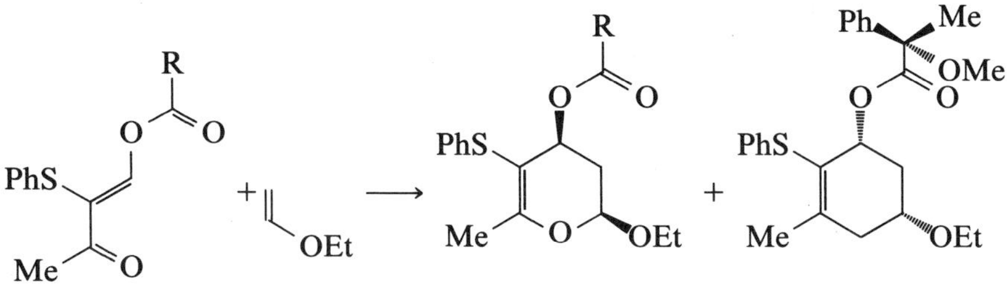

63 R = Me **65** R = Me **66**

64 R = 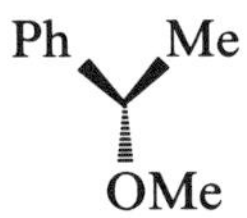

67 R = 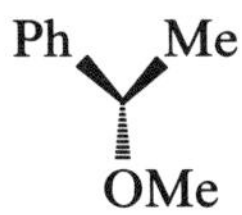

The α, β-unsaturated carbonyl compounds can behave as dienophiles when reacting with dienes and in the dimerization reaction. 1,3-Butadiene and acrolein give 4-formylcyclohexene, and this cycloaddition seems the rule with alkyl- or arylbutadienes.[486–488,494] With oxygenated dienes such as **68** the route to dihydropirans is preferred.[484,485] The dihydropyrone **69** has been used as starting material in the synthesis of liconsamine.[484,485] The dimerization reaction is high-activation-energy cycloaddition, which gives selectively the regioisomers **70** via a prevalent *endo* transition state.[499]

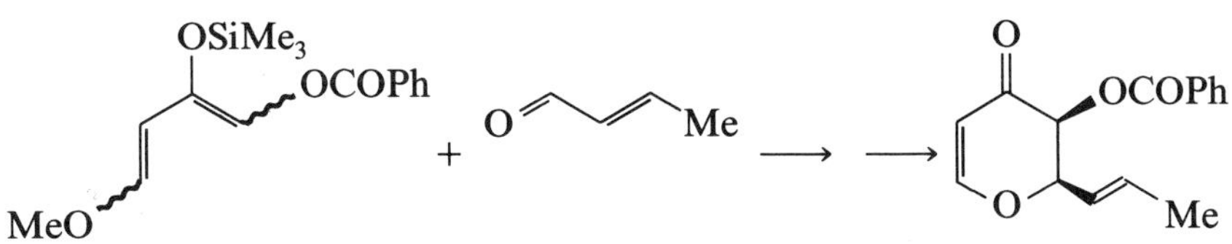

68 **69**

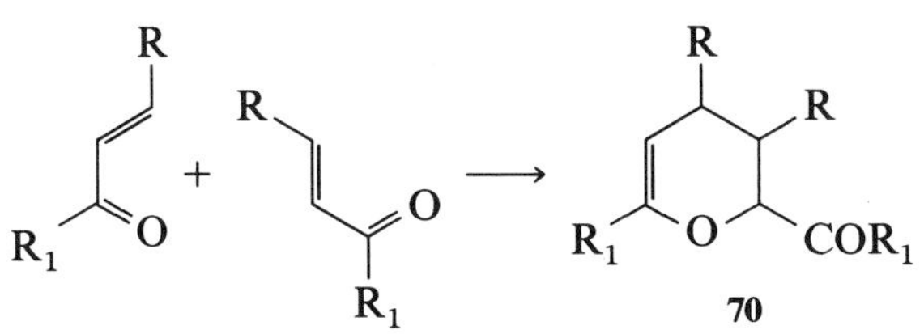

70

$R = R_1 = H$

$R = Me;\ R_1 = H$

$R = H;\ R_1 = Me, Et, Ph$

The azabutadienes have been explored in recent years[428, 436, 500] as useful reagents for the construction of nitrogen heterocyclic rings. 2-Aza-1,3-butadienes with electron-releasing substituents react with electron-poor dienophiles to give piperidine and pyridine derivatives (**71** → **72**, **73**). The reaction of electron-neutral 2-aza-1,3-butadienes with electron-rich dienophiles, which normally requires high temperature, is facilitated by Lewis acid catalysis[337, 436] (**74** → **75**; **76** → **77**).

Butadienes with two or three heteroatoms have attracted the attention of researchers because of their importance in heterocyclic chemistry and in natural product synthesis. Some recent typical reactions are depicted in Scheme 2.8.

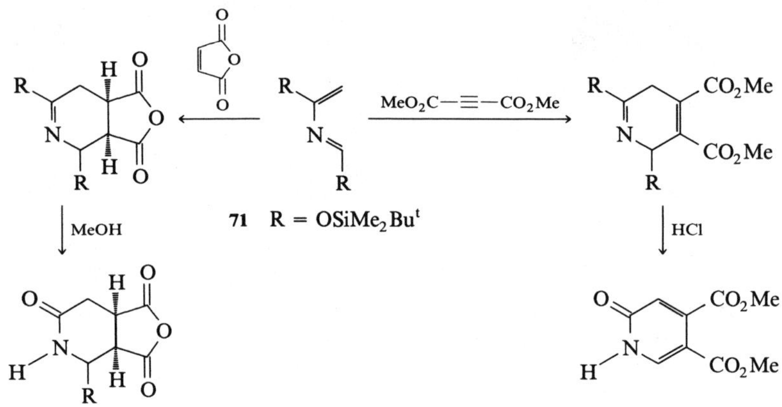

71 R = $OSiMe_2Bu^t$

72 **73**

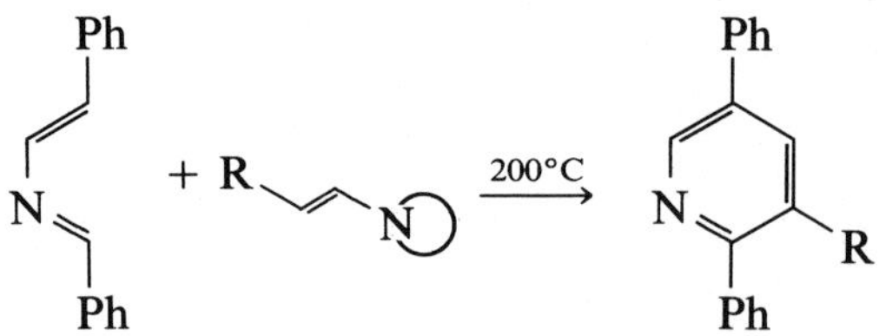

74 **75**

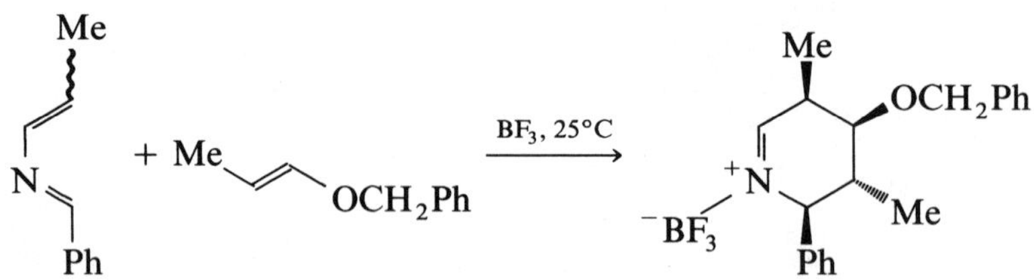

76 **77**

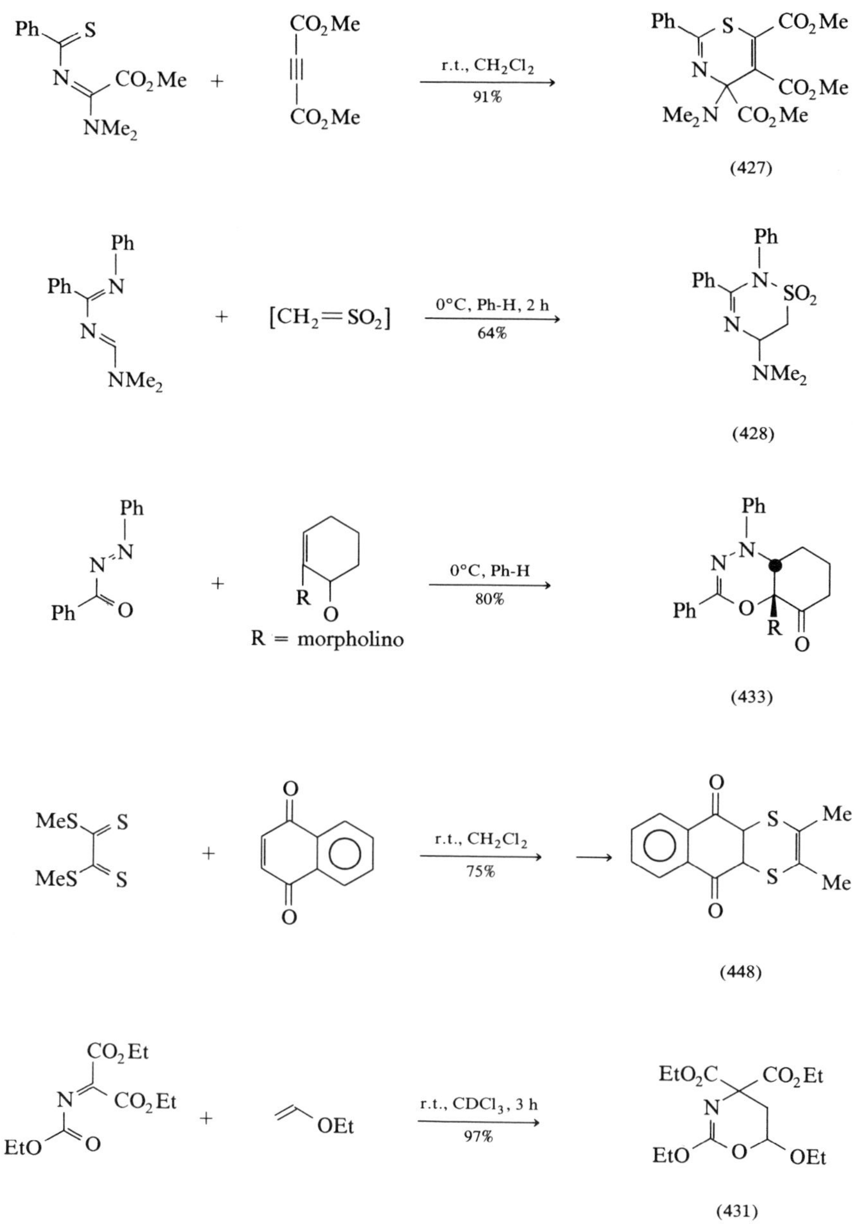

Scheme 2.8 (Reference numbers in parentheses)

Tables of Diels–Alder Reactions See Tables 2.34 to 21.37; see also the general remarks about classifications, abbreviations, and symbols in the Introduction.

TABLE 2.34 Diels–Alder Reactions of 1-Hetero-1,3-butadienes

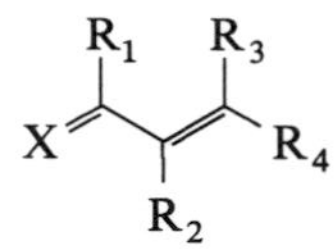

X	R_1	R_2	R_3	R_4		Dienophile	Yield (%)	Ref.
O	H	H	H	H	A8	Ethyl vinyl ether		418
					B2	α-Methylenetetrahydrofuran	82	425
					D1	Dihydropyrans	55	419
O	H	H	H	Me	A8	Ethyl vinyl ether	60–80	419
O	H	H	H	Ph	A8	Ethyl vinyl ether	60–80	419
O	H	Me	H	H	A8	Ethyl vinyl ether	60–80	419
					A10	Benzyl prop-1-en-1-ylether	58	443
					F9	Lithiated(fluorosilyl)diazasila-cyclopentenes	7–9	421
O	H	CHO	OH	H	A8	Ethyl vinyl ether	65	415
					D1	Dihydrofuran	73	415
					D6	Dihydrothiophen	30	415
O	H	CO_2Me	H	OH	C1	1-Methoxycyclopentene	82	440
					D1	Dihydrofuran	60	440
O	H	CO_2Me	H	OAc	D1	Dihydrofuran	100	440
O	H	CO_2Me	H	*O*-Camphanoyl (*)	D1	Dihydrofuran	55	440
O (R′ = H, OMe, OBn, Ph)	Me	Ac	H	R′	A8	Enol ethers	57–87	416, 422
					A8	Enediol ethers	60	422
					A10	(*E*)-1,2-Dimethoxyethene	71–84	416
O (R′ = OMe, OBn)	Me	CO_2Me	H	R′	A8	Ethyl vinyl ether	96	416
					A9	1,1-Diethoxyethene	80–92	422
					A10	(*E*)-1,2-Dimethoxyethene	69	416

O (R′ = Ph, OAc)	Me	SPh	H	R′	A8	Enol ethers	96–98	422
O (R′ = H, Bn; R″ = Me, Ph)	Me	SPh	H	NR′COR″	A8	Ethyl vinyl ether	89–93	452
O (R′ = Me, CH(OMe)Ph*)	Me	SPh	H	OCOR′	A8	Ethyl vinyl ether	60–90	423
O (R′ = Ph, OEt R″ = Me, Ph)	R″	COR′	H	H	A8	Vinyl ethers	96–99	420
O (R′ = H, Bn; R″ = Me, Ph, CO_2Me, C_6H_4CO)	CO_2Me	H	H	NR′COR″	A8	Vinyl ethers	76–98	452
O $CH(CF_3)_2$	$SCF(CF_3)_2$			—O—	A8	Styrene	33	348
					F4	Propionaldehyde	100	348
					H1	Methylthiocyanate	48	348
[S	Ph	H	H	Ph]	A1	Acrylamide	40	413
					A8	ACN	44	413
[S (R′ = Bu[t], Bn; R″ = H, Me, Cl)	R′	H	H	p-R″C_6H_4]	A8	ACN	37–96	424
					C2	Norbornene	27–84	424
S (R′ = Me, Ph; R″ = H, Me, Ph, Bn)	R′	H	NR″	Me	D3	MA	42–96	411
NC_6H_4R'—p (R′ = OMe, Me, NO_2, OH, OEt, Cl, Br)	H	H	Ph	H	A9	[Ketene]	16–90	417
N-NMe_2 (R′, R″ = H, Me, Et)	R′	R″	H	H	A1	MVK		414
					A13	1,3-Bis(ethoxycarbonyl)allene	35–52	288, 412
					C9	NQ	92	414

TABLE 2.35 Diels–Alder Reactions of 2-Aza-1,3-heterobutadienes

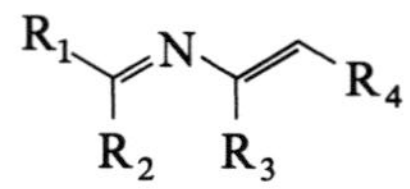

R_1	R_2	R_3	R_4		Dienophile	Yield (%)	Ref.
CH_2Me	Ph	Ph	Me	F4	Aldehydes	75–95	437
				I1	Dialkylazodicarboxylates	85–90	445
CH_2Et	Ph	Ph	Et	F1	Isothiocyanates	50–82	445
				F4	Benzaldehyde	85	437
CH_2Me	p-MeC_6H_4	p-MeC_6H_4	Me	F4	Aldehydes	72–80	437
				F4	Isocianates	75–80	445
$CHMe_2$	$N(CH_2)_4$	Ph	H	A1	AC	24	442
				A7	DMM		442
				A8	ACN	37	442
Ph	H	H	Me	A8	n-Butyl vinyl ether	27	436
				A8	Benzyl vinyl ether	30	436
				D1	Dihydropyran	5	436
p-PhC_6H_4	H	CO_2Me	H	A2	N-(Arylmethylene)dehydroalanine methyl ester	56	444
$=N-R'$ (R′ = Ph, Me, C_6H_{11}, p-MeC_6H_4, p-$OMeC_6H_4$, p-ClC_6H_4; R″ = H, Me)		Ph	p-$R''C_6H_4$	A12	TCNE	25–71	446
$OSiMe_2Bu^t$ (R′ = H, Me)	R′	$OSiMe_2Bu^t$	H	C9	NQ	44–74	435
				E3	DMAD	61–64	435

TABLE 2.36 Diels–Alder Reactions of 2-Aza-4-(oxa, aza, thio)-1,3-heterobutadienes

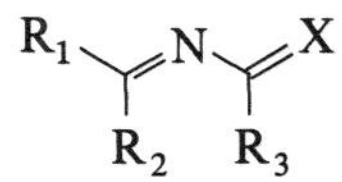

X	R_1	R_2	R_3		Dienophile	Yield (%)	Ref.
NAr	NMe_2	H	Ph	A12	Monophenylketene	84–92	447
				A12	Diphenylketene	85–96	447
				F1	Methanosulfonylchloride	60–70	428
NAr	NMe_2	H	SMe	F1	Methanosulfonylchloride	80–94	428
O	CO_2Et	CO_2Et	OEt	A8	Vinyl ethers	61–97	431
O	—S—		Ph	F7	Dimethyl-*N*-tolylketenimine	17	426
[O	—S—		1-Adamanthyl]	A1	1-Morpholinocyclohexene	50	449
[S	CF_3	CF_3	Ar]	A12	TCNE	53–60	429
S	NMe_2	R′	Ph	A1	MVK	80–98	427
	(R′ = Me, CO_2Et)			E3	DMAD	81	427
[S	NMe_2	H	1-Adamanthyl]	A1	MVK	36	449
S	NR'_2	H	Ar	A8	Acrylates		430
	(R'_2 = Me, $(CH_2)_5$, $(CH_2)_4O$)						
S	—O—		Ar	G4	2-Phenyl-1-azaspiro[2.2]pent-1-ene	52–70	441

TABLE 2.37 Diels–Alder Reactions of Miscellaneous 1,3-Heterobutadienes

Dienes					Dienophile	Yield (%)	Ref.
1-Adamanthyl 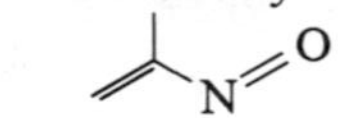				A14	2,3-Dimethyl-1,3-butadiene	59	449
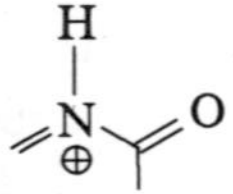 1-Adamanthyl				C1	Cyclohexene	33	449
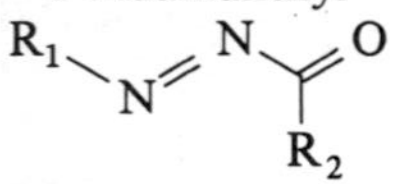	R_2	R_1					
	COPh	Ph		A1	Acrylates (k)		434
				A8	Styrenes (k)		434
	Ph	Ph		C4	2-Morpholinocyclo-2-hexen-1-one	80	433
	Ph	p-$NO_2C_6H_4$		C4	2-Morpholinocyclo-2-hexen-1-one	85	433
	p-$NO_2C_6H_4$	Ph		C4	2-Morpholinocyclo-2-hexen-1-one	95	433
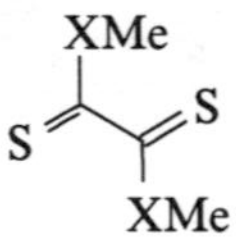	X =		O, S	C1	Cyclopentene	67–80	432
	X =		S	C1	Cycloheptatriene	30	448
	X =		O, S	C2	Norbornene	77–80	432
	X =		S	C9	NQ	75	448
	X =		O, S	E4	Phenylacetylene	4–80	432

REFERENCES

1. a. A. Oliva, J. I. Fernandez-Alonso, and J. Bertran, *Tetrahedron*, **34**, 2029 (1978).
   b. A. Oliva, J. I. Fernandez-Alonso, and J. Bertran, *An. Quim.*, **75**, 45 (1979).
   c. B. Blaise and O. Henri-Rousseau, *Nouv. J. Chim.*, **3**, 369 (1979).
   d. K. Jug and H. W. Krueger, *Theor. Chim. Acta*, **52**, 19 (1979).
   e. J. Pancir, *J. Am. Chem. Soc.*, **104**, 7424 (1982).
   f. N. L. Bauld, D. J. Bellville, R. Pabon, R. Chelsky, and G. Green, *J. Am. Chem. Soc.*, **105**, 2378 (1983).
   g. M. Ortega, A. Oliva, J. M. Lluch, and J. Bertran, *Chem. Phys. Lett.*, **102**, 317 (1983).
   h. F. K. Brown and K. N. Houk, *Tetrahedron Lett.*, **25**, 4609 (1984).
   i. F. Bernardi, A. Bottoni, M. A. Robb, M. J. Field, I. H. Hillier, and M. F. Guest, *J. Chem. Soc. Chem. Commun.*, 1051 (1985).
   j. A. Fernandez, *Theor. Chim. Acta*, **68**, 285 (1985).
   k. M. J. S. Dewar, S. Olivella, and J. J. P. Stewart, *J. Am. Chem. Soc.*, **108**, 5771 (1986).
2. T. Kortvelyesi, M. Gorgenyi, and L. Seres, *Acta Phys. Chem.*, **30**, 85 (1984).
3. P. Herczegh, M. Zsely, R. Bognar, and L. Szilagyi, *Tetrahedron Lett.*, **27**, 1509 (1986).
4. J. Jurczak and M. Tkacz, *Synthesis*, 42 (1979).
5. H. Morikawa, S. Kitazume, and S. Odaka, *Jpn. Kokai Tokyo Koho*, JP 80, 11,549 (1980); *Chem. Abstr.*, **92**, 214977x (1980).
6. P. Muller, R. Etienne, J. Pfyffer, N. Pineda, and M. Schipoff, *Helv. Chim. Acta*, **61**, 2482 (1978).
7. F. Farina, A. Galan, and J. L. Garcia Ruano, *An. Quim.*, **74**, 954 (1978).
8. T. Harayama, M. Takatani, and Y. Inubuschi, *Chem. Pharm. Bull.*, **27**, 726 (1979).
9. F. Fringuelli, F. Pizzo, A. Taticchi, and E. Wenkert, *Synth. Commun.*, 391 (1979).
10. R. Van Doorn, N. M. M. Wibbering, A. J. V. Ferrer-Correia, and K. R. Jennings, *Org. Mass Spectrom.*, **13**, 729 (1978).
11. K. Friedrich and M. Zamkanei, *Chem. Ber.*, **112**, 1867 (1979).
12. P. F. Newton and G. H. Whitham, *J. Chem. Soc. Perkin Trans. 1*, 3067 (1979).
13. Y. Kobayashi, M. Honda, Y. Hanzawa, I. Kumadaki, and A. Ohsawa, *J. Chem. Soc. Perkin Trans. 1*, 1743 (1979).
14. Y. Kobayashi, T. Nakano, R. Takahashi, and I. Kumadaki, *J. Chem. Soc. Perkin Trans. 1*, 2253 (1979).
15. J. Jurczak, *Bull. Chem. Soc. Jpn.*, **52**, 3438 (1979).
16. H. Labaziewicz and F. G. Riddell, *J. Chem. Soc. Perkin Trans. 1*, 2926 (1979).
17. G. R. Lenz, *J. Org. Chem.*, **44**, 4299 (1979).
18. M. F. Ahern and G. W. Gokel, *J. Chem. Soc. Chem. Commun.*, 1019 (1979).
19. S. Takenaka, K. Yamamoto, and T. Nishida, *Jpn. Kokai Tokyo Koho*, JP 79, 157,546 (1979); *Chem. Abstr.*, **92**, 180724v (1980).
20. T. Harayama, M. Takatani, and Y. Inubushi, *Tetrahedron Lett.*, 4307 (1979).

21. C. Rucker, D. Lang, J. Sauer, H. Friege, and R. Sustmann, *Chem. Ber.*,. **113**, 1663 (1980).
22. J. E. Lyons, H. K. Myers, and A. Schneider, *Ann. N.Y. Acad. Sci.*, **333**, 273 (1980).
23. G. Weber, K. Menke, and H. Hopfi, *Chem. Ber.*, **113**, 531 (1980).
24. S. Knapp, R. Lis, and P. Michna, *J. Org. Chem.*, **46**, 624 (1981).
25. a. W. R. Dolbier, Jr., and C. R. Burkholder, *Tetrahedron Lett.*, **21**, 785 (1980).
    b. W. R. Dolbier, Jr., and C. R. Burkholder, *J. Org. Chem.*, **49**, 2381 (1984).
    c. W. R. Dolbier, Jr., C. R. Burkholder, G. E. Wicks, G. J. Palenik, and M. Gawron, *J. Am. Chem. Soc.*, **107**, 7183 (1985).
26. P. Beslin and P. Metzner, *Tetrahedron Lett.*, **21**, 4657 (1980).
27. Y. Kobayashi, T. Yoshida, Y. Hanzawa, and I. Kumadaki, *Tetrahedron Lett.*, **21**, 4601 (1980).
28. A. V. Kamernitskii, V. N. Ignatov, I. S. Levina, and B. S. El'Yanov, *Izv. Akad. Nauk SSSR, Ser. Khim.*, 1902 (1980).
29. E. Kobayashi, S. Matsumura, and J. Furukawa, *Polym. Bull. (Berlin)*, **3**, 285 (1980).
30. T. Sano, J. Toda, N. Kashiwaba, Y. Tsuda, and Y. Iitaka, *Heterocycles*, **16**, 1151 (1981).
31. M. Peled and D. Ginsburg, *Heterocycles*, **15**, 1187 (1981).
32. K. C. Nicolaou and R. E. Zipkin, *Angew. Chem. (Int. Ed. Engl.)*, **20**, 785 (1981).
33. H. O. House, R. F. Sieloff, and D. Van Derveer, *J. Org. Chem.*, **46**, 4639 (1981).
34. D. N. Gupta, P. Hodge, and N. Khan, *J. Chem. Soc. Perkin Trans. 1*, 689 (1981).
35. I. M. Salimgareeva, N. G. Bogatova, O. Zh. Zhebarov, V. P. Yur'ev, and S. R. Rafikov, *Dokl. Akad. Nauk SSSR*, **261**, 118 (1981).
36. M. Oda and Y. Kanao, *Chem. Lett.*, 1547 (1981).
37. H. K. Hall, Jr., P. Nogues, J. W. Rhoades, R. C. Sentman, and M. Detar, *J. Org. Chem.*, **47**, 1451 (1982).
38. S. M. Makin, A. I. Pomogaev, T. N. Boikov, and A. P. Nikiforova, *Zh. Org. Khim.*, **17**, 2277 (1981).
39. Z. Kafka, L. Vodicka, and V. Galik, *Collect. Czech. Chem. Commun.*, **47**, 286 (1982).
40. H. K. Hall, Jr., R. C. Sentman, and P. Nogues, *J. Org. Chem.*, **47**, 3647 (1982).
41. F. Farina, J. C. Vega, and P. Prados, *An. Quim.*, **78**, 354 (1982).
42. a. F. Fringuelli, F. Pizzo, A. Taticchi, T. D. J. Hallas, and E. Wenkert, *J. Org. Chem.*, **47**, 5056 (1982).
    b. F. Fringuelli, F. Pizzo, A. Taticchi, and E. Wenkert, *J. Org. Chem.*, **48**, 2802 (1983).
    c. E. C. Angell, F. Fringuelli, F. Pizzo, A. Taticchi, and E. Wenkert, *J. Org. Chem.*, **53**, 1424 (1988).
    d. F. Fringuelli, L. Minuti, L. Radics, A. Taticchi, and E. Wenkert, *J. Org. Chem.*, **53**, 4607 (1988).
43. H. Link, K. Bernauer, and G. Englert, *Helv. Chim. Acta*, **65**, 2645 (1982).
44. R. O. Angus, Jr., and R. P. Johnson, *J. Org. Chem.*, **48**, 273 (1983).

45. F. Fringuelli, L. Minuti, F. Pizzo, A. Taticchi, T. D. J. Hallas, and E. Wenkert, *J. Org. Chem.*, **48**, 1810 (1983).
46. J. F. W. Keana, J. S. Bland, P. J. Boyle, M. Erion, R. Hartling, J. R. Husman, and R. B. Roman, *J. Org. Chem.*, **48**, 3627 (1983).
47. J. L. Primeau, R. C. Anderson, and B. Fraser-Reid, *J. Am. Chem. Soc.*, **105**, 5874 (1983).
48. K. Uneyama, K. Takano, and S. Torii, *Bull. Chem. Soc. Jpn.*, **56**, 2867 (1983).
49. N. Wiberg and G. Wagner, *Angew. Chem. (Int. Ed. Engl.)*, **22**, 1005 (1983).
50. E. S. Han, I. Lee, and B. D. Chang, *Bull. Korean Chem. Soc.*, **4**, 197 (1983).
51. R. S. Glass, K. Reineke, and M. Shanklin, *J. Org. Chem.*, **49**, 1527 (1984).
52. W. Oppolzer, C. Chapuis, and M. J. Kelly, *Helv. Chim. Acta*, **66**, 2358 (1983).
53. T. C. Sedergran, M. Yokoyama, and C. D. Dittmer, *J. Org. Chem.*, **49**, 2408 (1984).
54. Y. Tamaru, H. Satomi, O. Kitao, and Z. Yoshida, *Tetrahedron Lett.*, **25**, 2561 (1984).
55. G. A. Tolstikov, V. P. Kazakov, E. E. Shullts, R. G. Bulgakov, and R. G. Kantyukova, *Zh. Org. Khim.*, **20**, 313 (1984).
56. I. Lee, E. S. Han, and Y. J. Choi, *J. Comput. Chem.*, **5**, 606 (1984).
57. J. M. McIntosh and L. Z. Pillon, *Can. J. Chem.*, **62**, 2089 (1984).
58. D. L. Boger, M. Patel, and F. Takusagawa, *J. Org. Chem.*, **50**, 1911 (1985).
59. W. Oppolzer, C. Chapuis, and G. Bernardinelli, *Helv. Chim. Acta*, **67**, 1397 (1984).
60. Y. K. Rao and M. Nagarajan, *Synthesis*, 757 (1984).
61. S. C. Cooper and P. G. Sammes, *J. Chem. Soc. Perkin Trans. 1*, 2407 (1984).
62. B. Jousseaume, *J. Chem. Soc. Chem. Commun.*, 1452 (1984).
63. T. Watabe and M. Oda, *Chem. Lett.*, 1791 (1984).
64. P. G. Williard and S. E. de Laszlo, *J. Org. Chem.*, **50**, 3738 (1985).
65. J. F. Buzinkai, D. M. Hrubowchak, and F. X. Smith, *Tetrahedron Lett.*, **26**, 3195 (1985).
66. J.-L. Gras and A. Guerin, *Tetrahedron Lett.*, **26**, 1781 (1985).
67. S. Danishefsky and M. Bednarski, *Tetrahedron Lett.*, **26**, 2507 (1985).
68. T. Poll, A. Sobczak, H. Hartmann, and G. Helmchen, *Tetrahedron Lett.*, **26**, 3095 (1985).
69. a. E. C. Angell, F. Fringuelli, T. D. J. Hallas, F. Pizzo, B. Porter, A. Taticchi, A. P. Touris, and E. Wenkert, *J. Org. Chem.*, **50**, 4691 (1985).
    b. F. Fringuelli, F. Pizzo, A. Taticchi, V. F. Ferreira, E. L. Michelotti, B. Porter, and E. Wenkert, *J. Org. Chem.*, **50**, 890 (1985).
70. E. C. Angell, F. Fringuelli, L. Minuti, F. Pizzo, B. Porter, A. Taticchi, and E. Wenkert, *J. Org. Chem.*, **50**, 4686 (1985).
71. E. C. Angell, F. Fringuelli, F. Pizzo, B. Porter, A. Taticchi, and E. Wenkert, *J. Org. Chem.*, **50**, 4696 (1985).
72. V. V. Bazylchik, P. I. Fedorov, N. A. Klyuer, and E. Kh. Dank, *Zh. Org. Khim.*, **21**, 1450 (1985).

73. K. Mach, H. Antropiusova, L. Petrusova, F. Turecek, V. Hanus, P. Sedmera, and J. Schraml, *J. Organomet. Chem.*, **289**, 331 (1985).
74. D. Stephan, A. Gorgues, and A. Le Coq, *Tetrahedron Lett.*, **27**, 4295 (1986).
75. R. M. Ortuno, J. Corbera, and J. Font, *Tetrahedron Lett.*, **27**, 1081 (1986).
76. a. E. C. Angell, F. Fringuelli, F. Pizzo, B. Porter, A. Taticchi, and E. Wenkert, *J. Org. Chem.*, **51**, 2642 (1986).
    b. F. Fringuelli, F. Pizzo, A. Taticchi, and E. Wenkert, *Synth. Commun.*, **16**, 245 (1986).
    c. E. C. Angell, F. Fringuelli, L. Minuti, F. Pizzo, A. Taticchi, and E. Wenkert, *J. Org. Chem.*, **51**, 5177 (1986).
77. R. M. Dessau, *J. Chem. Soc. Chem. Commun.*, 1167 (1986).
78. V. L. Bell, A. B. Holmes, S.-Y. Hsu, G. A. Mock, and R. A. Raphael, *J. Chem. Soc. Perkin Trans. 1*, 1507 (1986).
79. D. L. Boger and C. E. Brotherton, *Tetrahedron*, **42**, 2777 (1986).
80. G. A. Kraus and S. H. Woo, *J. Org. Chem.*, **51**, 114 (1986).
81. W. R. Dolbier, Jr., M. Seabury, D. Daly, and B. E. Smart, *J. Org. Chem.*, **51**, 974 (1986).
82. S. Amin, K. Huie, N. Hussain, G. Balanikas, S. G. Carmella, and S. S. Hecht, *J. Org. Chem.*, **51**, 1206 (1986).
83. G. A. Kraus and Y.-S. Hon, *J. Org. Chem.*, **51**, 116 (1986).
84. T. Katada, S. Eguchi, and T. Sasaki, *J. Org. Chem.*, **51**, 314 (1986).
85. R. V. Honeychuck, P. V. Bonnesen, J. Farahi, and W. H. Hersh, *J. Org. Chem.*, **52**, 5293 (1987).
86. D. Tobia and Rickborn, *J. Org. Chem.*, **52**, 2611 (1987).
87. F. Jensen and C. S. Foote, *J. Am. Chem. Soc.*, **109**, 6376 (1987).
88. K. Furuta, K. Iwanaga, and H. Yamamoto, *Tetrahedron Lett.*, **27**, 4507 (1986).
89. N. Oda, K. Kobayashi, T. Ueda, and I. Ito, *Chem. Pharm. Bull.*, **26**, 2578 (1978).
90. M. Kakushima, *Can. J. Chem.*, **57**, 2564 (1979).
91. R. H. Fleming and B. M. Murray, *J. Org. Chem.*, **44**, 2280 (1979).
92. T. Saito and S. Motoki, *J. Org. Chem.*, **44**, 2493 (1979).
93. M. Kakushima and D. Scott, *Can. J. Chem.*, **57**, 1399 (1979).
94. E. Vedejs, M. J. Arnost, J. M. Dolphin, and J. Eustache, *J. Org. Chem.*, **45**, 2601 (1980).
95. J. S. Tou and W. Reusch, *J. Org. Chem.*, **45**, 5012 (1980).
96. A. B. Teck and H. K. Siong, *Occas. Pap. Nanyang Univ. Coll. Grad. Stud. Inst. Nat. Sci.*, 32 (1977); *Chem. Abstr.*, **90**, 203355a (1979).
97. E. J. Corey and H. Estreicher, *Tetrahedron Lett.*, **22**, 603 (1981).
98. I.-M. Tegmo-Larsson, M. D. Rozeboom, and K. N. Houk, *Tetrahedron Lett.*, **22**, 2043 (1981).
99. D. Liotta, M. Saindane, and C. Barnum, *J. Am. Chem. Soc.*, **103**, 3224 (1981).
100. M. Shen and A. G. Schultz, *Tetrahedron Lett.*, **22**, 3347 (1981).
101. T. J. Brocksom and M. G. Constantino, *J. Org. Chem.*, **47**, 3450 (1982).
102. S. K. Shakhova and B. S. El'yanov, *Izv. Akad. Nauk SSSR, Ser. Khim.*, 1194 (1982).

103. a. C. Carpanelli and G. Gaiani, *Gazz. Chim. Ital.*, **112**, 187 (1982).
b. C. Carpanelli and G. Gaiani, *Gazz. Chim. Ital.*, **112**, 191 (1982).

104. H.-J. Liu, N. C. E. Browne, and P. R. Pednekar, *Can. J. Chem.*, **60**, 921 (1982).

105. H.-J. Liu and T. Kongooi, *Synth. Commun.*, **12**, 715 (1982).

106. J.-L. Gras, *J. Chem. Res. (S)*, 300 (1982).

107. H.tom Dieck and R. Diercks, *Angew. Chem. (Int. Ed. Engl.)*, **22**, 778 (1983).

108. R. A. Pabon, D. J. Bellville, and N. L. Bauld, *J. Am. Chem. Soc.*, **105**, 5158 (1983).

109. J. Das, M. Kakushima, Z. Valenta, K. Jankowski, and R. Luce, *Can. J. Chem.*, **62**, 411 (1984).

110. J. B. Hendrickson and V. Singh, *J. Chem. Soc. Chem. Commun.*, 837 (1983).

111. K. Nakagawa, S. Iwase, Y. Ishii, S. Hamanaka, and M. Ogawa, *Rev. Roum. Chim.*, **24**, 613 (1979).

112. D. D. Weller and E. P. Stirchak, *J. Org. Chem.*, **48**, 4873 (1983).

113. F. Bronberger and R. Huisgen, *Tetrahedron Lett.*, **25**, 57 (1984).

114. D. A. Evans, K. T. Chapman, and J. Bisaha, *J. Am. Chem. Soc.*, **106**, 4261 (1984).

115. B. S. El'Yanov, S. K. Shakhova, B. D. Polkovnikov, and L. F. Rar, *J. Chem. Soc. Perkin Trans. 2*, 11 (1985).

116. V. Branchadell, A. Oliva, and J. Bertran, *J. Mol. Struct. (Theochem.)*, **21**, 85 (1985).

117. H. Felber, G. Kresze, R. Prewo, and A. Vasella, *Helv. Chim. Acta*, **69**, 1137 (1986).

118. E. C. Angell, F. Fringuelli, L. Minuti, F. Pizzo, B. Porter, A. Taticchi, and E. Wenkert, *J. Org. Chem.*, **51**, 2649 (1986).

119. J. W. Herndon, *J. Org. Chem.*, **51**, 2853 (1986).

120. N. Ono, A. Kamimura, and A. Kaji, *J. Org. Chem.*, **51**, 2139 (1986).

121. J. Nakayama, Y. Nakamura, and M. Hoshino, *Heterocycles*, **23**, 1119 (1985).

122. T. J. Brocksom and M. G. Constantino, *An. Acad. Bras. Cienc.*, **54**, 655 (1982).

123. a. J. Jurczak and B. Baranowski, *Pol. J. Chem.*, **52**, 1857 (1978).
b. J. Jurczak and M. Tkacz, *J. Org. Chem.*, **44**, 3347 (1979).
c. J. Jurczak, *Pol. J. Chem.*, **53**, 2539 (1979).

124. J. Jurczak, M. Chmielewski, and S. Filipek, *Synthesis*, 41 (1979).

125. M. C. Carreno, F. Farina, A. Galan, and J. L. Garcia Ruano, *J. Chem. Res.*, *(S)*, 296 (1979).

126. A. A. Broekhuis, J. W. Scheeren, and R. J. F. Nivard, *Recl. Trav. Chim. Pays-Bas*, **99**, 6 (1980).

127. M. Chmielewski and J. Jurczak, *J. Org. Chem.*, **46**, 2230 (1981).

128. P. J. Card, *J. Org. Chem.*, **47**, 2169 (1982).

129. M. C. Carreno, F. Farina, A. Galan, and J. L. Garcia Ruano, *J. Chem. Res. (S)*, 370 (1981).

130. G. A. Kraus and M. J. Taschner, *J. Am. Chem. Soc.*, **102**, 1974 (1980).

131. J. C. Carretero, J. C. Cuevas, A. Echavarren, F. Farina, and P. Prados, *J. Chem. Res. (S)*, 6 (1984).

132. a. J. Jurczak, T. Bauer, and S. Jarosz, *Tetrahedron Lett.*, **25**, 4809 (1984).
b. J. Jurczak, T. Bauer, and S. Jarosz, *Tetrahedron*, **42**, 6477 (1986).

133. J. Jurczak and T. Bauer, *Tetrahedron*, **42**, 5045 (1986).

134. Y. Apeloig, D. Arad, M. Kapon, and M. Wallerstein, *Tetrahedron Lett.*, **28**, 5917 (1987).

135. R. Aguilar, A. Reyes, J. Tamariz, and J.-L. Birbaum, *Tetrahedron Lett.*, **28**, 865 (1987).

136. C. Shin, M. Yamaura, E. Inui, Y. Ishida, and J. Yoshimura, *Bull. Chem. Soc. Jpn.*, **51**, 2618 (1978).

137. M. Hirama, Y. Koyama, Y. Shoji, and S. Ito, *Tetrahedron Lett.*, 2289 (1978).

138. a. T. R. Kelly and M. Montury, *Tetrahedron Lett.*, 4311 (1978).
b. T. R. Kelly and M. Montury, *Tetrahedron Lett.*, 4309 (1978).
c. T. R. Kelly, N. D. Parekh, and E. N. Trachtenberg, *J. Org. Chem.*, **47**, 5009 (1982).

139. E. McDonald, A. Suksamrarn, and R. D. Wylie, *J. Chem. Soc. Perkin Trans. 1*, 1893 (1979).

140. W. B. T. Cruse, I. Fleming, P. T. Gallagher, and O. Kennard, *J. Chem. Res.* (*S*), 372 (1979).

141. B. M. Trost and J. M. Fortunak, *J. Am. Chem. Soc.*, **102**, 2841 (1980).

142. R. E. Banks, J. A. Miller, N. J. Nunn, P. Stanley, T. J. R. Weakley, and Z. Ullah, *J. Chem. Soc. Perkin Trans. 1*, 1096 (1981).

143. S. D. Darling and S. J. Brandes, *J. Org. Chem.*, **47**, 1413 (1982).

144. N. Ono, H. Miyake, and A. Kaji, *J. Chem. Soc. Chem. Commun.*, 33 (1982).

145. R. G. Hall and S. Trippett, *Tetrahedron Lett.*, **23**, 2603 (1982).

146. B. M. Trost, C. G. Caldwell, E. Murayama, and D. Heissler, *J. Org. Chem.*, **48**, 3252 (1983).

147. P. F. Gordon, *Brit. U.K. Pat. Appl.*, 2,138,419 (1984); *Chem. Abstr.*, **102**, 131717m (1985).

148. M. D. Johnson and C. I. Hoody, *J. Chem. Soc. Perkin Trans. 1*, 71 (1985).

149. P. A. Grieco, R. Lis, R. E. Zelle, and J. Finn, *J. Am. Chem. Soc.*, **108**, 5908 (1986).

150. T. R. Kelly, A. Whiting, and N. S. Chandrakumar, *J. Am. Chem. Soc.*, **108**, 3510 (1986).

151. G. A. Kraus and M. J. Taschner, *J. Org. Chem.*, **45**, 1174 (1980).

152. T. Oida, S. Tanimoto, T. Sugimoto, and M. Okano, *Synthesis*, 131 (1980).

153. J. Bélanger, N. L. Landry, J. R. J. Paré, and K. Jankowski, *J. Org. Chem.*, **47**, 3649 (1982).

154. S. M. Makin, T. F. Nguyen, O. A. Shavrygina, B. M. Arshava, and I. A. Romanova, *Zh. Org. Khim.*, **19**, 727 (1983); Engl. Ed., 640 (1983).

155. M. G. Voronkov, S. N. Adamovich, V. I. Rakhlin, and R. G. Mirskov, *Zh. Obshch. Khim.*, **54**, 475 (1984).

156. K. Huyakawa, H. Nishiyama, and K. Kanematsu, *J. Org. Chem.*, **50**, 512 (1985).

157. K. Maruoka, M. Sakurai, J. Fujiwara, and H. Yamamoto, *Tetrahedron Lett.*, **27**, 4895 (1986).

158. P. Cano, F. Farina, M. D. Parellada, and P. Prados, *J. Chem. Soc. Perkin Trans. 1*, 1923 (1986).

159. a. A. Echavarren, F. Farina, and P. Prados, *J. Chem. Res. (S)*, 366 (1986).
b. S. V. Ley, W. L. Mitchell, T. V. Radhakrishnan, and D. H. R. Barton, *J. Chem. Soc. Perkin Trans. 1*, 1582 (1981).

160. G. A. Kraus and S. H. Woo, *J. Org. Chem.*, **52**, 4841 (1987).

161. V. Yedidia and C. C. Leznoff, *Can J. Chem.*, **58**, 1144 (1980).

162. T. Tokoroyama, Y. Fukuyama, T. Kubota, and K. Yokotani, *J. Chem. Soc. Perkin Trans. 1*, 1557 (1981).

163. a. W. G. Dauben and W. R. Baker, *Tetrahedron Lett.*, **23**, 2611 (1982).
b. W. G. Dauben and R. A. Bunce, *Tetrahedron Lett.*, **23**, 4875 (1982).

164. A. Rougny, H. Fillion, C. Laharotte, and M. Daudon, *Tetrahedron Lett.*, **25**, 829 (1984).

165. R. R. Schmidt and K. Vogt, *Synthesis*, 799 (1983).

166. a. R. Z. Andriamialisoa, N. Langlois, and Y. Langlois, *J. Org. Chem.*, **50**, 961 (1985).
b. Y. Langlois, A. Pouilhes, D. Genin, R. Z. Andriamialisoa, and N. Langlois, *Tetrahedron*, **39**, 3755 (1983).

167. D. Genin, R. Z. Andriamialisoa, N. Langlois, and Y. Langlois, *J. Org. Chem.*, **52**, 353 (1987).

168. S. Danishefsky and F. M. Hershenson, *J. Org. Chem.*, **44**, 1180 (1979).

169. S. Tanimoto, T. Oida, T. Kokubo, T. Sugimoto, and M. Okano, *Bull. Inst. Chem. Res. Kyoto Univ.*, **58**, 289 (1980).

170. T. Oida, S. Tanimoto, H. Ikehira, and M. Okano, *Bull. Inst. Chem. Res. Kyoto Univ.*, **60**, 336 (1982).

171. M. Chigr, H. Fillion, and A. Rougny, *Tetrahedron Lett.*, **28**, 4529 (1987).

172. H. Tanaka, Y. Nagai, H. Irie, S. Uyeo, and A. Kuno, *J. Chem. Soc. Perkin Trans. 1*, 874 (1979).

173. J. Drexler, R. Lindermayer, M. A. Hassan, and J. Sauer, *Tetrahedron Lett.*, **26**, 2555, 2559 (1985).

174. J. E. Baldwin, A. K. Forrest, S. Monaco, and R. J. Young, *J. Chem. Soc. Chem. Commun.*, 1586 (1985).

175. M. M. Guseinov, M. G. Veliev, and S. A. Mamedov, *Issled. Obl. Sint. Polim. Monomernykh. Prod.*, 3 (1977).

176. G. L. Vaughn, *U.S. Pat.*, 4,110,353 (1978); *Chem. Abstr.*, **90**, 103723a (1979).

177. F. Gavina, P. Gil, and B. Palazon, *Tetrahedron Lett.*, 1333 (1979).

178. T. Hagihara, H. Tsuruta, and T. Yoshida, *Jpn. Kokai Tokyo Koho*, JP 79, 154,733 (1979); *Chem. Abstr.*, **92**, 214974u (1980).

179. B. M. Trost, D. O'Krongly, and J. L. Belletire, *J. Am. Chem. Soc.*, **102**, 7595 (1980).

180. M. J. Carter, I. Fleming, and A. Percival, *J. Chem. Soc. Perkin Trans. 1*, 2415 (1981).

181. V. V. Russkikh, A .D. Marushina, A. M. Detsina, and E. P. Fokin, *Izv. Sib. Otd. Akad. Nauk SSSR, Ser. Khim. Nauk*, 147 (1982).

182. P. A. Grieco, K. Yoshida, and P. Garner, *J. Org. Chem.*, **48**, 3137 (1983).

183. a. V. Branchadell, A. Oliva, and J. Bertran, *Chem. Phys. Lett.*, **97**, 378 (1983).
b. L. Pardo, V. Branchadell, A. Oliva, and J. Bertran, *J. Mol. Struct.* (*Theochem.*), **10**, 255 (1983).
c. J. Bertran, V. Branchadell, M. Duran, and A. Oliva, *J. Mol. Struct.* (*Theochem.*), **16**, 191 (1984).
d. V. Branchadell, A. Oliva, and J. Bertran, *Chem. Phys. Lett.*, **113**, 197 (1985).
184. P. A. Grieco, P. Garner, K. Yoshida, and J. C. Huffman, *Tetrahedron Lett.*, **24**, 3807 (1983).
185. A. P. Kozikowski, K. Hiraga, J. P. Springer, B. C. Wang, and Z.-B. Xu, *J. Am. Chem. Soc.*, **106**, 1845 (1984).
186. K. Yoshida and P. A. Grieco, *J. Org. Chem.*, **49**, 5257 (1984).
187. S. Huber, P. Stamouli, and R. Neier, *J. Chem. Soc. Chem. Commun.*, 533 (1985).
188. Y. Naruta, Y. Nishigaichi, and K. Maruyama, *Chem. Lett.*, 1703 (1986).
189. S. Huber, P. Stamouli, T. Jenny, and R. Neier, *Helv. Chim. Acta*, **69**, 1898 (1986).
190. P. A. Grieco, P. Galatsis, and R. F. Spohn, *Tetrahedron*, **42**, 2847 (1986).
191. A. Lubineau and Y. Queneau, *J. Org. Chem.*, **52**, 1001 (1987).
192. W. Oppolzer, L. Bieber, and E. Francotte, *Tetrahedron Lett.*, 4537 (1979).
193. a. R. K. Boeckman, Jr., T. M. Dolak, and K. O. Culos, *J. Am. Chem. Soc.*, **100**, 7098 (1978).
b. F. Farina and P. Prados, *Tetrahedron Lett.*, 477 (1979).
c. K. Krohn, *Tetrahedron Lett.*, **21**, 3557 (1980).
d. A. Echavarren, P. Prados, and F. Farina, *J. Chem. Res.* (*S*), 364 (1986).
e. W. B. Manning, T. R. Kelly, and G. M. Muschik, *Tetrahedron Lett.*, **21**, 2629 (1980).
f. K. Krohn and E. Broser, *J. Org. Chem.*, **49**, 3766 (1984).
194. a. M. Chandler and R. J. Stoodley, *J. Chem. Soc. Chem. Commun.*, 997 (1978).
b. M. Chandler and R. J. Stoodley, *J. Chem. Soc. Perkin Trans. 1*, 1007 (1980).
195. B. S. El'Yanov and M. E. Gonikberg, *J. Chem. Soc. Faraday Trans. 1*, 172 (1979).
196. G. A. Olah, D. Meidar, and A. P. Fung, *Synthesis*, 270 (1979).
197. R. V. C. Carr and L. A. Paquette, *J. Am. Chem. Soc.*, **102**, 853 (1980).
198. A. Hosomi, M. Saito, and H. Sakurai, *Tetrahedron Lett.*, **21**, 355 (1980).
199. G. Märkl and D. Rudnick, *Tetrahedron Lett.*, **21**, 1405 (1980).
200. L. N. Kryukov, L. Yu. Kryukova, and A. F. Kolomiets, *Zh. Org. Khim.*, **18**, 1873 (1982); E. E., 1638 (1982).
201. L. Stella and J. L. Boucher, *Tetrahedron Lett.*, **23**, 953 (1982).
202. T. Matsui, T. Inoue, M. Nakayama, and J. D. White, *Bull. Chem. Soc. Jpn.*, **56**, 647 (1983).
203. R. C. Gupta, D. A. Jackson, and R. J. Stoodley, *J. Chem. Soc. Chem. Commun.*, 929 (1982).
204. A. D. Buss and S. Warren, *Tetrahedron Lett.*, **24**, 111 (1983).
205. C. G. Richards and D. E. Thurston, *Tetrahedron*, **39**, 1817 (1983).
206. W. D. Wulff and D. C. Yang, *J. Am. Chem. Soc.*, **105**, 6726 (1983).
207. H. Sakurai, A. Hosomi, M. Saito, K. Sasaki, H. Iguchi, J.-L. Sasaki, and Y. Araki, *Tetrahedron*, **39**, 883 (1983).

208. R. V. C. Carr, A. V. Williams, and L. A. Paquette, *J. Org. Chem.*, **48**, 4976 (1983).
209. J. Das, Z. Valenta, H.-J. Liu, and T. K. Ngooi, *Can. J. Chem.*, **62**, 481 (1984).
210. T. Hayama, S. Tomoda, T. Yoshito, and Y. Nomura, *J. Org. Chem.*, **49**, 3235 (1984).
211. D. Hermeling and H. J. Schafer, *Angew. Chem.* (*Int. Ed. Engl.*), **23**, 233 (1984).
212. I. Lee and E. S. Han, *Haksurwon Nonmunjip, Cha'yon Kwahak P'yon*, **22**, 43 (1983).
213. S. D. Larsen and P. A. Grieco, *J. Am. Chem. Soc.*, **107**, 1768 (1985).
214. M. E. Jung and K. R. Buszek, *J. Org. Chem.*, **50**, 5440 (1985).
215. J. L. Boucher and L. Stella, *Tetrahedron*, **41**, 875 (1985).
216. J.-L. Boucher and L. Stella, *Tetrahedron*, **42**, 3871 (1986).
217. R. M. Ortuno, R. Batllori, M. Ballesteros, M. Monsalvatje, J. Corbera, F. Sanchez-Ferrando, and J. Font, *Tetrahedron Lett.*, **28**, 3405 (1987).
218. C. R. Johnson and J. F. Kadow, *J. Org. Chem.*, **52**, 1493 (1987).
219. P. G. Gassman, D. A. Singleton, J. J. Wilwerding, and S. P. Chavan, *J. Am. Chem. Soc.*, **109**, 2182 (1987).
220. M. E. Jung and C. A. McCombs, *Org. Synth.*, **58**, 163 (1978).
221. S. Danishefsky and M. Kahn, *Tetrahedron Lett.*, **22**, 489 (1981).
222. M. E. Jung, C. A. McCombs, Y. Takeda, and Y.-G. Pan, *J. Am. Chem. Soc.*, **103**, 6677 (1981).
223. M. E. Jung, K. Shishido, L. Light, and L. Davis, *Tetrahedron Lett.*, **22**, 4607 (1981).
224. I. Ojima, M. Yatabe, and T. Fuchikami, *J. Org. Chem.*, **47**, 2051 (1982).
225. W. D. Wulff and C. D. Yang, *J. Am. Chem. Soc.*, **106**, 7565 (1984).
226. T.-K. Yin, J. G. Lee, and W. T. Borden, *J. Org. Chem.*, **50**, 531 (1985).
227. P. N. Preston, T. Winwick, and J. O. Morley, *J. Chem. Soc. Perkin Trans. 1*, 39 (1985).
228. J. P. Vacca, *Tetrahedron Lett.*, **26**, 1277 (1985).
229. B. M. Trost and S. Mignani, *J. Org. Chem.*, **51**, 3435 (1986).
230. A. V. R. Rao, V. H. Deshpande, and N. L. Reddy, *Tetrahedron Lett.*, **21**, 2661 (1980).
231. R. N. Ferguson, J. F. Whidby, E. B. Sanders, R. J. Levins, T. Katz, J. F. De Bardeleben, and W. N. Einolf, *Tetrahedron Lett.*, 2645 (1978).
232. D. G. Batt and B. Ganem, *Tetrahedron Lett.*, 3323 (1978).
233. a. E. Vedejs, T. H. Eberlein, and D. L. Varie, *J. Am. Chem. Soc.*, **104**, 1445 (1982).
    b. E. Vedejs, D. A. Perry, K. N. Houk, and N. G. Rondan, *J. Am. Chem. Soc.*, **105**, 6999 (1983).
234. a. K. Steliou, Y. Gareau, and D. N. Harpp, *J. Am. Chem. Soc.*, **106**, 799 (1984).
    b. K. Steliou, P. Salama, D. Brodeur, and Y. Gareau, *J. Am. Chem. Soc.*, **109**, 926 (1987).
235. D. Martina and F. Brion, *Tetrahedron Lett.*, **23**, 865 (1982).
236. Y. Kimura, M. Suzuki, T. Matsumoto, R. Abe, and S. Terashima, *Chem. Lett.*, 473 (1984).

237. O. Tsuge, E. Wada, S. Kanemasa, and H. Sakoh, *Chem. Lett.*, 469 (1984).
238. C. L. Liotta and J. W. Verbicky, Jr., *Tetrahedron Lett.*, **26**, 1395 (1985).
239. S. Kanemasa, H. Sakoh, E. Wada, and O. Tsuge, *Bull. Chem. Soc. Jpn.*, **59**, 1869 (1986).
240. O. Tsuge, E. Wada, S. Kanemasa, and H. Sakoh, *Chem. Lett.*, 709 (1984).
241. M. Taddei and A. Mann, *Tetrahedron Lett.*, **27**, 2913 (1986).
242. T. Minami, T. Chikugo, and T. Hanamoto, *J. Org. Chem.*, **51**, 2210 (1986).
243. E. Wada, N. Nagasaki, S. Kanemasa, and O. Tsuge, *Chem. Lett.*, 1491 (1986).
244. T. Chou, H. H. Tso, Y. T. Tao, and L. C. Lin, *J. Org. Chem.*, **52**, 244 (1987).
245. J. E. Backrall and S. K. Juntunen, *J. Am. Chem. Soc.*, **109**, 6396 (1987).
246. R. Kaiser and D. Lamparsky, *Helv. Chim. Acta*, **67**, 1184 (1984).
247. H. C. J. G. van Balen, A. A. Broekhuis, J. W. Scheeren, and R. J. F. Nivard, *Recl. Trav. Chim. Pays-Bas*, **98**, 36 (1979).
248. W. E. Bottomley and G. V. Boyd, *J. Chem. Soc. Chem. Commun.*, 790 (1980).
249. F. Zutterman and A. Krief, *J. Org. Chem.*, **48**, 1135 (1983).
250. A. D. Josey, *Angew. Chem.* (*Int. Ed. Engl.*), **20**, 686 (1981).
251. P. A. Grieco, G. Vidari, S. Ferrino, and R. C. Haltiwanger, *Tetrahedron Lett.*, **21**, 1619 (1980).
252. V. Kvita, H. Sauter, and G. Rihs, *Helv. Chim. Acta*, **66**, 2769 (1983).
253. P. V. Alston, M. D. Gordon, R. M. Ottenbrite, and T. Cohen, *J. Org. Chem.*, **48**, 5051 (1983).
254. *Jpn. Kokai Tokyo Koho*, JP 59, 212,444 (1984); *Chem. Abstr.*, **102**, 203769e (1985).
255. G. A. Kraus and M. E. Krolski, *J. Org. Chem.*, **51**, 3347 (1986).
256. B. Potthoff and E. Breitmaier, *Synthesis*, 584 (1986).
257. M. E. Garst and P. Arrhenius, *Synth. Commun.*, **11**, 481 (1981).
258. T. Cohen and Z. Kosarych, *J. Org. Chem.*, **47**, 4005 (1982).
259. P. A. Grieco, P. Garner, and Z. He, *Tetrahedron Lett.*, **24**, 1897 (1983).
260. T. Ibuka, H. Minakata, M. Hashimoto, L. E. Overman, and R. L. Freerks, *Heterocycles*, **22**, 485 (1984).
261. S. Danishefsky, T. Kitahara, C. F. Yan, and J. Morris, *J. Am. Chem. Soc.*, **101**, 6996 (1979).
262. S. Danishefsky, T. Harayama, and R. K. Singh, *J. Am. Chem. Soc.*, **101**, 7008 (1979).
263. S. Danishefsky, J. Morris, and L. A. Clizbe, *J. Am. Chem. Soc.*, **103**, 1602 (1981).
264. a. L. A. Paquette and W. A. Kinney, *Tetrahedron Lett.*, **23**, 131 (1982).
b. L. A. Paquette and W. A. Kinney, *Tetrahedron Lett.*, **23**, 5127 (1982).
265. Y. Tsuda, T. Ohshima, T. Sano, and J. Toda, *Heterocycles*, **19**, 2053 (1982).
266. P. Pellon, Y. Y. C. Yeung Lam Ko, P. Cosquer, J. Hamelin, and R. Carrié, *Tetrahedron Lett.*, **27**, 4299 (1986).
267. J. Jurczak, A. Goebiowski, and A. Rahm, *Tetrahedron Lett.*, **27**, 853 (1986).
268. S. Danishefsky and R. Hungate, *J. Am. Chem. Soc.*, **108**, 2486 (1986).
269. S. Danishefsky, J. Morris, and L. A. Clizbe, *Heterocycles*, **15**, 1205 (1981).

270. M. Bednarski and S. Danishefsky, *J. Am. Chem. Soc.*, **105**, 3716 (1983).
271. T. Sano, J. Toda, and Y. Tsuda, *Chem. Pharm. Bull.*, **31**, 356 (1983).
272. S. Danishefsky, T. Kitahara, and P. F. Schuda, *Org. Synth.*, **61**, 147 (1983).
273. M. A. Wantz and P. L. Fuchs, *J. Org. Chem.*, **52**, 5298 (1987).
274. I. Fleming and A. Percival, *J. Chem. Soc. Chem. Commun.*, 178 (1978).
275. T. R. Kelly and W.-G. Tsang, *Tetrahedron Lett.*, 4457 (1978).
276. B. Fraser-Reid, Md. Abdur Rahman, D. R. Kelly, and R. M. Srivastava, *J. Org. Chem.*, **49**, 1835 (1984).
277. a. B. M. Trost, P. G. McDougal, and K. J. Haller, *J. Am. Chem. Soc.*, **106**, 383 (1984).
b. B. M. Trost and P. G. McDougal, *J. Am. Chem. Soc.*, **104**, 6110 (1982).
278. T. R. Kelly, L. Ananthasubramanian, K. Borah, J. W. Gillard, R. N. Goerner, Jr., P. F. King, J. M. Lyding, W.-G. Tsang, and J. Vaya, *Tetrahedron*, **40**, 4569 (1984).
279. P. G. Lenhert, C. M. Lukehart, and L. A. Sacksteder, *J. Am. Chem. Soc.*, **108**, 793 (1986).
280. D. L. Flynn and D. E. Nies, *Tetrahedron Lett.*, **27**, 5075 (1986).
281. R. E. Ireland and D. M. Obrecht, *Helv. Chim. Acta*, **69**, 1273 (1986).
282. M. Vaultier, F. Truchet, B. Carboni, R. W. Hoffmann, and I. Denne, *Tetrahedron Lett.*, **28**, 4169 (1987).
283. R. C. Cookson, M. C. Cramp, and P. J. Parsons, *J. Chem. Soc. Chem. Commun.*, 197 (1980).
284. a. P. Dowd and W. Weber, *Tetrahedron Lett.*, **23**, 2155 (1982).
b. P. Dowd and W. Weber, *J. Org. Chem.*, **47**, 4774 (1982).
285. T. Mandai, K. Osaka, T. Wada, M. Kawada, and J. Otera, *Tetrahedron Lett.*, **24**, 1171 (1983).
286. T. Mandai, K. Osaka, M. Kawagishi, M. Kawada, and J. Otera, *Synth. Commun.*, **14**, 797 (1984).
287. T. Mandai, K. Osaka, M. Kawagishi, M. Kawada, and J. Otera, *J. Org. Chem.*, **49**, 3595 (1984).
288. K. T. Potts, D. Bhattacharjee, and E. B. Walsh, *J. Chem. Soc. Chem. Commun.*, 114 (1984).
289. R. P. Joyce, J. A. Gainor, and S. M. Weinreb, *J. Org. Chem.*, **52**, 1177 (1987).
290. R. C. Gupta, P. A. Harland, and R. J. Stoodley, *J. Chem. Soc. Chem. Commun.*, 754 (1983).
291. R. A. Baylouny, R. P. Danna, and K. H. Toth, *J. Org. Chem.*, **44**, 204 (1979).
292. D. Cooper and S. Trippett, *J. Chem. Soc. Perkin Trans. 1*, 2127 (1981).
293. A. P. Kozikowski, T. Konoike, and T. R. Nieduzak, *J. Chem. Soc. Chem. Commun.*, 1350 (1986).
294. R. C. Gupta, A. M. Z. Slawin, R. J. Stoodley, and D. J. Williams, *J. Chem. Soc. Chem. Commun.*, 668 (1986).
295. H.-J. Liu and W. M. Feng, *Synth. Commun.*, **16**, 1485 (1986).
296. M. Bednarski and S. Danishefsky, *J. Am. Chem. Soc.*, **105**, 6968 (1983).
297. N. L. Agarwal and H. W. Scheeren, *Chem. Lett.*, 1057 (1982).
298. E. Wada, S. Kanemasa, and O. Tsuge, *Bull. Chem. Soc. Jpn.*, **59**, 2451 (1986).

299. U. Gruseck and M. Heuschmann, *Tetrahedron Lett.*, **28**, 2681 (1987).
300. S. M. Bennett and D. L. J. Clive, *J. Chem. Soc. Chem. Commun.*, 878 (1986).
301. R. S. Garigipati, A. J. Freyer, R. R. Whittle, and S. M. Weinreb, *J. Am. Chem. Soc.*, **106**, 7861 (1984).
302. G. A. Kraus, S. Yue, and J. Sy, *J. Org. Chem.*, **50**, 283 (1985).
303. H. Natsugari, R. R. Whittle, and S. M. Weinreb, *J. Am. Chem. Soc.*, **106**, 7867 (1984).
304. R. Gree, J. Kessabi, P. Mosset, J. Martelli, and R. Carrié, *Tetrahedron Lett.*, **25**, 3697 (1984).
305. S. A. Harkin, O. Singh, and E. J. Thomas, *J. Chem. Soc. Perkin Trans. 1*, 1489 (1984).
306. P. A. Grieco, K. Yoshida, and Z.-M. He, *Tetrahedron Lett.*, **25**, 5715 (1984).
307. J. F. W. Keana and D. D. Ward, *Synth. Commun.*, **13**, 729 (1983).
308. H. J. Reich, J. M. Kelly, R. E. Olson, and R. C. Holtan, *Tetrahedron*, **39**, 949 (1983).
309. R. R. Schmidt and A. Wagner, *Tetrahedron Lett.*, **24**, 4661 (1983).
310. K. T. Potts and D. Bhattacharjee, *Synthesis*, 31 (1983).
311. H.-J. Altembach, B. Voss, and E. Vogel, *Angew. Chem. (Int. Ed. Engl.)*, **22**, 410 (1983).
312. M. Koreeda and M. A. Ciufolini, *J. Am. Chem. Soc.*, **104**, 2308 (1982).
313. W. Abele and R. R. Schmidt, *Tetrahedron Lett.*, **22**, 4807 (1981).
314. I. Angelova, *God. Sofii. Univ. Fak.*, 77 (1980); *Chem. Abstr.*, **94**, 15350d (1981).
315. a. H. Irikawa, T. Koyama, and Y. Okumura, *Bull. Chem. Soc. Jpn.*, **52**, 637 (1979).
     b. K. Hayakawa, K. Ueyama, and K. Kanematsu, *J. Chem. Soc. Chem. Commun.*, 71 (1984).
316. C. Larsen and D. N. Harpp, *J. Org. Chem.*, **45**, 3713 (1980).
317. A. P. Davis and G. H. Whitham, *J. Chem. Soc. Chem. Commun.*, 639 (1980).
318. P. Dowd and W. Weber, *J. Org. Chem.*, **47**, 4777 (1982).
319. R. Ideses, A. Shani, and J. T. Klug, *Chem. Ind. (London)*, 409 (1982).
320. F. P. Ballistreri, E. Maccarone, G. Perrini, G. A. Tomaselli, and M. Torre, *J. Chem. Soc. Perkin Trans. 2*, 273 (1982).
321. I. Hughes and R. A. Raphael, *Tetrahedron Lett.*, **24**, 1441 (1983).
322. L. E. Overman, C. B. Petty, T. Ban, and G. T. Huang, *J. Am. Chem. Soc.*, **105**, 6335 (1983).
323. J. F. W. Keana, A. P. Guzikowski, C. Morat, and J. J. Volwerk, *J. Org. Chem.*, **48**, 2661 (1983).
324. H. Hiranuma and S. I. Miller, *J. Org. Chem.*, **48**, 3096 (1983).
325. G. Revial, M. Blanchard, and J. D'Angelo, *Tetrahedron Lett.*, **24**, 899 (1983).
326. M. L. Durrant and E. J. Thomas, *J. Chem. Soc. Perkin Trans. 1*, 901 (1984).
327. M. D. Fryzuk, G. S. Bates, and C. Stone, *J. Org. Chem.*, **52**, 2334 (1987).
328. C. Ferroud, G. Revial, and J. D'Angelo, *Tetrahedron Lett.*, **26**, 3981 (1985).
329. R. W. Franck, S. Argade, C. S. Subramaniam, and D. M. Frechet, *Tetrahedron Lett.*, **26**, 3187 (1985).

330. K. Hartke, T. Kissel, J. Quante, and G. Henssen, *Angew. Chem. (Int. Ed. Engl.)*, **17**, 953 (1978).
331. E. D. Weiler and J. J. Brennan, *J. Heterocycl. Chem.*, **15**, 1299 (1978).
332. K. Laping and M. Hanack, *Tetrahedron Lett.*, 1309 (1979).
333. J.-L. Ripoll and M.-C. Lasne, *Tetrahedron Lett.*, 5201 (1978).
334. H. Tomisawa, H. Kato, R. Fujita, and H. Hongo, *Chem. Pharm. Bull.*, **27**, 810 (1979).
335. A. Gorgues and A. Le Coq, *Tetrahedron Lett.*, 4829 (1979).
336. P. A. T. W. Porskamp, R. C. Haltiwanger, and B. Zwanenburg, *Tetrahedron Lett.*, **24**, 2035 (1983).
337. M. Komatsu, S. Takamatsu, M. Uesaka, S. Yamamoto, Y. Ohshio, and T. Agawa, *J. Org. Chem.*, **49**, 2691 (1984).
338. H. Gnichtel, C. Gumprecht, and P. Luger, *Liebigs Ann. Chem.*, 1531 (1984).
339. H. J. Bestmann and R. Schobert, *Angew. Chem. (Int. Ed. Engl.)*, **24**, 790 (1985).
340. R. W. Saalfrank and W. Rost, *Angew. Chem. (Int. Ed. Engl.)*, **24**, 855 (1985).
341. A. Schwobel and G. Kresze, *Liebigs Ann. Chem.*, 453 (1985).
342. B. E. Smart and W. J. Middleton, *J. Am. Chem. Soc.*, **109**, 4982 (1987).
343. A. Corsico Coda, G. Desimoni, A. Gamba Invernizzi, P. P. Righetti, and G. Tacconi, *Tetrahedron Lett.*, **26**, 3137 (1985).
344. R. Fujita, H. Hongo, and H. Tomisawa, *Annu. Rep. Tohoku Coll. Pharm.*, 51 (1978); *Chem. Abstr.*, **92**, 76252d (1980).
345. A. V. Kamernitskii, V. N. Ignatov, I. S. Levina, B. S. El'Yanov, *Izv. Akad. Nauk SSSR, Ser. Khim.*, 1902 (1980).
346. G. de Lauzon, C. Charrier, H. Bonnard, F. Mathey, J. Fisher, and A. Mitschler, *J. Chem. Soc. Chem. Commun.*, 1272 (1982).
347. J.-L. Gras, *J. Chem. Res. (S)*, 300 (1982).
348. D. C. England, *J. Org. Chem.*, **46**, 153 (1981).
349. L. A. G. M. van den Broek, P. A. T. W. Porskamp, R. Curtis Haltiwanger, and B. Zwanenburg, *J. Org. Chem.*, **49**, 1691 (1984).
350. R. Appel, F. Knoch, and R. Zimmermann, *Chem. Ber.*, **118**, 814 (1985).
351. K.-H. Scholz, H.-G. Heine, and W. Hartmann, *Org. Synth.*, **62**, 149 (1982).
352. P. Laszlo and I. Lucchetti, *Tetrahedron Lett.*, **25**, 2147 (1984).
353. B. G. Lenz, H. Regeling, and B. Zwannenburg, *Tetrahedron Lett.*, **25**, 5947 (1984).
354. G. Augelmann, J. Streith, and H. Fritz, *Helv. Chim. Acta*, **68**, 95 (1985).
355. F. Farina, M. C. Paredes, and V. Stefani, *Tetrahedron*, **42**, 4309 (1986).
356. H. Fisher, U. Gerbing, J. Riede, and R. Benn, *Angew. Chem. (Int. Ed. Engl.)*, **25**, 78 (1986).
357. K. Furuta, S. Hayashi, Y. Miwa, and H. Yamamoto, *Tetrahedron Lett.*, **28**, 5841 (1987).
358. D. P. G. Hamon and P. R. Spurr, *Synthesis*, 873 (1981).
359. P. J. Proteau and P. B. Hopkins, *J. Org. Chem.*, **50**, 141 (1985).
360. P. J. Garratt and A. Tsotinis, *Tetrahedron Lett.*, **27**, 2761 (1986).
361. A. Hosomi, K. Otaka, and H. Sakurai, *Tetrahedron Lett.*, **27**, 2881 (1986).

362. H. D. Scharf, H. Plum, J. Fleischhauer, and W. Schleker, *Chem. Ber.*, **112**, 862 (1979).
363. M. T. Reetz and G. Neumeier, *Chem. Ber.*, **112**, 2209 (1979).
364. Y. Ishino, I. Nishiguchi, F. Takihira, and T. Hirashima, *Tetrahedron Lett.*, **21**, 1527 (1980).
365. B. M. Trost, W. C. Vladuchick, and A. J. Bridges, *J. Am. Chem. Soc.*, **102**, 3554 (1980).
366. D. Bellus, H. Sauter, and C. D. Weis, *Org. Synth.*, **58**, 67 (1978).
367. B. M. Trost and M. Shimizu, *J. Am. Chem. Soc.*, **104**, 4299 (1982).
368. A. J. Bridges and J. W. Fischer, *J. Org. Chem.*, **49**, 2954 (1984).
369. L. S. Hegedus, N. Kambe, Y. Ishii, and A. Mori, *J. Org. Chem.*, **50**, 2240 (1985).
370. H. Hopf, I. Bohm, and J. Kleinschroth, *Org. Synth.*, **60**, 41 (1981).
371. R. V. Bonnert and P. R. Jenkins, *Tetrahedron Lett.*, **28**, 697 (1987).
372. P. G. Gassman and D. A. Singleton, *Tetrahedron Lett.*, **28**, 5969 (1987).
373. P. G. McDougal, J. G. Rico, and D. Van Derveer, *J. Org. Chem.*, **51**, 4492 (1986).
374. T. Cuvigny, C. Hervé du Penhoat, and M. Julia, *Tetrahedron*, **42**, 5329 (1986).
375. S. Danishefsky, J. Aubé, and M. Bednarski, *J. Am. Chem. Soc.*, **108**, 4145 (1986).
376. Y. Mori, M. Inaba, and M. Suzuki, *Chem. Pharm. Bull.*, **34**, 3488 (1986).
377. a. O. Tsuge, E. Wada, S. Kanemasa, and H. Sakoh, *Bull. Chem. Soc. Jpn.*, **57**, 3221 (1984).
b. O. Tsuge, S. Kanemasa, H. Sakoh, and E. Wada, *Bull. Chem. Soc. Jpn.*, **57**, 3234 (1984).
c. O. Tsuge, E. Wada, and S. Kanemasa, *Chem. Lett.*, 1525 (1983).
d. O. Tsuge, E. Wada, and S. Kanemasa, *Chem. Lett.*, 239 (1983).
378. P. G. Gassman and D. A. Singleton, *J. Am. Chem. Soc.*, **106**, 7993 (1984).
379. T. Schmidlin, R. Gamboni, P. Strazewski, and C. Tamm, *Helv. Chim. Acta*, **66**, 1796 (1983).
380. D. W. Cameron, G. I. Feutrill, and P. G. McKay, *Aust. J. Chem.*, **35**, 2095 (1982).
381. L. A. Paquette and G. D. Crouse, *J. Org. Chem.*, **48**, 141 (1983).
382. M. Fink, H. Gaier, and H. Gerlach, *Helv. Chim. Acta*, **65**, 2563 (1982).
383. T. Ibuka, Y. Mitsui, K. Hayashi, H. Minakata, and Y. Inubushi, *Tetrahedron Lett.*, **22**, 4425 (1981).
384. A. P. Kozikowski, K. Sugiyama, and E. Huie, *Tetrahedron Lett.*, **22**, 3381 (1981).
385. D. W. Cameron, G. I. Feutrill, and P. Perlmutter, *Tetrahedron Lett.*, **22**, 3273 (1981).
386. a. A. P. Kozikowski and K. Sugiyama, *Tetrahedron Lett.*, **21**, 4597 (1980).
b. A. P. Kozikowski, K. Sugiyama, and J. P. Springer, *Tetrahedron Lett.*, **21**, 3257 (1980).
387. S. Danishefsky and F. J. Walker, *J. Am. Chem. Soc.*, **101**, 7018 (1979).
388. S. Danishefsky, C.-F. Yan, R. K. Singh, R. B. Gammill, P. M. McCurry, Jr., N. Fritsch, and J. Clardy, *J. Am. Chem. Soc.*, **101**, 7001 (1979).
389. D. W. Cameron, G. I. Feutrill, P. G. Griffiths, and D. J. Hodder, *J. Chem. Soc. Chem. Commun.*, 688 (1978).

390. M. Gillard, C. T. Kint, E. Sonveaux, and L. Ghosez, *J. Am. Chem. Soc.*, **101**, 5837 (1979).
391. S. Danishefsky and S. J. Etheredge, *J. Org. Chem.*, **43**, 4604 (1978).
392. C. Boan and L. Skattebol, *J. Chem. Soc. Perkin Trans. 1*, 1568 (1978).
393. K. Inomata, H. Kinoshita, H. Takemoto, Y. Murata, and H. Kotake, *Bull. Chem. Soc. Jpn.*, **51**, 3341 (1978).
394. V. M. Berestovitskaya, M. V. Titova, and V. V. Perekalin, *Zh. Org. Khim.*, **16**, 891 (1980).
395. S. V. Attwood, A. G. M. Barrett, and J. C. Florent, *J. Chem. Soc. Chem. Commun.*, 556 (1981).
396. G. Schon and H. Hopf, *Liebigs Ann. Chem.*, 165 (1981).
397. A. P. Kozikowski, K. Sugiyama, and J. P. Springer, *J. Org. Chem.*, **46**, 2426 (1981).
398. N. B. Kaz'mina, E. I. Mysov, M. Yu. Antipin, A. I. Akhmedov, and Yu. T. Struchkov, *Izv. Akad. Nauk SSSR, Ser. Khim.*, 842 (1981).
399. R. Gompper and M. Sramek, *Synthesis*, 649 (1981).
400. S. Danishefsky and T. A. Craig, *Tetrahedron*, **37**, 4081 (1981).
401. R. S. Garigipati, J. A. Morton, and S. M. Weinreb, *Tetrahedron Lett.*, **24**, 987 (1983).
402. a. P. Laszlo and J. Lucchetti, *Tetrahedron Lett.*, **25**, 1567 (1984).
    b. D. W. Reynolds, K. T. Lorenz, H.-S. Choiu, D. J. Bellville, R. A. Pabon, and N. L. Bauld, *J. Am. Chem. Soc.*, **109**, 4960 (1987).
403. a. O. Tsuge, S. Kanemasa, H. Sakoh, and E. Wada, *Chem. Lett.*, 277 (1984).
    b. O. Tsuge, S. Kanemasa, H. Sakoh, and E. Wada, *Chem. Lett.*, 273 (1984).
404. R. L. Snowden, *Tetrahedron Lett.*, **25**, 3835 (1984).
405. S. Castellino and J. J. Sims, *Tetrahedron Lett.*, **25**, 2307 (1984).
406. L. A. Paquette, G. J. Wells, and G. Wickham, *J. Org. Chem.*, **49**, 3618 (1984).
407. C. M. Angelov, D. M. Mondeshka, and T. N. Tancheva, *J. Chem. Soc. Chem. Commun.*, 647 (1985).
408. A. P. Kozikowski and Y. Xia, *J. Org. Chem.*, **52**, 1375 (1987).
409. A. B. Reitz, A. D. Jordan, and B. E. Maryanoff, *J. Org. Chem.*, **52**, 4800 (1987).
410. R. Gompper and R. Sobotta, *Tetrahedron Lett.*, 921 (1979).
411. G. Adiwidjaja, T. Proll, and W. Walter, *Tetrahedron Lett.*, **22**, 3175 (1981).
412. Y. Tamura, T. Tsugoshi, Y. Nakajima, and Y. Kita, *Synthesis*, 930 (1984).
413. T. Karakasa and S. Motoki, *J. Org. Chem.*, **43**, 4147 (1978).
414. B. Serckx-Poncin, A. M. Hesbain-Frisque, and L. Ghosez, *Tetrahedron Lett.*, **23**, 3261 (1982).
415. L. F. Tietze, K. H. Gluesenkamp, K. Harms, G. Remberg, and G. M. Sheldrick, *Tetrahedron Lett.*, **23**, 1147 (1982).
416. R.R. Schmidt, and M. Maier, *Tetrahedron Lett.*, **23**, 1789 (1982).
417. K. Krishan, A. Singh, B. Singh, and S. Kumar, *Synth. Commun.*, **14**, 219 (1984).
418. Daicel Chem. Ind., *Jpn. Kokai Tokyo Koho*, JP 59, 108,734 (1984); *Chem. Abstr.*, **101**, 170708h (1984).
419. S. Danishefsky and M. Bednarski, *Tetrahedron Lett.*, **25**, 721 (1984).

420. M. Yamauchi, S. Katayama, O. Baba, and T. Watanabe, *J. Chem. Soc. Chem. Commun.*, 281 (1983).

421. W. Clegg, O. Graalmann, M. Haase, U. Klingebiel, G. M. Sheldrick, P. Werner, G. Henkel, and B. Krebs, *Chem. Ber.*, **116**, 282 (1983).

422. M. Maier and R. R. Schmidt, *Liebigs Ann. Chem.*, 2261 (1985).

423. R. R. Schmidt and M. Maier, *Tetrahedron Lett.*, **26**, 2065 (1985).

424. H. Tanaka, S. Kametani, T. Saito, and S. Motoki, *Bull. Chem. Soc. Jpn.*, **58**, 667 (1985).

425. R. E. Ireland and D. Habich, *Tetrahedron Lett.*, **21**, 1389 (1980).

426. G. L'Abbé, J.-P. Dekerk, J.-P. Declercq, G. Germain, and M. van Meerssche, *Tetrahedron Lett.*, 3213 (1979).

427. C. T. Gokou, J.-P. Pradere, and H. Quiniou, *J. Org. Chem.*, **50**, 1545 (1985).

428. S. N. Mazumdar, M. Sharma, and M. P. Mahajan, *Tetrahedron Lett.*, **28**, 2641 (1987).

429. a. K. Burger and H. Goth, *Angew. Chem. (Int. Ed. Engl.)*, **19**, 810 (1980).
b. K. Burger, H. Goth, W. Schöntag, and J. Firl, *Tetrahedron*, **38**, 287 (1982).

430. J.-C. Meslin and H. Quiniou, *Bull. Soc. Chim. Fr.*, 347 (1979).

431. H. K. Hall, Jr., and D. L. Miniutti, *Tetrahedron Lett.*, **25**, 943 (1984).

432. K. Hartke, T. Kissel, J. Quante, and G. Henssen, *Angew. Chem. (Int. Ed. Engl.)*, **17**, 953 (1978).

433. M. Forchiassin, G. Pitacco, C. Russo, and E. Valentin, *Gazz. Chim. Ital.*, **112**, 335 (1982).

434. F. Bassam and R. G. Jones, *J. Chem. Soc. Chem. Commun.*, 917 (1979).

435. F. Sainte, B. Serckx-Poncin, A.-M. Hesbain-Frisque, and L. Ghosez, *J. Am. Chem. Soc.*, **104**, 1428 (1982).

436. Y.-S. Cheng, E. Ho, P. S. Mariano, and H. L. Ammon, *J. Org. Chem.*, **50**, 5678 (1985).

437. J. Barluenga, J. Joglar, S. Fustero, V. Gotor, C. Kruger, and M. J. Romao, *Chem. Ber.*, **118**, 3652 (1985).

438. J. J. Gajewski, K. B. Peterson, and J. R. Kagel, *J. Am. Chem. Soc.*, **109**, 5545 (1987).

439. S. Wolfe and J. R. Cambell, *Synthesis*, 117 (1979).

440. L. F. Tietze and K.-H. Glusenkamp, *Angew. Chem. (Int. Ed. Engl.)*, **22**, 887 (1983).

441. O. Tsuge, T. Ohnishi, and H. Watanabe, *Heterocycles*, **16**, 2085 (1981).

442. Y. Nomura, Y. Takeuchi, S. Tomoda, and M. M. Ito, *Bull. Chem. Soc. Jpn.*, **54**, 2779 (1981).

443. Y. Yamamoto, H. Suzuki, and Y. Moro-oka, *Chem. Lett.*, 73 (1986).

444. G. Wulff and H. Bohnke, *Angew. Chem. (Int. Ed. Engl.)*, **25**, 90 (1986).

445. J. Barluenga, F. J. Gonzalez, S. Fustero, and V. Gotor, *J. Chem. Soc. Chem. Commun.*, 1179 (1986).

446. T. Saito, M. Nakane, M. Endo, H. Yamashita, Y. Oyamada, and S. Motoki, *Chem. Lett.*, 135 (1986).

447. S. N. Mazumdar, I. Ibnusaud, and M. P. Mahajan, *Tetrahedron Lett.*, **27**, 5875 (1986).
448. T. Gillmann and K. Hartke, *Chem. Ber.*, **119**, 2859 (1986).
449. T. Sasaki, K. Shimizu, and M. Ohno, *Chem. Pharm. Bull.*, **32**, 1433 (1984).
450. *Jpn. Kokai Tokyo Koho*, JP 58, 216,188 (1983); *Chem. Abstr.*, **100**, 209875s (1984).
451. W. Ando, Y. Hanyu, T. Takata, T. Sakurai, and K. Kobayashi, *Tetrahedron Lett.*, **25**, 1483 (1984).
452. L. F. Tietze and E. VoB, *Tetrahedron Lett.*, **27**, 6181 (1986).
453. G. W. Kirby, *Chem. Soc. Rev.*, **6**, 1 (1977).
454. S. M. Weinreb and R. R. Staib, *Tetrahedron*, **38**, 3087 (1982).
455. S. M. Weinreb and J. I. Levin, *Heterocycles*, **12**, 949 (1979).
456. J. Sauer and R. Sustmann, *Angew. Chem. (Int. Ed. Engl.)*, **19**, 779 (1980).
457. J. J. Gaiewski, *J. Am. Chem. Soc.*, **101**, 4393 (1979).
458. B. S. El'yanov, S. K. Shakhova, B. D. Polkovnikov, and L. F. Rar, *J. Chem. Soc. Perkin Trans. 2*, 11 (1985).
459. S. D. Kahn, C. F. Pau, L. E. Overman, and W. J. Hehre, *J. Am. Chem. Soc.*, **108**, 7381 (1986).
460. Yu. A. Titov, *Russ. Chem. Rev.*, **11**, 267 (1962).
461. a. R. Breslow and U. Maitra, *Tetrahedron Lett.*, **25**, 1239 (1984).
b. S. Danishefsky and J. F. Kerwin, Jr., *J. Org. Chem.*, **47**, 3183 (1982).
462. H. J. Liu and W. M. Feng, *Synth. Commun.*, **17**, 1777 (1987).
463. S. Danishefsky, M. P. Prisbylla, and S. Hiner, *J. Am. Chem. Soc.*, **100**, 2918 (1978).
464. S. Danishefsky, T. Kitahara, R. McKee, and P. F. Schuda, *J. Am. Chem. Soc.*, **98**, 6715 (1976).
465. M. Petrizilka and J. I. Graison, *Synthesis*, 753 (1981).
466. S. Danishefsky, C. F. Yan, and P. M. McCurry, *J. Org. Chem.*, **42**, 1819 (1977).
467. S. Danishefsky, *Acc. Chem. Res.*, **14**, 400 (1981).
468. S. Danishefsky, J. F. Kerwin, Jr., and S. Kobayashi, *J. Am. Chem. Soc.*, **104**, 358 (1982).
469. A. Hosomi, H. Iguchi, J. Sasaki, and H. Sakurai, *Tetrahedron Lett.*, **23**, 551 (1982).
470. G. J. Janz, in "1,4-Cycloaddition Reactions: The Diels–Alder Reaction in Heterocyclic Synthesis", J. Hamer, Ed., Academic Press, New York, p. 97 (1967).
471. N. J. Leonard, A. Playtis, F. Skoog, and R. Y. Schmidz, *J. Am. Chem. Soc.*, **93**, 3056 (1971).
472. S. M. Weinreb, R. S. Garigipati, and G. A. Gainor, *Heterocycles*, **21**, 309 (1984).
473. S. S. Novikov, G. A. Shvekhgeimer, and A. A. Dudinskaya, *Russ. Chem. Rev.*, **29**, 79 (1960).
474. S. David, A. Lubienau, and A. Thieffry, *Tetrahedron*. **34**, 299 (1978).
475. S. David and A. Lubienau, *Nouv. J. Chim.*, **1**, 375 (1977).
476. S. David, J. Eustache, and A. Lubienau, *J. Chem. Soc. Perkin Trans. 1*, 1795 (1979).

477. a. S. L. Paquette, "Asymmetric Synthesis", Academic Press, New York, vol. 3, p. 445 (1984).
b. S. D. Kahn and W. J. Hehre, *J. Am. Chem. Soc.*, **109**, 663 (1987).
478. K. Alder, and H. von Bracher, *Liebigs Ann. Chem.*, **608**, 195 (1957).
479. G. F. Woods and A. Viola, *J. Am. Chem. Soc.*, **78**, 4380 (1956).
480. H. Fleischacker and F. Wood, *J. Am. Chem. Soc.*, **78**, 3436 (1956).
481. A. S. Onischenko, "Diene Synthesis", Israel Program of Scientific Translation, Jerusalem (1964).
482. B. K. Rammash, C. M. Gladstone, and J. L. Wong, *J. Org. Chem.*, **46**, 3036 (1981).
483. M. G. Veliev, M. M. Guseinov, L. A. Yanovskaya, and K. Ya. Burstein, *Tetrahedron*, **41**, 749 (1985).
484. E. Larson and S. Danishefsky, *J. Am. Chem. Soc.*, **105**, 6715 (1983).
485. S. Danishefsky, E. R. Larson, and J. P. Springer, *J. Am. Chem. Soc.*, **107**, 1274 (1985).
486. J. P. Schirmann, G. Bonnard, and F. Weiss, *Bull Soc. Chim. Fr.*, 3326 (1968).
487. J. Soula, D. Lumbroso, M. Hellin, and F. Coussemant, *Bull. Soc. Chim. Fr.*, 2059 (1966).
488. G. Büchi and J. E. Powell, Jr., *J. Am. Chem. Soc.*, **89**, 4559 (1967).
489. G. Desimoni and G. Tacconi, *Chem. Rev.*, **75**, 651 (1975).
490. C. W. Smith, D. G. Norton, and S. A. Ballard, *J. Am. Chem. Soc.*, **73**, 5267, 5270 (1951).
491. H. Fiessermann and J. Ribka, *Chem. Ber.*, **89**, 27 (1956).
492. B. P. Mundy, R. D. Otzenberger, and A. R. Debarnardis, *J. Org. Chem.*, **36**, 2390 (1971).
493. R. N. Schut and T. M. H. Liu, *J. Org. Chem.*, **30**, 2845 (1965).
494. G. Opitz and H. Holtmann, *Liebigs Ann. Chem.*, **684**, 79 (1965).
495. D. L. Boger and K. D. Robarge, *J. Org. Chem.*, **53**, 3373 (1988).
496. J. Castells, F. Camps, and F. S. Ferrando, *An. R. Soc. Esp. Fis. Quim., Ser. B*, **66**, 175 (1970).
497. R. I. Longley and W. S. Emerson, *J. Am. Chem. Soc.*, **72**, 3079 (1950).
498. S. Danishefsky and M. Bednarsky, *Tetrahedron Lett.*, **25**, 3175 (1984).
499. J. Colonge and G. Descontes, in "1,4-Cycloaddition Reactions: The Diels–Alder Reaction in Heterocyclic Syntheses", J. Hamer, Ed., Academic Press, New York (1967).
500. D. L. Boger, *Tetrahedron*, **39**, 2869 (1983).

3

OUTER-RING DIENES

Reactivity, Selectivity, Synthesis Several types of outer-ring dienes are known (i.e, **1**–**8**) in dependence on the presence of one or two diene moieties and their nature (carbodiene or heterodiene), and the type of ring (carbocyclic and/or heterocyclic). The reactivity of *o*-carbodimethylenes has been related to the distance between the methylene carbons (1,4-distance) and is strongly influenced by the presence of heteroatoms in the ring. The reaction rate of type **1** compounds with MA (Table 3.1) is higher as the 1,4-distance is shorter.[1,88] The large 1,4-distance of **1** ($n = 0$) obstructs [4 + 2] cycloadditions. Indeed, in the reaction of **1** ($n = 0$) with 1,1-difluoroallene, a 3 : 1 ratio of adducts is derived from [4 + 2] and [2 + 2] reactions.[2] The 1,3-dioxolanes **2** do not react with MA; they only react with the more reactive dienophiles (Table 3.1), and therefore, in comparison with compounds **1**, they show a lower diene reactivity.

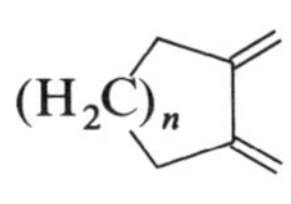

1

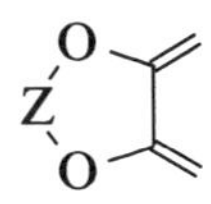

2

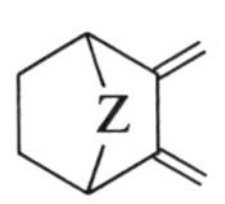

3

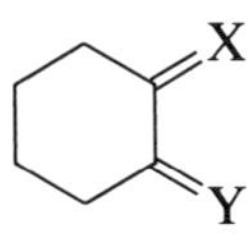

4

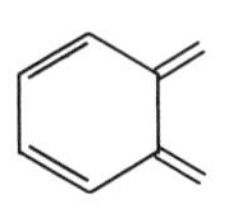

5

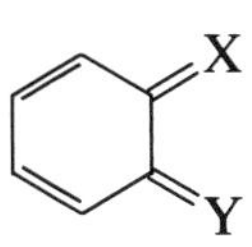

6

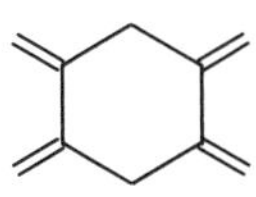

7

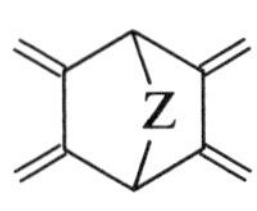

8

o-Carbodimethylenes have been used to construct polycyclic structures through multiple [4 + 2] cycloadditions. In principle (Scheme 3.1), the cyclo-

TABLE 3.1 Rate Constants of Reactions of 1,2-Dimethylenes with 4-*t*-Butyl-1,2,4-triazoline-3,5-dione (BTAD) in Ethyl Acetate at 30°C and Maleic Anhydride (MA) in Dioxane at 30°C and 1,4-Diene Distance

Diene		Dienophile	$k_2 \times 10^5$ ($L \cdot mol^{-1} \cdot s^{-1}$)	$r_{1,4}$ (Å)
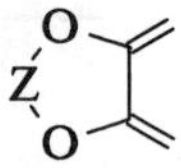	Z = CO	BTAD	4,200	3.41
	CH_2	BTAD	171,300	3.39
	CMe_2	BTAD	62,300	3.38
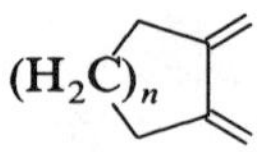	*n* = 0	MA	20.8	3.34
	1	MA	3,590	3.14
	2	MA	960	3.17
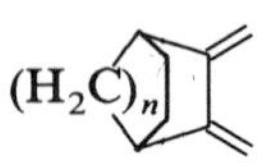	*n* = 1	MA	17.3	3.17
	2	MA	316	3.08
	3	MA	1,190	3.03
		MA	6,780	2.36

addition of butatriene to 1,3-butadiene gives 1,2-dimethylenecyclohex-4-ene which, can react with a second molecule of butatriene, and so on. The butatriene cannot be used because of its weak dienophilicity and its tendency to oligomerizate. Diethylacetilenedicarboxylate has been used as a butatriene synthetic equivalent (Scheme 3.1), and dehydroaromatic esters have been prepared by a repetition of a four-step sequence.[16]

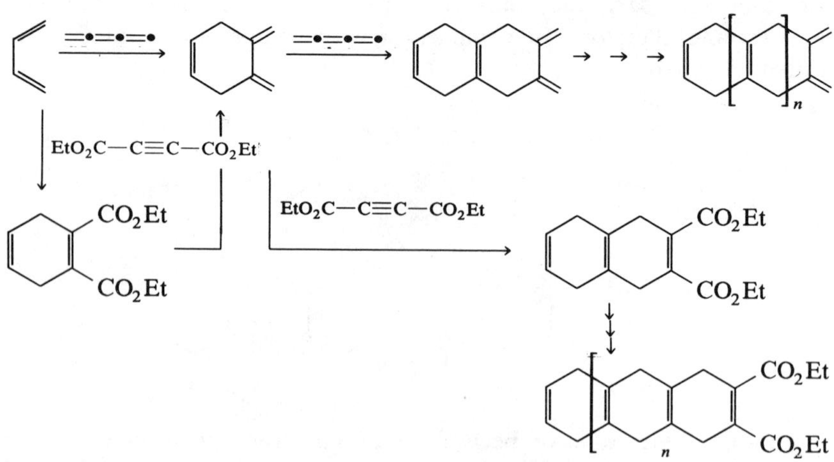

Scheme 3.1

Extensive studies on the regio- and diastereoselectivity of *o*-dimethylenes such as **3** have been carried out by Vogel et al.[57,61,64] The regioselectivity of **3** (Z = CH_2) is affected by substituents at C-2. The dimethylene-2-norbornanol **9** reacts with MP to give preferentially the *meta* adduct **10**, while its oxidation derivative, the dimethylene-2-norbornanone, yields the *para* adduct as predominant regioisomer.

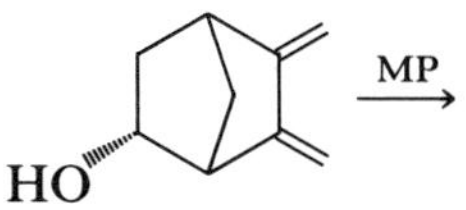

9

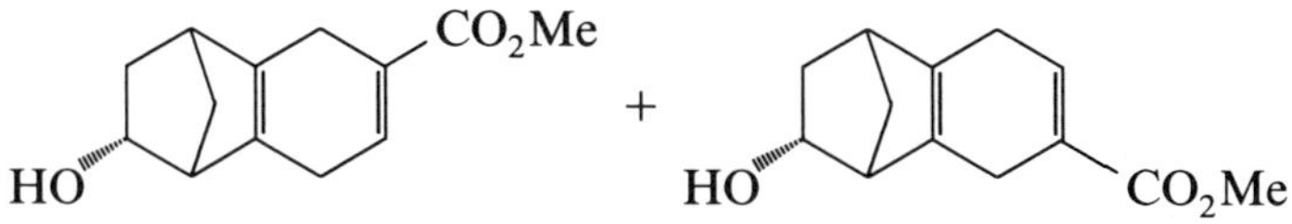

para *meta*

10

The facial-diastereoselectivity of these compounds is of theoretical and synthetic interest. Whereas the 7-oxabicyclo[2.2.1]heptanes (i.e., **3** and derivatives, Z = O) preferentially give Diels–Alder addition *syn* to oxygen bridge, the cycloadditions of bicyclo[2.2.2]octenes **11**–**14** occur on the side opposite the ethano bridge (*anti* addition).[61,62] The Diels–Alder reactions of **14** (a mixture of stereoisomers) with MA give the adducts **15** and **16**, which are the result of four attacks; *anti* attack is preferred in the *exo* addition, while no facial selectivity is observed[62] in the *endo* addition.

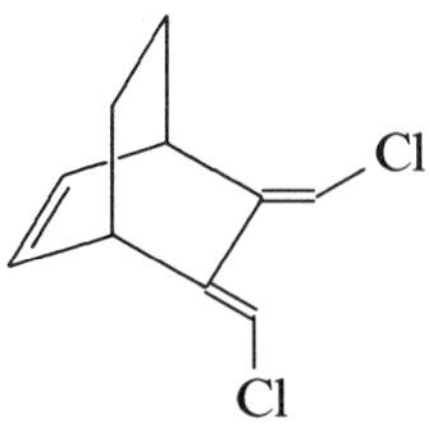

11

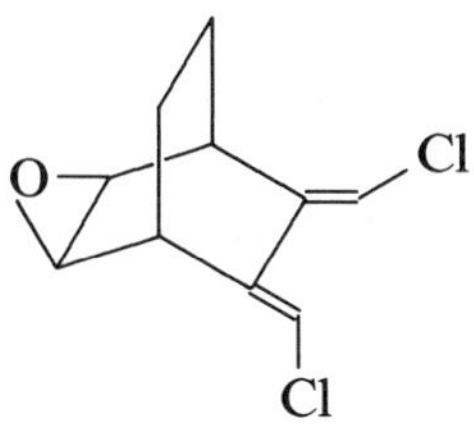

12

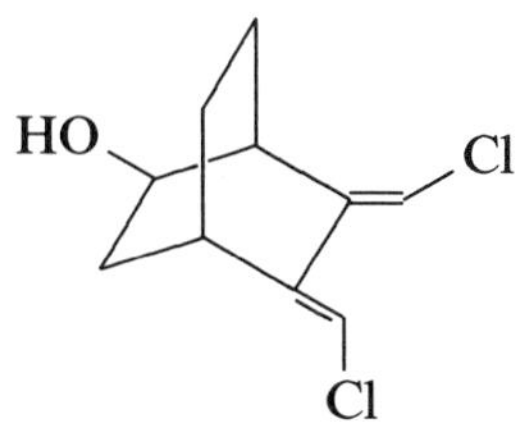

13

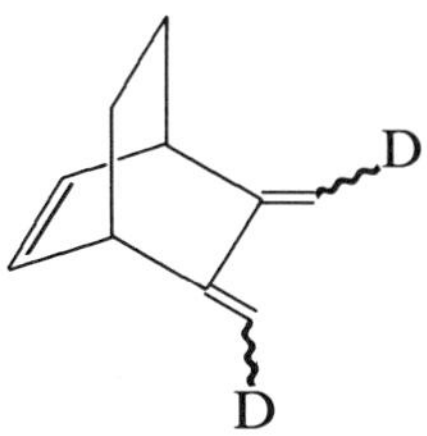

14

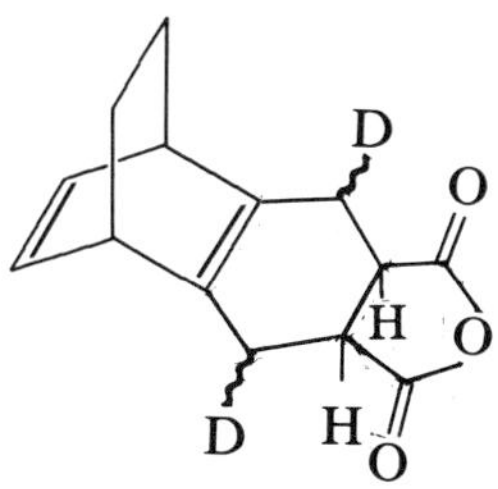

15 { *anti-exo* 36%, *syn-endo* 24% }

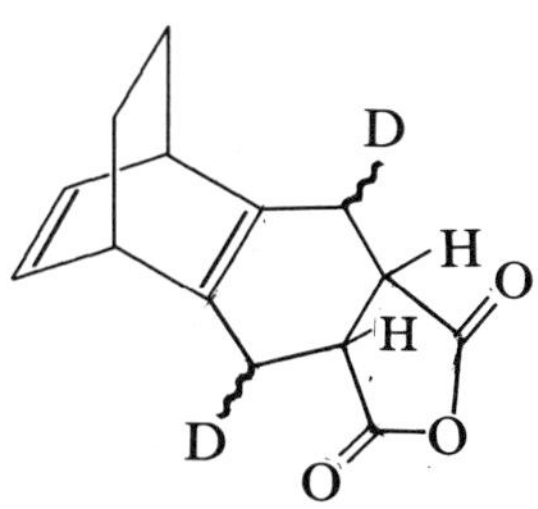

16 { *anti-endo* 24%, *syn-exo* 16% }

The *o*-quinodimethane **5** and its derivatives are very reactive and unstable dienes (they are usually prepared in situ) which undergo Diels–Alder reactions with the exocyclic diene moiety and have been used as intermediates in the synthesis of alkaloids, steroids, and anthracyclines.

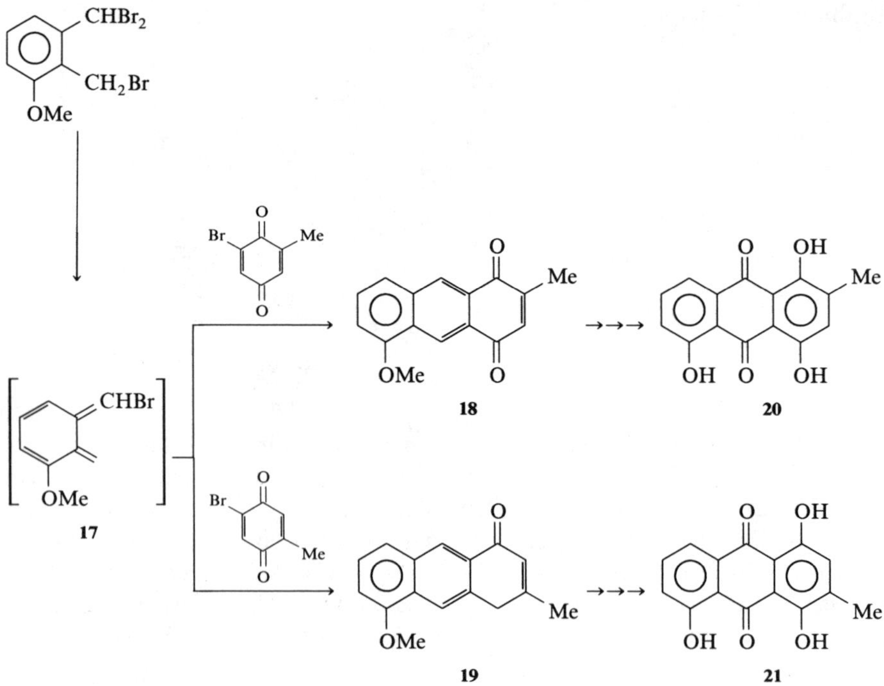

Examples are the regioselective cyclizations of **17** with 6-bromo- and 5-bromotoluquinone to anthraquinones **18** and **19**, which have been converted to islandicin (**20**) and digitopurpone (**21**), respectively, by a multistep sequence.[89]

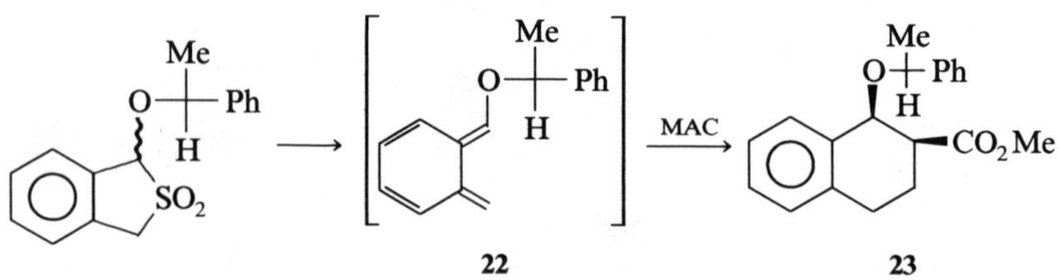

o-Quinodimethanes with chiral α-alkoxy groups (i.e., **22**) react *endo*, *ortho*, and *enantio* selectively with acrylates to give optically active tetrahydronaphthalenes (i.e., **23**),[31] while the α-phenyl- and α-phenyl-α'-methoxy derivatives

24 react with DMF and DMM via an *exo* transition state, indicating that the phenyl group has a stereo-directing effect.[34a]

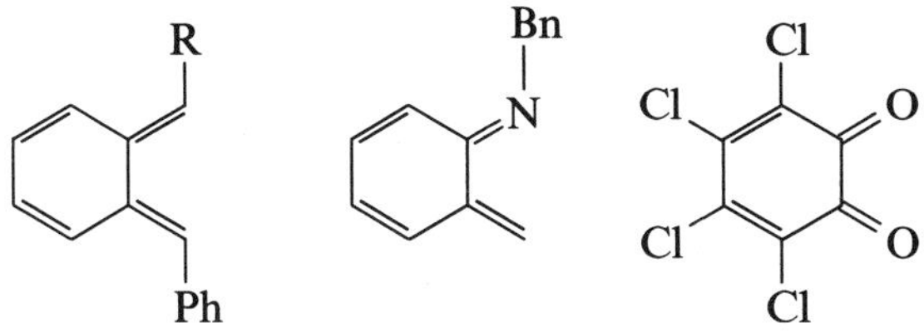

24 R = H, OMe **25** **26**

Exocyclic heterodienes **4** and **6** allow the building of heterocyclic rings. Thus quinone methide-*N*-alkylimines (i.e., **25**) react with the electron-deficient dienophiles to give tetrahydroquinolines via a preferential *endo-ortho* transition state.[49] The tetrachlorobenzoquinone (*o*-chloranil) **26** gives Diels–Alder reactions with an inverse electron demand with heterocyclic dienophiles such as 1-ethoxycarbonyl-1*H*-diazepine[55b] and oxazoles,[55a] affording heteroannulated 1,4-dioxins.

Tetraenes **27** react with dienophiles faster than do their monoadducts **28** ($k_1 > k_2$) and are therefore attractive starting materials to synthetize polycyclic systems **29** by timed Diels–Alder reactions.[65,68,71,80] A ratio $k_1/k_2 = 376$ was measured for the cycloadditions of TCNE with **27** and **28** (Z = O, R = H) and a similarly high ratio was observed for **27** and **28** [Z = O, R = $CH(OMe)_2$]. Lower ratios ($k_1/k_2 \simeq 10$) were found in the cycloadditions of **27** and **28** (Z = CH_2CHOH; Z = CH_2CO; R = H) with TCNE, while no difference ($k_1/k_2 \simeq 1$) was observed in the two successive Diels–Alder reactions with MP. The oxabicyclo **27** (Z = O, R = H) has been used to prepare both the anthracyclinone skeleton by timed Diels–Alder reactions with two different dienophiles and linear epoxypolyacenequinones by repetitive quinone Diels–Alder reactions–oxidation reactions (Scheme 3.2).

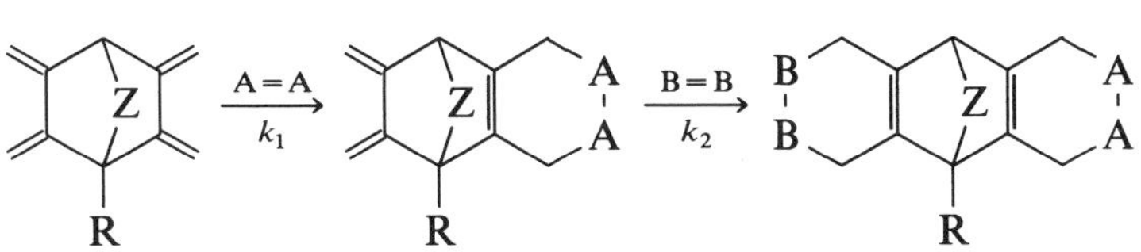

27 **28** **29**

The tetraene **30** permits control of both the tandem cycloadditions and their regio- and stereoselectivity.[80] With butynone and MVK the adducts **31** and **32** are obtained as a result of an *ortho* addition *anti* to the oxygen bridge. With TCNE the addition is *syn* selective, as generally occurs for 7-oxabicyclo[2.2.1]heptanes.

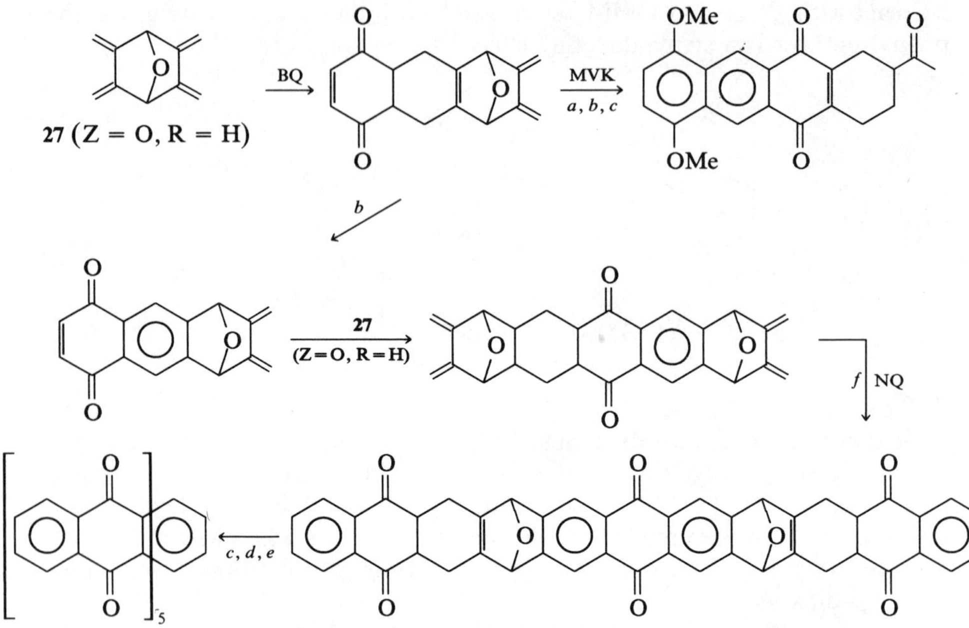

a = MeI; b = K_2CO_3/DDQ; c = TFA; d = KOH; e = HClO; f = Al_2O_3

Scheme 3.2

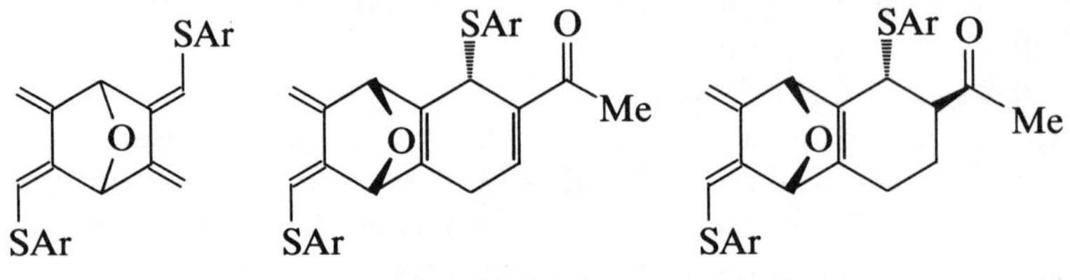

30 **31** **32**

The more complex dienes **33** and **34** favor the *para*-cycloaddition of the dienophile.[64]

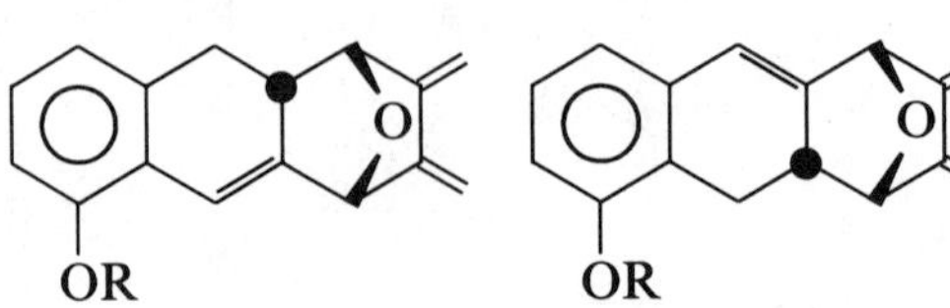

33 **34**

On the contrary, the cycloadditions of tetraenes **27** [Z = O; R = $CH(OMe)_2$; R = CH_2OH] with MP and butynone are not greatly regioselective and the selectivity is not improved by Lewis acid catalysts.[71]

Tables of Diels–Alder Reactions See Tables 3.2 to 3.11; see also the general remarks about classifications, abbreviations, and symbols in the Introduction.

TABLE 3.2 Diels–Alder Reactions of 1,2-Dimethylene Cycloalkanes and 1,2-Dimethylene Cycloalkenes

Diene	Dienophile		Yield (%)	Ref.
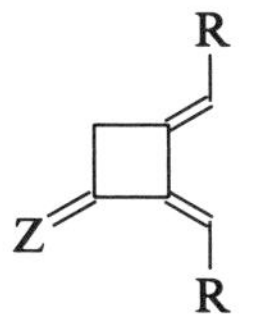				
R = H, Z = H_2	A13	1,1-Difluoroallene	60	2
	D3	MA (k)	62	1
R = H, Z = CH_2	I5	NPTAD		4
R = Ph, Z = H_2	D5	NPM	74	3
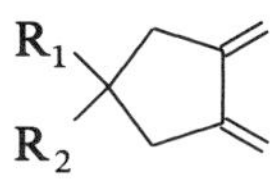				
$R_1 = R_2 =$ H	A12	TCNE (k)		81
	D3	MA (k)	50	1
$R_1 =$ H, $R_2 =$ Me	D5	NPM	87	5
$R_1 = CO_2Me$, $R_2 =$ H	D3	MA	80	6
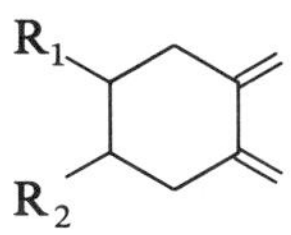				
$R_1 = R_2 =$ H	C9	Quinoline-5,8-quinones	82–97	7
	C9	Quinoxaline-5,8-quinone	84	7
	C9	1,4-Naphthoquinones	87–98	8
	D3	MA (k)	80	1
[$R_1 =$ H, $R_2 =$ Ac]	D5	NPM	60	9
[$R_1 =$ H, $R_2 = CO_2Me$]	A1	MVK	66	9
[$R_1 = R_2 = CO_2Me$]	D5	NPM	45	9
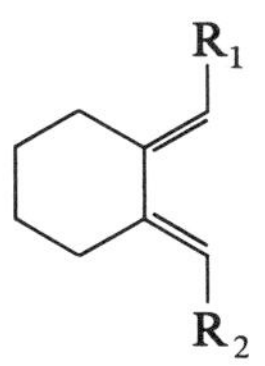				
$R_1 =$ H, $R_2 =$ OAc	I5	NPTAD	47	10
$R_1 =$ H, $R_2 = S(CH_2)_3Me$	I5	NPTAD	19	10
$R_1 = R_2 =$ Me	D5	NPM	90	5
$R_1 =$ Me, $R_2 = Pr^i$	D3	MA	88	5
$R_1 = R_2 =$ Et	D5	NPM	86	5

TABLE 3.2 (*Continued*)

Diene		Dienophile	Yield (%)	Ref.
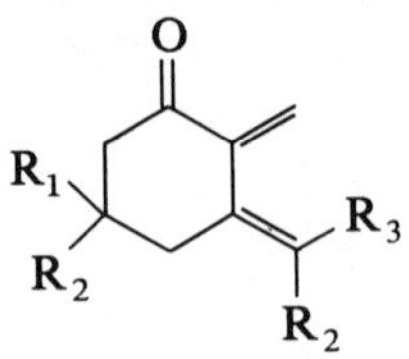				
[R_1 = H, R_2 = R_3 = OMe]	C9	1,4-Naphthoquinones	70–88	14
[R_1 = H, R_2 = R_3 = OEt]	C9	1,4-Naphthoquinones	22–88	14
[R_1 = Me, R_2 = OEt, R_3 = H]	E3	DMAD	88	11
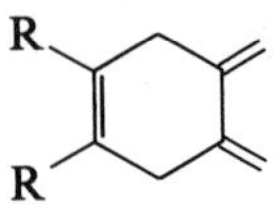				
R = H	C1	Tetrachlorocyclopropene	87	12
	C1	1,1-Dichloro-3,3-difluorocyclopropene	85	12
	E3	Diethylacetylene dicarboxylate	94	16
R = CO_2Me	D5	NPM	45	9
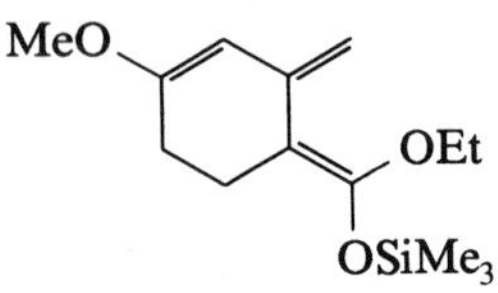	C9	1,4-Naphthoquinones	36–61	13
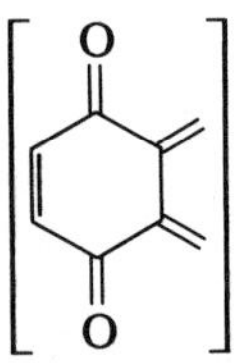	C9	NQ	11	15
	D5	NPM	29	15
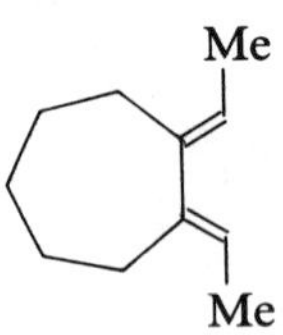	D5	NPM	82	5

TABLE 3.3 Diels–Alder Reactions of *o*-Quinodimethanes

Diene			Dienophile	Yield (%)	Ref.
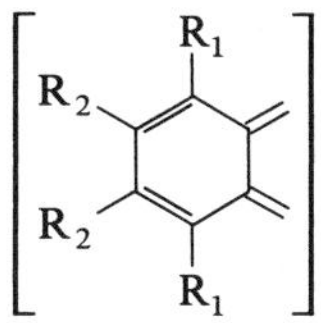					
R_1	**R_2**				
H	H	A2	3-(Trimethylsilyloxy)-3-buten-2-one	26	24
		A8	Ethene		(19)
		A8	Vinyltrichlorosilane	55	20
		C1	Cyclopentene	6–20	17
		C2	Norbornene	71	17
		C3	Dihydronaphthalenes	67–73	23
		C8	Quinone		22
		D2	D-Glucose-based enone (*)	70	18
		E3	DMAD (k)		21
H	OMe	A10	Ethyl cinnamate	55	25
		A11	Benzalmalonitriles	27–41	25
		D5	NPM	78	25
H	Cl	D3	MA		26
Me	H	A1	MAC	73	27
		A1	MVK	74	27
		D3	MA	70	27
OME	H	A2	3-(Trimethylsilyloxy)-3-buten-2-one	17	24
—CH=CH—CH=CH—		D2	D-Glucose-based enone (*)	20–30	18
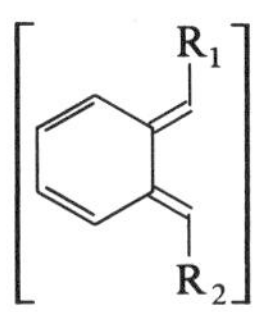					
R_1	**R_2**				
H	CN	D1	Dihydropyran	70	28
H	OH	C5	Dimethylcyclohexa-2,5-dien-1-one	76	29
		C5	4,4-Ethylenedioxycyclohexa-2,5-dien-1-one		30
		C9	1,4-Naphthoquinones		30
[H	$OCHMeCH_2NMe_2$]	A1	MAC	90	33
Ph	OMe	A7	DMM		34a
		A7	DMF		34a
Ph	OAc	A7	DMM	61	34
		A7	DMF	90	34
		D3	MA	90	34
OCHMePh	H (*)	A7	DMF	100	31
Br	Br	C3	[1,2-Dibromobenzo-cyclobut-1-ene]	71	32

TABLE 3.3 (*Continued*)

Diene			Dienophile	Yield (%)	Ref.
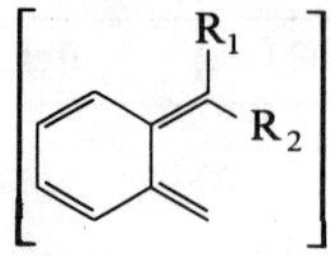					
R_1	**R_2**				
Me	Me	D3	MA	48	35
Me	Et	D3	MA	24	35
Et	Me	D3	MA	40	35
[—N(Me)—CMe_2—CH_2O—]		A8	ACN	89	33
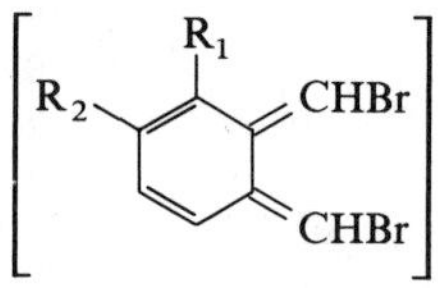					
R_1	**R_2**				
H	PhCO	C3	[1,2-Dibromobenzo-cyclobut-1-ene]	95	32
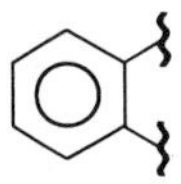		C3	[1,2-Dibromobenzo-cyclobut-1-ene]	28	32
OAc	H	C7	BQ		39
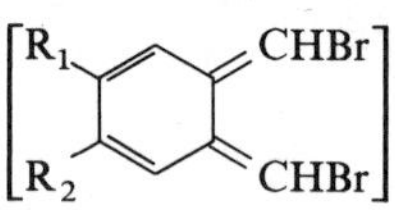					
R_1	**R_2**				
N(Ac)—CO—N(Ac)		D3	MA	82	56
N=CH—NCOPh		D3	MA		56
N=CH—NCO_2Ph		D3	MA		56
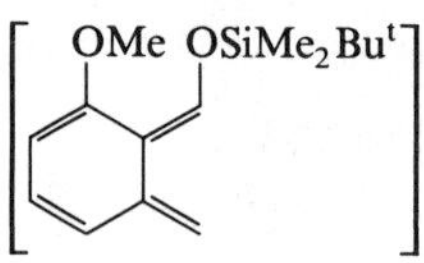		A3	β-Substituted ethyl acrylates (*)	71	36
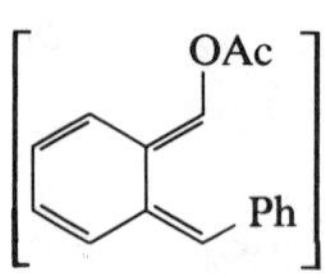		A7	DMF	94	34b
		A7	DMM	41	34b
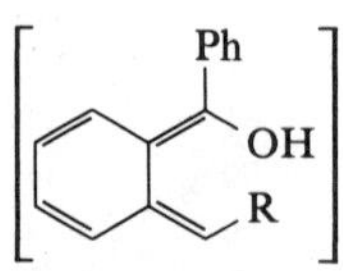		A1	MAC		38
		A7	DMF		37
(R = Me, Ph)		D3	MA		37

TABLE 3.4 Diels–Alder Reactions of 1,2-Dimethylene Heterocycles

Diene	Dienophile		Yield (%)	Ref.
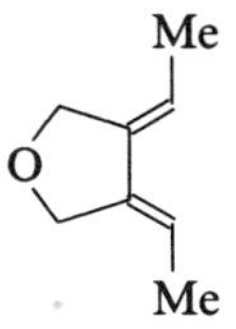	C9	NQ	91	5
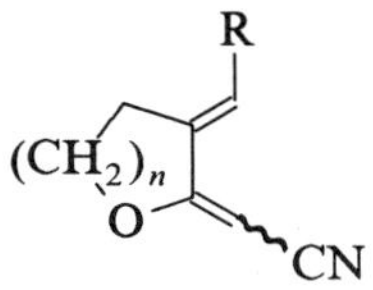 ($n = 1$, R = H, Et, Ph, OMe; $n = 2$, R = Ph)	A1	MVK	23–40	40
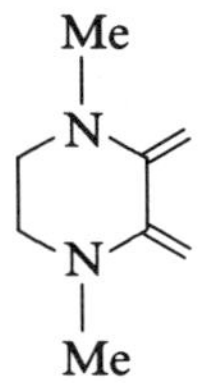	A1	MVK	90	41
	A1	AC	61	41
	A7	DMF	60	41
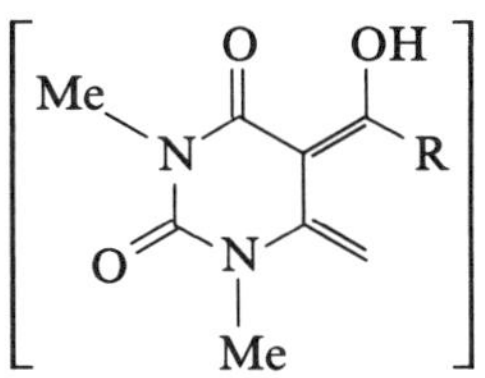 (R = H, Me, OEt)	A7	DMM	56–63	42
	A7	DMF	47–70	42
	E3	DMAD	49–83	42
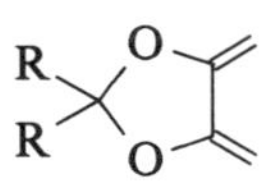 (R = H, Me)	I5	NPTAD (k)	96	1
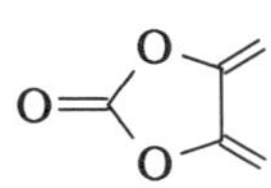	I5	NPTAD (k)	64	1
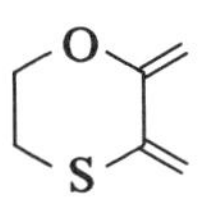	A1	MVK	60	43
	D5	NPM	62	43
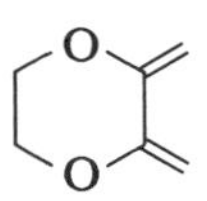	D3	MA	65	1

TABLE 3.5 Diels–Alder Reactions of 1,2-Dimethylene Fused Polycyclic Compounds

Diene		Dienophile	Yield (%)	Ref.
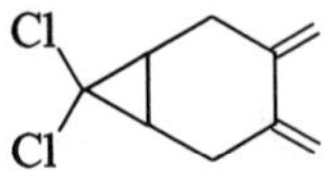	C7	1,4-Benzoquinones	57–69	44
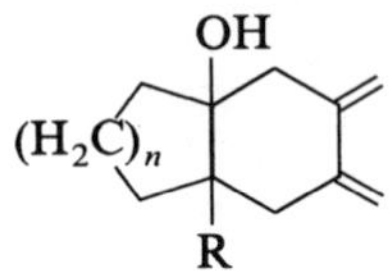 $n = 1$, R = CO_2Me	A8	Phenylvinylsulfone	80	45
$n = 1$, R = SO_2Ph	E3	DMAD	95	45
$n = 2$, R = SO_2Ph	D5	NPM	66	45
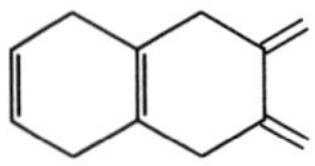	E3	DMAD	55	16
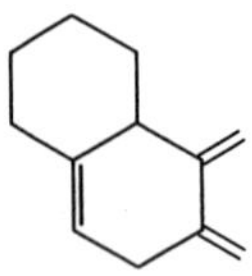	C1	1-Bromo-2-chloro-cyclopropene	57	46
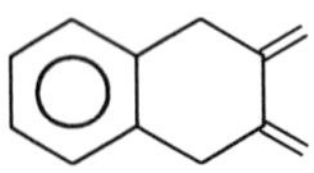	C1	1-Bromo-2-chloro-cyclopropene	76	47
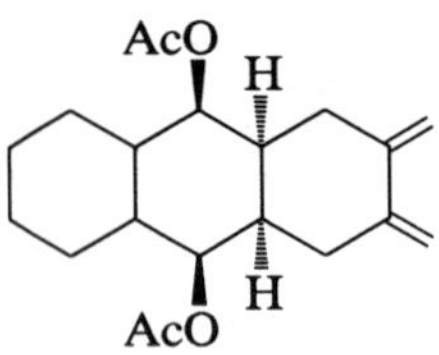	A1	MVK		9
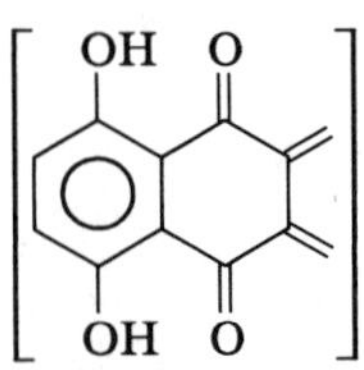	A1	MVK	80	48
	A2	Methyl methacrylate	62	48
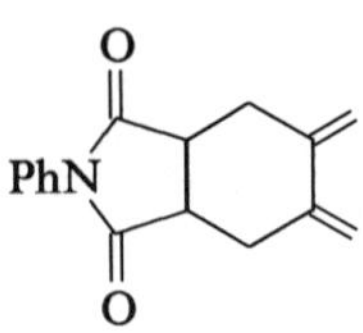	A1	MVK	59	9

TABLE 3.6 Diels–Alder Reactions of 1,2-Dimethylene Bridged Bicyclo Compounds

Diene		Dienophile		Yield (%)	Ref.
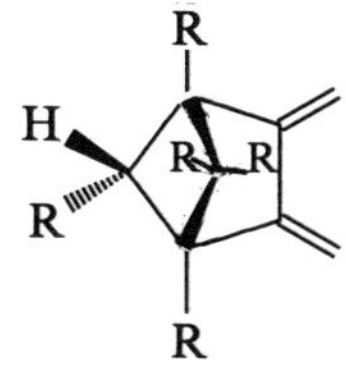	R = H	A12	TCNE (k)		81
	R = Me	A12	TCNE (k)		81
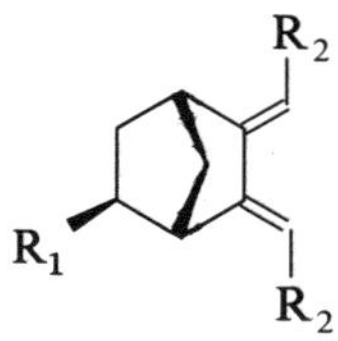	R_1 = H, R_2 = H	E1	MP	95	57, 58
	R_1 = H, R_2 = Me	C7	BQ	82	58
		D5	NPM	92	58
		I5	NMTAD	100	58
	R_1 = OH, R_2 = H	A1	MVK	89	57
		E1	MP	92	57
	R_1 = OAc, R_2 = H	A1	MVK	82	57
		E1	MP	88	57
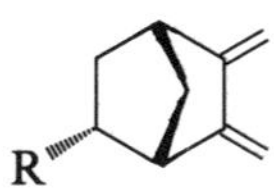	R = OH, OAc, OBs	E1	MP	92–97	59
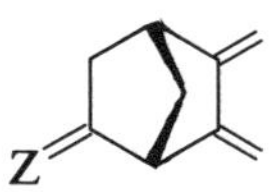	Z = CH_2	A12	TCNE	86	64
		E1	MP	100	64, 76
	Z = $C(CN)_2$	E1	MP	47	76
	Z = O	A1	MVK	85	57
		E1	MP	92–97	59, 76
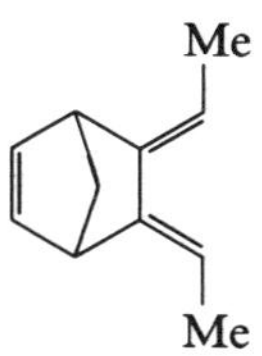		C7	BQ	85	58
		D5	NPM	100	58
		E3	DMAD	100	58
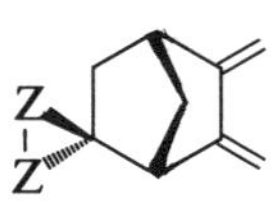 (Z = CH_2, 0)		E1	MP	90	60

TABLE 3.6 (*Continued*)

Diene		Dienophile		Yield (%)	Refs.
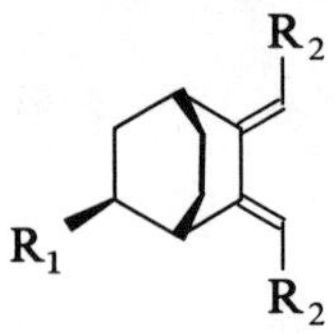	$R_1 = R_2 = H$	E1	MP	90	57
	$R_1 = OH, R_2 = H$	E1	MP	88	57
	$R_1 = OH, R_2 = Cl$	A12	TCNE	95	61
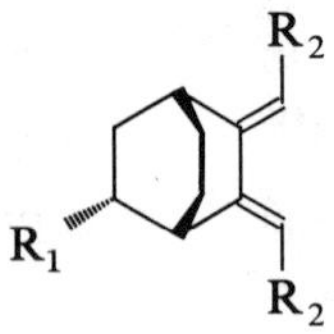	$R_1 = OH, R_2 = H$	E1	MP	87	57
	$R_1 = OH, R_2 = Cl$	A12	TCNE	> 95	61
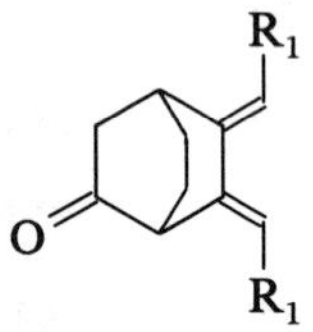	$R_1 = H$	E1	MP	90	57
	$R_1 = Cl$	A12	TCNE	> 95	61
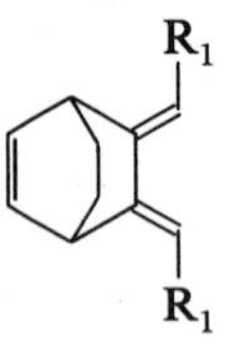	$R_1 = D$	A12	TCNE	95	62
		E3	DMAD	95	62
		I5	NPTAD	95	62
	$R_1 = Cl$	A12	TCNE	95	61, 63
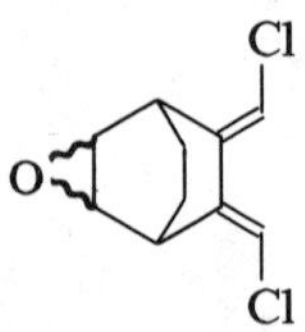	*Endo*; *exo*	A12	TCNE	95	61, 63

TABLE 3.7 Diels–Alder Reactions of 1,2-Heterodimethylenes

Diene		Dienophile		Yield (%)	Ref.
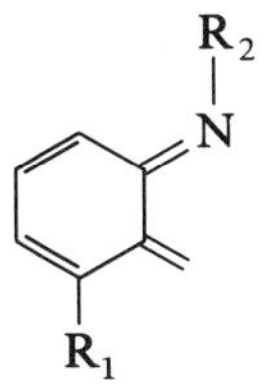	R_1 = H, R_2 = Et	A7	DMF	83	49
		A7	DMM	40	49
	R_1 = H, R_2 = Bn	A7	DMF	85	49
		A7	DMM	50	49
	R_1 = OMe, R_2 = Pr	A7	DMF	80	49
		A7	DMM	61	49
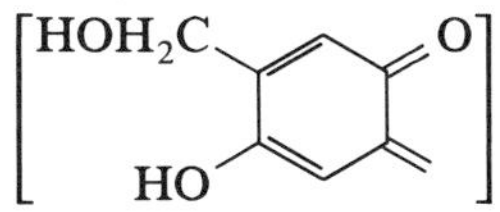		A8	Styrene	49	50
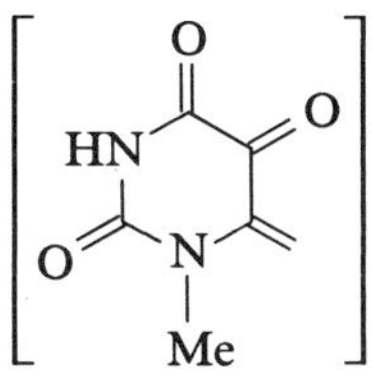		A9	5-Oxo-6-methylene-pyrimidine	80	51
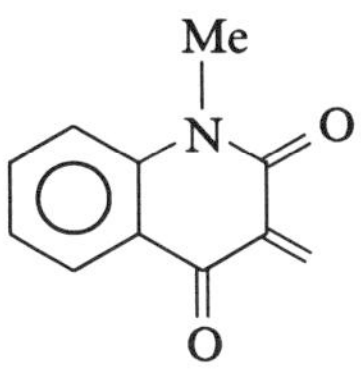		D1	*N*-Methylflindersine		52
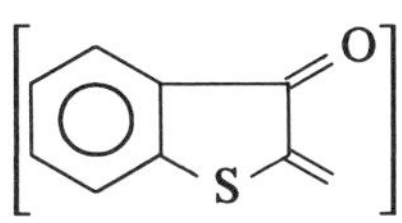		A12	2-Methylene-benzo[*b*]thiophen-3(2*H*)-one		53
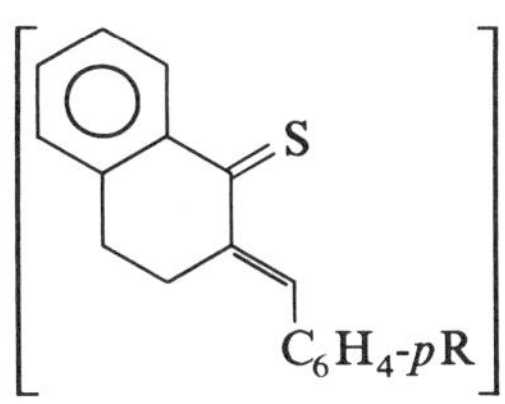 (R = H, OMe, Cl)		A8	ACN	68–90	54
		A8	Styrene	72–77	54
		C2	Norbornene	27–53	54
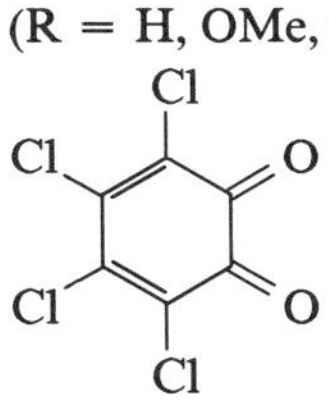		D4	1-Ethoxycarbonyl-1*H*-azepine	8	55
		D8	2-(*N*-Benzyl-*N*-methylamino) oxazole	87	55

TABLE 3.8 Diels–Alder Reactions of 2,3-Dimethylene-7-oxanorbornanes

Diene	Dienophile		Yield (%)	Ref.
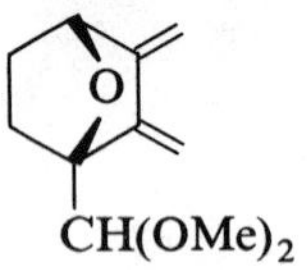	A1	MVK	90–95	70
	12	TCNE	80	71
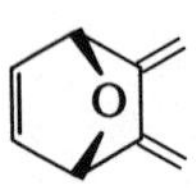	E3	DMAD	75–90	72
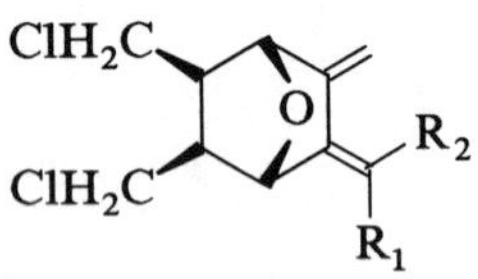				
R_1 = H, R_2 = OMe	A12	TCNE (k)	90	73
R_1 = H, R_2 = Cl	A12	TCNE (k)	10	73
R_1 = Cl, R_2 = H	A12	TCNE (k)	90	73
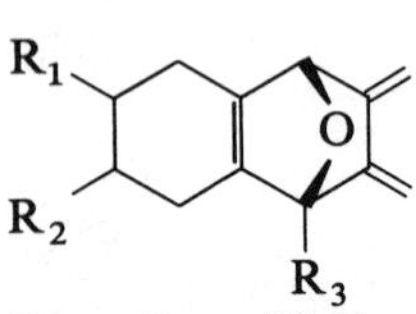				
$R_2 = R_3$ = H, R_1 = Ac	E6	[Benzyne]	50	75
$R_1 = R_2 = (CN)_2$, $R_3 = CH(OMe)_2$	A12	TCNE		74
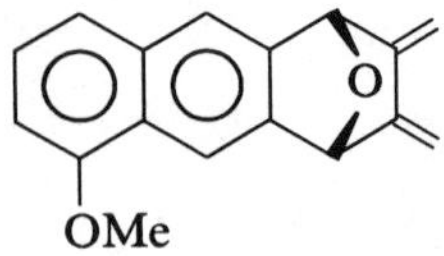	A1	MVK		76
	E1	MP		76

TABLE 3.8 (*Continued*)

Diene	Dienophile		Yield (%)	Ref.
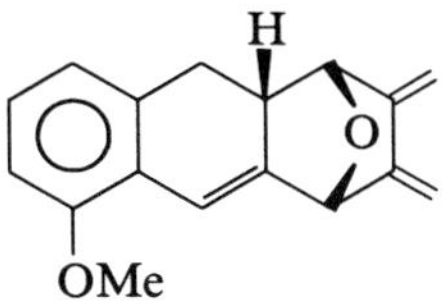	A1	MVK	80	76
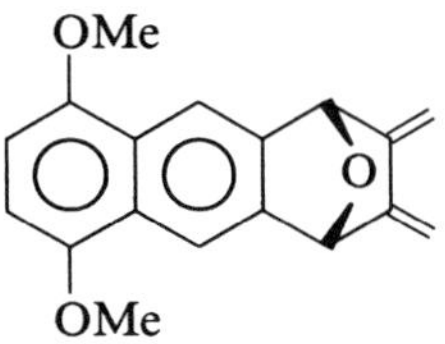	A1	MVK	93	77
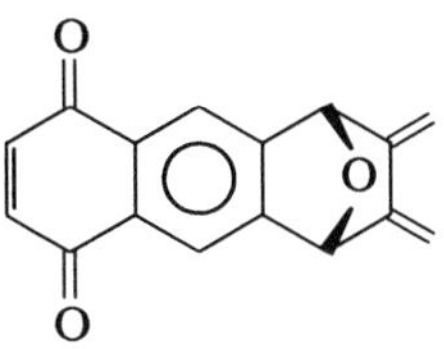	C7 C9	BQ NQ	15 39	65 65
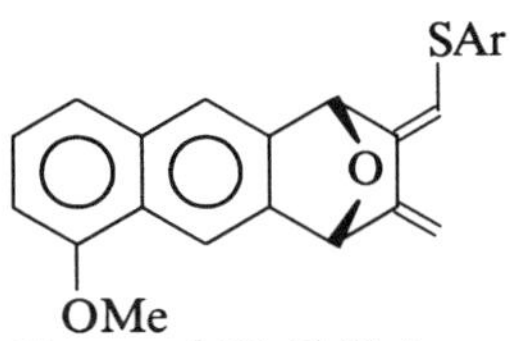 (Ar = o-$NO_2C_6H_4$)	A1	MVK	88	78
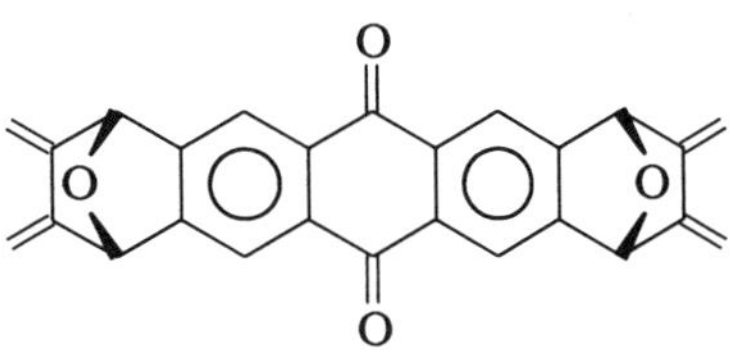	C7 C9	BQ NQ	62 59	65 65

TABLE 3.9 Diels–Alder Reactions of Tetramethylene Cycloalkanes

Diene	Dienophile		Yield (%)	Ref.
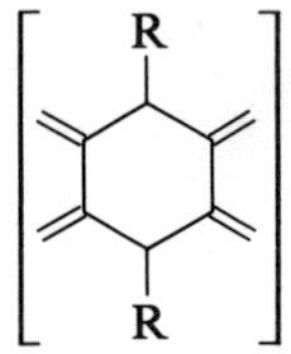 (R = OMe, OAc)	C7	2,3-Dimethylbenzoquinone	18–21	65
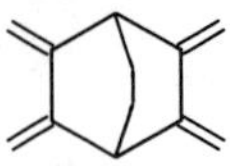	A12	TCNE (k)	80	66
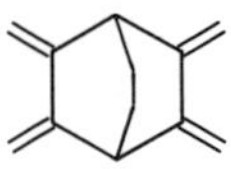	A12	TCNE	80	67
	E3	DMAD (k)		67
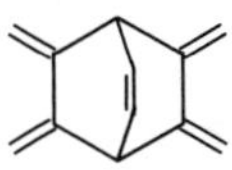	A12	TCNE	95	67
	E3	DMAD (k)		67
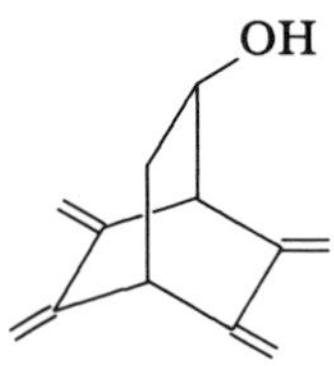	A12	TCNE	93	68
	E1	MP	71	68
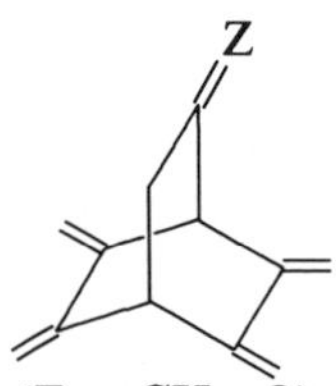 (Z = CH_2, O)	A12	TCNE	85–93	64, 68
	E1	MP	35–70	64, 68
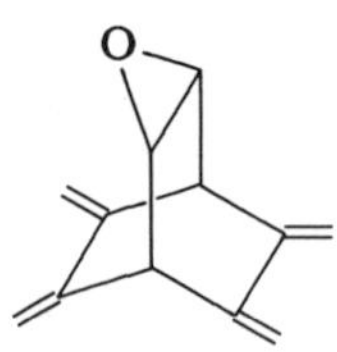	A12	TCNE	100	69

TABLE 3.10 Diels–Alder Reactions of Tri- and Tetramethylene-7-oxanorbornanes

Diene		Dienophile		Yield (%)	Ref.
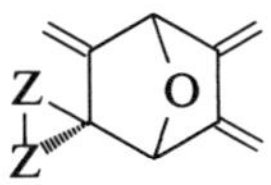	Z = CH_2, O	E1	MP	90	60
		E1	Butynone	89–93	60
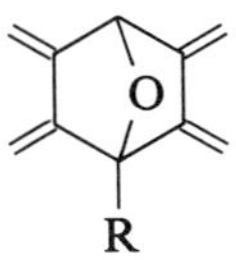	R = H	A12	TCNE (k)	94	66
		C6	[Benzyne]	26	75
		C7	BQ	22–91	65, 79
		C9	NQ	82	65, 79
	R = $CH(OMe)_2$	A12	TCNE	70	71, 74
		E1	Butynone	83	71, 74
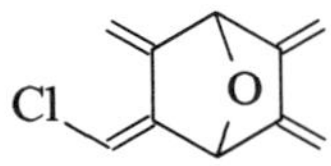		A12	TCNE	85	73
Ar = o-$NO_2C_6H_4$ 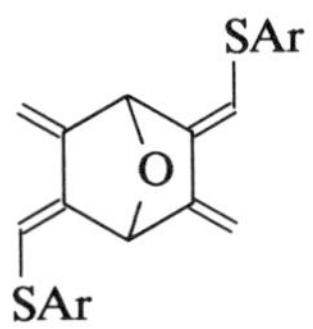		A1	MVK	95	80
		A1	MAC	94	80
		C6	[Methoxybenzyne]	46	78
		E1	Butynone	84	80

TABLE 3.11 Diels–Alder Reactions of Miscellaneous Outer-Ring Dienes

Diene	Dienophile		Yield (%)	Ref.
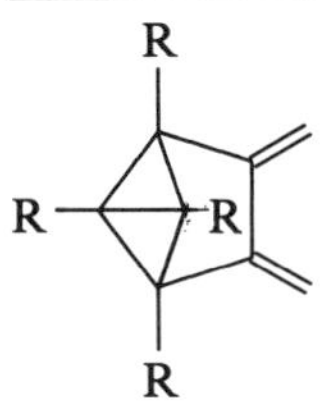 (R = H, Me)	A12	TCNE (k)	70	81
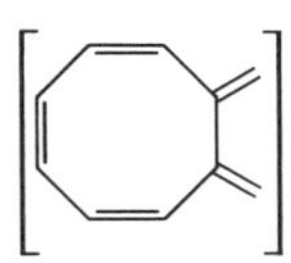	C3	[Benzocyclobutadiene]	31–40	82
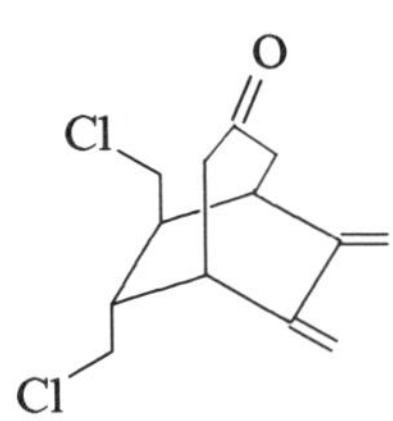	A1	MVK	69	83

TABLE 3.11 (*Continued*)

Diene		Dienophile	Yield (%)	Ref.
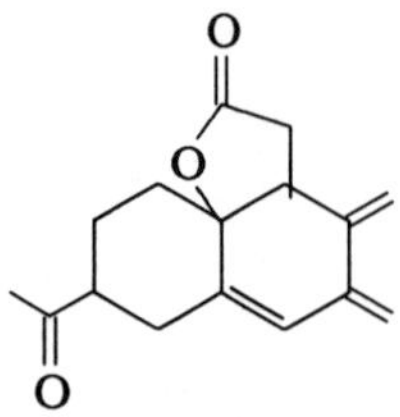	C6	[2,3-Didehydroanisole]	35	83
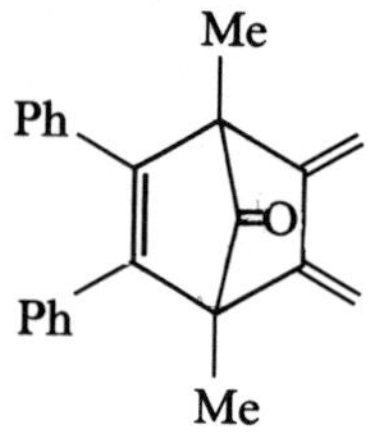	C1	[Dewar benzene]		84
	I5	*N-p*-Tolyltriazolinedione		84
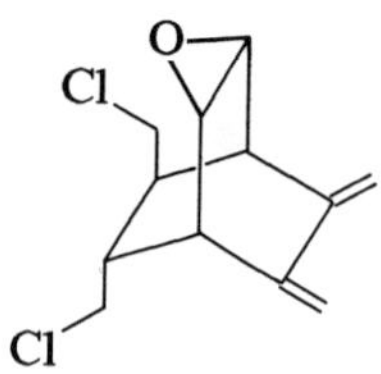	A12	TCNE	96	69
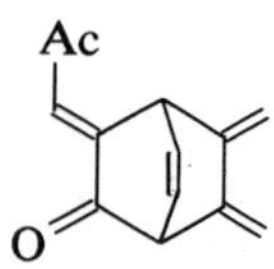	E1	MP	53	85
	E3	DMAD	85	85
	I5	NPTAD	85	85
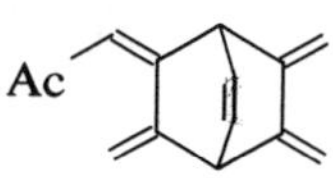	E1	MP	> 25	85
	E3	DMAD	85	85
	I5	NPTAD	85	85
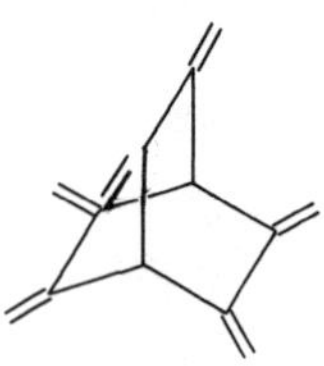	A12	TCNE (k)	92	86
	E3	DMAD(k)	83	67, 86
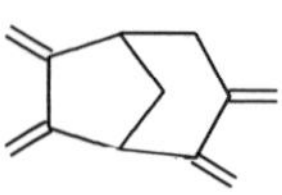	A12	TCNE	93	87

REFERENCES

1. H.-D. Scharf, H. Plum, J. Fleischhauer, and W. Schleker, *Chem. Ber.*, **112**, 862 (1979).
2. W. R. Dolbier, Jr., and C. R. Burkholder, *J. Org. Chem.*, **49**, 2381 (1984).
3. a. T. Minami, Y. Taniguchi, and I. Hirao, *J. Chem. Soc. Chem. Commun.*, 1046 (1984).
   b. T. Minami, N. Harui, and Y. Taniguchi, *J. Org. Chem.*, **51**, 3572 (1986).
4. H.-D. Martin and B. Mayer, *Tetrahedron Lett.*, 2351 (1979).
5. W. A. Nugent and J. C. Calabrese, *J. Am. Chem. Soc.*, **106**, 6422 (1984).
6. L. S. Hegedus, N. Kambe, Y. Ishii, and A. Mori, *J. Org. Chem.*, **50**, 2240 (1985).
7. N. Oda, K. Kobayoshi, T. Ueda, and I. Ito, *Heterocycles*, **15**, 857 (1981).
8. N. Oda, S.-I. Nagai, and I. Ito, *Chem. Pharm. Bull.*, **27**, 2229 (1979).
9. B. M. Trost and M. Shimizu, *J. Am. Chem. Soc.*, **104**, 4299 (1982).
10. M. P. Johnson and C. J. Moody, *J. Chem. Soc. Perkin Trans. 1*, 71 (1985).
11. A. J. Barker, M. J. Begley, A. M. Birch, and G. Pattenden, *J. Chem. Soc. Perkin Trans. 1*, 1919 (1983).
12. P. Muller and H. C. Nguyen Thi, *Isr. J. Chem.*, **21**, 135 (1981).
13. J. P. Gesson, J. C. Jacquesy, and M. Mondon, *Tetrahedron Lett.*, **21**, 3351 (1980).
14. R. K. Boeckman, Jr., M. H. Delton, T. M. Dolak, T. Watanabe, and M. D. Glick, *J. Org. Chem.*, **44**, 4396 (1979).
15. Y. Kanao and M. Oda, *Bull. Chem. Soc. Jpn.*, **57**, 615 (1984).
16. R. O. Angus, Jr., and R. P. Johnson, *J. Org. Chem.*, **48**, 273 (1983).
17. T. Tuschka, K. Naito, and B. Rickborn, *J. Org. Chem.*, **48**, 70 (1983).
18. S. Chew and R. J. Ferrier, *J. Chem. Soc. Chem. Commun.*, 911 (1984).
19. T. Kametani, T. Honda, Y. Ebisawa, and H. Ichikawa, *Tetrahedron*, **41**, 3643 (1985).
20. T. Kohei, Y. Jun-Ichi, A. Munetaka, S. Yoshihiro, I. Takahisa, and K. Makoto, *Bull. Chem. Soc. Jpn.*, **55**, 255 (1982).
21. W. H. Hersh and R. G. Bergman, *J. Am. Chem. Soc.*, **105**, 5846 (1983).
22. M. J. Broadhurst, C. H. Hassall, and G. J. Thomas, *J. Chem. Soc. Chem. Commun.*, 158 (1982).
23. L. A. Levy and L. Pruitt, *J. Chem. Soc. Chem. Commun.*, 227 (1980).
24. R. J. Ardecky, F. A. J. Kerdesky, and M. P. Cava, *J. Org. Chem.*, **46**, 1483 (1981).
25. J. Afzal, M. Vairamani, B. G. Hazra, and K. G. Das, *Synth. Commun.*, **10**, 843 (1980).
26. L. A. Levy, *Synth. Commun.*, **13**, 639 (1983).
27. G. M. Rubotton and J. E. Wey, *Synth. Commun.*, **14**, 507 (1984).
28. R. W. Franck and T. V. John, *J. Org. Chem.*, **45**, 1170 (1980).
29. M. C. Carreno, F. Farina, A. Galan, and J. L. Garcia Ruano, *J. Chem. Res. (S)*, 296 (1979).
30. M. C. Carreno, F. Farina, J. L. Garcia Ruano, and L. Puebla, *J. Chem. Res. (S)*, 288 (1984).

31. a. J. L. Charlton, *Tetrahedron Lett.*, **26**, 3413 (1985).
    b. E. Charlton, *Can. J. Chem.*, **64**, 720 (1986).
32. J. W. Barton, M. K. Shepherd, and R. J. Willis, *J. Chem. Soc. Perkin Trans. 1*, 967 (1986).
33. Y. Ito, Y. Amino, M. Nakatsuka, and T. Saegusa, *J. Am. Chem. Soc.*, **105**, 1586 (1983).
34. a. T. Durst, E. C. Kozma, and J. L. Charlton, *J. Org. Chem.*, **50**, 4829 (1985).
    b. J. L. Charlton, M. M. Alauddin, and G. H. Peuner, *Can. J. Chem.*, **64**, 793 (1986).
35. J. M. Hornback and R. D. Barrows, *J. Org. Chem.*, **47**, 4285 (1982).
36. R. W. Franck, T. V. John, K. Olejniczak, and J. F. Blount, *J. Am. Chem. Soc.*, **104**, 1106 (1982).
37. M. Pfau, S. Combrisson, J. E. Rowe, Jr., and N. D. Heindel, *Tetrahedron*, **34**, 3459 (1978).
38. M. Pfau, J. E. Rowe, Jr., and N. D. Heindel, *Tetrahedron*, **34**, 3469 (1978).
39. F. Farina, J. Primo, and T. Torres, *Chem. Lett.*, 77 (1980).
40. A. W. Murray, N. D. Murray, and R. G. Reid, *J. Chem. Soc. Chem. Commun.*, 1230 (1986).
41. H. Ahlbrecht, M. Dietz, and W. Raab, *Synthesis*, 231 (1983).
42. S. Senda, T. Asao, I. Sugiyama, and K. Hirota, *Tetrahedron Lett.*, **21**, 531 (1980).
43. B. M. Trost, W. C. Vladuchick, and A. J. Bridges, *J. Am. Chem. Soc.*, **102**, 3554 (1980).
44. L. K. Bee and P. J. Garratt, *J. Chem. Res. (S)*, 368 (1981).
45. B. M. Trost and R. Remuson, *Tetrahedron Lett.*, **24**, 1129 (1983).
46. W. E. Billups, L.-J. Lin, B. E. Arney, Jr., W. A. Rodin, and E. W. Casserly, *Tetrahedron Lett.*, **25**, 3935 (1984).
47. W. E. Billups, E. W. Casserly, and B. E. Arney, Jr., *J. Am. Chem. Soc.*, **106**, 440 (1984).
48. T. Watabe and M. Oda, *Chem. Lett.*, 1791 (1984).
49. Y. Ito, E. Nakajo, and T. Saegusa, *Synth. Commun.*, **16**, 1073 (1986).
50. P. Lanteri, R. Longeray, and J. Royer, *J. Chem. Res. (S)*, 168 (1981).
51. I. M. Sasson, R. P. Gagnier, and B. A. Otter, *J. Heterocycl. Chem.*, **20**, 753 (1983).
52. M. F. Grundon, V. M. Ramachandran, and B. M. Sloan, *Tetrahedron Lett.*, **22**, 3105 (1981).
53. Y. Tamura, Y. Takebe, S. M. Bayomi, C. Mukai, M. Ikeda, M. Murase, and M. Kise, *J. Chem. Soc. Perkin Trans. 1*, 1037 (1981).
54. a. H. Tanaka, S. Kametani, T. Saito, and S. Motoki, *Bull. Chem. Soc. Jpn.*, **58**, 667 (1985).
    b. T. Karakasa and S. Motoki, *J. Org. Chem.*, **43**, 4147 (1978).
55. a. A. Dondoni, M. Fogagnolo, A. Mastellari, P. Pedrini, and F. Ugozzoli, *Tetrahedron Lett.*, **27**, 3915 (1986).
    b. K. Saito, T. Mukai, and S. Iida, *Bull. Chem. Soc. Jpn.*, **59**, 2485 (1986).
56. T. M. Stevenson and N. J. Leonard, *J. Org. Chem.*, **49**, 2158 (1984).
57. M. Avenati, P. A. Carrupt, D. Quarroz, and P. Vogel, *Helv. Chim. Acta*, **65**, 188 (1982).

58. L. A. Paquette, A. G. Shaefer, and J. F. Blount, *J. Am. Chem. Soc.*, **105**, 3642 (1983).
59. P. A. Carrupt, M. Avenati, D. Quarroz, and P. Vogel, *Tetrahedron Lett.*, 4413 (1978).
60. C. Mahaim, L. Schwager, P.-A. Carrupt, and P. Vogel, *Tetrahedron Lett.*, **24**, 3603 (1983).
61. M. Avenati and P. Vogel, *Helv. Chim. Acta*, **65**, 204 (1982).
62. M. Avenati and P. Vogel, *Helv. Chim. Acta*, **66**, 1279 (1983).
63. M. Avenati, J.-P. Hagenbuch, C. Mahaim, and P. Vogel, *Tetrahedron Lett.*, **21**, 3167 (1980).
64. G. Burnier, L. Schwager, and P. Vogel, *Helv. Chim. Acta*, **69**, 1310 (1986).
65. A. D. Thomas and L. L. Miller, *J. Org. Chem.*, **51**, 4160 (1986).
66. O. Pilet and P. Vogel, *Helv. Chim. Acta*, **64**, 2563 (1981).
67. O. Pilet, J.-L. Birbaum, and P. Vogel, *Helv. Chim. Acta*, **66**, 19 (1983).
68. R. Gabioud and P. Vogel, *Helv. Chim. Acta*, **66**, 1134 (1983).
69. R. Gabioud and P. Vogel, *Tetrahedron*, **36**, 149 (1980).
70. J.-L. Metral and P. Vogel, *Tetrahedron Lett.*, **25**, 5387 (1984).
71. J.-L. Metral, J. Lauterwein, and P. Vogel, *Helv. Chim. Acta*, **69**, 1287 (1986).
72. P. Vioget, M. Bonivento, R. Roulet, and P. Vogel, *Helv. Chim. Acta*, **67**, 1630 (1984).
73. C. Mahaim and P. Vogel, *Helv. Chim. Acta*, **65**, 866 (1982).
74. J.-L. Metral and P. Vogel, *Helv. Chim. Acta*, **68**, 334 (1985).
75. Y. Bessiere and P. Vogel, *Helv. Chim. Acta*, **63**, 232 (1980).
76. J. Tamariz, L. Schwager, J. H. A. Stibbard, and P. Vogel, *Tetrahedron Lett.*, **24**, 1497 (1983).
77. P.-A. Carrupt and P. Vogel, *Tetrahedron Lett.*, 4533 (1979).
78. J.-M. Tornare and P. Vogel, *Helv. Chim. Acta*, **68**, 1069 (1985).
79. L. L. Miller, A. D. Thomas, C. L. Wilkins, and D. A. Weil, *J. Chem. Soc. Chem. Commun.*, 661 (1986).
80. J.-M. Tornare and P. Vogel, *J. Org. Chem.*, **49**, 2510 (1984).
81. F. Lanzendorfer and M. Christl, *Angew. Chem.* (*Int. Ed. Engl.*), **22**, 871 (1983).
82. E. Vogel, H. W. Engels, S. Matsumoto, and J. Lex, *Tetrahedron Lett.*, **23**, 1797 (1982).
83. B. Demarchi and P. Vogel, *Tetrahedron Lett.*, **28**, 2239 (1987).
84. R. N. Warrener, R. A. Russell, and I. G. Pitt, *J. Chem. Soc. Chem. Commun.*, 1675 (1984).
85. E. Tagliaferri, U. Hanisch, R. Roulet, P. Vogel, and K. J. Schenk, *Helv. Chim. Acta*, **68**, 1362 (1985).
86. U. Hanisch, E. Tagliaferri, R. Roulet, and P. Vogel, *Helv. Chim. Acta*, **66**, 2182 (1983).
87. R. Gabioud and P. Vogel, *J. Org. Chem.*, **51**, 2385 (1986).
88. R. Sustmann, M. Böhm, and J. Sauer, *Chem. Ber.*, 883 (1979).
89. J. R. Wiseman, J. J. Penderi, C. A. Otto, and K. G. Chiong, *J. Org. Chem.*, **45**, 516 (1980).

4

INNER–OUTER-RING DIENES

Reactivity, Selectivity, Synthesis The behavior of inner–outer-ring dienes in Diels–Alder reactions depends on the type of ring (carbocyclic, heterocyclic, cyclodienic, aromatic) that bears the vinyl group.

The regioselectivity of cycloadditions of 1-vinylcycloalkenes **1** is influenced by the strain of the ring, the presence of alkyl substituents, and the nature of dienophile. In the reaction of **1** ($R = R_1 = H$) with MP the amount of regioisomer **2** increases as the ring size increases (i.e., as the ring strain decreases): $n = 0$, **2/3** = 0.98; $n = 1$, **2/3** = 1.39; $n = 2$, **2/3** = 2.06.[8] The presence of a geminal dimethyl group (**1**, $n = 2$, $R = H$, $R_1 = Me$) increases the *ortho* isomer **2** (**2/3** = 4), but the presence of another methyl group (**1**, $n = 2$, $R = R_1 = Me$) makes the reaction completely regiounselective.[27] Cycloadditions of 1-vinylcyclohexenes with dienophiles of ethylenic nature such as nitrocrotonates are poorly regioselective.[27] On the contrary, complete *ortho* regioselectivity is observed in the reactions of **1** ($n = 2$, $R = R_1 = H$) with 3- and 4-carbomethoxy-1,2-benzoquinones. These cycloadditions are also highly *endo*-diastereoselective and the adducts **4** and **5** are useful intermediates in the synthesis of diterpenes.[15]

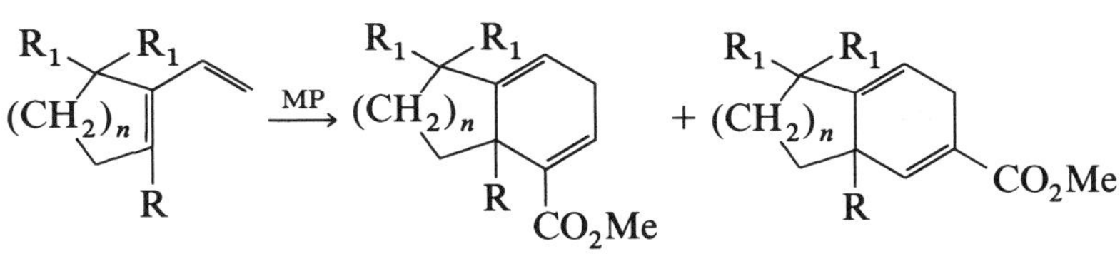

1 **2** **3**

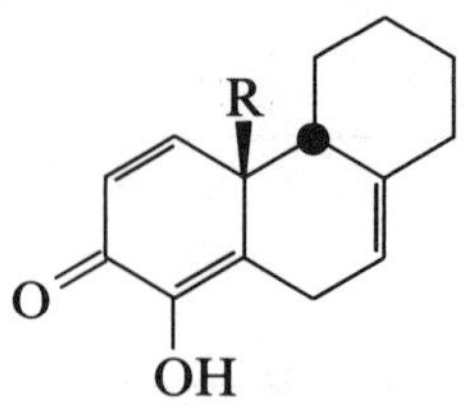

4 R = CO_2Me

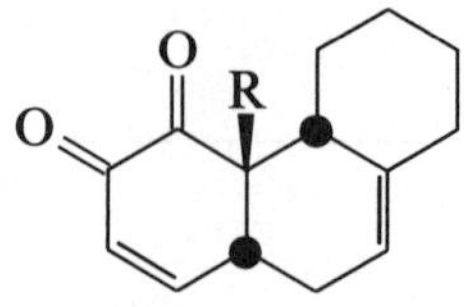

5 R = CO_2Me

The 3-, 4-, and 6-substituted vinylcyclohexenes **6**, **8**, and **11** are "flexible" dienes that allow evaluations of the facial selectivity of conformational isomers. 3- and 6-acetoxy dienes **6** and **11** with DMAD give a single adduct **7** and nearly a 1 : 1 mixture of diastereoisomers **12** and **13**, respectively.[5] The 4-methylderivative **8** reacts with MA, affording a 2 : 1 mixture of **9** and **10**.[26] The results can be explained on the basis of an energetically favored antiparallel attack.[61] Thus the antiparallel *endo-syn* attack on the less abundant but more reactive[26] axial conformer **15** affords **10**, and the antiparallel *endo-anti* attack on the more abundant but less reactive **14** produces the predominant diastereoisomer **9**. The observed ratio of the adducts is a function of the reactivity of conformers and their conformational equilibrium.

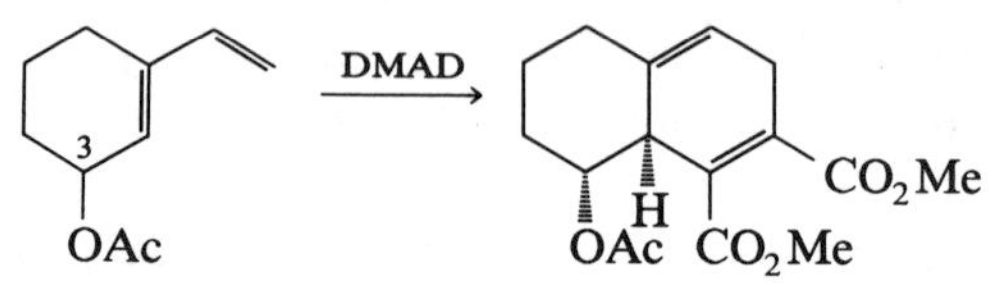

6 **7**

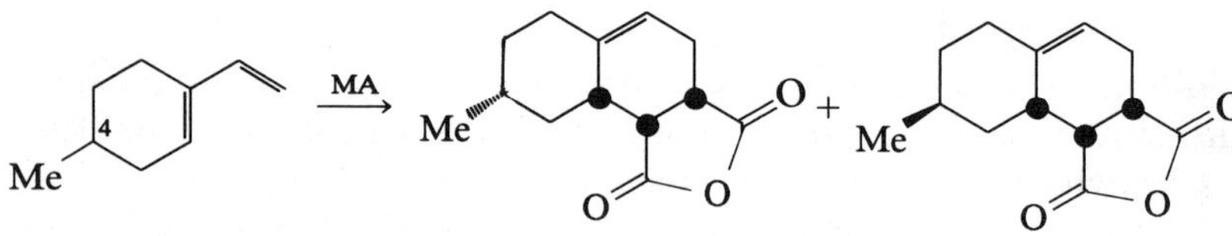

8 **10** **9**

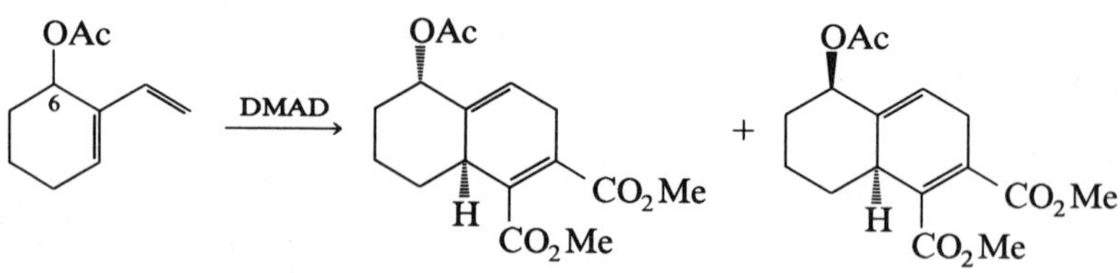

11 **12** **13**

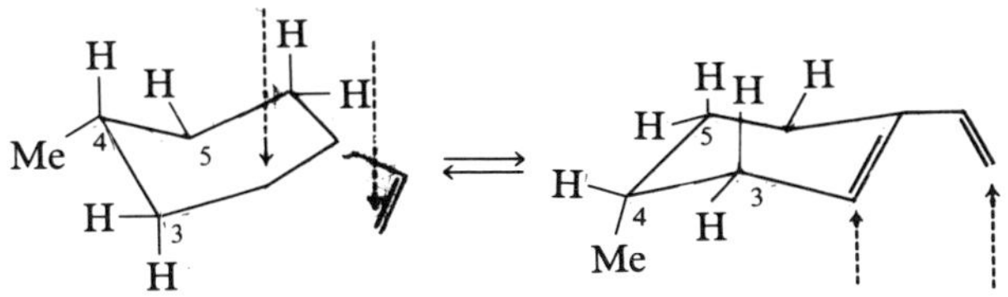

14 15

Introduction of a vinyl group on a cyclodiene system increases the possible sites for a Diels–Alder reaction and can influence the reactivity of the parent diene. The 1- and 2-vinylcyclopentadienes **16** and **18** have two sites for Diels–Alder reactions, and the diene system including the vinyl double bond is the less reactive. The cycloadditions therefore primarily involve the cyclopentadiene moiety. This is evident in the reactions of **16** and **18** with DMAD.[62] The former gives only monoadduct **17**, while its valence isomer **18** affords the bisadduct **19**.

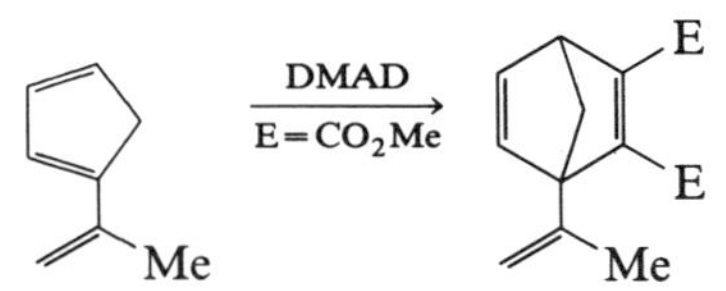

16 17

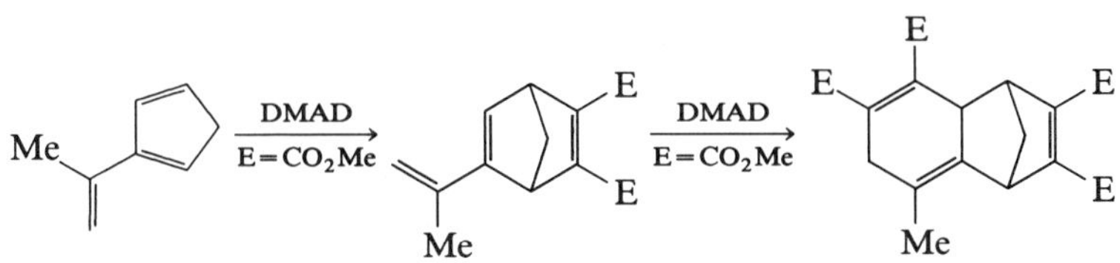

18 19

A vinyl-substituted cycloheptatriene–norcaradiene system has on the whole four or five sites for the Diels–Alder reaction.[60] The 1- and 2-vinyl compounds **20** and **21** react exclusively at the diene part including the vinyl group of the cycloheptatriene form, but the 3-vinylderivative **22** reacts from the cycloheptatriene or norcaradiene form, depending on the electron affinity of dienophiles.[60] No bisadduct has been observed.

The [4 + 2] cycloadditions of styrenes **23** (R, R_1 = H, Me) with BQ, NQ, and DMAD require high temperature and long reaction times, and the yields are generally low. Styrenes **23** (R = H, Me; R_1 = H, Cl, NO_2, Bu^t) react with very reactive dienophiles such as NPTAD and NMTAD at room temperature and give bisadduct **24** and Diels–Alder-ene adducts **25**.[32]

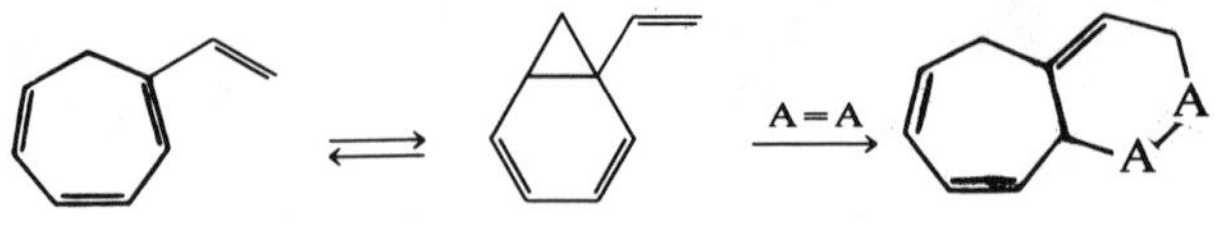

20

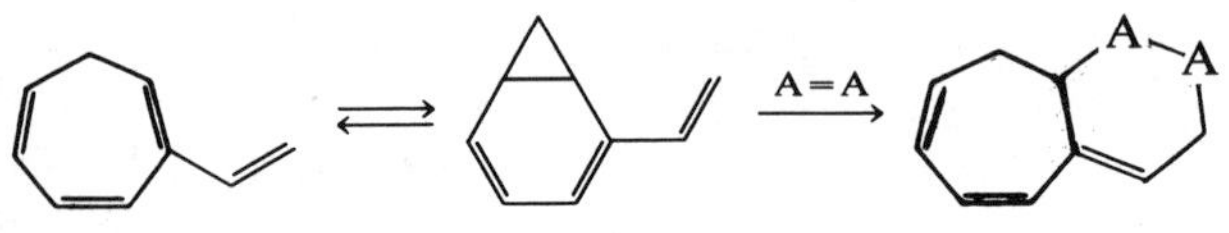

21

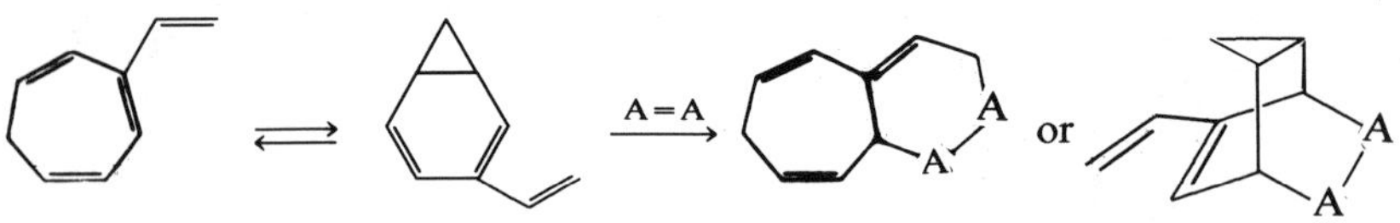

22

The styrenes **23** (R = H, R_1 = $OSiMe_3$) and **26** do not give Diels–Alder reactions with BQ and DMAD, but the protected dihydroxystyrene **27** reacts with some dienophiles, affording a simple method to prepare *peri*-hydroxylated polycyclic aromatic compounds **28**, which have a high biological interest.[43]

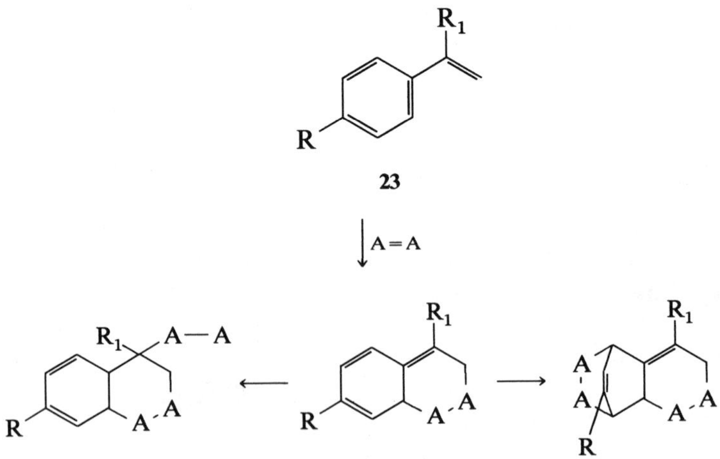

23

25 **24**

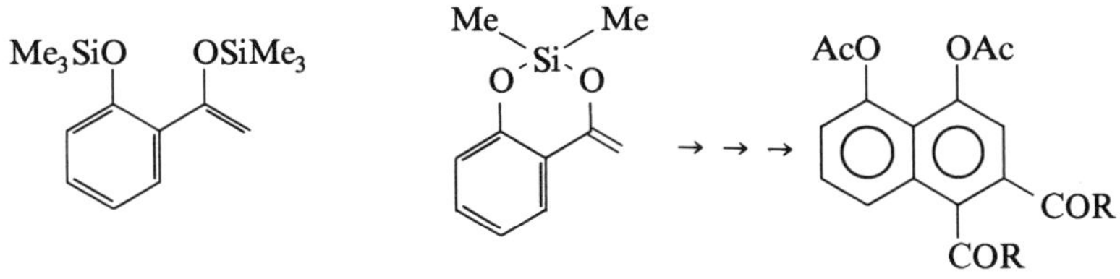

26 27 28 R = OMe, Ph

1-Vinylnaphthalenes react more easily than styrenes and have been used to synthesize steroidlike compounds. The cycloaddition of **29** with optically active substituted carbomethoxyvinylphosphine oxides is regiospecific but not diastereospecific. The diastereoisomers **30** are converted into optically pure 17-phosphasteroids **31**, which generally have specific pharmacological properties.[58]

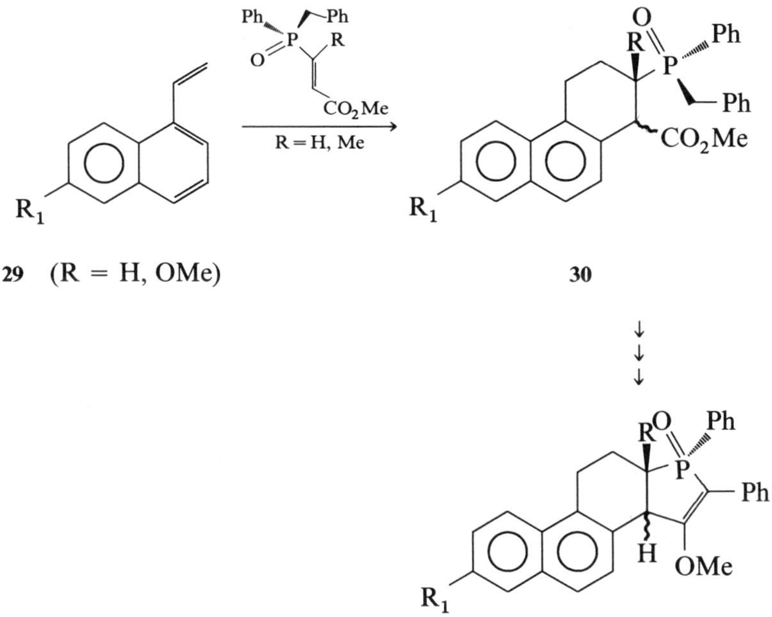

29 (R = H, OMe) **30**

31

The Diels–Alder reactions of vinylheterodienes involve synthetic, mechanistic, and theoretical aspects. 2-Vinylfuran **32** (X = O, R = H) and 2-vinylthiophen **32** (X = S, R = H) show a different behavior with respect to that of analogous cyclopentadienes. They react with MA to give adducts such as **33**. 2-Vinylfuran also reacts with DMAD and PM to give adducts that dehydrogenate easily to **34** and **35**.[75]

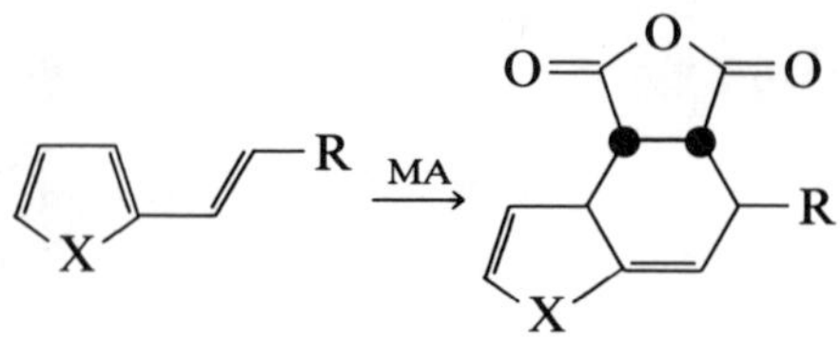

32 33 (R = H)

An electron-withdrawing substituent on the β-carbon of the vinyl group (**32**: X = O, R = CHO, COOH, NO_2; **32**: X = S, R = CO_2Me) prevents reactions with MA and DMAD. On the contrary, 2-vinylpyrroles having electron-withdrawing substituents (i.e., **37**: R = H, R_1 = CO_2Me; **37**: R = R_1 = CO_2Me) react with MP and DMAD to give fully dehydrogenated mono adducts. This result can be explained in terms of the greater electron-releasing ability of the nitrogen atom, with respect to oxygen and sulfur, in the heterocyclic ring.[75] 2-(1-Trimethylsilyloxyvinyl)furan **36** (X = O) is six times less reactive than the analogous thiophene derivative **36** (X = S) in the cycloaddition reaction with NPM, while furan is known to give [4 + 2] cycloadditions much more readily than thiophene.[65]

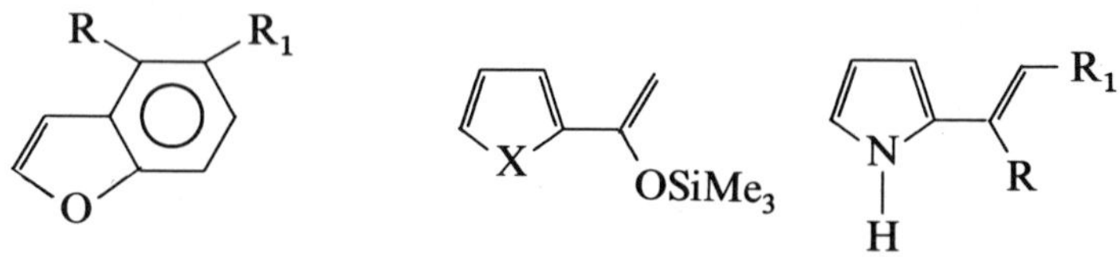

34 R = R_1 = CO_2Me
35 R = CO_2Me; R_1 = H

36 **37**

Sesquiterpenes α-methylene lactones have been synthesized starting from chiral vinyl dienes **38**, which react with MA *endo* selectively and *anti* with respect to the *O*-isopropylidene ring.[85] Diels–Alder reactions of isoxazoles (i.e., **39**) are of synthetic interest to functionalize cyclic compounds in view of the easy opening of the heterocyclic ring.[67]

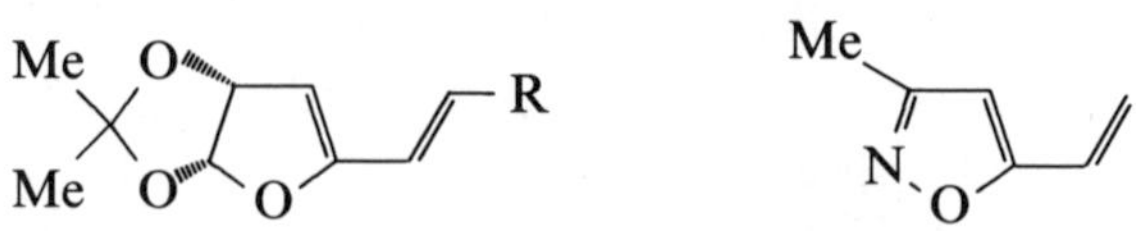

38 R = H, $CH_2SiMe_2Bu^t$ **39**

Tables of Diels–Alder Reactions See Tables 4.1 to 4.9; See also the general remarks about classifications, abbreviations, and symbols in the Introduction.

TABLE 4.1 Diels–Alder Reactions of Vinylcarbomonocycloalkenes

Diene		Dienophile	Yield (%)	Ref.
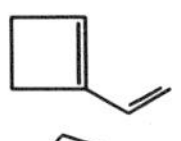	D3	MP	10–40	8
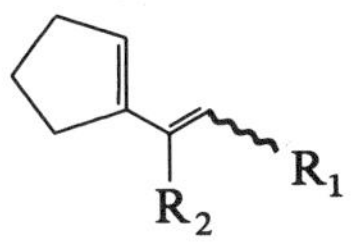				
R_1 **R_2**				
H H	E1	MP	10–40	8
H OAc	C4	2-Cyclopentenones	62–67	3
H $OSiMe_3$	C4	2-Cyclopentenones		3
H $OSiMe_2Bu^t$	C4	2-Carboxymethyl-2-cyclohexenones	20–62	4
NMe_2 H (*E*)	A1	MAC	81	2
	E1	MP	89	1
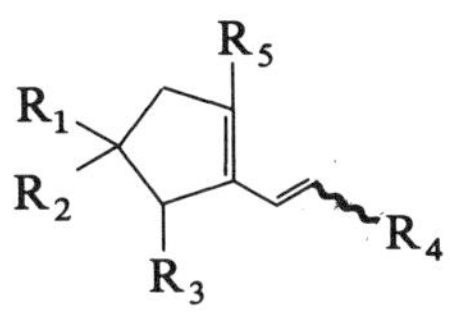				
$R_1 = R_2 = R_4 = R_5 = H$, $R_3 = OAc$	E3	DMAD	31	5
$R_1 = R_2 = R_3 = H$, $R_4 = NMe_2$, $R_5 = Me$ (*E*)	A1	MAC	82	2
$R_1 = R_2 = Me$, $R_3 = R_5 = H$, $R_4 = CH_2OH$ (*E*)	D3	Bromomethylmaleic anhydride	58	6
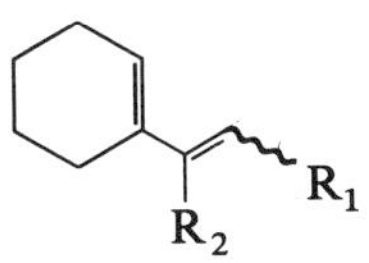				
R_1 **R_2**				
H H	C4	2-Methyl-2-cyclopentenone	70	11
	C4	2-Acetyl-4-methyl-4-cyclopenten-1,3-dione	71	12, 14
	C7	2-Carbomethoxy-1,4-benzoquinone	92	(7), 9
	C7	Carbomethoxy-1,2-benzoquinones	41–55	15
	C7	1,4-Benzoquinones		16

TABLE 4.1 (*Continued*)

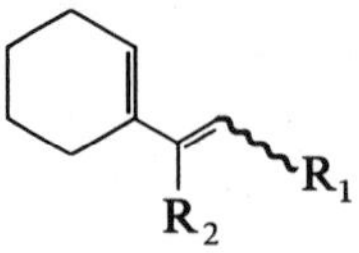

R_1	R_2		Dienophile	Yield (%)	Ref.
H	H	C10	6-Dehydrotestosterone acetate		18
		D3	Chloromethylmaleic anhydride	72	10, 17
		D3	Citraconic anhydride	47	12, 17
		D3	Aconitic anhydride	26	17
		E1	MP	10–40	8
		I4	[3-Pyrazolones]		13
H	OAc	C7	Methoxycarbonylbenzoquinones	70–95	19
H	$OSiMe_3$	A10	2-Nitrostyrene	80	23
		C4	Bridgehead enones	52–97	24
		F7	Benzylideneaniline	50–72	22
H	$OSiMe_2Bu^t$	C4	2-Methyl-2-cyclopentenone	64	11
		D3	MA	82	21
NMe_2	H	A1	MAC	83	2
		E1	MP	76	1
OAc	H	A1	Ethyl acrylate	64	20
		C7	Benzoquinones	20–75	19
		D3	MA	72	20
		D5	NPM	91	20
SPh	OMe (*E*)	A1	MVK	80	25
		C4	2-Cyclohexenone	85	25

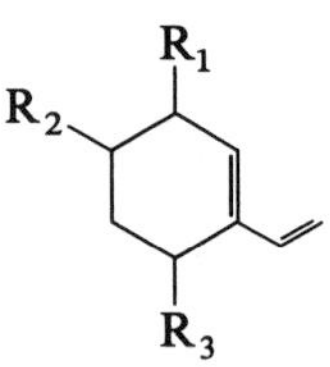

R_1	R_2	R_3		Dienophile	Yield (%)	Ref.
H	H	Ac	E3	DMAD	42	5
H	Me	H	A12	TCNE	68	26
			D3	MA		26
H	Bu^t	H	D3	MA	41	26
OAc	H	H	E3	DMAD	30	5

TABLE 4.1 (*Continued*)

Diene			Dienophile		Yield (%)	Ref.
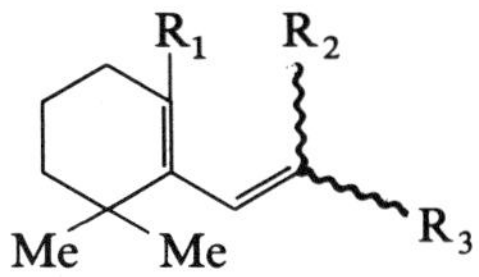						
R_1	**R_2**	**R_3**				
H	H	H	A2	(*E*)-Ethyl-*β*-nitroacrylate	78	27
			C6	3-Isopropyl-*o*-benzoquinone	28–30	28
			E1	Ethyl propiolate	80	27
			E3	DMAD	95	27
H	H	NMe_2 (*E*)	A1	MAC	76	2
			E1	MP	82	1
Me	H	H	A2	(*E*)-Ethyl-*β*-nitroacrylate	66	27
			E1	Ethyl propiolate	55	27
			E3	DMAD	87–94	27, 29, 31
Me	NMe_2	H (*E*)	A1	MAC	84	2
Me	$N(CH_2)_4$	H (*E*)	A7	DMF	95	30
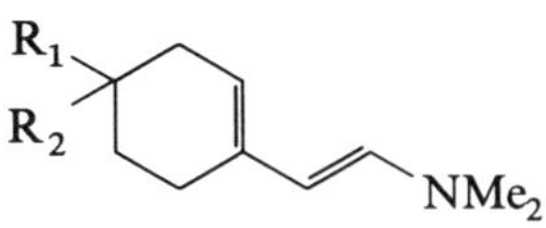						
R_1	**R_2**					
H	Bu^t		E1	MP	81	1
Me	Me		A1	MAC	84	1
			E1	MP	83	1
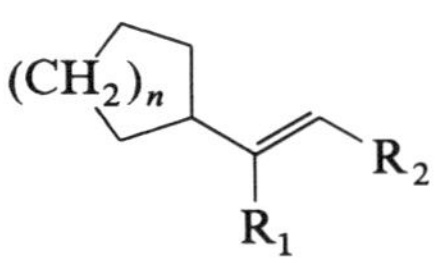						
R_1	**R_2**					
H (n = 3, 8)	NMe_2		D3	MA	86	2
			E1	MP	52–78	1
CH_2SO_2Ph (n = 1, 2, 3, 8)	H		E3	DMAD	44–100	44

TABLE 4.2 Diels–Alder Reactions of Styrenes

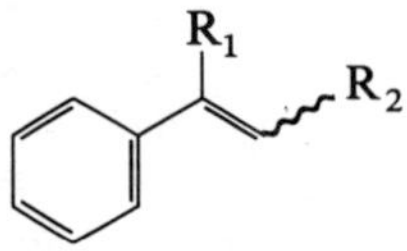

R_1	R_2			Dienophile	Yield (%)	Ref.
H	H		A12	TCNE (k)		36
			C7	BQ	31	35
			C9	1,4-Naphthoquinones	48–58	33
			D3	Carbomethoxymaleic anhydride		34
			I5	NPTAD		32
H	Ph (*E*, *Z*)		A12	TCNE (k)		36
Me	H		A12	TCNE (k)		36
Ph	H		A12	TCNE (k)		36
			D3	Carbomethoxymaleic anhydride	20	34
			D3	MA		37

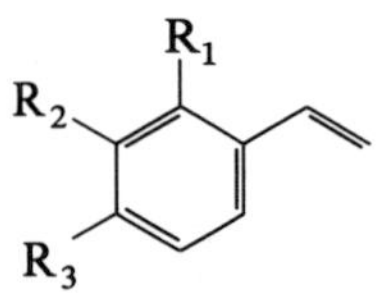

R_1	R_2	R_3		Dienophile	Yield (%)	Ref.
H	H	Me	A12	TCNE (k)		39
H	H	$CH{=}CH_2$	D3	MA		38
H	H	OMe	C7	BQ	42	35
			D3	Carbomethoxymaleic anhydride	67	34
H	H	Cl	A12	TCNE (k)		39
H	H	Br	A12	TCNE (k)		39
H	OMe	H	C7	BQ	26	35
			C9	NQ	40	40
H	OMe	OMe	C7	BQ	31	35
OMe	H	H	C7	BQ	50	35
OMe	OMe	H	C7	BQ	63	35
			C9	NQ	27	40

TABLE 4.2 (*Continued*)

Diene				Dienophile	Yield (%)	Ref.
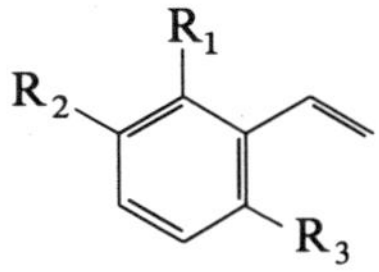						
R_1	**R_2**	**R_3**				
H	OMe	OMe	C7	BQ	37	35
Cl	H	Cl	I5	NPTAD		32
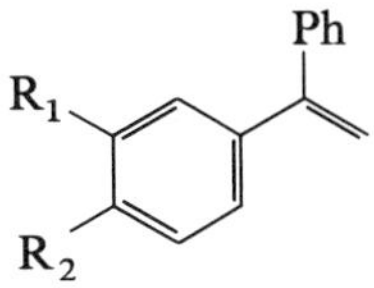						
R_1	**R_2**					
H	$CH{=}CH_2$		D3	MA		37
H	$CMe{=}CH_2$		D3	MA		37
OH	OMe		C7	2-Methyl-1,4-benzoquinone	40	41
OMe	OMe		C7	2-Methoxy-1,4-benzoquinone	45	41
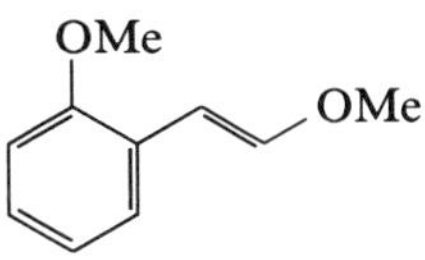			I3	[Singlet oxygen]	100	42
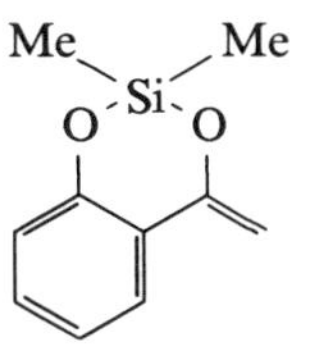			C9	NQ	52	43
			E1	Benzoylacetylene	43	43
			E3	DMAD	16	43

TABLE 4.3 Diels–Alder Reactions of Vinylcarbobicycloalkenes

Diene			Dienophile	Yield (%)	Ref.
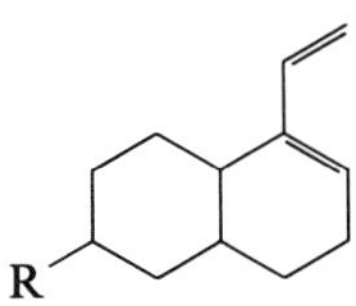	R = H, OMe	I4	Diazolones		45
		I5	NPTAD		45
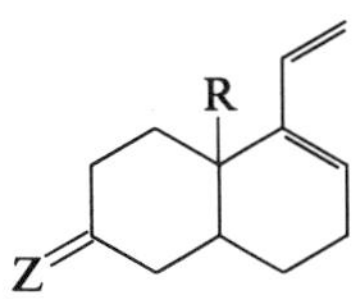	R = H, Me; Z = H_2, O	I4	Diazolones		45
		I5	NPTAD		45

TABLE 4.3 (*Continued*)

Diene	Dienophile	Yield (%)	Ref.
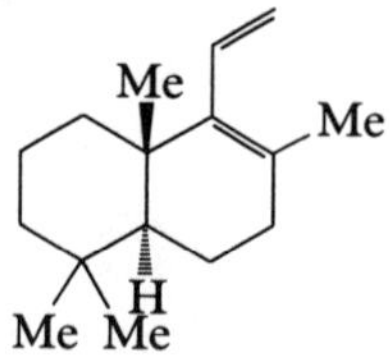	F4 Diethyl mesoxalate	35	46
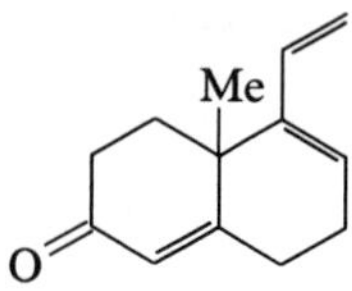	A4 (*E*)-Ethyl-2-methyl-4-oxo-2-butenoate		47
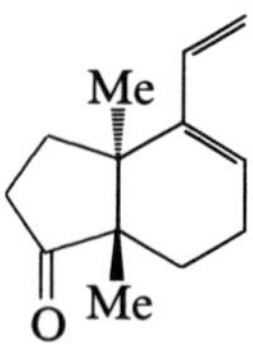	C7 2-Methoxy-5-methyl-1,4-benzoquinone	40–75	48
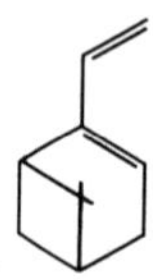	A8 Phenylvinylsulfone	98	50
	I5 NPTAD		49
	I5 NMTAD		49
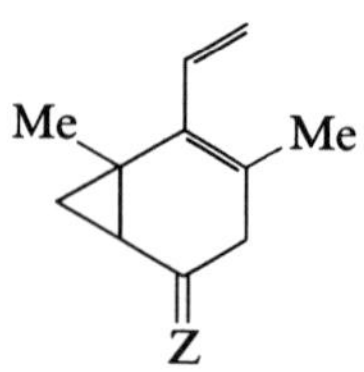 Z = H_2, $(OCH_2)_2$	C7 2,5-Dichloro-1,4-benzoquinone	100	51
	C7 2-Chloro-6-methoxy-1,4-benzoquinone	100	51
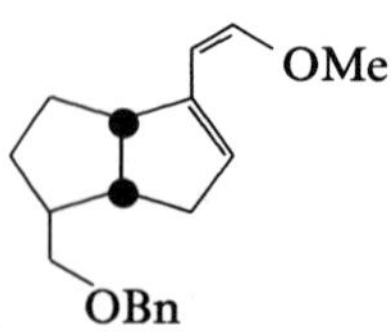	E1 MP	65	52
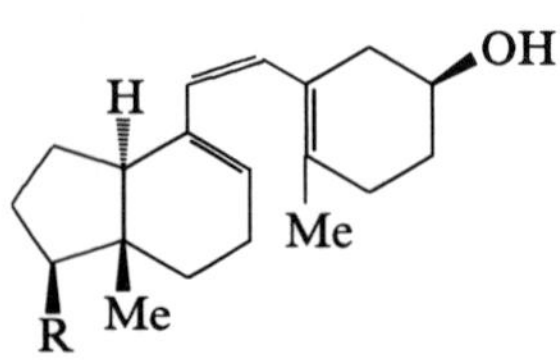 R = C_8H_{17}	I5 NPTAD	50	53

TABLE 4.4 Diels–Alder Reactions of Vinylnaphthalenes, Vinyldihydronaphthalenes, and Vinylphenanthrenes

Diene		Dienophile	Yield (%)	Ref.
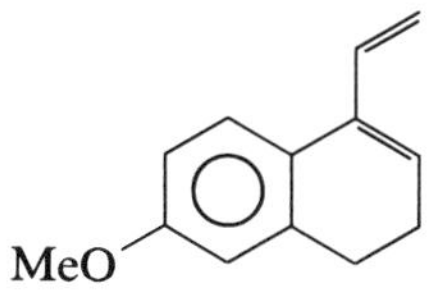	A10	(E)-Nitroarylethenes		55
	C7	Dimethyl-1,4-benzoquinones	52–82	56
	C7	2-Acetyl-1,4-benzoquinone	90	56
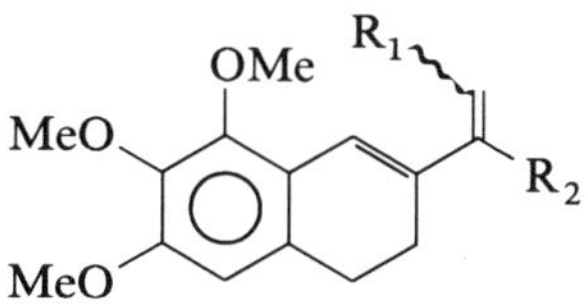 $\{R_1 = H, Me, CO_2Et;$ $R_2 = H, OAc\}$	D3	MA	90–92	57
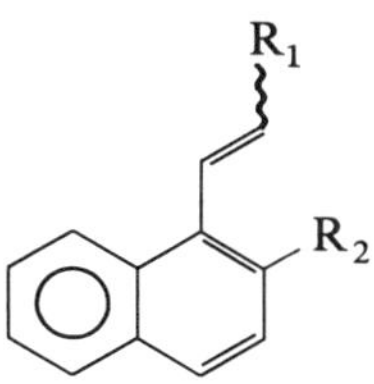				
$R_1 = R_2 = H$	A3	*trans*-Benzylphenyl[β-(carbomethoxy)-vinyl]phosphine oxide (*)	70	58
$\{R_1 = H\ R_2 = Me$	A12	TCNE	43	59
$R_1 = R_2 = OMe\ (E, Z)\}$	I3	[Singlet oxygen]	86–98	42
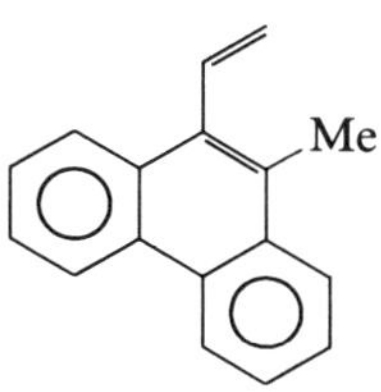	A12	TCNE	32	59

TABLE 4.5 Diels–Alder Reactions of Vinylcycloheptadienes

Diene	Dienophile		Yield (%)	Ref.
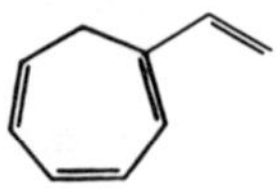	A12	TCNE	80	60
	E3	DMAD	82	60
	I5	NPTAD	97	60
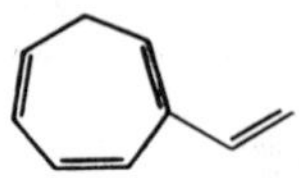	A12	TCNE	70	60
	E3	DMAD	88	60
	I5	NPTAD	60	60
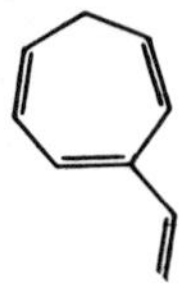	A12	TCNE	75	60
	E3	DMAD	98	60
	I5	NPTAD	95	60

TABLE 4.6 Diels–Alder Reactions of Vinyl Five-Membered Heterocycles

Diene	Dienophile		Yield (%)	Ref.
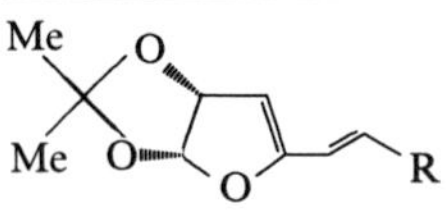				
R = H (*)	D3	MA	86	64, 85
R = $OSiMe_2Bu^t$ (*)	D3	MA	86	85
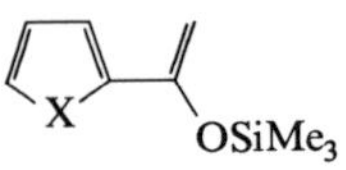 X = O, S	D5	NPM	43–85	65
X = S	A11	Triethylethylene carboxylate	73	65
	I4	Diethylazo dicarboxylate	47	65
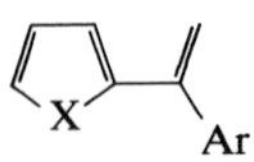	A8	ACN	5–79	77
	A9	2-Chloroacrylonitrile	56–76	77
	C2	Norbornene	60–80	77
(X = O, S; Ar = Ph, $Me_3C_6H_2$)	D3	MA	12–69	77
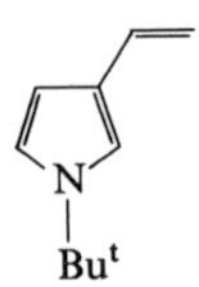	D3	MA	75	74
	E1	MP	66	74

TABLE 4.6 (*Continued*)

Diene			Dienophile	Yield (%)	Ref.
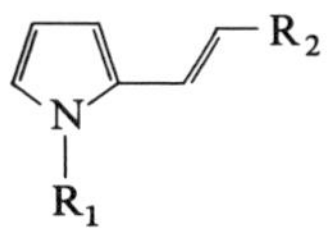					
R_1	**R_2**				
H	CO_2Me	E1	MP	35	75
Me	H	D3	MA	54	74
		E3	DMAD	67	74
Me	Me	E3	DMAD		76
Me	CO_2Me	E3	DMAD	37	75
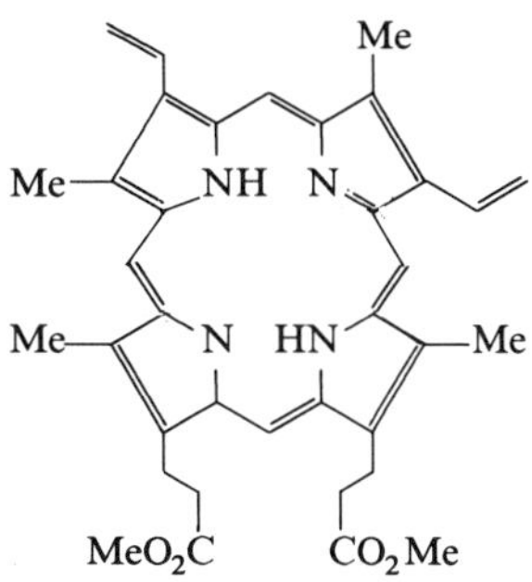		E2	Methyl-β-(phenylsulfonyl)-propiolate	41	68
		E3	DMAD	20	68
		E3	Diethylacetylene-dicarboxylate	39	68
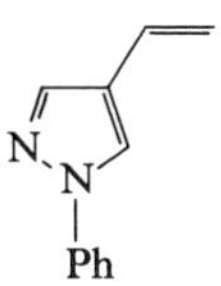		D5	NPM	35	66
		E1	MP	65	66
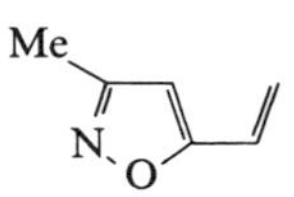		A1	MVK	25	67
		A1	MAC	6	67

TABLE 4.7 Diels–Alder Reactions of Vinyl Six-Membered Heterocycles

Diene		Dienophile	Yield (%)	Ref.
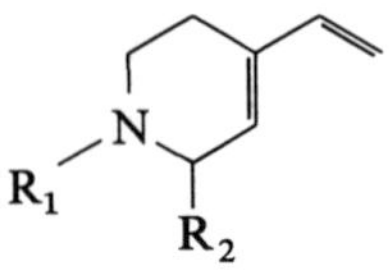				
R_1 = Me, R_2 = H	D5	NPM	63	71
R_1 = Bn, R_2 = H	E3	Dibenzoylacetylene	11	72
	E5	Perfluoro-2-butyne	62	72
R_1 = Me, R_2 = p-OMe$C_6H_4CH_2$	D5	NPM	64	71
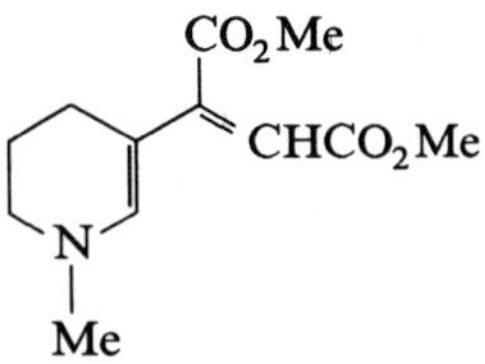	A1	MAC		73
	E3	DMAD		73
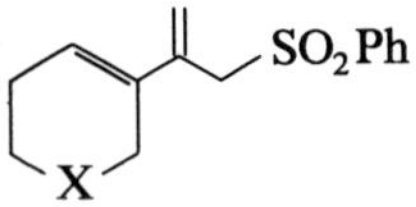 X = NMe	E3	DMAD	11	44
X = S	E3	DMAD	44	44
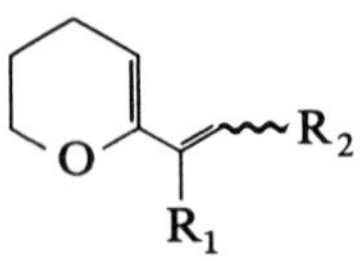				
R_1 = Me, Et, Ph, Bn	D3	MA		70
R_2 = H, Me	E3	DMAD		70
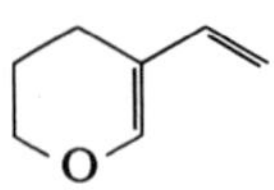	A10	β-Nitrostyrene	83	69
	A12	TCNE	33	69
	I5	NPTAD	84	69
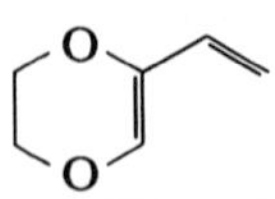	A12	TCNE	28	69
	I4	Diethylazo dicarboxylate	59	69

TABLE 4.8 Diels–Alder Reactions of Vinylindoles, Vinylbenzo[*b*]furans, and Vinylbenzo[*b*]thiophenes

Diene		Dienophile		Yield (%)	Ref.
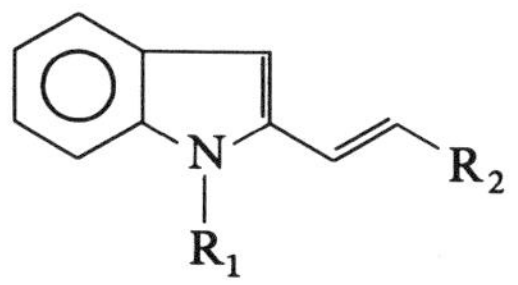					
R_1	**R_2**				
H	NO_2	A1	MAC	10	78
		A8	ACN	36	78
Me	Me	D3	MA	31	79
Me	CO_2Et	D3	MA	70	79
Me	OMe	A1	MAC	27	78
		A8	ACN	50	78
		D3	MA	72	79
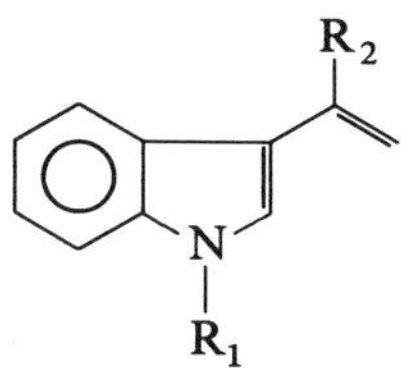					
R_1	**R_2**				
[H	OMe]	E3	DMAD	11	83
Me	H	C7	BQ	77	81
		C9	NQ	82	81
Me	Me	C9	NQ	68	81
[Me	Et]	E3	DMAD	6	83
Bn	H	C7	BQ	86	80
		D3	MA	61	80
SO_2Ph	H	C7	BQ	50	82
		D5	NPM	47	82
Ts	H	C9	NQ	54	81
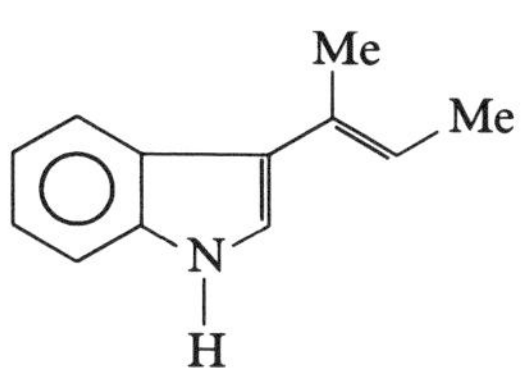		A7	Maleic acid	24	84
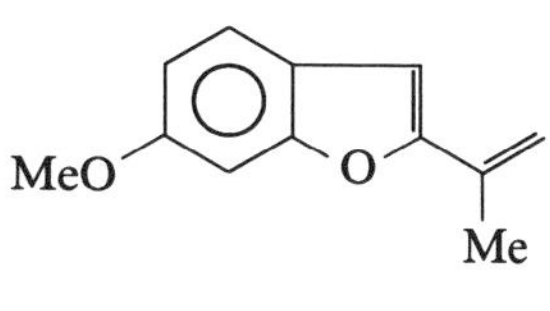		E1	MP	20	86

TABLE 4.8 (*Continued*)

Diene		Dienophile		Yield (%)	Ref.
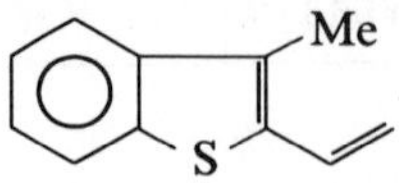		A12	TCNE	24	59
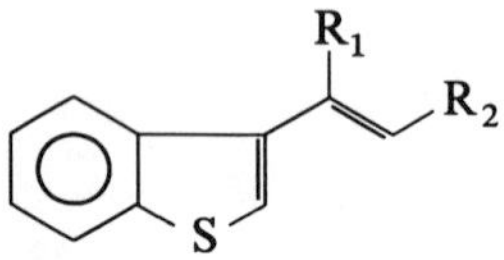					
R_1	**R_2**				
H	H	A12	TCNE	72	59
		C9	NQ		87
H	Me	A12	TCNE	52	59
		C9	NQ	68	87
Me	H	A12	TCNE	63	59
		C9	NQ	73	87

TABLE 4.9 Diels–Alder Reactions of Miscellaneous Inner–Outer-Ring Dienes

Diene	Dienophile		Yield (%)	Ref.
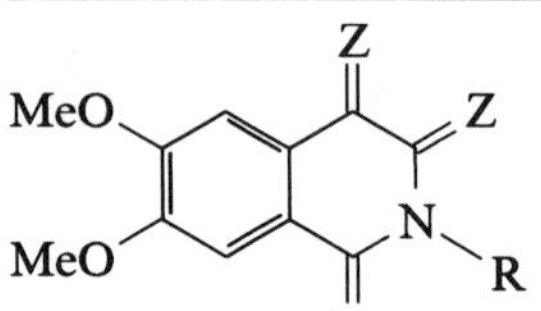 ($Z = H_2$, O; R = H, COMe)	E6	[Benzyne]	40	88
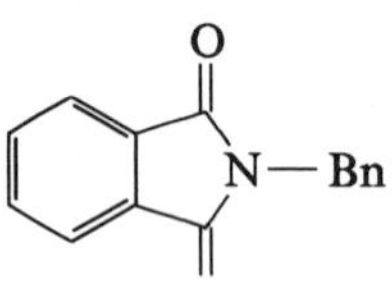	E6	[Benzyne]	22	88
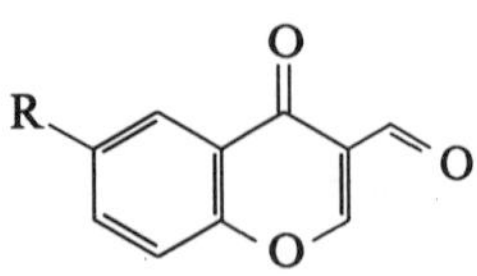				
R = Me	A9	2-Methoxypropene		89
R = H, OAc, OMe, OH, NO_2	A8	Ethoxyethene	66–94	90

TABLE 4.9 (*Continued*)

Diene		Dienophile	Yield (%)	Ref.
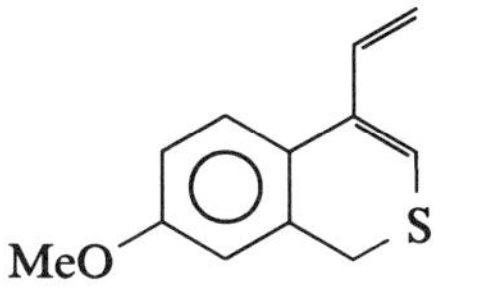	D3	MA	65	91
	D3	Citraconic anhydride	65	91
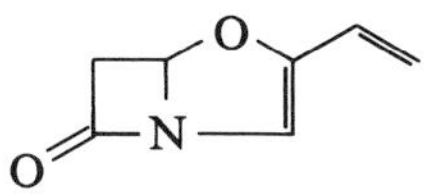 (*)	A12	TCNE		92
	D3	MA		92
	J1	Nitrosobenzene		92
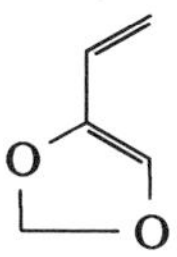	A1	AC	50	93
	D3	MA	65	93
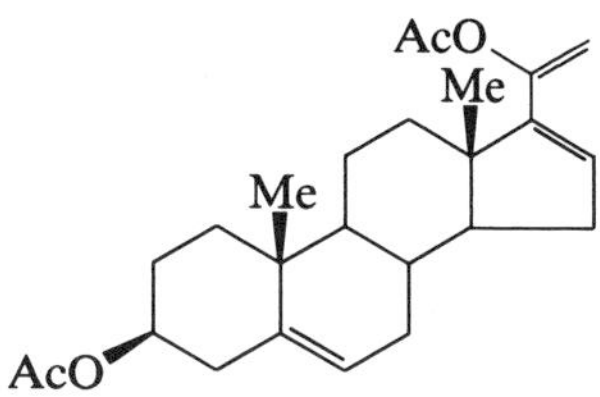	F4	Ethyl glyoxylate		54
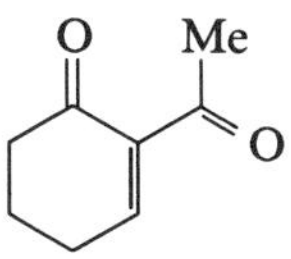	A8	Ethyl vinyl ether	75	63
	A10	Propenyl ethyl ether (E and Z)	70	63

REFERENCES

1. R. L. Snowden and M. Wüst, *Tetrahedron Lett.*, **27**, 703 (1986).
2. R. L. Snowden and M. Wüst, *Tetrahedron Lett.*, **27**, 699 (1986).
3. D. Caine, C. R. Harrison, and D. G. Van Derveer, *Tetrahedron Lett.*, **24**, 1353 (1983).
4. J. Orban and J. V. Turner, *Tetrahedron Lett.*, **24**, 2697 (1983).
5. T. R. Hoye and M. J. Rother, *J. Org. Chem.*, **44**, 458 (1979).
6. W. J. Greenlee and R. B. Woodward, *Tetrahedron*, **36**, 3367 (1980).
7. D. Pitea, *J. Mol. Struct.* (*Theochem.*), **10**, 347 (1983).
8. J. H. Markgraf, E. W. Greeno, M. D. Miller, W. J. Zaks, and G. A. Lee, *Tetrahedron Lett.*, **24**, 241 (1983).

9. F. Orsini, F. Pelizzoni, D. Pitea, E. Abbondanti, and A. Mugnoli, *J. Org. Chem.*, **48**, 2866 (1983).
10. T. Tokoroyama, K. Matsuo, H. Kotsuki, and R. Kanazawa, *Tetrahedron*, **36**, 3377 (1980).
11. E. R. Ireland, W. J. Thompson, N. S. Mandel, and G. S. Mandel, *J. Org. Chem.*, **44**, 3583 (1979).
12. V. A. Danisenko, V. L. Novikov, and G. B. Elyakov, *Zh. Org. Khim.*, **15**, 251 (1979).
13. M. F. Brana, J. M. Castellano, E. Gonzalez, M. Lora-Tamayo, and J. L. Soto, *An. Quim.*, **75**, 927 (1979); *Chem. Abstr.*, **93**, 26336j (1980).
14. V. L. Novikov, O. P. Shestak, A. V. Kamernitskii, and G. B. Elyakov, *Izv. Akad. Nauk SSSR, Ser. Khim.*, 1390 (1980).
15. D. Pitea, M. Gastaldi, F. Orsini, F. Pelizoni, A. Mugnoli, and E. Abbondanti, *J. Org. Chem.*, **50**, 1853 (1985).
16. J. B. Hendrickson and V. Singh, *J. Chem. Soc. Chem. Commun.*, 837 (1983).
17. T. Tokoroyama, K. Matsuo, and T. Kubota, *Tetrahedron*, **34**, 1907 (1978).
18. G. R. Lenz, *J. Org. Chem.*, **44**, 4299 (1979).
19. M. Forte, F. Orsini, F. Pelizzoni, and G. Ricca, *Gazz. Chim. Ital.*, **115**, 41 (1985).
20. R. C. Cookson, M. C. Cramp, and P. J. Parsons, *J. Chem. Soc. Chem. Commun.*, 197 (1980).
21. M. Ihara, Y. Ishida, K. Fukumoto, and T. Kametani, *Chem. Pharm. Bull.*, **33**, 4102 (1985).
22. C. Veyrat, L. Wartski and J. Seyden-Penne, *Tetrahedron Lett.*, **27**, 2981 (1986).
23. F. Richter and H.-H. Otto, *Tetrahedron Lett.*, **28**, 2945 (1987).
24. G. A. Kraus and Y.-S. Hon, *J. Org. Chem.*, **51**, 116 (1986).
25. T. Cohen and Z. Kosarych, *J. Org. Chem.*, **47**, 4005 (1982).
26. L. D. Quinn and J. E. MacDiarmid, *J. Org. Chem.*, **47**, 3248 (1982).
27. S. P. Tanis and Y. M. Abdallah, *Synth. Commun.*, **16**, 251 (1986).
28. S. Knapp and S. Sharma, *J. Org. Chem.*, **50**, 4996 (1985).
29. S. C. Howell, S. V. Ley, M. Mahon, and P. A. Worthington, *J. Chem. Soc. Chem. Commun.*, 507 (1981).
30. R. L. Snowden, *Tetrahedron Lett.*, **25**, 3835 (1984).
31. D. M. Hollinshead, S. C. Howell, S. V. Ley, M. Mahon, and N. M. Ratcliffe, *J. Chem. Soc. Perkin Trans. 1*, 1579 (1983).
32. Yu.-C. Lai, S. E. Mallakpour, G. B. Butler, and G. J. Palemic, *J. Org. Chem.*, **50**, 4378 (1985).
33. W. B. Manning, *Tetrahedron Lett.*, 1661 (1979).
34. H. K. Hall, Jr., P. Nogues, J. W. Rhoades, R. C. Sentman, and M. Detar, *J. Org. Chem.*, **47**, 1451 (1982).
35. W. B. Manning, T. R. Kelly, and G. M. Muschik, *J. Org. Chem.*, **45**, 2535 (1980).
36. Y. Uosaki, M. Nakahara, and J. Osugi, *Bull. Chem. Soc. Jpn.*, **54**, 2569 (1981).
37. Sh. S. Vezirov, S. M. Aliev, and E. S. Zeinalova, *Zh. Prikl. Khim.* (*Leningrad*), **58**, 1582 (1985); *Chem. Abstr.*, **104**, 6206j (1986).
38. Z. Nippon, *Jpn. Kokai Tokyo Koho*, JP **60**, 23,383 [JP 85, 23, 383] (1985); *Chem. Abstr.*, **103**, 37468e (1985).

39. Y. Uosaki, M. Nakahara, and J. Osugi, *Bull. Chem. Soc. Jpn.*, **54**, 3681 (1981).
40. W. B. Manning and D. J. Wilbur, *J. Org. Chem.*, **45**, 733 (1980).
41. T. O. Criodain, M. O'Sullivan, M. J. Meegan, and D. M. X. Donnelly, *Phytochemistry*, **20**, 1089 (1981).
42. M. Matsumoto, K. Kuroda, and Y. Suzuki, *Tetrahedron Lett.*, **22**, 3253 (1981).
43. Y. Kita, H. Yasuda, O. Tamura, and Y. Tamura, *Tetrahedron Lett.*, **25**, 1813 (1984).
44. K. Hayakawa, M. Takewaki, I. Fujimoto, and K. Kanematsu, *J. Org. Chem.*, **51**, 5100 (1986).
45. M. F. Brana, J. M. Castellano, M. Lora-Tamayo, and J.-L. Soto, *An. Quim.*, **75**, 931 (1979).
46. W. M. Daniewski, E. Kubak, and J. Jurczak, *J. Org. Chem.*, **50**, 3963 (1985).
47. M. Kakushima, L. Allain, R. A. Dickinson, P. S. White, and Z. Valenta, *Can. J. Chem.*, **57**, 3354 (1979).
48. a. L. Kolaczkowski and W. Reusch, *J. Org. Chem.*, **50**, 4766 (1985).
b. J. S. Tou and W. Reusch, *J. Org. Chem.*, **45**, 5012 (1980).
49. P. Ashkenazi, R. Benn, and D. Ginsburg, *Helv. Chim. Acta*, **67**, 583 (1984).
50. R. V. C. Carr, R. V. Williams, and L. A. Paquette, *J. Org. Chem.*, **48**, 4976 (1983).
51. M. Sakaino and J. Meinwald, *Tetrahedron Lett.*, **28**, 3201 (1987).
52. K. Shimoji and M. Hayashi, *Tetrahedron Lett.*, **21**, 1255 (1980).
53. L. J. Vanmaele, P. J. De Clercq, and M. Vandewalle, *Tetrahedron Lett.*, **23**, 995 (1982).
54. M. Kocor and W. Wojciechowska, *Pol. J. Chem.*, **54**, 1907 (1980).
55. P. M. Kelkar, S. N. Rastogi, and N. Anand, *Indian J. Chem., Sect. B*, **18 B**, 71 (1979).
56. J. Das, R. Kubela, G. A. MacAlpine, Z. Stojanac, and Z. Valenta, *Can. J. Chem.*, **57**, 3308 (1979).
57. N. S. Narasimhan and T. Mukhopadhyay, *Indian J. .Chem., Sect. B*, **22 B**, 1096 (1983).
58. R. Bodalsky, J. Koszuk, H. Krawczyk, and K. M. Pietrusiewicz, *J. Org. Chem.*, **47**, 2219 (1982).
59. W. H. Cherry, J. T. Craig, and Q. N. Porter, *Aust. J. Chem.*, **32**, 133 (1979).
60. M. Aga, K. Okada, and M. Oda, *Tetrahedron Lett.*, **27**, 5653 (1986).
61. E. C. Angell, F. Fringuelli, F. Pizzo, B. Porter, A. Taticchi, and E. Wenkert, *J. Org. Chem.*, **51**, 2642 (1986).
62. H. M. R. Hoffmann and O. Koch, *J. Org. Chem.*, **51**, 2939 (1986).
63. B. B. Snider, *Tetrahedron Lett.*, **21**, 1133 (1980).
64. K. M. Sun and B. Fraser-Reid, *J. Am. Chem. Soc.*, **104**, 367 (1982).
65. T. Sasaki, Y. Ishibashi, and M. Ohno, *Heterocycles*, **20**, 1933 (1983).
66. M. Medio Simon and J. Sepulveda Arques, *Tetrahedron*, **42**, 6683 (1986).
67. A. Brandi, F. DeSarlo, A. Guarna, and A. Goti, *Heterocycles*, **23**, 2019 (1985).
68. V. S. Pangka, A. R. Morgan, and D. Dolphin, *J. Org. Chem.*, **51**, 1094 (1986).
69. B. Potthoff and E. Breitmaier, *Chem. Ber.*, **119**, 3204 (1986).
70. A. Lebouc, J. Delanney, and O. Riobe, *J. Chem. Res. (S)*, 187 (1980).

71. J. R. L. Smith, R. O. C. Norman, M. E. Rose, A. C. W. Curran, and J. W. Lewis, *J. Chem. Soc. Perkin Trans. 1*, 2868 (1979).

72. J. A. Finkelstein and C. D. Perchonock, *Tetrahedron Lett.*, **21**, 3323 (1980).

73. R. M. Achenson and G. Paglietti, *Studi Sassar.*, *Sez. 2*, **57**, 439 (1979); *Chem Abstr.*, **94**, 121038t (1981).

74. R. A. Jones, M. T. P. Marriotti, W. P. Rosenthal, and J. S. Arques, *J. Org. Chem.*, **45**, 4515 (1980).

75. W. E. Noland, C. K. Lee, S. K. Bae, B. Y. Chung, C. H. Hahn, and K. J. Kim, *J. Org. Chem.*, **48**, 2488 (1983).

76. C. K. Lee, *Bull. Korean Chem. Soc.*, **5**, 50 (1984).

77. H. Ohmura and S. Motoki, *Bull. Chem. Soc. Jpn.*, **57**, 1131 (1984).

78. N. S. Narasimhan and R. S. Kusurkar, *Indian J. Chem.*, *Sect. B*, **22 B**, 846 (1983).

79. N. S. Narasimhan, R. S. Kusurkar, and D. D. Havale, *Indian J. Chem.*, *Sect. B*, **22 B**, 1004 (1983).

80. J. D. Lambert and Q. N. Porter, *Aust. J. Chem.*, **34**, 1483 (1981).

81. R. Bergamasco, Q. N. Porter, and C. Yap, *Aust. J. Chem.*, **31**, 1841 (1978).

82. B. Saroja and P. C. Srinivasan, *Synthesis*, 748 (1986).

83. U. Pindur and L. Pfeuffer, *Tetrahedron Lett.*, **28**, 3079 (1987).

84. W. E. Noland and S. R. Wann, *J. Org. Chem.*, **44**, 4402 (1979).

85. K.-M. Sun, R. M. Giuliano, and B. Fraser-Reid, *J. Org. Chem.*, **50**, 4774 (1985).

86. R. T. Scannel and R. Stevenson, *J. Chem. Soc. Chem. Commun.*, 1103 (1980).

87. W. H. Cherry and Q. N. Porter, *Aust. J. Chem.*, **32**, 145 (1979).

88. L. Castedo, E. Guitián, J. M. Saa, and R. Suau, *Tetrahedron Lett.*, **23**, 457 (1982).

89. F. M. Dean, M. Al-Sattar, and D. A. Smith, *J. Chem. Soc. Chem. Commun.*, 535 (1983).

90. S. T. Saengchantara and T. W. Wallace, *J. Chem. Soc. Perkin Trans. 1*, 789 (1986).

91. S. Kano, K. Tanaka, S. Hibino, and S. Shibuya, *J. Org. Chem.*, **44**, 1580 (1979).

92. E. Hunt and I. I. Zomaya, *J. Chem. Soc. Perkin Trans. 1*, 1509 (1982).

93. T. V. Sotnichenko, R. I. Kruglikova, and B. V. Unkovskii, *Zh. Org. Khim.*, **15**, 471 (1979).

5

ACROSS-RINGS DIENES

Reactivity, Selectivity, Synthesis Conjugated dienes with a diene moiety in two rings have been used to prepare unusual compounds for mechanistic studies and in synthesis of natural compounds. Cycloadditions of bicyclodienes **1** with 1-bromo-2-chlorocyclopropene (**2**) give the adducts **3**, which are easily dehydrogenated to spectroscopically and theoretically interesting tris-annulated benzenes **4**.[1]

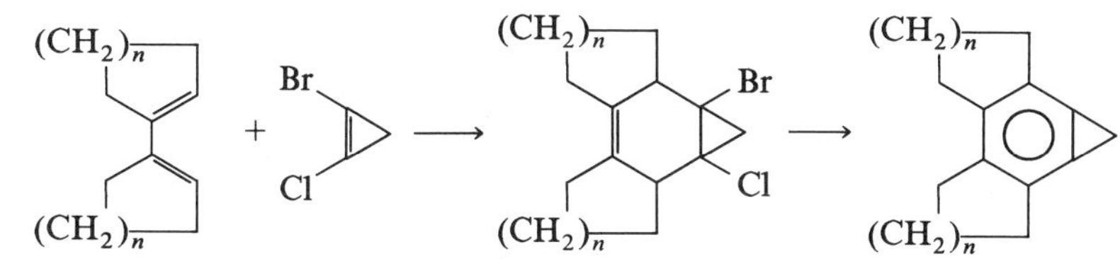

1 ($n = 0, 1, 2$) **2** **3** **4**

Cation radical cycloadditions of 1,1′-bicyclopentenyl **1** (n = 1) with vinyl ethers $CH_2{=}CH{-}R$ (R = OEt, OPh, SPh), under photosensitized electron-transfer conditions or by the tris(*p*-bromophenyl)aminium hexachlorostibnate procedure, yield preferentially cyclobutane adducts **5** (69–98%) along with minor (2–31%) Diels–Alder adducts **6**.[5] This result contrasts with that of cation radical diene–diene cycloadditions and with those of additions of 1,3-cyclohexadiene with phenoxy and thiophenoxy vinyl ethers, all of which show a high Diels–Alder chemoselection.[5]

5 6

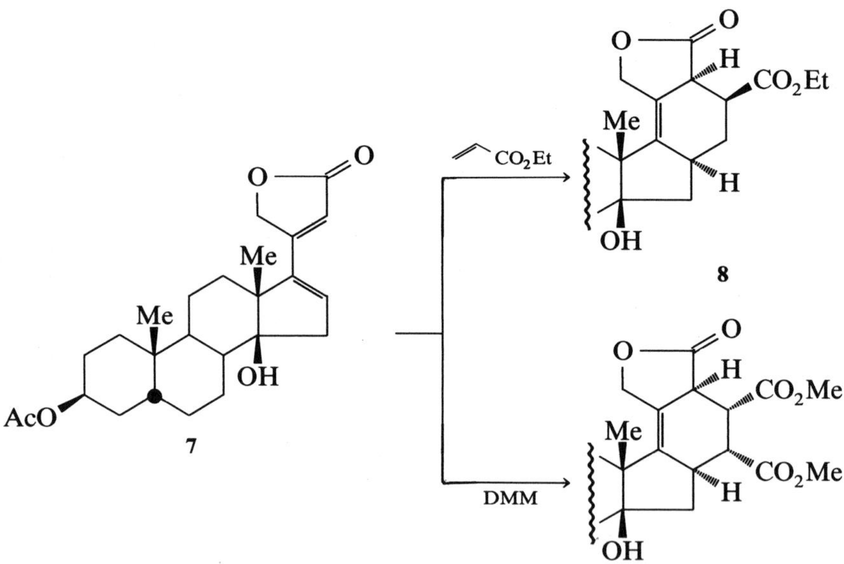

7 8 9

In the course of studies on digitalis glycosides, it was observed that the Diels–Alder reaction of 16,17-dehydrodigitoxigenin-3-acetate (**7**) with ethyl acrylate gives the ester **8** via an *endo* addition *syn* to angular substituents (*β*-face of the molecule), while the reaction with DMM occurs with *syn-exo* selectivity affording the adduct **9**. DMF gives the expected adduct from *syn-endo* addition.[12a]

Tables of Diels–Alder Reactions See Tables 5.1 and 5.2; see also the general remarks about classifications, abbreviations, and symbols in the Introduction.

TABLE 5.1 Diels–Alder Reactions of Carbobicyclo Across-Rings Dienes

Diene		Dienophile	Yield (%)	Ref.
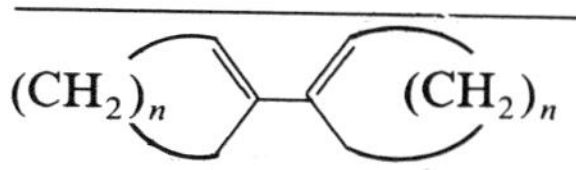				
$n = 1$	C1	1-Bromo-2-chlorocyclopropene	19	1
	C1	Cyclobut-1-ene-1,2-dicarbonitrile	65	2
	C1	Dimethylcyclobutene-1,2-dicarboxylate	47	3
$n = 2$	A8	Ethyl vinyl ether	71	5
	A8	Phenyl vinyl ether		5
	A8	Phenyl vinyl thioether		5
	A8	(β-Chloroethyl)vinyl ether		5
	C1	1-Bromo-2-chlorocyclopropene		1
	C1	1,1′-Dicyclopentenyl	50	6
	E6	[3,4 : 5,6-Dicyclopentabenzyne]	11	4
$n = 3$	C1	1-Bromo-2-chlorocyclopropene		1
	I2	[Singlet sulfur]	50–70	7
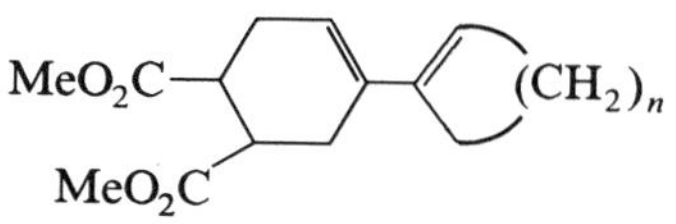				
$n = 2$	D5	NMM	73	8
	E3	DMAD	63	8
$n = 3$	E3	DMAD	83	8
$n = 4$	E3	DMAD	77	8

TABLE 5.2 Diels–Alder Reactions of Miscellaneous Across-Rings Dienes

Diene		Dienophile	Yield (%)	Ref.
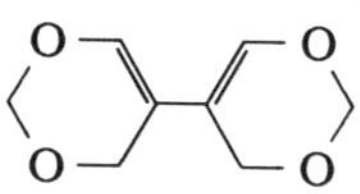	D3	MA	80	9
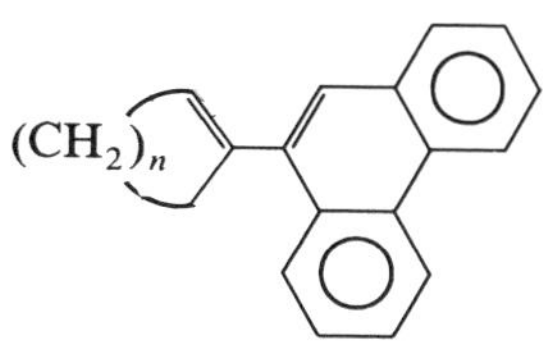 ($n = 2, 3$)	D3	MA	85–97	10

TABLE 5.2 (*Continued*)

Diene		Dienophile	Yield (%)	Ref.
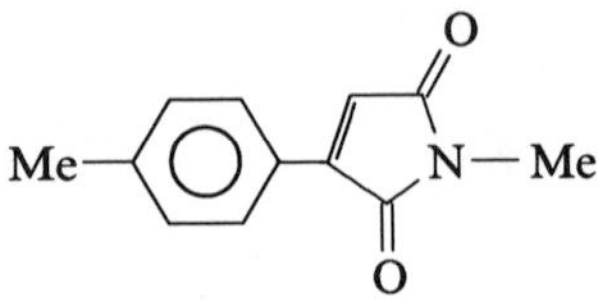	D5	*p*-Tolyl-*N*-methylmaleimide	94	11
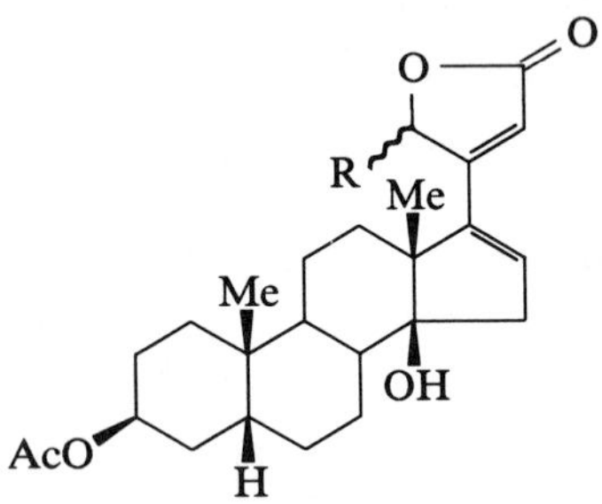 (R = H, OEt)	A1 C1	Ethyl acrylate 16,17-Dehydrodigitoxygenin-3-acetate	38	12a 12b
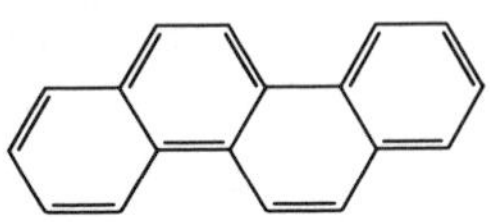	D3	MA (k)		13
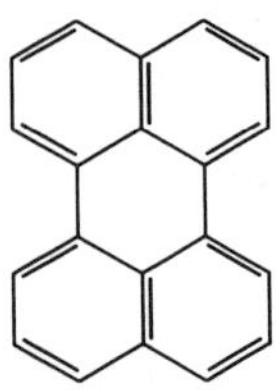	I4	Diethyldiazenedicarboxylate	38–40	14
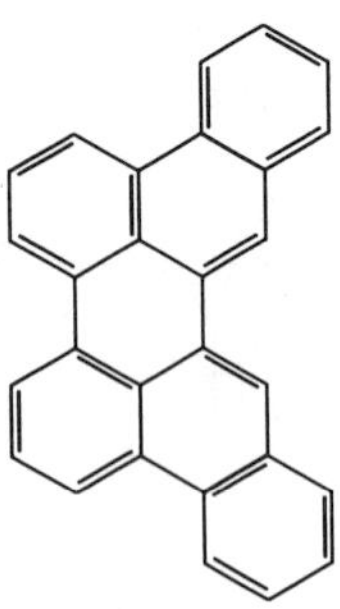	G4	Dehydrohydantoins	13–96	15

REFERENCES

1. W. E. Billups, B. E. Arney, Jr., and L.-J. Lin, *J. Org. Chem.*, **49**, 3436 (1984).
2. G. McMullen, M. Lutterbeck, H. Fritz, H. Prinzbach, and C. Krueger, *Isr. J. Chem.*, **22**, 19 (1982).
3. W. Nutakul, R. P. Thummel, and A. D. Taggart, *J. Am. Chem. Soc.*, **101**, 770 (1979).
4. R. P. Thummel and T. Fuchigami, *J. Org. Chem.*, **46**, 5217 (1981).
5. R. A. Pabon, D. J. Bellville, and N. L. Bauld, *J. Am. Chem. Soc.*, **106**, 2730 (1984).
6. D. J. Bellville, D. D. Wirth, and N. L. Bauld, *J. Am. Chem. Soc.*, **103**, 718 (1981).
7. a. K. Steliou, P. Salama, D. Brodeur, and Y. Gareau, *J. Am. Chem. Soc.*, **109**, 926 (1987).
   b. K. Steliou, Y. Gareau, and D. N. Harpp, *J. Am. Chem. Soc.*, **106**, 799 (1984).
8. O. Tsuge, F. Wada, S. Kanemasa, and H. Sakoh, *Chem. Lett.*, 709 (1984).
9. V. V. Zorin, V. N. Trifonova, S. S. Zlot-Skii, and D. L. Rakhmankulov, *Zh. Org. Khim.*, **20**, 1590 (1984); Engl. Ed., 1447 (1985).
10. E. G. Lekveishvili, L. D. Melikadze, N. I. Tabashidze, and E. V. Kartvelishvili, *Soobshch. Akad. Nauk Gruz. SSR*, **109**, 305 (1983).
11. J. W. Epstein, T. C. McKenzie, M. F. Lovell, and N. A. Perkinson, *J. Chem. Soc. Chem. Commun.*, 314 (1980).
12. a. T. Hashimoto, Y. Kato, M. Shiro, Y. Takahashi, Y. Jinbo, and D. Satoh, *Chem. Pharm. Bull.*, **30**, 494 (1982).
    b. T. Hashimoto, Y. Kato, H. Shibahara, K. Toyooka, T. Ohta, and D. Satoh, *Chem. Pharm. Bull.*, **27**, 2975 (1979).
13. H. Karpf, O. E. Polansky, and M. Zauder, *Monatsh. Chem.*, **112**, 659 (1981).
14. S. Tokita, K. Hiruta, K. Kitahara, and H. Nishi, *Synthesis*, 229 (1982).
15. S. Tokita, K. Hiruta, Y. Yaginuma, S. Ishikawa, and H. Nisni, *Synthesis*, 270 (1984).

6

INNER-RING DIENES

Dienes, in which dienic functionality is part of a ring (inner-ring dienes), are numerous and are markedly different structurally, depending on the type and dimension of the ring and the nature and number of dienic functionalities present in the ring. They can be divided into three subclasses: carbocycles, aromatic hydrocarbons, and heterocycles.

6.1 CARBOCYCLIC DIENES

Four- and Five-Membered Rings

Cyclobutadienes Cyclobutadienes are poorly stable but very reactive compounds and their [4 + 2] cycloadditions afford strained adducts. Kinetically stabilized cyclobutadienes **1** react[413] with phosphalkynes **3** ($R_2 = Pr^i$, Bu^t, etc.) to give the regioisomers Dewar phosphinines **4** and **5** with a high prevalence of *ortho* adduct **4**.

The tetra-*tert*-butylcyclobutadiene **2** has been cyclized with DMAD and dicyanoacetylene to prepare stable Dewar benzenes[150, 423] **6**. The formation of adduct **6** (R = CN) has been hypothesized to occur via radical ion-pair intermediates.

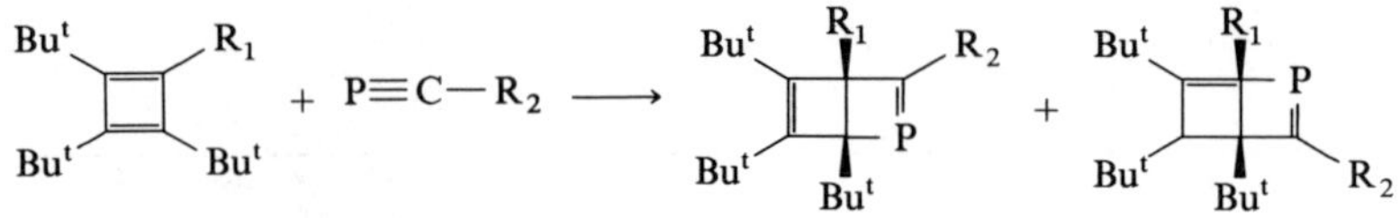

1 $R_1 = CO_2Me, CO_2Bu^t$ **3** **4** **5**

2 $R_1 = Bu^t$

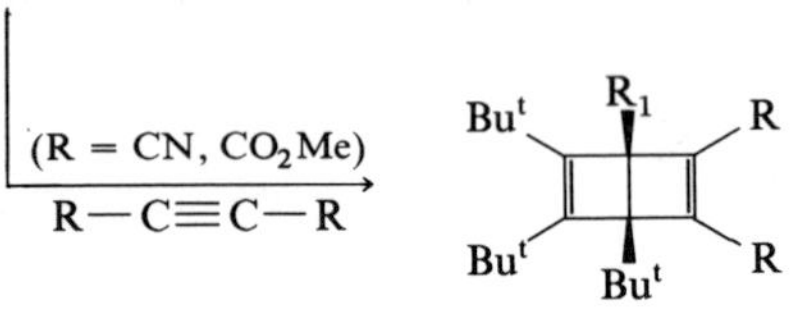

6

The cycloadducts **8**, prepared from the isolable cyclobutadiene **7**, are converted[149] into tetra-*tert*-butylbenzenes **9** (Z = CH, CPh) and tetra-*tert*-butylpyridines **9** (Z = N), which are difficult to prepare by other procedures.

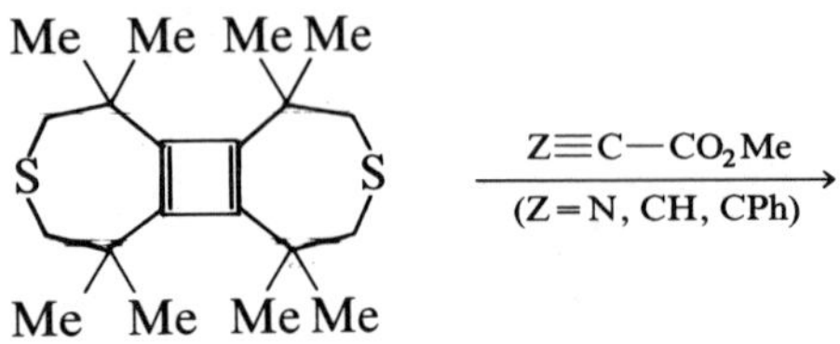

7

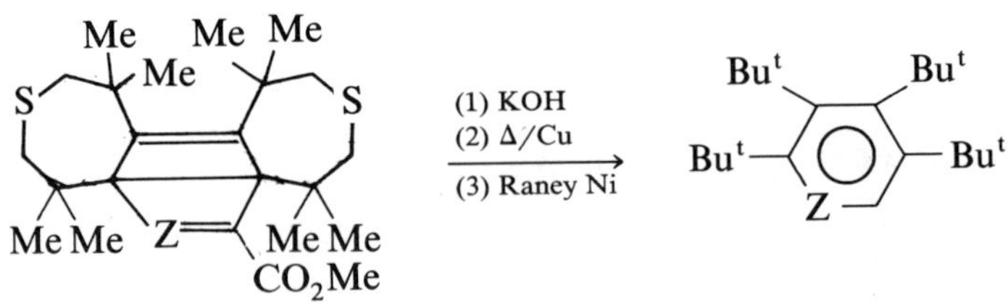

8 **9**

Cyclopentadienes Cyclopentadienes are widely used to study mechanistic and regio- and stereochemical aspects of the Diels–Alder reaction and for synthetic purposes. The parent compound **10** (R = H) is highly reactive (Table 6.1) if compared with other cyclic and aliphatic hydrocarbon dienes; this is due mainly to the geometry of the ring, which has the most favorable 1,4-dienic distance.[63d,f]

TABLE 6.1 Relative Rate Constants of Various Dienes with TCNE

Diene	k_{rel} (20°C, CH_2Cl_2)
Cyclopentadiene	2,109,000
1,3-Cyclohexadiene	811
1,3-Cycloheptadiene	2.6
1,3-Butadiene	1

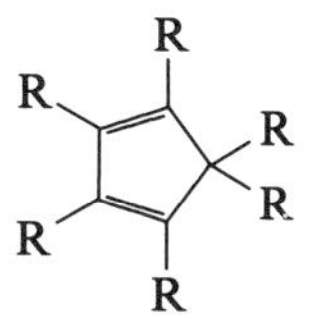

10

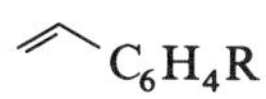

12

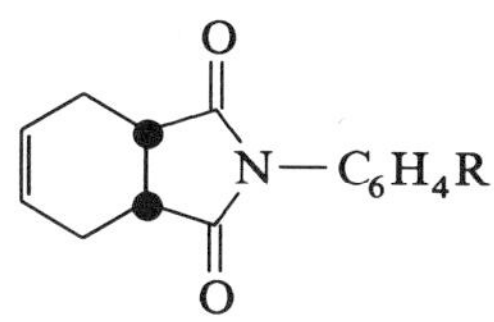

13

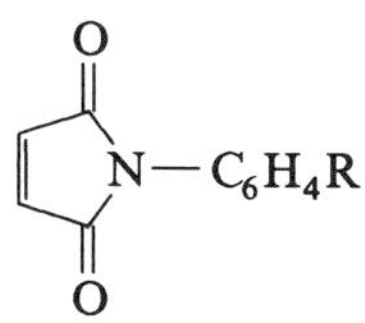

11

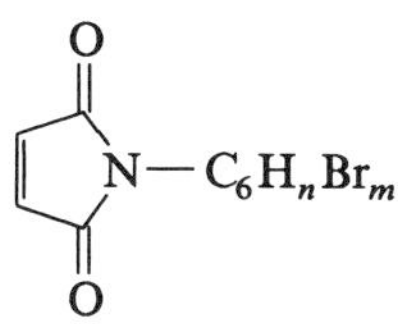

15

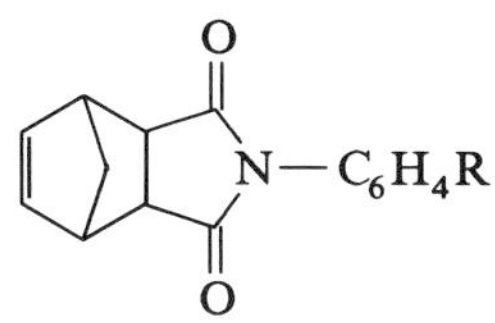

14

Kinetic studies have proven that while the cycloaddition of cyclopentadiene (**10**: R = H) with *N*-arylmaleimides **11** is a normal Diels–Alder reaction,[424] the cyclizations of hexachlorocyclopentadiene (**10**: R = Cl) with styrenes **12**[425] and *N*-arylamides **13** and **14**,[181a, b, e] as well as the additions of hexabromocyclopentadiene (**10**: R = Br) with *N*-polybromoarylmaleimides **15**,[426] are Diels–Alder reactions with inverse electron demand.

Stereochemical studies have pointed out that the Diels–Alder reaction of cyclopentadienes is prevalently *endo*-diastereoselective; the *endo/exo* ratio can be solvent, temperature, and catalyst dependent. Sometimes an *endo-exo* isomerization occurs.

Under mild conditions cyclopentadiene reacts with maleic and citraconic anhydrides **16** and **17** to give *endo* adducts **18** and **19**, which at 190°C isomerize to *exo* isomers **20** and **21**, respectively.[427]

The *endo/exo* ratio of the cycloaddition of crotonic acid to cyclopentadiene changes from 85% to 40% by changing the reaction temperature from −10°C to 170°C.[428] Similarly, the *endo* preference (81%) of cycloaddition with cyclopentene at 200°C is greatly reduced (29%) at 300°C.[86]

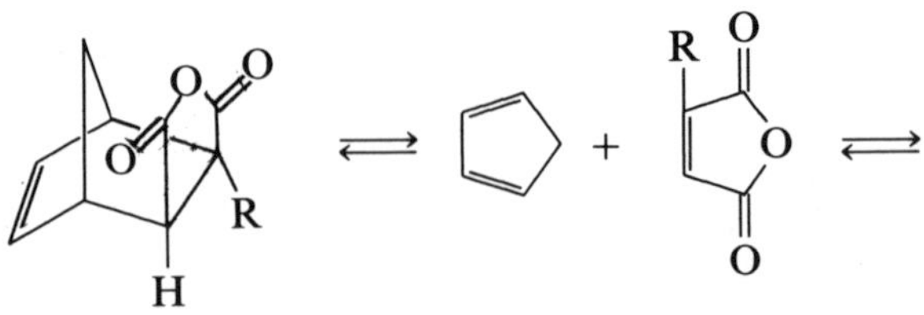

20 R = H
21 R = Me

10

16 R = H
17 R = Me

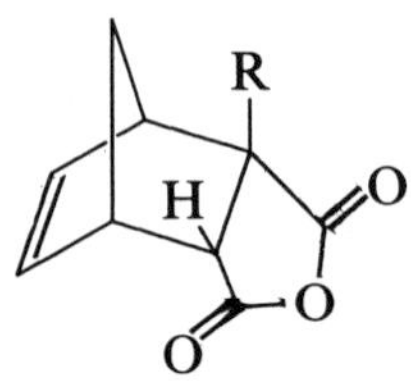

18 R = H
19 R = Me

A detailed study[429] of the diastereoselectivity of the reactions between cyclopentadiene and monosubstituted ethenes and their α- and β-methyl derivatives (Table 6.2) shows that the *endo* selectivity decreases in the order: *cis*-β-Me > H > *trans*-β-Me > α-Me. This indicates that the Me group possesses an appreciable *exo*-orienting ability. This methyl effect has also been shown with other dienophiles, such as *cis*- and *trans*-3-chloro-2-methyl propenal[133] and 2-cyclohexenones.[430] Bulky geminal groups in 5,5-dimethoxytetrachlorocyclopentadiene (**22**) interfere sterically in the transition state with the large carbomethoxymethyl group of methyl methacrylate, and the *endo* adduct **23** (R = CO_2Me) is the sole reaction product, while the steric interaction with the slim linear nitrile group of acrylonitrile (R = CN) is greatly reduced and the *exo* addition is preferred (77%).[201]

TABLE 6.2 Endo Percentage of Diels–Alder Reaction of Cyclopentadiene with Various Dienophiles at 25°C

R	$CH_2{=}CHR$	$CH_2{=}C(Me){-}R$	(Me)(H)C=C(H)(R)	(H)(Me)C=C(H)(R)
CN	57.5	12.0	28.1	79.5
COMe	79.4	47.0	64.6	74.7
CO_2Me	74.3	30.1	50.9	79.6
CHO	74.4	17.0	63.1	—
COOH	80.2	29.2	62.4	83.7

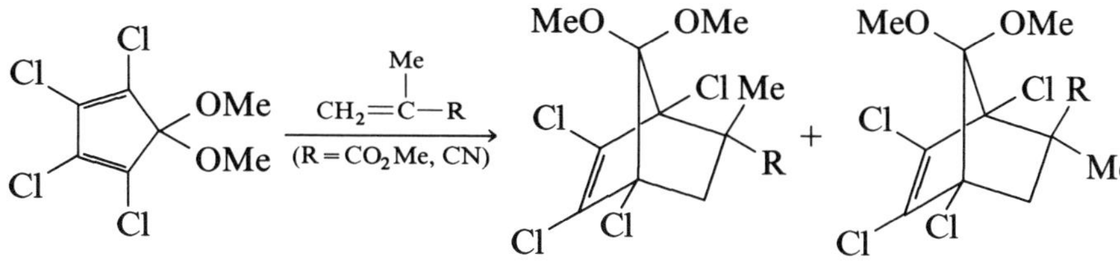

22 **23**

Steric effects, van der Waals interactions, secondary orbital overlap, the relationship between reactivity and selectivity, and/or a combination of these have been used to explain the *endo-exo* diastereoselectivity.[86,133,429,431] The importance of steric interactions versus electronic interactions is always a debated topic. Recent MNDO and AM1 semiempirical calculations show that the steric effects are as important as secondary orbital interactions in controlling the *endo-exo* selectivity.[86]

The cycloadditions of hexachlorocyclopentadiene with *trans*-alkenes **24** (Scheme 6.1) are examples of nonstereospecific Diels–Alder reactions. Experimental evidence in the reaction with **24** (R = Cl) excludes the possibility that the presence of unexpected **26** (R = Cl) is the consequence of isomerizations of **24** (R = Cl) or **25** (R = Cl). It has been hypothesized that steric and electronic interactions obstruct a synchronous [4 + 2] cycloaddition and favor a biradical intermediate[431,432] which can give ring closure to **26** (R = Cl) or β-scission to hexachlorocyclopentadiene and *cis*-1,2-dichloroethene, which produce the adduct **26** (R = Cl).

The π-facial diastereoselectivity of Diels–Alder cycloadditions of unsymmetrical cyclopentadienes has been extensively studied experimentally and theoretically.[171,433] The directionality of the addition depends on both the

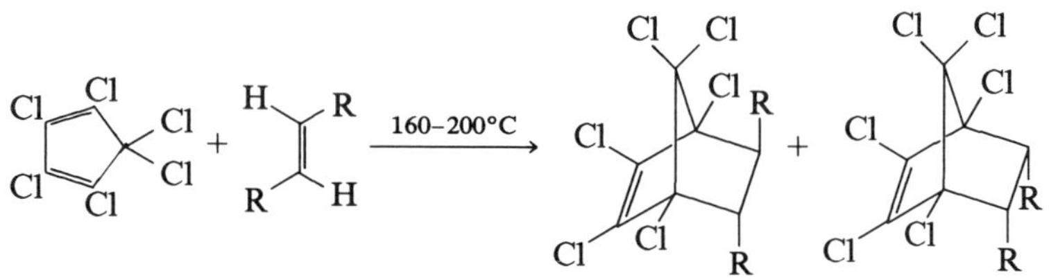

24 **25** **26**

R	Percentage **25**	Percentage **26**
Cl	5	95
CN	57	43
Ph	71	29
CO_2Me	73	27

Scheme 6.1

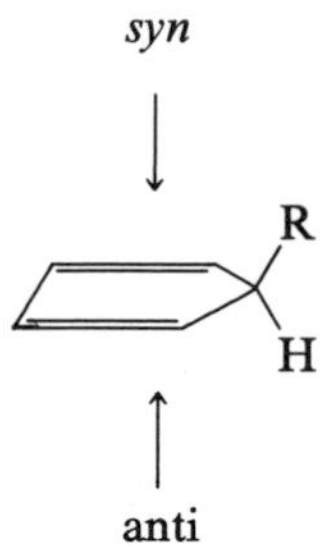

R	Dienophile	*anti* Addition	Ref.
CH_2OCH_2Ph	MA	100%	157
CH_2OMe	$CH_2{=}CHNO_2$	100%	120*a*
$SiMe_3$	$CH_2{=}CHNO_2$	100%	120*a*
OAc	$CH_2{=}CH_2$	0%	436
Me	NPM	50%	436

Scheme 6.2

stereoelectronic nature of the diene and dienophile. Scheme 6.2 and Table 6.3 summarize some results of reactions of mono- and polysubstituted cyclopentadienes. The results of cycloaddition of 5-acetoxycyclopentadiene and ethene are unexpected; the diene approaches exclusively from the sterically more hindered face. The reactions of pentachlorocyclopentadiene with MA, BQ, and MAC are the same unexpected if they are compared with those of the pentamethyl analog.[171,434–437]

Examples of facial stereoselectivity of norbornyl–norbornenyl fused cyclopentadienes and their analogs are reported in Table 6.4. The dienes **28**–**30** react with triazolinediones from the *endo* face (*syn* to ethano or etheno bridge). In contrast, the isodicyclopentadiene **27** reacts with the same dienophiles from the *exo* face (*anti* to the ethano or etheno bridge). DMAD prefers the *endo* face of dienes **27** and **29** but the *exo* face of diene **30**. Crossover in stereoselection also occurs in the reactions of **27**, **28**, and **29** with BQ. Below-plane additions occur exclusively in the Diels–Alder reactions of dehydroisodicyclopentadiene **28** with all dienophiles except (*E*)-1,2-bis(phenylsulfonyl)ethene.[93,171,246,247a,250,254,258,260,438]

To date there is no agreement on the origin of the observed π-facial stereoselectivity, which has been thought to result from attractive interactions due to dipole–dipole, dipole–induced dipole, and London dispersion forces between the diene and the dienophile,[434,435] secondary orbital interactions,[436]

TABLE 6.3 Percentage of *Anti* Addition in the Diels–Alder Reactions of 1,2,3,4,5 Pentamethyl- and 1,2,3,4,5 Pentachlorocyclopentadiene

Dienophile	Pentamethylcyclopentadiene	Pentachlorocyclopentadiene
MA	79	9.2
BQ	83	39.9
ACN		28.3
MAC	82	46.8

TABLE 6.4 Facial Stereoselectivity of Norbornyl–Norbornenyl Fused Cyclopentadienes and Analogs

Diene	Addition from: Endo Face		Exo Face	
exo face ↓ 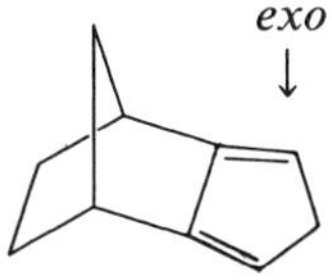 ↑ *endo* face **27**	BQ	100%	NPTAD	100%
	DMAD	100%	NMTAD	100%
	MAC	100%	(*E*)PSE[a]	100%
	NPM	> 90%		
	(*Z*)PSE[a]	55%		
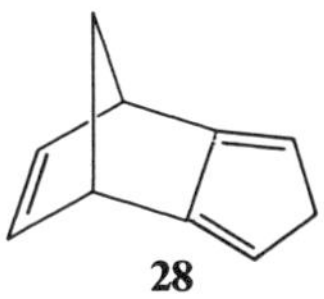 **28**	NMTAD	100%	(*E*)PSE[a]	90%
	NPTAD	100%		
	MAC	100%		
	MA	100%		
	(*Z*)PSE[a]	100%		
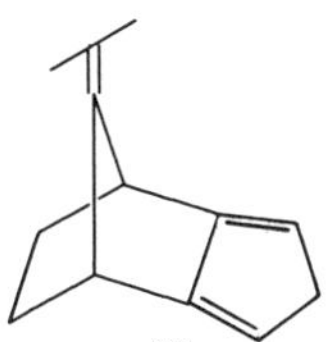 **29**	NMTAD	100%	BQ	63%
	NPM	50%	NPM	50%
	DMAD	100%		
exo face ↓ 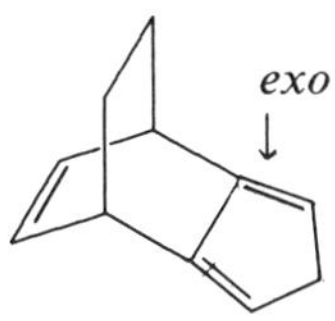 ↑ *endo* face **30**	NMTAD	100%	DMAD	86%
	NPM	80%	MP	79%
	MA	80%	Benzyne	81%

[a](*E*) and (*Z*) 1,2-bis(phenylsulfonyl)ethene.

π-orbital tilting associated with σ/π interactions,[437] combination of torsional effects,[171] and product stabilities.[439]

Study of regiochemistry achieved on 1- or 2-monosubstituted cyclopentadienes are complicated by the 1,5-sigmatropic transpositions.[152,164,165,440] 1-Methylcyclopentadiene **31** interconverts to a mixture of **31**–**33** at room temperature. However, the low equilibrium percentage of **33** permits[440] a careful study of the regiochemistry of additions of ACN and MAC to **31** and

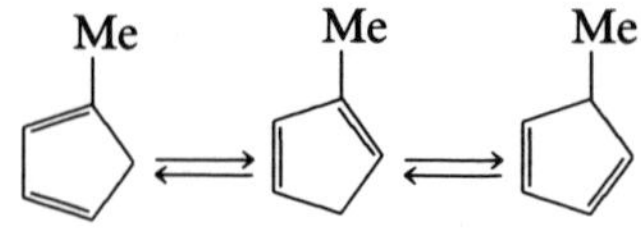

31 **32** **33**

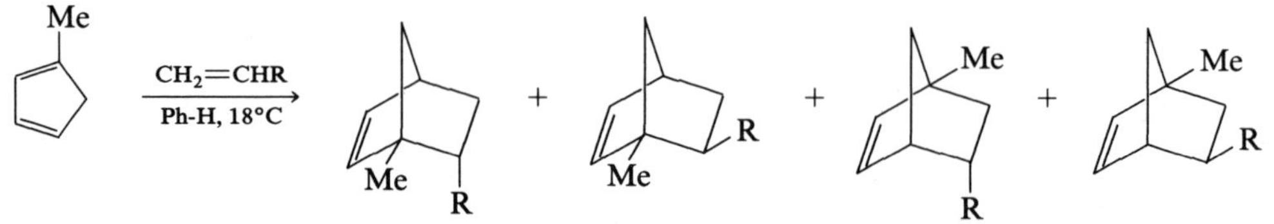

31 R	*ortho-endo*	*ortho-exo*	*meta-endo*	*meta-exo*
CN	34%	53.2%	7.9%	4.9%
CO_2Me	65.6%	22.6%	8.4%	3.4%

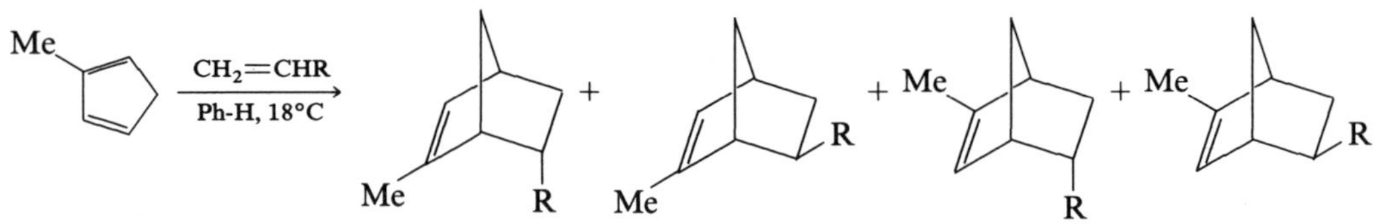

32 R	*meta-endo*	*meta-exo*	*para-endo*	*para-exo*
CN	21.2%	7.7%	39.1%	32%
CO_2Me	21.4%	4.3%	60.3%	14%

Scheme 6.3

32. The results (Scheme 6.3) follow those of 1-methyl- and 2-methylbutadiene. 1-Methylcyclopentadiene gives prevalently *ortho* adducts, while the 2-methyl isomer shows a prevalence toward the *para* addition.

The methylcyclopentadienyltrimethylsilanes (**34**, **35**) are interesting molecules that allow the examination of facial and *endo-exo* diastereoselectivity and regioselectivity at the same time.[165b] A rapidly equilibrating mixture of **34** and **35** reacts with MAC in the presence of $BF_3 \cdot Et_2O$ and produces an 83 : 17 mixture of **36** (*para-endo-anti* addition) and **37** (*ortho-endo-anti* addition) in 90% yield.

Cyclopentadienes are versatile reagents for organic synthesis. The cycloaddition of imine **38** and cyclopentadiene gives the adduct **39**, which was used as a key starting material for the synthesis of antibiotic aristeromycin (**40**).[443] The azabicyclo **41**, which is the first example of Diels–Alder cycloadduct obtained by a nitrile, has been converted into puromycin analog **42**[441] and into the natural product **43**.[442]

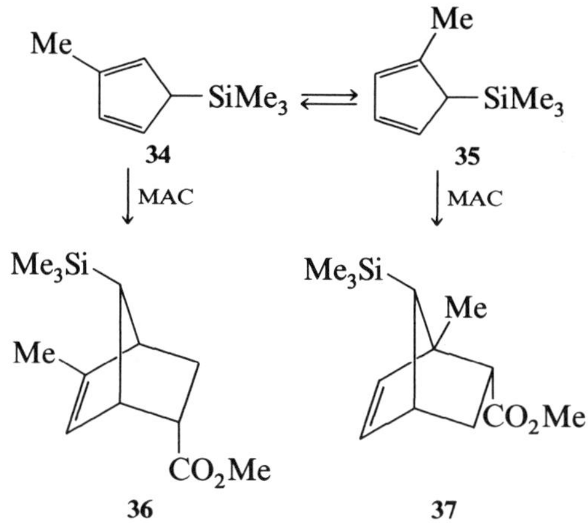

An example of an asymmetric reaction of cyclopentadienes with dienophiles that have a removable chiral auxiliary is the cycloaddition with the chiral allenic ester **44**, which allows a high enantioselective synthesis of (-)β-santalene (**45**).[444]

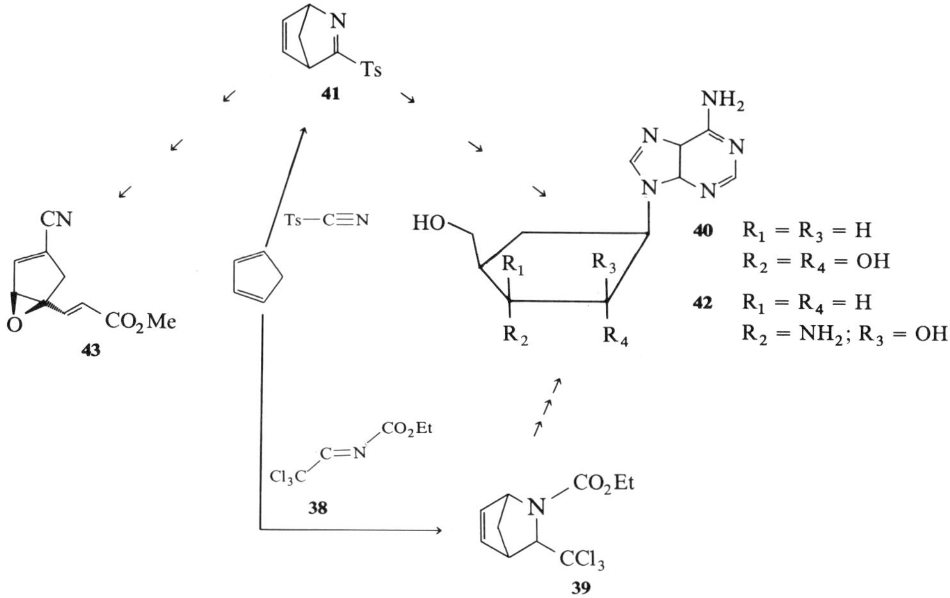

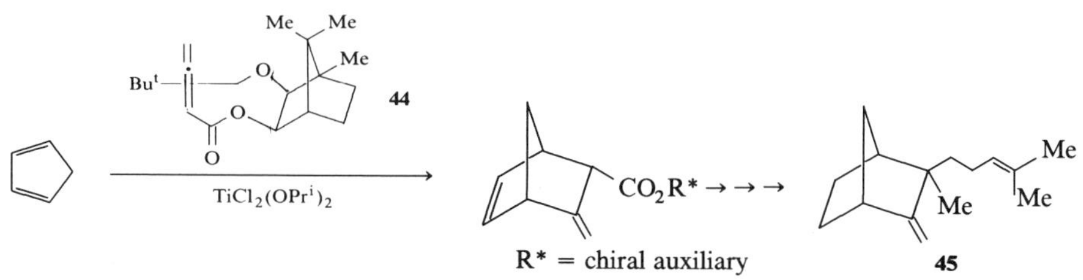

Fulvenes Fulvenes have considerable synthetic potential and have been studied in great detail.[445] They have three conjugated double bonds but only the pair in *cisoid* form give Diels–Alder adducts. In the field of Diels–Alder reactions, they are distinguishable from cyclopentadienes because the cycloadditions are generally less *endo*-selective and the adducts easily give a retro Diels–Alder reaction.

The diastereoselectivity of the cycloaddition depends on the substituents of the external methylene and temperature. Fulvene (**46**: R = H) and 6,6-dimethylfulvene (**46**: R = Me) react with MA under kinetic control to give a mixture of *endo* and *exo* adducts in which the *endo* adduct slightly predominates. Under thermodynamic control the more stable *exo* adduct is prevalent (ca. 85%). Similar results are obtained with the anhydride **47**,[32] but with the dienophiles **48**, the 6,6-dimethylfulvene (**46**: R = Me) gives exclusively *exo* addition.[243] The 6,6-diphenylfulvene (**46**: R = Ph) with MA and, under kinetic control, gives only *endo* adduct, which isomerizes to the more stable *exo* under equilibrating conditions.[446]

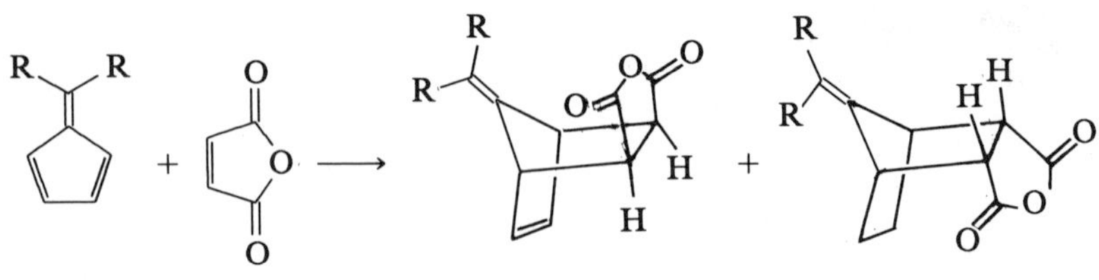

46 *exo* *endo*

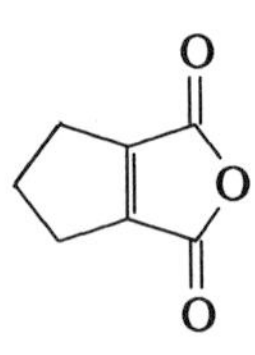

47

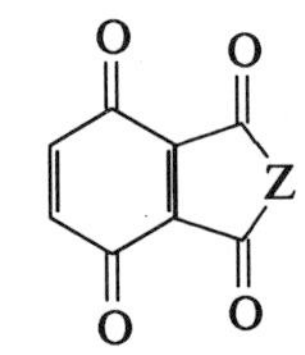

48 Z = O, NPh

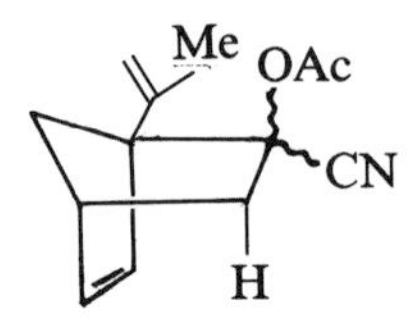

49

The higher *endo* selectivity of 6,6-diphenylfulvene (**46**: R = Ph) with respect to 6,6-dimethyl analog (**46**: R = Me) was also found in the cycloaddition reaction with aliphatic dienophiles.[34a]

An unexpected result is the cycloaddition of **46** (R = Me) with 1-acetoxyacrylonitrile. In addition to the normal *endo* and *exo* adducts, the unexpected isomer **49** was isolated.[244] Experimental evidence seems to indicate that **49** is produced by the 1-isopropenylcyclopentadiene, which is the result of an acid-catalyzed isomerization of **46** (R = Me).

Cyclopentadienones (Cyclones) Cyclopentadienones react with ethylenic and acetylenic dienophiles to give mainly *endo* adducts, which, when heated, have a tendency to eliminate the carbonyl bridge with the formation of cyclohexadienes or benzenes. Secondary adducts are sometimes produced as a consequence of the competitive cycloaddition of cyclohexadiene derivative with the dienophile.

Cyclopentadienone (**51**) is an unisolable compound trapped by dienophiles.[209,210] When **51** is generated slowly by the simple heating of **50** in the presence of an excess of ethyl acrylate and in the absence of a base, only the *endo* adduct **52** is isolated in 20% yield and utilized as starting material for the synthesis of sarkomycin **53**.[210]

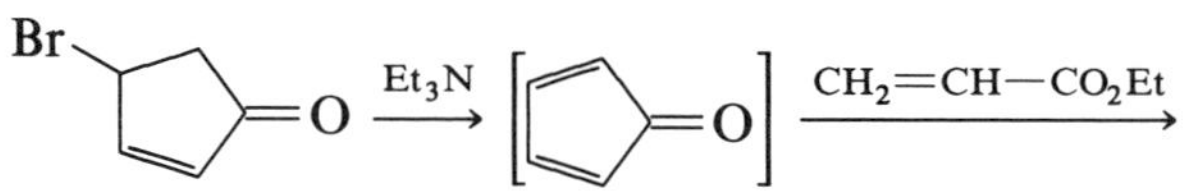

50 **51**

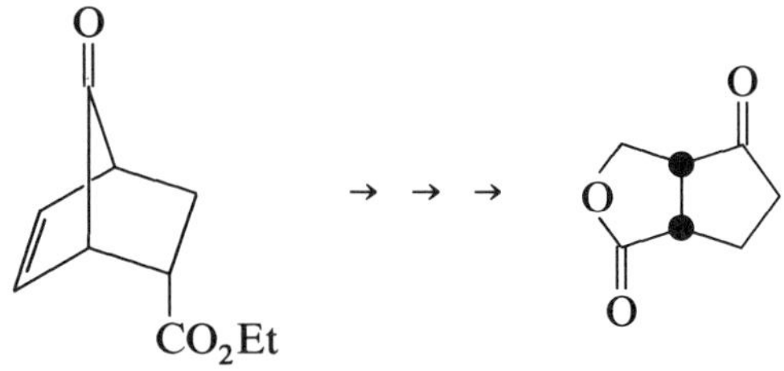

52 **53**

Mixtures of *endo* and *exo* adducts are obtained in the cycloaddition of 2,5-dimethyl-3,4-diphenylcyclopentadienone (**54**) with cyclopentene (**55**: $n =$ 1) and cyclohexene (**55**: $n = 2$) in benzene solution at 105°C. The *endo/exo* product ratios (0.91, $n = 1$ and 2.6, $n = 2$) are the same as those found in liquid-crystalline solvents.[222]

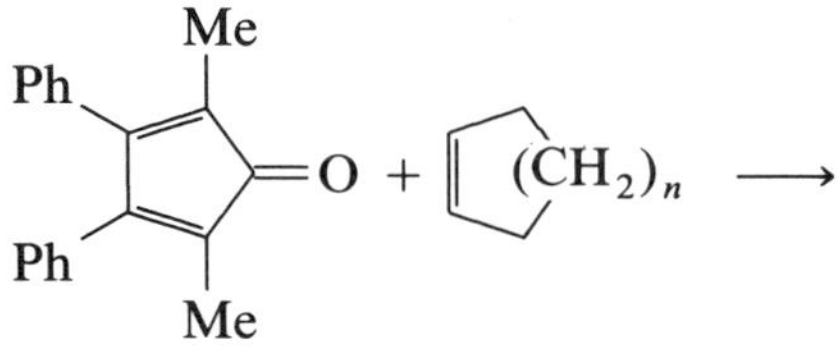

54 **55**

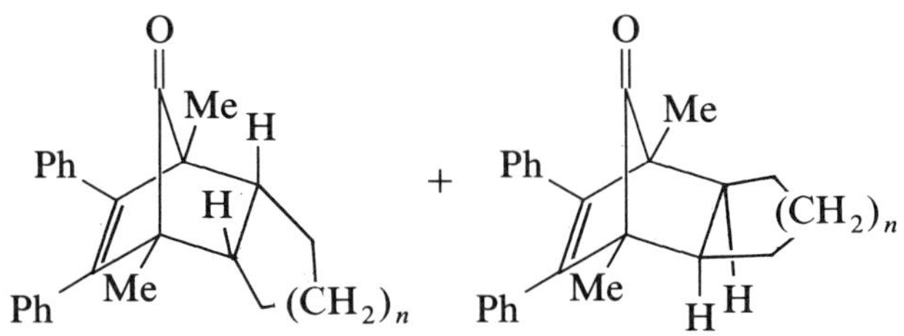

endo *exo*

An example of regioselectivity is observed[221] in the reaction of cyclopentadienone **56** with 2-phenyl-1-azaspiro[2.2]pent-1-ene (**57**). The reaction proceeds via the initial formation of Diels–Alder adducts, followed by loss of CO to give primarily 2*H*-azepines, which rearrange to more stable 3*H*-azepines **58** and **59**. The heterocyclic compounds **58** and **59** are formed in a 1.6 : 1 ratio, and the prevalence of **58** is ascribable to electronic factors which overcome the steric ones in the energetic balance of the transition state.

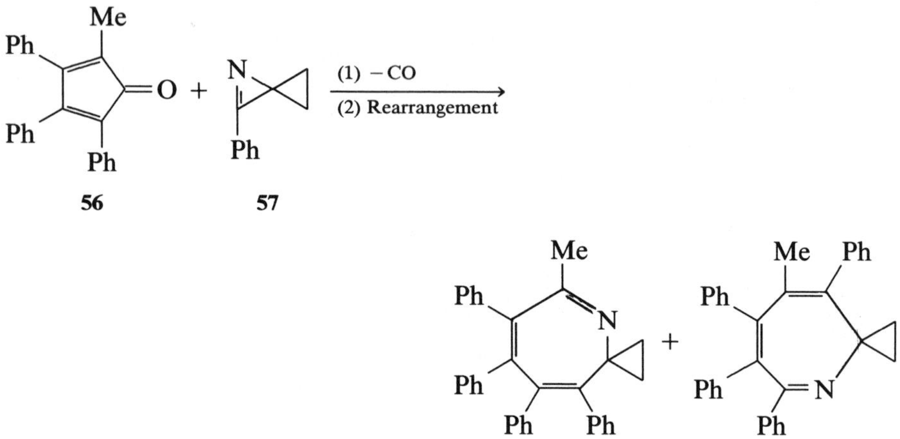

56 **57** **58** **59**

Cyclones are versatile compounds used to study mechanistic aspects of the Diels–Alder reaction. Neutral-type Diels–Alder reactions were observed[424,431] in the studies of cycloadditions of tetracyclones **60** (R = Ph, Ar), phencyclones **61** (R = H, Ar), and acecyclone **62** (R = H) with styrenes and arylacetylenes.

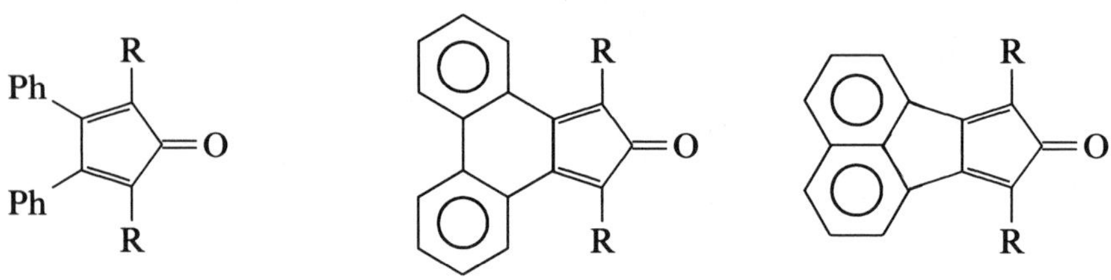

60 (R = Me, Ph, Ar, CO_2Et) **61** (R = H, OMe, Cl, Ph, Ar) **62** (R = H, CO_2Et)

Tetracyclone **60** (R = Ph) has a tendency to give neutral type Diels–Alder reactions when it reacts with *N*-arylmaleimides in the presence of $AlCl_3$.[447] Similarly, the cyclopentadienone **60** (R = Me) shows good reactivity with 1-aryl-5-morpholino- and 1-aryl-5-pyrrolidino-4-methylene-*v*-triazolines; on the basis of FMO theory, these cycloadditions also can be rationalized as being "neutral."[217]

In neutral Diels–Alder reactions, both electron-donating and electron-withdrawing substituents increase the stabilization of the transition state and therefore increase the reactivity of the system. Thus the dependence curve of log k versus the Hammett σ-constants of the substituents should be U-shaped. Substituted phencyclones **61** give neutral-type Diels–Alder reactions when they react with p-substituted styrenes and when log k/k_0 is plotted against σ^+ values,[424,431] a U-shaped curve with a minimum for H as substituent is observed.

A transition from neutral to normal and inverse Diels–Alder reactions is possible by the introduction of appropriate substituents into the diene. Kinetic data and FMO theory indicate that the reactions of biscarboxyethylcyclones **60** (R = CO_2Et) and **62** (R = CO_2Et) with p-substituted styrenes can be explained as Diels–Alder reactions with inverse electron demand.[229]

Tables of Diels–Alder Reactions See Tables 6.5 to 6.16; see also the general remarks about classifications, abbreviations, and symbols in the Introduction.

Six-Membered Rings

1,3-Cyclohexadienes 1,3-Cyclohexadienes are versatile compounds in organic synthesis. Cycloaddition of 1-methoxy-1,3-cyclohexadienes **65** with acetylenes is a simple route to phthalates, benzoates, and phthalides.[301] When the reaction is carried out above 100°C, the Diels–Alder adducts **67** lose ethene and aromatic compounds (i.e., **66**) are obtained in good yield. The dienes **65** can be prepared in situ by isomerization of 1,4-isomers **64**, which can easily be

TABLE 6.5 Diels–Alder Reactions of Cyclobutadienes

Diene		Dienophile	Yield (%)	Ref.
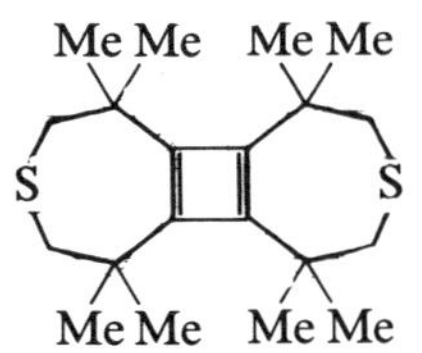	E1	MP	> 48	149
	E2	Phenylcarboxymethylacetylene	> 53	149
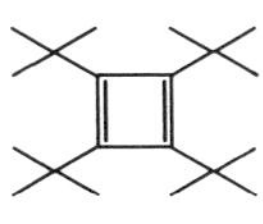	E5	Dicyanoacetylene	63	150
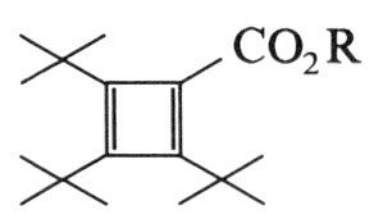 (R = Me, Bu^t)	H3	Phosphaalkynes		413

TABLE 6.6 Diels–Alder Reactions of Cyclopentadiene

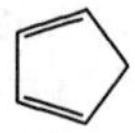

	Dienophile	Yield (%)	Ref.
A1	5-Methyl-2-(1-methyl-1-phenylethyl)-cyclohexyl acrylate (*)	81	16
A1	Methyl vinyl ketones	42–95	31
A1	Acrylates (*)	36–98	36
A1	AC, Cat. (*)	65–100	1c, 59
A1	Acryloyl chloride		144
A2	MTA, Cat. (*)	36–83	1
A2	α,β-Unsaturated acyliron complexes (*)	84–99	10
A2	Alkyl-*N*-acetyl-α,β-dehydroalaninates	56–79	67
A2	α-Acetamido acrylates	56	75
A2	α-Acrylmethyl acrylates	44–53	108
A3	Diethylocta-2,6-dien-4,5-isopropylidenedioxy-1,8-dicarboxylate (*)	90	5
A3	*N*-(2-Alkylacryloyl)1,3-oxazolidin-2-one, Cat. (*)	72–93	25
A3	3-Phenylsulfinyl-prop-2-enoic acid (*E* and *Z*)	89–100	30
A3	1,1,1-Trifluoro-2-penten-4-one	80	11
A3	*N*-Acyl sultam (*)	30–98	37
A3	Methyl-(*E*)-2,3-dideoxyarabinohept-2-enoates (*)	40–79	43
A3	(*E*)-6,6-Dimethoxy-3-hexen-2-one	60	56
A3	Cinnamic acids	25–99	60
A3	Methyldecan-8-en-10-oxo-oate	73	65
A3	α-Methyl-β-chloromethyl acrolein	90	82
A3	(*E*)-4-Oxo-2-pentenoic acid	100	83
A3	(*E*)-4-Benzyloxybut-2-enal	57–100	84
A3	(*Z*)-Methyl-*p*-tolylsulfinyl-3-propenoate (*)	100	104
A3	Crotonate imides (*)	78–88	124
A3	Carbomethoxyvinyltrialkylammonium salts (*E* and *Z*)	78–100	148
A4	2-(4-Chlorophenylseleno)-2-butenal	76–83	46
A4	3-Chloro-2-methylpropenal (*E* and *Z*)	32–68	133
A4	α-Methylcrotonaldehydes	64	137
A7	Menthylphenethyl fumarate (*)	93	3
A7	Diethyl (and dimethyl) fumarate (k)	70–100	23
A7	Dibutyl maleate	86	55
A7	Poly(ethene maleate)		105
A8	1-Alkenylsilatranes	60–80	39
A8	*p*-Tolylvinyl sulfoxide (*)	95	42
A8	*S*-*p*-Tolyl-*S*-vinyl-*N*-sulfoximines	81–95	49
A8	3,3,3-Trifluoropropene	70	52

TABLE 6.6 (*Continued*)

	Dienophile	Yield (%)	Ref.
A8	Vinylsulfone	80–100	78
A8	Ethene		(91), (136)
A8	Vinyltriphenylphosphonium bromide	100	107
A8	Allylacetylenes	66–69	116
A8	Nitroethenes	94–100	120
A8	Pentacarbonylmethoxyvinyl chromium and tungsten complexes	78–93	130
A8	Acrylic monomers, Cat. (*)		142
A9	Pentacarbonyl[isopropenyl(methoxy)carbene] chromium	68	6
A9	1,1-Bis(*p*-tolylsulfinyl)ethene (*)	98	8
A9	1-Acyl-1(trimethylsilyloxy)ethene	48–58	33
A9	Acrylonitriles	60–92	132, (136)
A10	1-Sulfonyl-2-sulfinylethenes (*)	> 90	12
A10	1-Sulfonyl-2-nitroethene	82	13
A10	*N*-Vinylpyridinium tetrafluoroborates	65–80	22
A10	*E*-1-Nitro-2-(phenylsulfinyl)ethene	97	24
A10	Phenylsulfonylethenes (*E* and *Z*)	30–89	48
A10	(*E*)-1-Benzenesulfonyl-2-(trimethylsilyl)ethene	89	112
A10	*O*,*O*-Bis(*β*-chloroethyl)-2-nitrovinylphosphonate		40
A10	1-Trifluoromethyl-2-sulfonylfluoride-ethene	70	117
A10	3-Trimethylsilyacrylic acid	78	123
A10	1-*p*-Anisylpropene	74	128
A11	Acetoxymethylene dimethylmalonate	80	7
A11	3-(Pentafluoroisopropenyl)-5,5-bis(trifluoromethyl)-1,4,2-dioxazoline	84	38
A11	Nitrostyrenes	47–90	90
A11	4-Methyl-3-penten-2-oles	52–56	119
A12	Alkenesulfonyl fluorides (*E* and *Z*)	75–87	94
A12	Tetranitroethene	85	97
A12	1,1-Dicyano-2-chloroethenes	22–95	109
A12	Cyanoethenes (k)		63d–g, (136)
A13	1,2-Butadiene	57	26
A13	Arylsulfonylallenes (*)	32–90	34
A13	Allenes (*)	83–98	53
A13	Allenic esters	50–90	113
A13	Fluoroallenes	99	122
A14	(*E*)-2-Benzenesulfonyl-2,4-pentadiene	55	15
A14	1,3-Butadiene		101

TABLE 6.6 (*Continued*)

	Dienophile	Yield (%)	Ref.
A14	[2-Formyl-1,3-butadiene]	100	131, 147
A14	[2-Carboxyethyl-1,3-butadiene]	50	147
B1	2,2-Difluoromethylene cyclopropane	55	14
B1	Methyl(2-chloro-2-cyclopropylidene) acetate	91	111
B2	Substituted-5-methylene barbituric acids	84–88	71
B2	5-Methylene-2,2-dimethyl-1,3-dioxane-4,6-dione		50
C1	Tricyclo[4.2.2.2^{2,5}]dodeca-1,5-diene	61	45
C1	Spiropentene	100	64
C1	1,2,3-Tris(trifluoromethyl)-3-trifluoroacetylcyclopropene	100	66
C1	*cis*-3,4-Dichlorocyclobutene	85	74
C1	Cyclobutenedione		95
C1	3-Methyl-3-cyanocyclopropene		100
C1	Hexachlorocyclopentadiene		115
C1	Dichlorobistrifluoromethylcyclopropenes	84–95	118
C1	Cyclopropenone ketal	57	139
C2	Hexakis(trifluoromethyl)bicyclo[2.2.0]hexa-2,5-diene	52	73
C2	Hexakis(trifluoromethyl)benzvalene	100	79
C2	1,2-Bis(phenylsulfonyl)norbornadiene	91	93
C2	Bicycloalkenes (k)		121
C4	Cycloalkenones	20–99	54
C5	2,5-Cyclohexadienones	63–84	72
C6	*p*-Tropoquinone	> 80	85
C6	1,2-Benzoquinones		102
C7	1,4-Benzoquinones	40–100	28, 63g
C8	1,2-Dihydro-4,7-dihydroxycyclobuta[*b*]-naphthalene-3,8-dione	98	35
C8	1-Ethyl-4,5,8-trioxo-1,4,5,8-tetrahydro-3-quinolincarboxylic acid	89	50
C8	Pyrrolo[2,1*c*][1,2,4]benzotriazine-6,9-dione	70	69
C8	4*a*,9*a*-Epoxy-4*a*,9*a*-dihydroanthracene-1,4,9,10-tetrones	85	81
C8	Diquinones	100	138
C9	1,4-Naphthoquinones	75–95	19
C12	[1-Phenylphosphole oxide]	68	47
D1	Hexenopiranoside (*)	30	126
D2	Butenolides (*)	55–85	4

TABLE 6.6 (*Continued*)

	Dienophile	Yield (%)	Ref.
D2	Levuglucosenone (*)	81	44
D2	Glutaconic anhydride	70	70
D2	Acylbutenolides	59–77	108
D2	3-Carbethoxy-5,5-dimethyl-2(5*H*)furanone	100	129
D3	3-Oxabicyclo[3.3.0]oct-1(5)-ene-2,4-dione	88	32
D3	Maleic anhydrides (k)	80–90	63a–f, (86)
D3	Norbornene-2,3-dicarboxylic anhydride		110
D4	Δ^3-Pyrrolin-2-ones (*)	58–67	29
D4	2-Ethoxycarbonylhexafluoro-2-azabicyclo[3.2.0]hepta-3,6-diene	98	57
D4	2,3,4,5,5-pentachloro-1-azacyclopentadiene	90	134
D5	[*N*-Methylacylmaleimides]	30–74	125
D6	2-Thiolene-1,1-dioxides (k)		141
D6	3-Thiolen-2-one	63	146
D10	1,4-Benzodithiin oxides	95	96
E2	Methyl(phenylthio)propiolate	90	51
E2	Alkynoyl chlorides	78–84	106
E2	Methyl-3-bromopropiolate	89	135
E3	[Acetylendicarbonyl dichloride]	66	58
E3	Acetylendicarbaldehyde	72	76
E4	[Diphenylpropargyl cation]	80	2
E4	Propargylic alcohol	72	98
E4	Ethynil-*p*-tolylsulfone	100	145
E5	(Triisopropylsilyl)nitroacetylene	90	87
E5	Pentacarbonyl(methoxyethynylcarbene) chromium and tungsten complexes	85–91	88, 127
E6	Cyclooctyne	17	41
F1	(*Z*)-Alkanethial-*S*-oxides	67	27
F1	[Sulfines and sulfenes]	71	99
F1	Cyanothioformamides	31–94	143
F2	Adamantane thione	91	18
F7	[*C*-Acyliminium ions]	65–86	21
F7	[Methylenemethyliminium chloride]	100	40
F7	2-(Acylimino)perfluoropropane	91–93	61
F7	[Diacylimines]	55–81	62
F7	[*N*-Carbobenzyloxy-2-iminoacetate]	56	92
F10	[Selenoaldehydes]	39–89	9
F10	Pentacarbonyl(selenocarbonyles)tungsten	53–85	20

TABLE 6.6 (*Continued*)

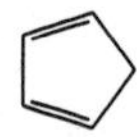

	Dienophile	Yield (%)	Ref.
I1	Arylazocyanides	58–100	77
I1	Bis(2,2,2-trichloroethyl)azodi-carboxylate	90	89
I4	2,3-Diazabicyclo[2.2.1]hept-2-ene	37	17
I4	[6,7-Dimethoxyphthalazine-1,4-dione]	52	68
I4	2,3-Diazabicyclo[2.2.1]hept-2-ene-3H^+		103
I5	1,2,4-Triazolin-3,5-diones	63	80

TABLE 6.7 Diels–Alder Reactions of Monosubstituted Cyclopentadienes

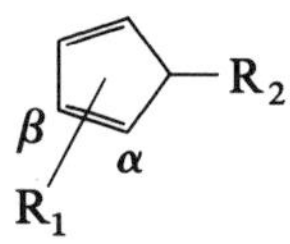

R_1	R_2		Dienophile	Yield (%)	Ref.
Me	H	A1	Acrylates		151
		A1	MVK	65	160
		A1	1-Octen-3-one	64	160
		A3	Crotonaldehyde		162
		A9	α-Chloroacrylonitrile	64	152
		C7	BQ	94	28b
$CH{=}C{=}CH_2$	H	E5	Hexafluorobutyne	26	161
$CMe{=}CH_2$	H	E3	DMAD	70	153
$CEt{=}CH_2$	H	E3	DMAD	70	153
α-COR	H	E3	DMAD	11–15	155
(R = Me, Pr_i, Bu^t, n-C_7H_{15})					
CO_2Me	H	A8	Vinyl acetate	34	154
		A9	α-Chloroacrylonitrile	20	154
$C(OMe)_3$	H	C7	BQ		164
		E3	DMAD		164
β-OMe	H	A1	MAC	70	159
$SiMe_3$	H	A2	Acrylic acids		156
		D3	MA		156
		D5	NPM	90	165
H	CH_2OBn	D3	MA	60	157
H	$SiMe_3$	A7	DMF		166
H	Nb	E5	Hexafluorobutyne	10	158
H	Pt	E5	Hexafluorobutyne		163

TABLE 6.8 Diels–Alder Reactions of Polysubstituted Cyclopentadienes

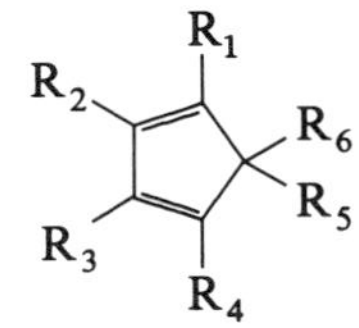

R_1	R_2	R_3	R_4	R_5	R_6		Dienophile	Yield (%)	Ref.
				Disubstituted					
$-(CH{=}CH)_2-$		H	H	H	H	A10	(*E*)-1,2-Bis(phenyl-sulfonyl)ethene	22	48b, 382
$-(CH_2)_6-$		H	H	H	H	E5	Hexafluorobut-2-yne		263
$-CH{=}CH(-CH_2-)_4$		H	H	H	H	E5	Hexafluorobut 2-yne	90	263
p-RC_6H_4	H	H	*p*-RC_6H_4	H	H	I3	[Singlet oxygen] (k)		169
Me	H	H	H	H	$SiMe_3$	A1	MAC	16	165b
						D5	NPM		165a
[H	$-(CH{=}CH)_2-$		H	H	H]	D3	MA		264
[H	Ph	Ph	H	H	H]	A12	TCNE	65	168
H	$OSiMe_3$	$OSiMe_3$	H	H	H	A7	Diethylfumarate	87	208
						C1	Cyclobutenedione		95
						D3	MA	87	208
H	Me	H	H	H	$SiMe_3$	A1	MAC	74	165b
						D5	NPM		165a
[H	H	H	$-(CH_2)_2-$		H]	A1	Acryloyl chloride	70	172
H	H	H	H	$-(CH_2)_2-$		C1	[Tricyclo[4.1.0.0^{2,7}]hept-1(7)ene]	38	240
						C4	Bicyclo[3.2.2]nona-3,6-dien-2-one	58	54a

TABLE 6.8 (*Continued*)

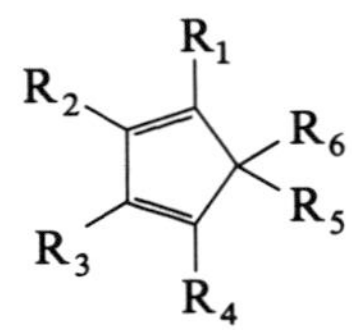

R_1	R_2	R_3	R_4	R_5	R_6		Dienophile	Yield (%)	Ref.
				Disubstituted (Continued)					
H	H	H	H	CN	CN	A1	MAC	63	167
						C1	5,5-Dicyanocyclopentadiene	91	167
H	H	H	H	Ph	Ph	D3	MA	79	168
H	H	H	H	OEt	OEt	A7	Fumaroyl chloride	> 15	166
						C7	BQ		171
						D3	MA		171
						D5	NPM		171
						C7	BQ	100	170, 171
						D3	MA	95	170, 171
						D5	NPM	100	170, 171
				Trisubstituted					
[CO_2H	$-(CH{=}CH)_2-$		H	H	H]	A7	DMF	19	264
						D3	MA	71	264
						D5	NPM	46	264
[CO_2Me	$-(CH{=}CH)_2-$		H	H	H]	D3	MA	43	264
						D5	NPM	50	264

				Tetrasubstituted					
Me	Ph	Ph	Me	H	H	E3	DMAD		175
Ph	Ph	Ph	Ph	H	H	A10	*trans*-Stilbene (k)		174
						C2	Norbornadiene (k)		174
Cl	Cl	Cl	Cl	H	H	Cl	Fulvenes	43–51	173
H	Me	Me	H	Me	Me	E3	DMAD		175
H	Me	*p*-$MeOC_6H_4$	H	Me	Me	E3	DMAD		175
H	*p*-$MeOC_6H_4$	*p*-$MeOC_6H_4$	H	Me	Me	E3	DMAD		175
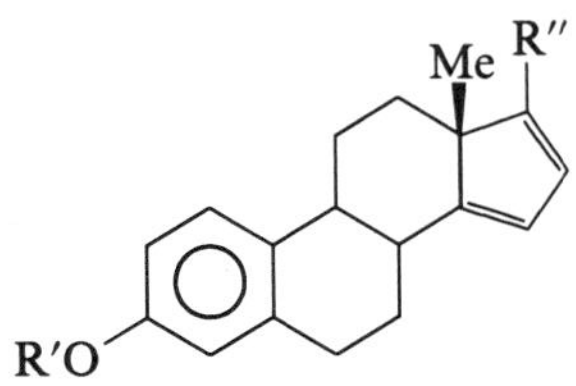	R′ = Ac, R″ = OAc					A8	Phenylvinylsulfone	92	265
	R′ = Me, R″ = Ph					E1	Propargylic aldehyde	94	266
						E1	Butynone	91	266
				Pentasubstituted					
Me	Me	Me	Me	Me	H	A1	MAC	> 90	177
						C7	BQ	> 90	177
						D3	MA	> 90	177
Cl	Cl	Cl	Cl	Cl	H	A8	Allyl-*o*-chlorobenzoate	87	176
				Hexasubstituted					
Cl	Cl	OMe	Cl	OMe	OMe	C7	BQ	95	194b
						C9	NQ	60	194
						D3	MA	86	194b
Cl	Cl	OMe	Cl	—$(CH_2)_2O$—		C7	BQ	71	194b
						C9	NQ	90	194b
						D3	MA	62	194b

TABLE 6.8 (*Continued*)

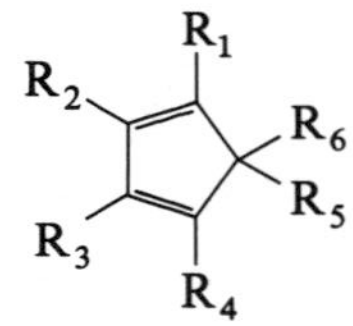

R_1	R_2	R_3	R_4	R_5	R_6		Dienophile	Yield (%)	Ref.
				Hexasubstituted (*Continued*)					
Cl	Cl	Cl	Cl	OMe	OMe	A1	MAC	97	201
						A2	Methyl methacrylate	86	201
						A8	Allyl-*p*-chlorobenzoate	78	198
						A8	Allylacetylenes	78–95	203
						A8	Allylchloro acetates	30–80	204
						A9	Methacrylonitrile	73	201
						A11	Isobutenyl acetate	38	179
						A12	Aliphatic alkenes		197
						C1	Fulvenes	35–68	173
						C1	*N*-Aryltetrahydro-phthalimides	75–90	199
						C3	7-Methylenebenzo-norbornadiene		200
						D2	Levuglucosenone (*)	68	44b
						D7	2-(*N*-Benzyl-*N*-methylamino)oxazole	85	190
						E2	Ethyl(diethoxy-phosphinyl)propynoate	100	196
						E4	Acetylenes	77–90	202

Cl	Cl	Cl	Cl	—$O(CH_2)_2O$—		A12	Aliphatic alkenes		197
Cl	Cl	Cl	Cl	Cl	Br	D3	MA		193
Cl	Cl	Cl	Cl	Cl	Cl	A5	4-Chlorohexa-3,5-dien-2-one	45	205
						A8	Allyl benzoate	88	178
						A8	*o*-Chlorophenyl allyl ether	95	185
						A8	Silatranes	60–80	189
						A8	Allylacetylenes	78–95	203
						A8	Allyl(β-cyanoethyl) ether	97	207
						A9	Styrenes (k)		187
						A10	*cis*-But-2,3-en-1,4-diole	83	184
						A10	(*E*)-β-Chloroacrylonitrile		195
						C1	*cis*-3-Methyl-4-cyclohexene-1,2-dicarboxylic anhydride (k)		180
						C1	Cycloalkenes (k)		188
						C1	Fulvenes	38–51	173
						C1	3-Methyl-*cis*-*cis*-4-cyclohexene-1,2-dicarboxylic acid-*N*-(4-pyridyl)imine	92	206
						C2	*N*-Arylimides (k)		181
						C2	*N*-Alkylimides (k)		183
						C2	Norbornadienes (k)		186
						D7	2-(*N*-Benzyl-*N*-methylamino)oxazole	86	190
						D9	4-Isothiazolin-3-one-1-oxide	33	182

TABLE 6.8 (*Continued*)

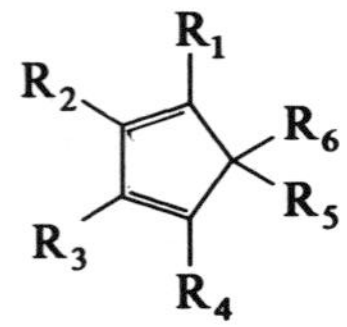

R_1	R_2	R_3	R_4	R_5	R_6		Dienophile	Yield (%)	Ref.
				Disubstituted					
						E4	Propargyl alcohol	72	98
						E4	Acetylenes	77–90	202
Br	Br	Br	Br	OMe	OMe	C2	*N*-Arylmaleimides	89–99	191d
Br	Br	Br	Br	Br	Br	A1	Acrylates (k)	85–87	192
						A1	Acrylic acid (k)	89	192
						A8	ACN (k)	43	192
						C2	*N*-Arylmaleimides (k)	89–99	191
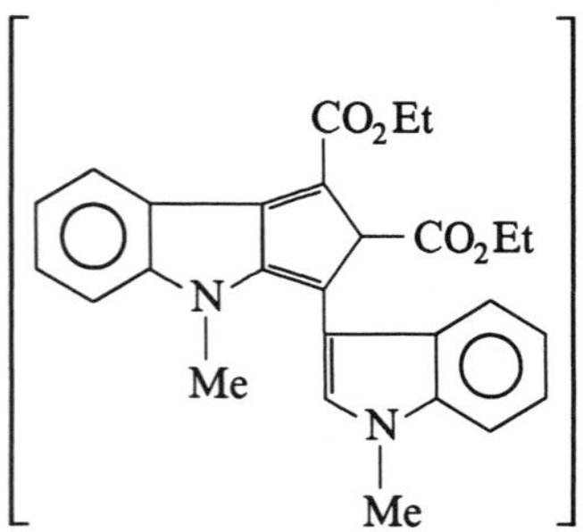						A8	ACN	47	421
						D3	MA	100	421
						D5	Maleimide	93	421

TABLE 6.9 Diels–Alder Reactions of Tricyclo[5.2.1.0^{2,6}]deca- and Tricyclo[5.2.2.0^{2,6}]undeca-2,5,8-trienes

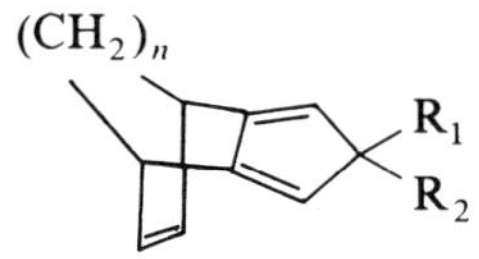

Dienophile		Yield (%)	Ref.
	$n = 1$ $R_1 = R_2 = H$		
A3	(*Z*)-Methyl-β-(phenylsulfonyl) acrylate	100	247a
A10	1,2-Bis(phenylsulfonyl)ethene (*E* and *Z*)	70	93
D3	MA	91	247
C7	BQ	85–88	247b, 248
E1	MP	83	247, 248
E4	*p*-Tolylsulfonylacetylene	> 98	250
E6	[Benzyne]	71	247b
I5	NMTAD	86	246, 248
	$n = 1$ $R_1 = H$ $R_2 = Me$		
D3	MA	66	249
	$n = 1$ $R_1 = Me$ $R_2 = H$		
D3	MA	66	249
	$n = 1$ $R_1 = R_2 = Me$		
D5	NPM	12	249
I5	NMTAD	47	249
	$n = 1$ $R_1, R_2 = -(CH_2)_2-$		
C7	BQ	32	252
D5	NPM	22	252
E3	DMAD	12	252
	$n = 1$ $R_1, R_2 = -(CH_2)_4-$		
D3	MA	94	252
E3	DMAD		252
I5	NMTAD	42	252
	$n = 2$ $R_1 = R_2 = H$		
D3	MA	48	251
D5	NPM	55	251
E1	MP	42	247b
E3	DMAD	62–94	247b, 251
I5	NMTAD		246

TABLE 6.10 Diels–Alder Reactions of Tricyclo[5.2.1.0^{2,6}]deca- and Tricyclo[5.2.2.0^{2,6}]undeca-2,5-dienes

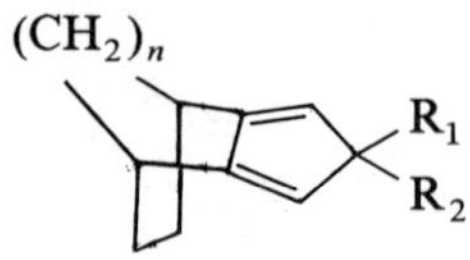

Dienophile		Yield (%)	Ref.
	$n = 1$ $R_1 = R_2 = H$		
A1	MAC	94	247b, 248
A8	Phenylvinylsulfone	91	248
A8	Thiophenylethene	69	253
A8	Phenylvinyl sulfoxide	77	253
A8	(Methoxyvinylcarbene)tungsten-pentacarbonyl	88	254
A12	TCNE	89	255
C1	[Cyclobutadiene]	42	254
C7	BQ	94	247b, 248
D3	MA	90–98	247b, c, 248, 249
D5	NPM	99	249
E1	MP	93	247b, 248
E3	DMAD	61	249
E5	Hexafluoro-2-butyne	48	254
E6	[Benzyne]	71	247b, 248
E6	Cyclooctyne	10	254
I5	NMTAD	82	246
I5	NPTAD	86	246b, 255
	$n = 1$ $R_1 = H$ $R_2 = Me$		
A1	MAC	24	249
A8	Phenylvinylsulfone	38	249
D3	MA	71	249
E3	DMAD	84	249
I5	NMTAD	58	249
	$n = 1$ $R_1 = H$ $R_2 = SiMe_3$		
A12	TCNE	51	256
D3	MA	35	256
D5	NPM	79	256
E3	DMAD	70	256
I5	NMTAD	23	256
	$n = 1$ $R_1 = SiMe_3$ $R_2 = H$		
D5	NPM	68	256
E3	DMAD	77	256
I5	NMTAD	40	256

TABLE 6.10 (*Continued*)

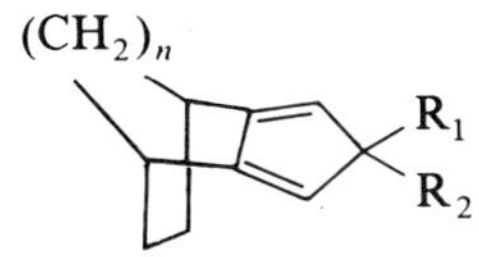

Dienophile		Yield (%)	Ref.
	$n = 1$ $R_1 = R_2 = Me$		
D3	MA	100	249
D5	NPM	83	249
I5	NMTAD	100	249
	$n = 1$ $R_1, R_2 = -(CH_2)_2-$		
A8	Phenylvinylsulfone	69	252
C7	BQ	45	252
E3	DMAD	84	252
	$n = 1$ $R_1, R_2 = -(CH_2)_4-$		
C7	BQ	65	252
D3	MA	47	252
D5	NPM	40	252
E3	DMAD	21	252
	$n = 1$ $R_1 = R_2 = SiMe_3$		
E3	DMAD	100	256
I5	NMTAD	60	256
I5	NPTAD	42	256
	$n = 2$ $R_1 = R_2 = H$		
I5	NMTAD	97	246a

TABLE 6.11 Diels–Alder Reactions of Isodicyclopentadiene-Related Compounds

Diene			Dienophile	Yield (%)	Ref.
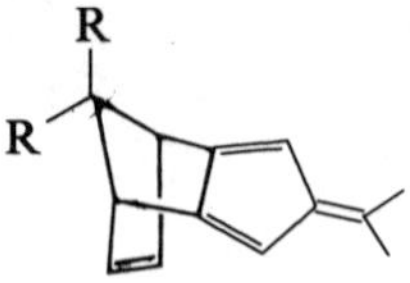	R = H	D3	MA		257
		D5	NPM	53	257
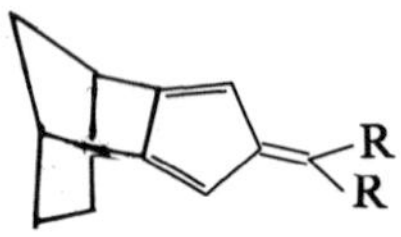	R = Me	E3	DMAD	90	257
		D3	MA	50	257
		D5	NPM	95	257
		E1	MP	54	257
	R = Ph	D5	NPM	87	258
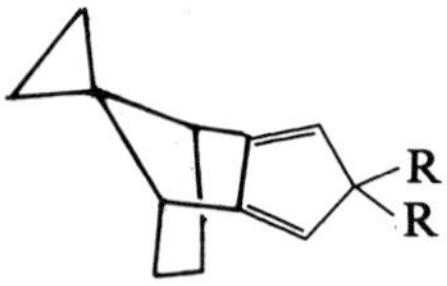	R = H	C7	BQ	27	259
		D5	NPM	63	259
		E3	DMAD	16	259
	R = Me	E3	DMAD	64	259
		I5	NPTAD	15	259
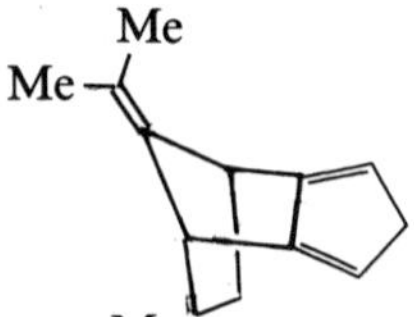		A8	Phenylvinylsulfone	92	260
		C7	BQ	92	260
		D5	NPM	63	260
		E3	DMAD	38	260
		I5	NMTAD	92	260
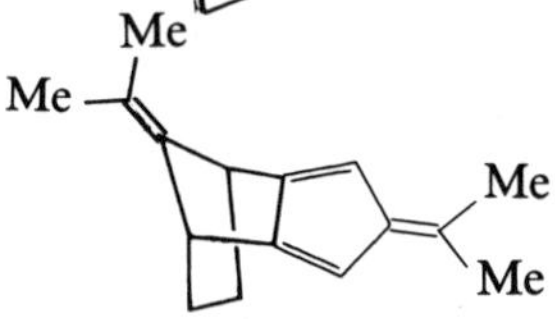		D5	NPM	96	258
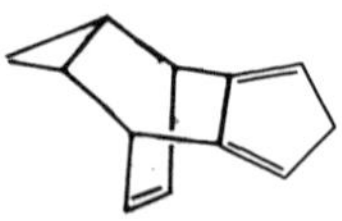		A8	Phenylvinylsulfone	68	251
		D5	NPM	61	251
		E3	DMAD	73	251
		I5	NMTAD	83	251
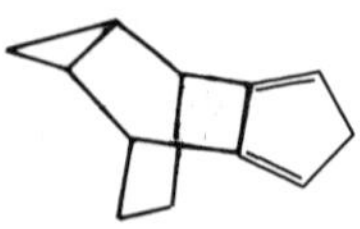		A8	Phenylvinylsulfone	66	251
		D5	NPM	72	251
		E3	DMAD	98	251
		I5	NMTAD		251
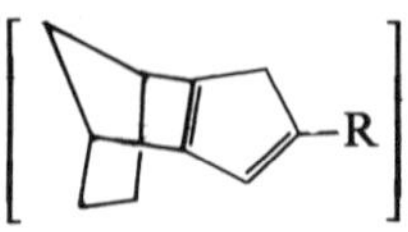	R = H	A8	Thiophenylethene	69	253
		A8	Phenylvinyl sulfoxide	77	253
		A10	(*E*)-1-(Phenylsulfonyl)-2-(trimethy-lsilyl)ethene	51	253
		A10	(*E*)-1,2-Dichloroethene	38	253
		A12	TCNE	30	255
	R = Me	A8	Phenlyvinylsulfone	66	253

TABLE 6.11 (*Continued*)

Diene		Dienophile	Yield (%)	Ref.
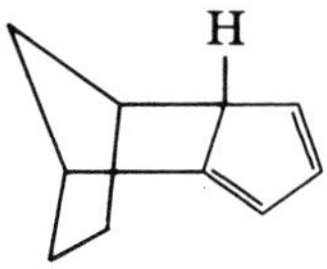	A12	TCNE	72	255
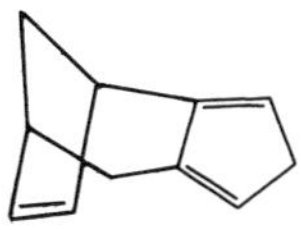	D3	MA		261
	E3	DMAD		261
	I5	NPTAD		261
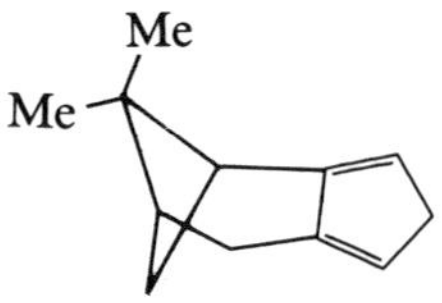 (*)	A10	(*Z*)-1,2-Bis(phenylsulfonyl)ethene	60	262
	C7	BQ	43	262
	D5	NPM	49	262
	E3	DMAD	88	262

TABLE 6.12 Diels–Alder Reactions of Fulvenes

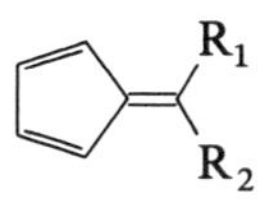

R_1	R_2		Dienophile	Yield (%)	Ref.
H	H	C1	Fulvene	65	241
H	Me	C1	6-Methylfulvene	63	241
H	$CH_2-\underset{nC_5H_{11}}{C}(CH_2O)_2$	A2	2-Chloroacryloylchloride	> 43	242a
		A9	2-Chloroacrylonitrile	88	242
Me	Me	A9	1-Acetoxyacrylonitrile	80	244
		A13	Phenylsulfonylpropadiene	87	34a
		C1	6,6-Dimethylfulvene	16	241
		C1	Cyclobutenedione		95
		C7	1,4-Benzoquinone-2,3-dicarboxylicanhydride	100	243
		D3	3-Oxabicyclo[3.3.0]oct-1(5)-ene-2,4-dione	92	32
Ph	Ph	A13	Phenylsulfonylpropadiene	83	34a
$-(CH_2)_2-O-(CH_2)_2-$		D3	MA		245
$-(CH_2)_2-S-(CH_2)_2-$		D3	MA		245

TABLE 6.13 Diels–Alder Reactions of Cyclopentadienones

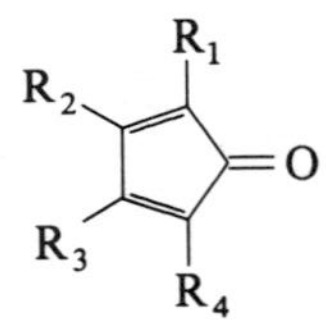

R_1	R_2	R_3	R_4		Dienophile	Yield (%)	Ref.
			Unsubstituted				
[H	H	H	H]	E3	Polimeric monoester of acetylene dicarboxylic acid		209
				A1	Ethyl acrylate	20	210
				A3	Methyl crotonate	13	210
			Disubstituted				
—CH=CH—CO—		H	H		Polimeric monoester of acetylene dicarboxylic acid	27	239
p—$OMeC_6H_4$	H	H	p-$OMeC_6H_4$	C2	Hexachlorotricyclononadiene	75	212
			Trisubstituted				
Me	Ph	H	Ph	C5	2-Methyl-3,5-diphenylcyclo-pentadien-1-one	23	213
Bu^t	Bu^t	H	Bu^t	C2	Hexachlorotricyclononadiene	57	212
Ph	H	Ph	Ph	C2	Hexachlorotricyclononadiene		212
Ph	Me	H	Ph	C5	3-Methyl-2,5-diphenylcyclo-pentadien-1-one	38	213
Ph	Ph	H	Ph	C5	2,3,5-Triphenylcyclopentadien-1-one	5	213

Tetrasubstituted

[Me	α-Fur	α-Fur	Me]	C1	1,3-Cyclohexadiene	72	228
				D3	MA	45	228
				E3	DMAD	63	228
[Me	α-Th	α-Th	Me]	D3	MA	55	228
				E3	DMAD	71	228
Ph	Me	Me	Ph	C2	Hexachlorotricyclononadiene	71	212
Ph	Me	Ph	Ph	C2	Hexachlorotricyclononadiene	71	212
Ph	p-MeOC$_6$H$_4$	Ph	p-MeOC$_6$H$_4$	C2	Hexachlorotricyclononadiene	68	212
Ph	p-MeOC$_6$H$_4$	p-MeOC$_6$H$_4$	Ph	C2	Hexachlorotricyclononadiene	78	212
p-MeOC$_6$H$_4$	p-MeOC$_6$H$_4$	Ph	Ph	C2	Hexachlorotricyclononadiene	75	212
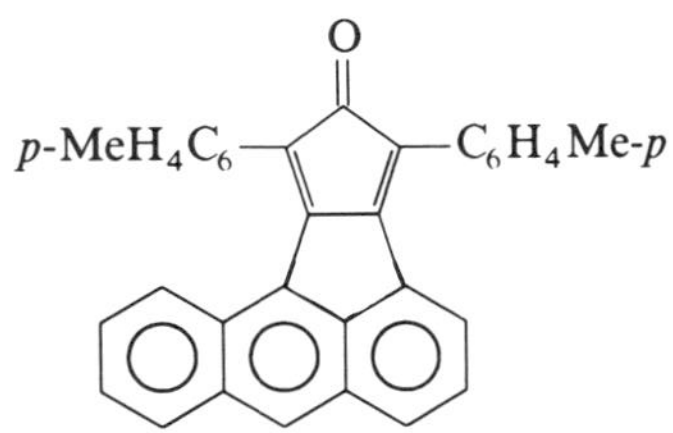				C3	Acenaphthylene		237
				D3	MA	80	237
				E5	Diphenylacetylene	55	237
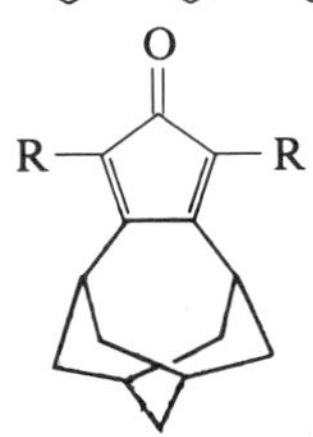				A12	Ethenes	44–83	238
				C1	1,5-Cyclooctadiene	58	238
				E5	Acetylenes	31–94	238
				I5	NPTAD	84	238
(R = Me, Ph, CO$_2$Me)							
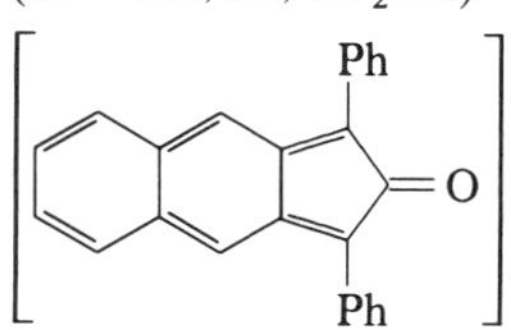				A7	DMF	27	420
				A7	DMM	28	420
				A10	But-2-ene (E and Z)	34–51	420
				C2	Norbornadiene	16	420
				D5	NPM	66	420

TABLE 6.14 Diels–Alder Reactions of 3,4-Diphenylcyclopentadienones

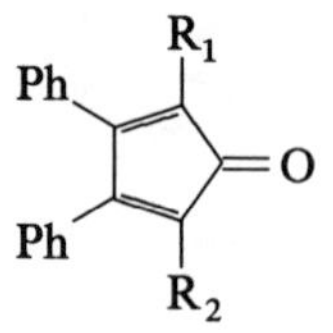

Dienophile		Yield (%)	Ref.
	$R_1 = R_2 = H$		
D5	*N*-(*p*-nitrophenyl)maleimide		211
	$R_1 = R_2 = Me$		
B2	5-Morpholino-4-methylene-*v*-triazoline	18	217
C1	Hexachlorotricyclononadiene	97	212
C1	*cis*-3-Chloro-4-methoxycyclobutene		220
C1	Cyclopentene	91	222
C1	Cycloheptene	80	222
C3	7-Methylenebenzonorbornadiene		200
C3	Indene	77	222
C3	Acenaphthylene	53	222
C8	1,2,3,6-Tetrahydrobenzocyclobutene-3,6-dione	84	28c
D10	1,4-Dithiine	95	223
G4	2-Phenyl-1-azaspiro[2.2]pent-1-ene	76	221
	$R_1 = R_2 = Et$		
C2	Norbornene (k)	37	226
D3	MA (k)	99	226
E3	DMAD	99	226
	$R_1 = Me \quad R_2 = Ph$		
C1	*cis*-3,4-Dichlorocyclobutene		220
	$R_1 = R_2 = Ph$		
A2	Acrylonitriles	18–92	216
A2	3-[(Trimethylsilyl)oxy]-3-buten-2-one	52	31c
B2	5-Morpholino-4-methylene-*v*-triazolines	30–47	217
C1	Cyclobutene		214
C1	[1,2-Dibromobenzocyclobutene]	55	219
C1	*cis*-3,4-Dichlorocyclobutene		220
C1	Hexachlorotricyclononadiene	87	212
C3	Acenaphthylenes	70–90	235
C8	1,2,3,6-Tetrahydrobenzocyclobutene-3,6-dione	81	28c
E2	Methylmethoxypropiolate	60	51

TABLE 6.14 (*Continued*)

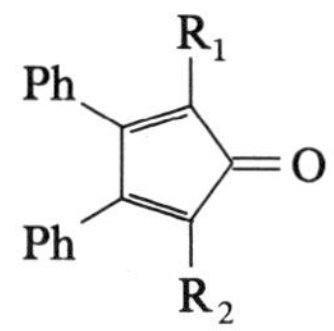

Dienophile		Yield (%)	Ref.
	$R_1 = R_2 = Ph$		
E6	[5,6,9,10-Tetrahydrobenzocyclooctene]		215
E6	[4,4′-Methylene-bis(benzyne)]	51	218
G4	2-Phenyl-1-azaspiro[2.2]pent-1-ene	86	221
	$R_1 = R_2 = p\text{-}OMeC_6H_4$		
C2	Hexachlorotricyclononadiene	74	212
	$R_1 = R_2 = p\text{-}BrC_6H_4$		
E4	Phenylacetylene		227
	$R_1 = R_2 = 3,4\text{-}Me_2C_6H_3$		
C3	Acenaphthylenes	70–90	235
E6	[Benzyne]	75	236
	$R_1 = R_2 = p\text{-}RC_6H_4$ (R = Me, Et, OMe)		
C3	Acenaphthylenes	70–90	235
	$R_1 = R_2 = CO_2Me$		
C1	Cyclooctatetraene (k)	9–55	224, 225
C2	Norbornene (k)	95	226
D3	MA (k)	66	226
E3	DMAD (k)	84	226

TABLE 6.15 Diels–Alder Reactions of 1,3-Disubstituted 2-Oxo-2*H*-cyclopenta[*i*]acenaphthylenes (Acecyclones)

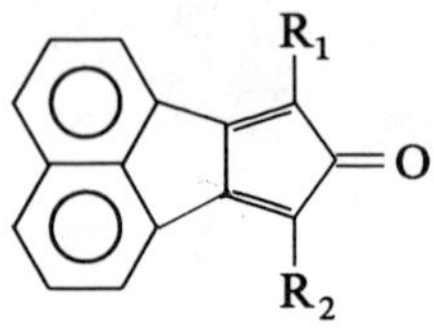

R_1	R_2		Dienophile	Yield (%)	Ref.
Me	Me	C2	Dimethyltricyclo[4.2.0.0^{2,5}]deca-3,7,9-triene-7,8-dicarboxylate		230
Me	Pr^i	C2	Dimethyltricyclo[4.2.0.0^{2,5}]deca-3,7,9-triene-7,8-dicarboxylate		230
Me	Ph	C2	Dimethyltricyclo[4.2.0.0^{2,5}]deca-3,7,9-triene-7,8-dicarboxylate	35	230
Et	CO_2Et	C2	Dimethyltricyclo[4.2.0.0^{2,5}]deca-3,7,9-triene-7,8-dicarboxylate	65	230
Pr^i	Pr^i	C2	Dimethyltricyclo[4.2.0.0^{2,5}]deca-3,7,9-triene-7,8-dicarboxylate		230
Ph	Ph	C2	Dimethyltricyclo[4.2.0.0^{2,5}]deca-3,7,9-triene-7,8-dicarboxylate		230
		C3	Acenaphthylenes	80–90	235
		G4	2-Phenyl-1-azaspiro[2.2]pent-1-ene	42	221
3,4-$Me_2C_6H_3$	3,4-$Me_2C_6H_3$	C2	Norbornadiene	85	236
		E6	[Benzyne]		236
p-RC_6H_4 (R = Me, Et, OMe)	*p*-RC_6H_4	C3	Acenaphthylenes	80–90	235
CO_2Et	CO_2Et	A9	*p*-Chlorostyrene (k)	67	229
		C1	1,3-Cyclohexadiene	89	229
		C2	Norbornene	71	229
		C2	Dimethyltricyclo[4.2.0.0^{2,5}]deca-3,7,9-triene-7,8-dicarboxylate	70	230
		D5	NPM	77	229

TABLE 6.16 Diels–Alder Reactions of 1,3-Disubstituted 2*H*-Cyclopent[1]-phenantren-2-ones (Phencyclones)

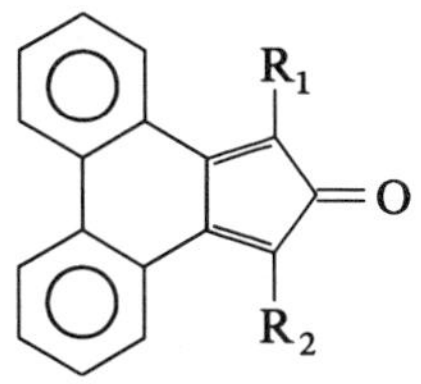

R_1	R_2		Dienophile	Yield (%)	Ref.
H	H	C1	1-Methyl-1-phenylcyclopropene		231
		C1	3,3-Diphenylcyclopropene		418
		D1	2,5-Dihydrofuran		419
H	Ph	D5	*N*-Arylmaleimides		211
Ph	Ph	A1	AC (k)		234
		A12	Cyanoethenes		(233)
		C1	*cis*-3,4-Dichlorocyclobutene		220
		C1	Cyclooctatetraene	85	225
		C3	1-Chloro-3,4-dihydronaphthalene		232
		C3	5,6-Dimethyl-1-chloro-1-3(*H*)-indene		232
		C3	Acenaphthylene		235
		D5	*N*-Arylmaleimides		211
		E5	Diphenylacetylene	78	236
		E6	[Benzyne]	73	236
		G4	2-Phenyl-1-azaspiro[2.2]pent-1-ene	96	221
p-MeC_6H_4	*p*-MeC_6H_4	A1	AC (k)		234
		C2	Norbornadiene	84	236
		C3	Acenaphthylene		235
		E5	Diphenylacetylene	67	236
		E6	[Benzyne]	67	236
p-EtC_6H_4	*p*-EtC_6H_4	C3	Acenaphthylene		235
p-$NO_2C_6H_4$	*p*-$NO_2C_6H_4$	A1	AC (k)		234
p-$MeOC_6H_4$	*p*-$MeOC_6H_4$	A1	AC (k)		234
		C2	Norbornadiene	80	236
		C3	Acenaphthylene		235
		C3	1-Chloroacenaphthylene		235
		E5	Diphenylacetylene	65	236
		E6	[Benzyne]	65	236
p-ClC_6H_4	*p*-ClC_6H_4	A1	AC (k)		234
3,4-$Me_2C_6H_3$	3,4-$Me_2C_6H_3$	C2	Norbornadiene	88	236
		C3	Acenaphthylene		235
		E5	Diphenylacetylene	75	236
		E6	[Benzyne]	75	236

obtained by Birch reduction of the corresponding anisoles **63**. Formally, the sequence is a substitution reaction in a benzene ring.

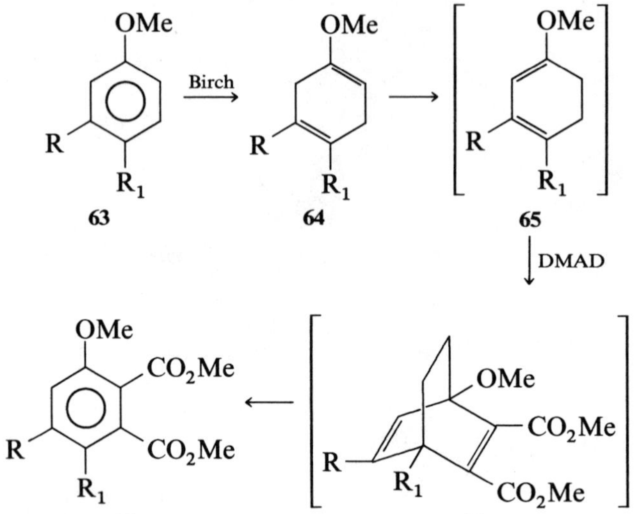

Trimethylsilyloxy-1,3-cyclohexadienes have been utilized to synthetize substituted phenols[448] (i.e., **68** → **69**) and to construct[449] complex carbon skeletons such as seychellene (**70**).

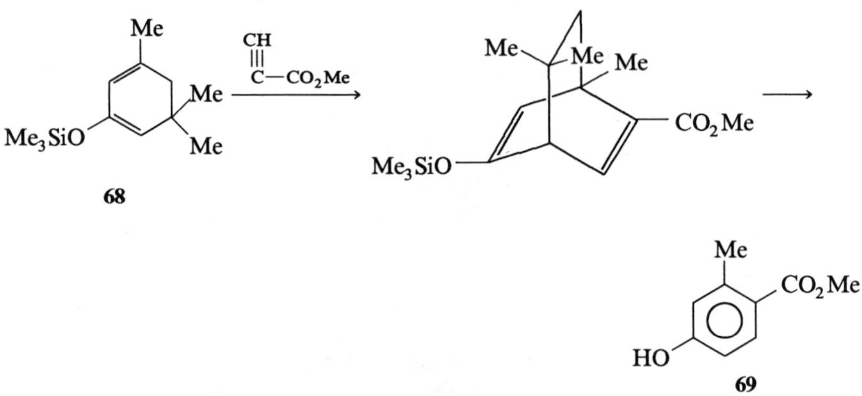

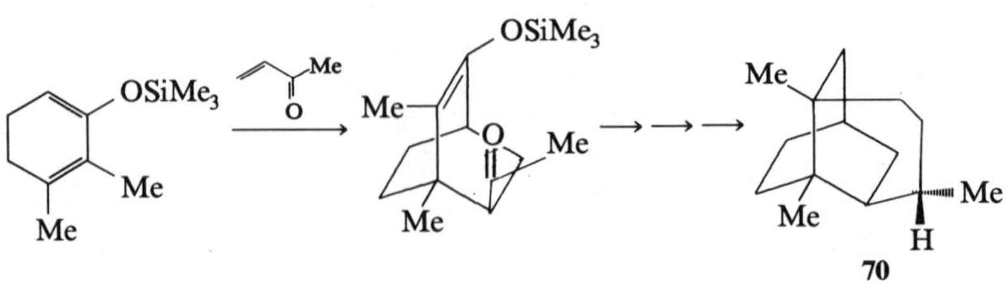

1,3-Cyclohexadiene (**71**: $R_1 = R_2 = H$) and 2,3-dialkilderivatives **71** readily undergo cyclization with MA *endo* selectively and adducts **72** are obtained in high yield.[316]

An example of π-facial diastereoselectivity is the cycloaddition of 4-substituted-5,6-dihydro-3,5-dimethylanisole **73** and MAC. The data of Scheme 6.4 show that the substituents influence the facial and *endo-exo* diastereoselection in the opposite way.[329]

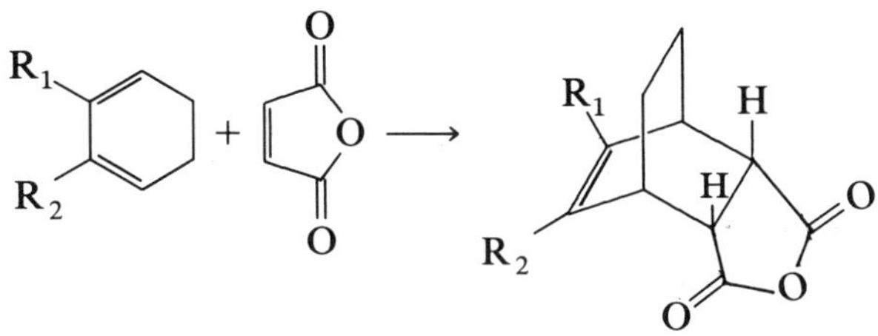

71 **72**

The regioselectivity does not always follow that of aliphatic dienes. The cycloaddition with nitrosodienophiles, which has been used in the stereospecific synthesis of tabtoxyn (**80**)[450] permits interesting observations. Both weakly electron-rich and electron-deficient 2-substituted-1,3-cyclohexadienes **74** and **75** with acylonitroso compounds afford *para-aza* adducts **76** predominantly, and the ratio of regioisomers **76** and **77** is independent of the reaction conditions.[310b,c] By contrast, isoprene with nitrosobenzene gives the expected *para-oxygen* adduct but with α-chloronitrosocyclohexane, a reversed orientation has been observed because of steric factors. In the cycloaddition of **78**, the observed adduct is the *ortho-oxygen* regioisomer **79**,[450] whereas the *ortho-aza* adduct is the main compound of condensation of piperylene with nitrosobenzene.[427,451]

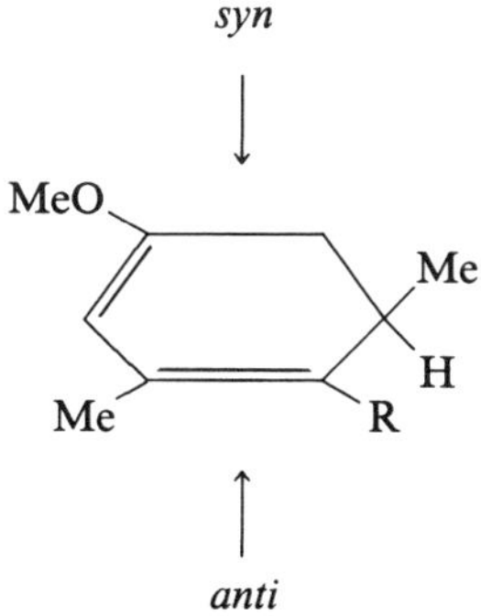

73

R	Addition *anti*	Addition *endo*
H	100%	75%
Me	100%	
Et	50%	80%
$(CH_2)_2OMe$	33%	88%
$(CH_2)_2OAc$	23%	100%

Scheme 6.4

74 $R_1 = CH_2SiMe_2Bu^t$
75 $R_1 = CO_2Me$

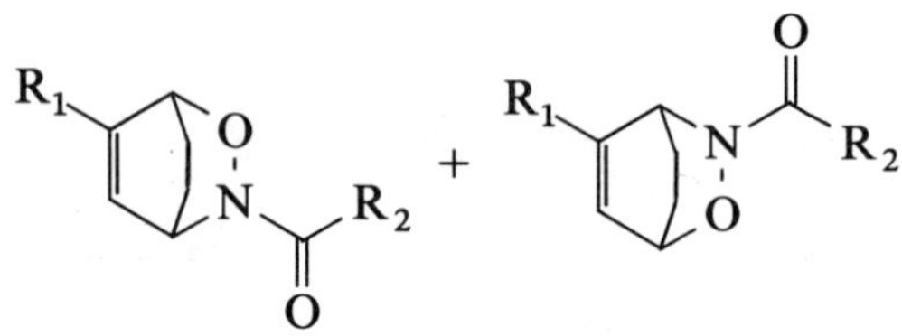

76 **77**

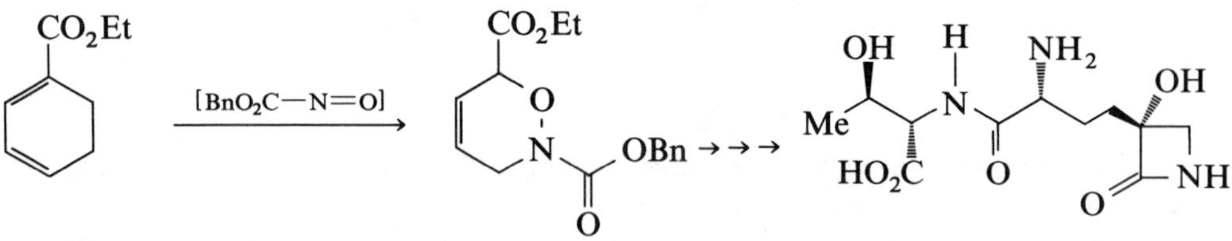

78 **79** **80**

The dimerization of 1,3-cyclohexadiene and its derivatives is of theoretical interest. Thermal reaction[277b] of **71** ($R_1 = R_2 =$ H) at 110°C gives a mixture of **81** (74.3%), **82** (11.4%), **83** (4.1%), **84** (2.5%), and **85** (7.7%) coming from [4 + 2] and [2 + 2] cycloadditions and [6 + 4] ene reaction. At temperatures above 150°C, only *endo* and *exo* adducts are observed. It has been hypothesized[277b] that the *endo* and *exo* adducts are formed by a concerted and nonconcerted mechanism, respectively.

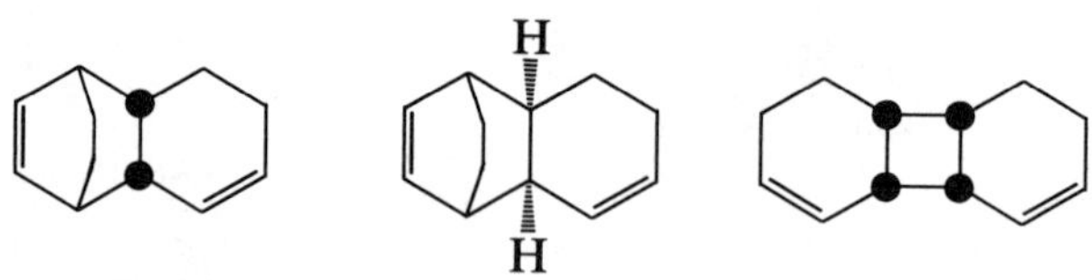

81 *endo* **82** *exo* **83** *syn*

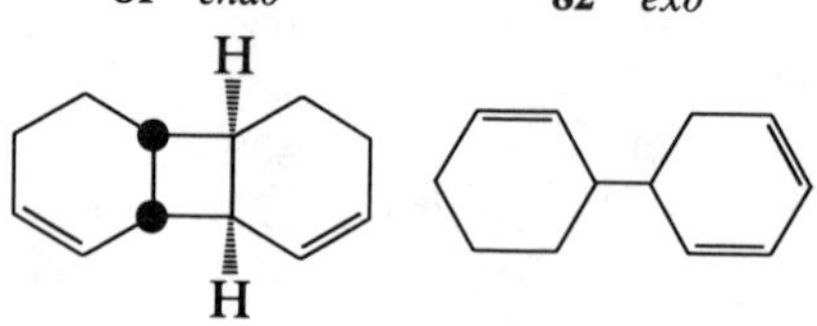

84 *anti* **85** *threo*

Radical cation dimerization of **71** ($R_1 = R_2 =$ H) induced upon photolysis by electron acceptors such as tris(*p*-bromophenyl)aminium hexachlorostibnate,[306] 2,4,6-triphenylpyryliumtetrafluoroborate,[452] 1,4-dicyanonaphthalene (in CH_3CN),[277e] and ketone–$LiClO_4$ mixtures[277h] give **81** and **82** almost exclusively in a variable ratio (from 1 : 1 to 6 : 1). The ratio between **81** and **82** is enhanced drastically (28 : 1) when the cation radical Diels–Alder reaction is induced electrochemically.[291a]

Diels–Alder reactions via radical cations generated photochemically or electrochemically also occur between electron-rich 1,3-cyclohexadienes and electron-rich dienophiles.[128,277h,291] These results seem to support the idea that the mechanism of the cation radical Diels–Alder reaction can also take place via the cation radical of the diene, not only via the cation radical of the dienophile.[277h,291a]

Cyclohexadienones Cyclohexa-2,4-dienones and cyclohexa-2,4-diendiones (*o*-benzoquinones) have received attention because of synthetic usefulness and theoretical interest. The cycloadditions of cyclohexadienones **86** and **87** with MA and BQ are highly stereo- and facial selective.[367] Only *endo* additions on the acetoxy and hydroxy faces have occurred (i.e., **88**).

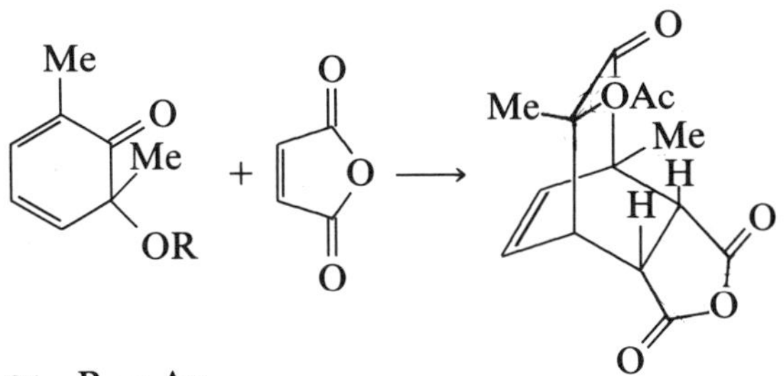

86 R = Ac
87 R = H

88

o-Chloranil (**89**) has been used to synthesize chlorinated aromatic systems (i.e., **91** and **93**) not readily available by traditional procedures.[379,381] The cycloadditions of **89** with alkynes and arynes occur mainly across the homodiene moiety (**89** → **90**, **92**) but can also occur across the dicarbonyl system[380] (i.e., **89** → **95**).

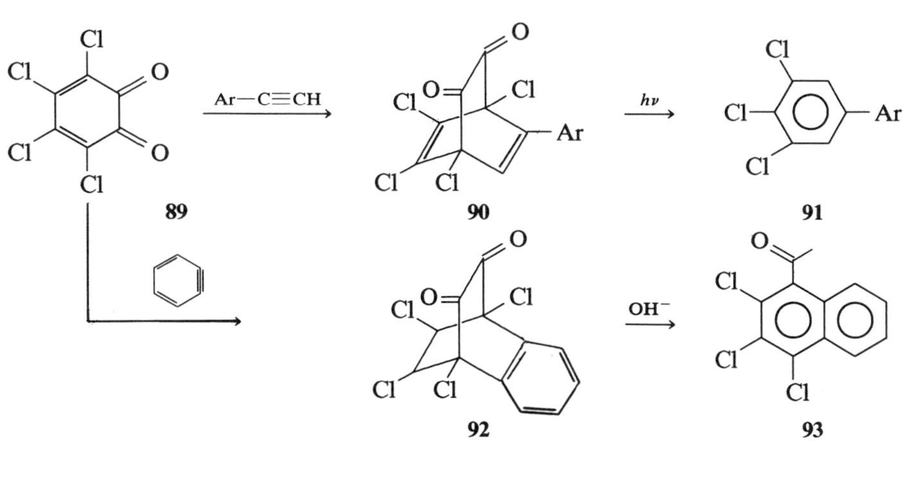

89 **90** **91**

92 **93**

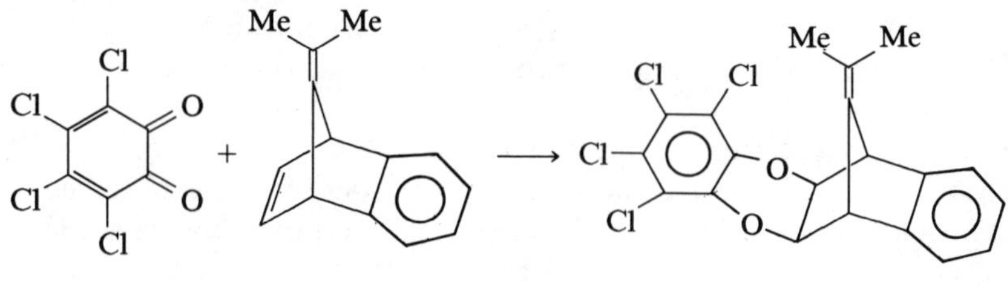

89 94 95

The tetrachloro-*o*-benzoquinone (**89**) should be an electron acceptor in relation to *p*-substituted styrenes, and therefore Diels–Alder with inverse electron demand is expected. However, it was found that neutral-type reactions occur with inversion of the reactivity at the *p*-chlorostyrene.[424]

Propellanes One or two cyclohexadiene rings are present in the tetraenic propellanes **96–100**. *Syn* and *anti* additions with respect to the hetero ring and mono and bis adducts (i.e., **101–104**: Y = O, CH_2; X = O, NMe_2) are observed dependently on the nature of the addends (—A=A—). Singlet oxygen attacks propellanes containing two CO groups (**97–99**) exclusively at the *syn* face. When only one or no CO group is present in the hetero ring (i.e., **96**, **100**), both *syn* and *anti* attack occur.[348] Similarly, NMTAD with **99** gives mainly the bis adducts **103** (Y = O, X = NMe) and with **100** exclusively the bis adduct **104** (Y = H_2, X = O).[356] The different facial selectivity of propellanes with respect to that of annulenes has been interpreted in terms of secondary orbital interactions.

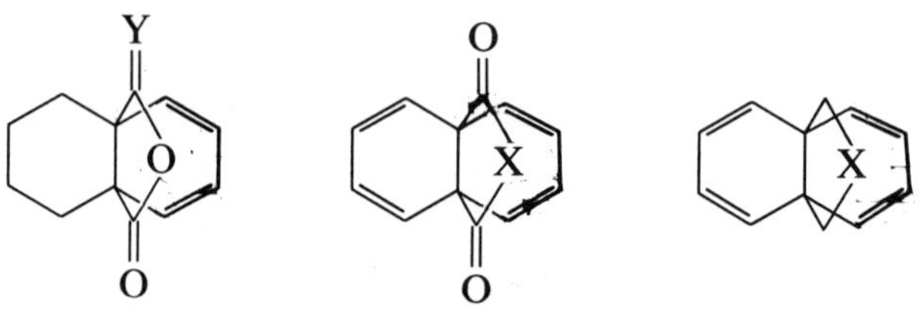

96 Y = CH_2
97 Y = O

98 X = O
99 X = NMe

100 X = O

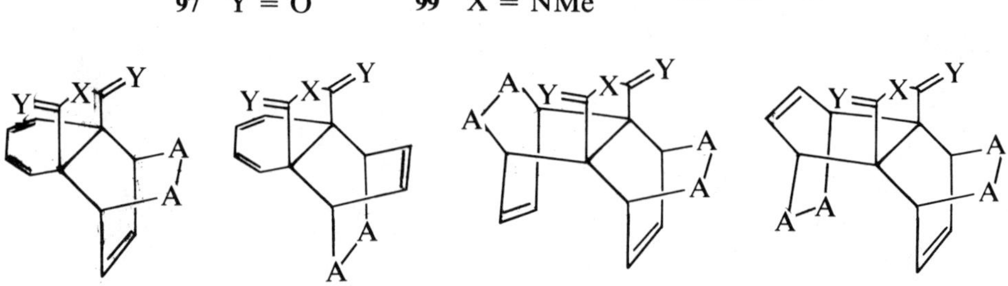

101 *syn* **102** *anti* **103** *syn-syn* **104** *anti-syn*

Tables of Diels–Alder Reactions See Tables 6.17 to 6.27; see also the general remarks about classifications, abbreviations, and symbols in the Introduction.

TABLE 6.17 Diels–Alder Reactions of 1,3-Cyclohexadiene

	Dienophile	Yield (%)	Ref.
A1	Vinyl aryl ketones		295
A3	(*E*)-β-Trimethylsilylacryloyl chloride	80	123
A3	3,4-Dichlorocinnamyl chloride		294b
A3	(*E*)-6,6-Dimethoxy-3-hexen-2-one	68	412
A7	Alkyl maleates	75–80	281
A8	3,3-Diethoxypropene	57	267
A8	Ethenes (k)		274
A8	Phenylvinylsulfone	90	48, 78a
A8	*S*-*p*-Tolyl-*S*-vinyl-*N*-(*p*-tolylsulfonyl)sulfoximine	95	49
A8	Pyrrolo[1,2,*a*]quinoxaline-6,9-dione	75	284
A8	Triethylallylammonium bromide		288a
A8	Styrenes	24	291
A9	Acrylonitriles	72–92	132a
A9	Methyl-α-methyl thioacrylate	81	132a
A10	1,2-Bis(phenylsulfonyl)ethene (*E* and *Z*)	42–97	48b, c, 282
A10	(*E*)-Nitrovinylsilane	61	120b
A10	Ethyne-bridged dimolybdenum complex	35	286, 296
A12	1,1-Dicyano-2,2-bis(trifluoromethyl)ethene		109
A12	TCNE		63f
C1	2,4-Hexadienes	36	269, 277f
C1	1,3-Cyclohexadiene (k)	45–80	277, 291b
C2	Hexakis(trifluoromethyl)benzvalene	85	79
C3	Indene	85	276
C4	Bicyclo[3.2.2]nona-3,6-dien-2-one	86	54a
C7	BQ		288b
C8	Dihydroanthracene-1,4,9,10-tetrones	67–78	81
C9	1,4-Anthraquinone	55	271
D2	2-Methoxy-5-oxo-5,6-dihydro-2*H*-pyran	60	290
D3	MA		63f
D9	4-Isothiazolin-3-one-1,1-dioxide	72	182
D9	2-*N*-Butyl-4-isothiazolin-3-one-1,1-dioxide	91	182
E2	Ethyl(diethoxyphosphinyl)propinoate	76	196
E2	Acetylenedicarbaldehyde monoacetal	90	268, 272
E2	Methyl methoxypropiolate	30	51
E2	Ethyl β-phenylsulfonylpropiolate	75	285
E2	Ethyl 4-aryl-4-hydroxybut-2-ynoates	41–49	294a
E3	Di-*tert*-butylacetylenedicarboxylate	51	270
E3	But-2-yne-dial	75	76a
E4	Acetylenehexacarbonyldicobalt complexes		289
E5	Trimethylsilylethynylmethoxypenta-carbonylchromium carbene complex	67	127

TABLE 6.17 (*Continued*)

	Dienophile	Yield (%)	Ref.
F4	Menthyl glyoxylate (*)	23–39	273
F4	*n*-Butyl glyoxylate		273
F7	[Methylenbenzyliminium chloride]	35	40
I1	Arylazocyanides	43	77
I4	2,3-Diazabicyclo[2.2.1]hept-2-ene	75	17
I5	*N*-Aryltriazolinediones	71	80
J1	17-β-Nitrosoandrostane (*)	69	279
J1	1-Nitroso-D-mannofuranose (*)	69	280
J1	Trifluoronitrosomethane	100	283
J1	[Acylnitroso compound]	65	287
J1	1-Chloro-1-nitrosocyclohexane	91	292b
J1	2,2-Chloronitrosopropane	100	293
J2	*N*-Sulfinylcarbamates (*)	55–77	275
J2	*N*,*N'*-Bistosylsulfurdiimide	42	278

TABLE 6.18 Diels–Alder Reactions of 1-Substituted 1,3-Cyclohexadienes

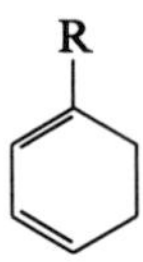

R		Dienophile	Yield (%)	Ref.
Me	A10	2-Methyl-2,4-hexadiene	75	(303)
	C1	1-Methyl-1,3-cyclohexadiene	80	269
CO_2Et	A1	Ethyl acrylate	77	304a, c
	A1	MVK	89	304a
	A8	ACN	79	304a
	A8	Nitroethene	62	304b
	J1	Benzylnitroso formate	93	305
$N[(CH_2CH_2)_2O]$	D3	MA		309
OMe	A7	DMF		307
	A7	Fumaryl chloride		307
	A8	Phenylvinylsulfone	68	48a, 78a
	A10	Fumaronitrile		300

TABLE 6.18 (*Continued*)

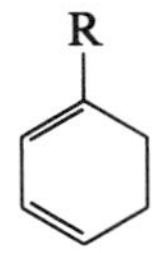

R		Dienophile	Yield (%)	Ref.
	A12	TCNE		299
	A13	Phenylsulphonylpropadiene	54	34a
	C1	1-Methoxy-1,3-cyclohexadiene	70	128, 269, 291
	C9	1,4-Naphthoquinones, Cat.(*)	93–98	297
	C9	Quinoline-5,8-diones	40–90	298
	F4	Ethyl mesoxalate	64	308
[OMe]	A12	TCNE		299
	E2	Methyl(tetrahydropyranyl)-2-pentynoate	65	302
	E2	Ethyl-4-aryl-4-hydroxybut-2-ynoate	41–52	294a
	E3	DMAD	91	301
OAc	A10	(*E*)-Anethole	28	306
	C1	1-Acetoxy-1,3-cyclohexadiene	14	291a

TABLE 6.19 Diels–Alder Reactions of 2-Substituted 1,3-Cyclohexadienes

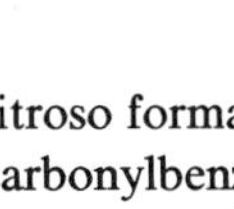

R		Dienophile	Yield (%)	Ref.
Ph	D3	MA	93	316
	D5	NPM	96	316
CO_2Me	J1	Methylnitroso formate	41	310c
	J1	Nitrosocarbonylbenzene	63	310c
$N(CF_3)_2$	J1	Trifluoronitrosomethane	67	311
[OAc]	C1	1-Acetoxy-1,3-cyclohexadiene	26–38	291b
$OSiMe_3$	F4	Ethyl mesoxalate	41	308
$OSiMe_2Bu^t$	D3	MA	74	310
	J1	Nitrosocarbonylbenzene	15–61	310b
	J1	Methylnitroso formate	72	310b

TABLE 6.20 Diels–Alder Reactions of Disubstituted 1,3-Cyclohexadienes

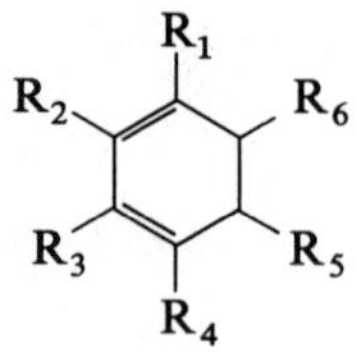

Substituents			Dienophile	Yield (%)	Ref.
			1, 2		
R_1	**R_2**				
Me	$OSiMe_2Bu^t$	D3	MA	77	310a
			1, 3		
R_1	**R_3**				
[OMe	Me]	E1	Ethyl propynoate	15	301
		E3	DMAD	45	301
OMe	OMe	C9	5-Hydroxy-7-methyl-1,4-naphthoquinone		297e
[OMe	OMe]	E1	Ethyl propynoate	55	301
		E3	DMAD	63	301
			1, 4		
R_1	**R_4**				
Me	Me	C1	2,5-Dimethyl-1,3-cyclohexadiene		312
Me	Pr^i	B2	Itaconic anhydride		314
Me	$N[(CH_2CH_2)_2O]$	D3	MA		309
[Me	OMe]	D3	Cyclopentenedicarboxylic-anhydride		313
		E2	Methyl-4-hydroxybut-2-ynoate	45	301
		E2	Methyl-5-(tetrahydropyranyloxy)-2-pentynoate	67	302
		E3	DMAD	45	301
Ph	Ph	D3	MA		416
CN	CN	D3	MA		416
[OMe	OMe]	E1	Ethyl propynoate	63	301
		E2	Methyl-4-hydroxybut-2-ynoate	39	301
		E2	Ethyl-4-aryl-4-hydroxybut-2-ynoate		294
		E3	DMAD	67	301
OEt	OEt	A9	Acrylonitriles		416
		D3	MA		416

TABLE 6.20 (*Continued*)

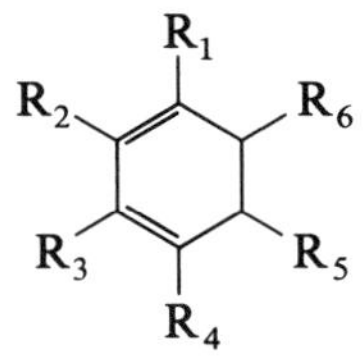

Substituents		Dienophile		Yield (%)	Ref.
		1, 6			
R_1	**R_6**				
$-C(=O)-O-CH_2-$		D3	MA	62	315
		D5	NPM	58	315
		2, 3			
R_2	**R_3**				
Me	Me	D3	MA	92	316
		D5	NPM	98	316
Me	Et	D3	MA	96	316
		D5	NPM	92	316
$OSiMe_2Bu^t$	Me	D3	MA	83	310a
$-(CH_2)_2-$		D3	MA	78	316
		D5	NPM	70	316
$-(CH_2)_3-$		D3	MA	88	316
		D5	NPM	78	316
$(CH_2)_2-CH(Me)$		D3	MA	80	316
		D5	NPM	86	316
$-(CH_2)_4-$		D3	MA	83	316
		D5	NPM	95	316
		2, 6			
R_2	**R_6**				
CO_2Me	OH	I5	NPTAD		317
		5, 6			
R_5	**R_6**				
CO_2R	CO_2R	J1	Nitroso compounds		292a
(R = H, Me, Et, Bu^t)					
$-[CH=CH]-$		I3	[Singlet oxygen]	89	320
$-[CH_2]_2-$		I3	[Singlet oxygen]	94	320
$>CHC(N_2)R$		I5	NPTAD	70–90	422
[R = $POPh_2$, $PO(OMe)_2$, CO_2Et]					

TABLE 6.20 (*Continued*)

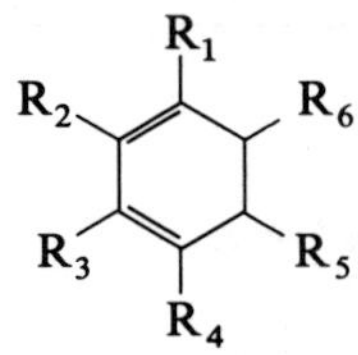

Substituents				Dienophile	Yield (%)	Ref.
				5, 6		
R_5	**R_6**					
OH	N_3	*t*	J1	1-Chloro-1-nitroso-cyclohexane	47	318
			J1	*p*-Chloronitrosobenzene	76	318
OH	OMe	*t*	J1	*p*-Chloronitrosobenzene	82	319
OAc	NHAc	*t*	J1	1-Chloro-1-nitroso-cyclohexane	68	318
			J1	*p*-Chloronitrosobenzene	62	318
OAc	OAc	*t*	J1	*p*-Chloronitrosobenzene	100	319
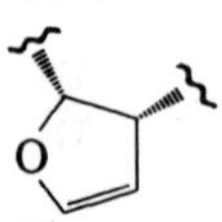			A12	TCNE		321
			D3	MA		321
			D5	NPM		321
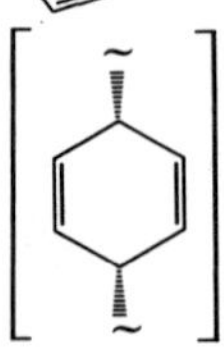			I5	NPTAD		322

TABLE 6.21 Diels–Alder Reactions of Trisubstituted 1,3-Cyclohexadienes

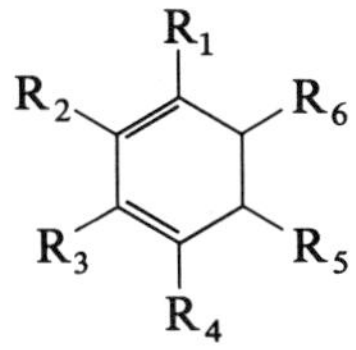

Substituents					Dienophile	Yield (%)	Ref.
				1, 2, 3			
R_1	**R_2**	**R_3**					
Me	Me	$OSiMe_3$		A1	2-Ethoxyethyl vinyl ketone	46	347
				A1	MVK	79–97	323, 324
				A1	AC	92	324
				A1	Divinyl ketone		347
				A8	ACN	95	324
Me	Me	$OSiMe_2Bu^t$		A1	AC		324
				A1	MVK		324
				A8	ACN		324
				1, 3, 5			
R_1	**R_3**	**R_5**					
OMe	Me	Me		A1	MAC	24	329
				1, 5, 5			
R_1	**R_5**	**R_5**					
$OSiMe_3$	Me	Me		A1	MVK	72	326
				1, 5, 6			
R_1	**R_6**	**R_5**					
$—C(=O)—O—CH_2—$		CO_2Me	*t*	A12	TCNE	65	315
				D3	MA	62	315
				D5	NPM	80	315

TABLE 6.22 Diels–Alder Reactions of Tetra- and Pentasubstituted 1,3-Cyclohexadienes

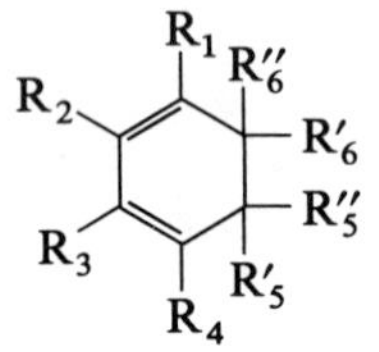

Substituents				Dienophile		Yield (%)	Ref.
			Tetrasubstituted				
R_1	**R_3**	**R'_5**	**R''_5**				
Me	Me	Me	Me	C7	BQ		327
R_1	**R_3**	**R_4**	**R'_5**				
OMe	Me	Me	Me	A1	MAC	58	329
OMe	Me	Et	Me	A1	MAC	25	329
OMe	Me	$CH_2CH(CH_2O)_2$	Me	A1	MAC	65	328
OMe	Me	$(CH_2)_2OMe$	Me	A1	MAC	22	329
OMe	Me	$CH_2CH(OC_2H_4O)$	Me	A1	MAC	65	329
				A9	2-Chloropropenenitrile	57	329
OMe	Me	$(CH_2)_2OAc$	Me	A1	MAC	46	329
R_1	**R_2**	**R''_5**	**R'_6**				
$SiMe_3$	$SiMe_3$	$—CH_2—CH{=}CH—$		E5	Bis(trimethylsilyl)-acetylene		330
R_2	**R_3**	**R''_5**	**R'_6**				
$SiMe_3$	$SiMe_3$	$—CH_2—CH{=}CH—$		E5	Bis(trimethylsilyl)-acetylene		330
R'_5	**R''_5**	**R'_6**	**R''_6**				
Me	Me	Me	Me	I5	NPTAD	36	331
OMe	OMe	Me	CO_2Me	D3	MA	5	366
				E3	DMAD	80	366
$—(CH_2)_2—$		$—(CH_2)_2—$		C7	BQ	70	331
				E6	[Benzyne]	4	331
		Pentasubstituted					
R_1	**R_3**	**R_4**	**R'_5 R''_5**				
$OSiMe_3$	Me	$OSiMe_3$	Me Me	A1	MAC	41	414
				A8	ACN	41	414

TABLE 6.23 Diels–Alder Reactions of Thebaines

Diene			Dienophile	Yield (%)	Ref.
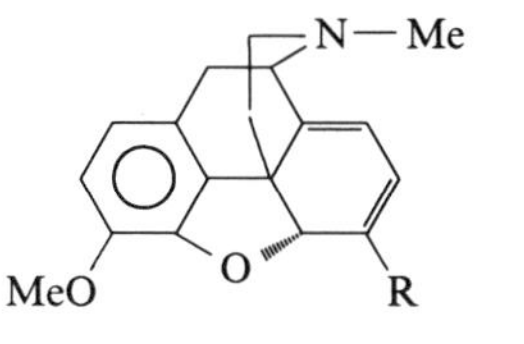	R = H	A1	Ethyl acrylate		333
		A1	MVK	51	334
		E3	DMAD	72	332
	R = OMe	A1	MVK	80	335
		D6	2-Thiolen-4-on-1,1-dioxide	85	337
		E1	MP	26–95	332, 336
		E3	DMAD	51–68	332, 336
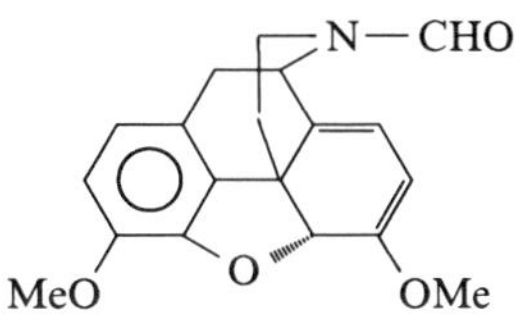		A8	Nitroethene	99	338
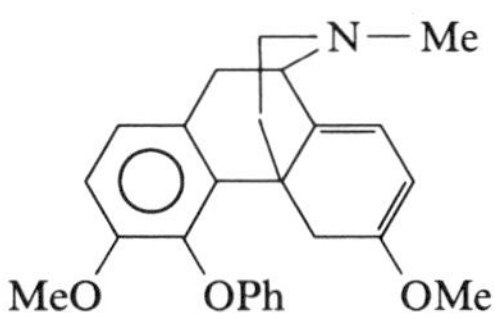		A1	MVK	80	335

TABLE 6.24 Diels–Alder Reactions of Substituted 2,4-Cyclohexadien-1-ones

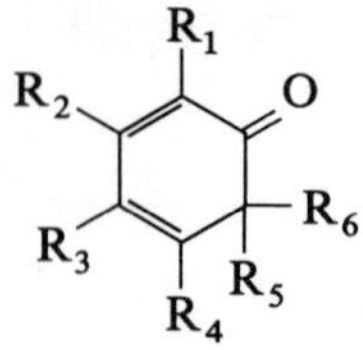

Substituents					Dienophile	Yield (%)	Ref.
					6,6-Disubstituted		
$\mathbf{R_5}$	$\mathbf{R_6}$						
Me	CO_2Me			E3	DMAD		366
Bn	CO_2Me			D3	MA	41	366
				E3	DMAD	56	366
[=CH—CHO]				A8	Isobutyl vinyl ether	68	369
				A10	(*E*)-2-Methoxystyrene	54	369
				D1	2,3-Dihydrofuran	13	369
					2,6,6-Trisubstituted		
$\mathbf{R_1}$	$\mathbf{R_5}$	$\mathbf{R_6}$					
Me	Me	OAc		D3	MA	53	367
				E1	Propiolic acid	46	367
Me	Me	Me		A1	MVK	96	368
				A1	AC		368
					3,6,6-Trisubstituted		
$\mathbf{R_2}$	$\mathbf{R_5}$	$\mathbf{R_6}$					
[OMe	=CH—CHO]			A8	Isobutyl vinyl ether	29	369
					4,6,6-Trisubstituted		
$\mathbf{R_4}$	$\mathbf{R_5}$	$\mathbf{R_6}$					
Me	Me	CO_2Me		A1	MAC		366
				E3	DMAD		366
					2,3,6,6-Tetrasubstituted		
$\mathbf{R_1}$	$\mathbf{R_2}$	$\mathbf{R_5}$	$\mathbf{R_6}$				
Pr	Pr	$—OCH_2COO—$		A1	Ethyl acrylate	45	373
				A8	Styrene	72	373
				A8	Ethyl vinyl ether	50	373
Pr	Pr	$—O(CH_2)_2O—$		E5	Diphenylacetylene		374
					2,4,6,6-Tetrasubstituted		
$\mathbf{R_1}$	$\mathbf{R_3}$	$\mathbf{R_5}$	$\mathbf{R_6}$				
Bu^t	Bu^t	$—OCH_2—$		E3	DMAD		372

TABLE 6.24 (*Continued*)

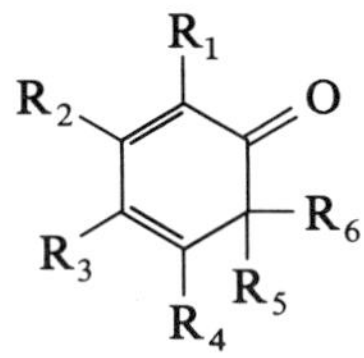

Substituents					Dienophile	Yield (%)	Ref.
		2,5,6,6-Tetrasubstituted					
R_1	**R_4**	**R_5**	**R_6**				
Pr	Pr	$—OCH_2COO—$		C1	Cycloheptatriene		370
				E5	Acetylenes	65–100	371
		3,4,6,6-Tetrasubstituted					
R_2	**R_3**	**R_5**	**R_6**				
CO_2Et	Me	Me	Me	I5	NPTAD	100	365
		3,5,6,6-Tetrasubstituted					
R_2	**R_4**	**R_5**	**R_6**				
[OMe	OBn	=CH—CHO]		A8	Isobutyl vinyl ether	39	369
		4,5,6,6-Tetrasubstituted					
R_3	**R_4**	**R_5**	**R_6**				
Pr	Pr	$—O(CH_2)_2O—$		E5	Acetylenes		374
Pr	Pr	$—OCH_2COO—$		A1	Ethyl acrylate	89	373
				A8	Styrene	95	373
				A8	Ethyl vinyl ether	59	373
				C1	1-Hexene	70	373
		2,3,4,5,6,6-Hexasubstituted					
$R_1 = R_2 = R_3 = R_4 = R_5 = R_6 = Me$							
				A9	Acrylonitriles	0–92	216

TABLE 6.25 Diels–Alder Reactions of *o*-Benzoquinones

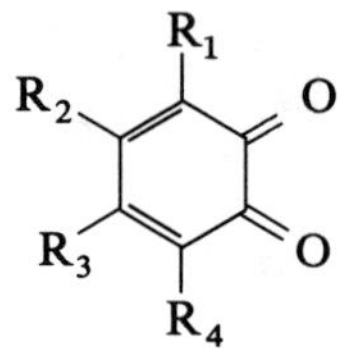

Substituents		Dienophile	Yield (%)	Ref.
$R_1 = R_2 = R_3 = R_4 = H$	C1	1-Methyl-1-phenyl-cyclopropene		231
$R_1 = R_3 = R_4 = H$; R_2 = CN, CHO, COMe, COEt, COPh, CO_2Me	C1	Cyclopentadiene		375
$[R_1 = R_4 = H, R_2 = R_3 = C_4H_4]$	C1	Cyclopentadiene	56	376
$R_3 = R_4 = H, R_1 = R_2 = Pr$	E3	DMAD		377
$R_1 = R_2 = R_3 = R_4 = Cl$	A8	Vinylethynyldimethyl-carbinol	85	98
	C1	1-Methyl-1-cyano-cyclopropene		378
	C3	7-Isopropylidenebenzo-norbornadiene	18	380
	E4	*p*-Chlorophenylacetylene	> 76	379
	E4	*m*-Chlorophenylacetylene	81	379
	E4	2,5-Dichlorophenyl-acetylene	37	379
	E4	Phenylacetylene	90	381
	E6	[Benzyne]	53	381

TABLE 6.26 Diels–Alder Reactions of Propellanes

Diene			Dienophile	Yield (%)	Ref.
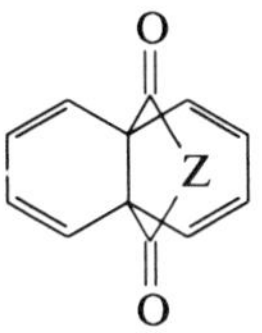	Z = NMe	I3	[Singlet oxygen]	64	348
		I5	NPTAD		356
		J1	Nitrosobenzenes		349
	Z = NCH(R)CO_2R′ (R = H, Me; R′ = Me, Bu^t)	I5	NPTAD	64–100	350
	Z = NCRR′CO_2Me (R = H, Me; R′ = H, Me, CH_2CO_2Me)	I5	NPTAD	82	350
	Z = NPh	I5	NPTAD	100	351
	Z = N-2,6-$Me_2C_6H_3$	I5	NPTAD		351
	Z = NCH_2CMe_3	I5	NPTAD	> 75	351
	Z = O	C7	BQ	68	352
		D5	NPM	85	352
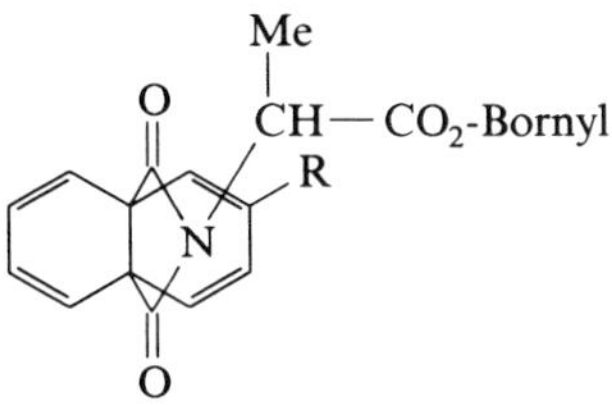	(*) R = H, Br	I5	NPTAD	80	355
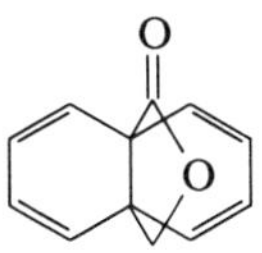		I5	NPTAD		353
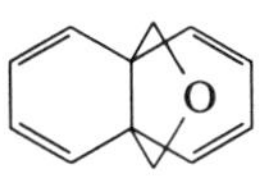		I3	[Singlet oxygen]	29–73	348
		I5	NMTAD		356
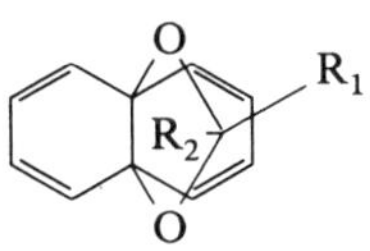	R_1 = R_2 = H, Me	I5	NMTAD		354
	R_1, R_2 = O	I5	NMTAD		354
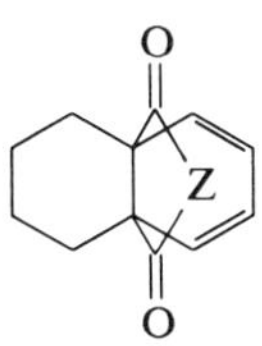	Z = O	I3	[Singlet oxygen]	76	348
	Z = NMe	I3	[Singlet oxygen]	77	348
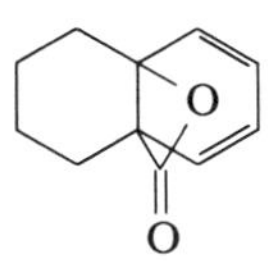		I3	[Singlet oxygen]	51	348

TABLE 6.26 (*Continued*)

Diene		Dienophile	Yield (%)	Ref.
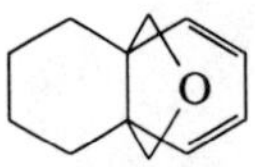		I3 [Singlet oxygen]	22	348
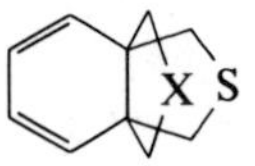	X = O, S	I5 NPTAD	100	356
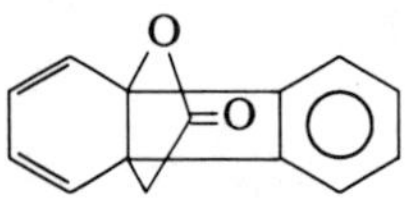		I5 NPTAD		359
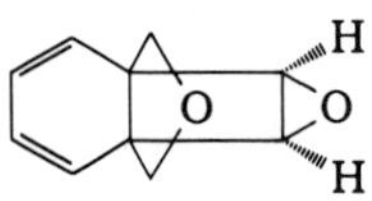		I5 NPTAD		358
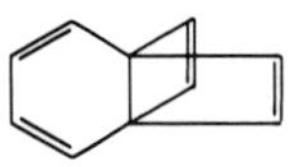		D3 MA E1 MP E3 DMAD	100 100	360 360 360
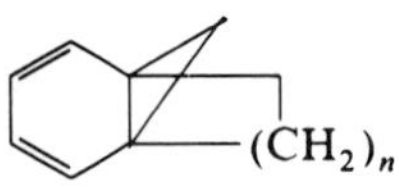	(n = 1, 2)	I5 NPTAD		357

TABLE 6.27 Diels–Alder Reactions of Miscellaneous 1,3-Cyclohexadiene-Related Compounds

Diene	Dienophile	Yield (%)	Ref.
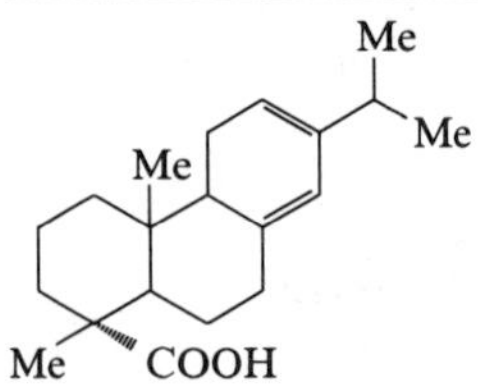	F1 Methylcyanodithioformate F1 *N*-Benzoyl-*N*-phenyl-cyanothioformamide	93 66	339 339
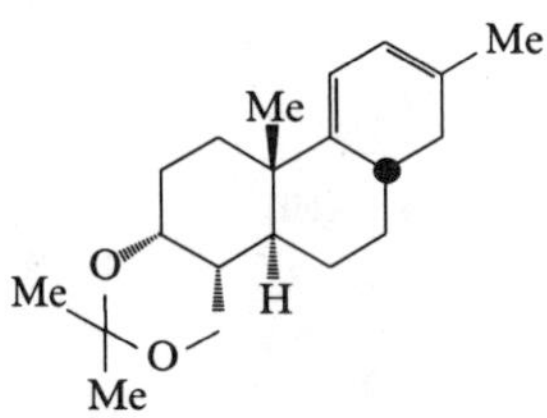	A2 α-Chloroacryloyl chloride A8 Nitroethene	 50	340 340

TABLE 6.27 (*Continued*)

Diene	Dienophile	Yield (%)	Ref.
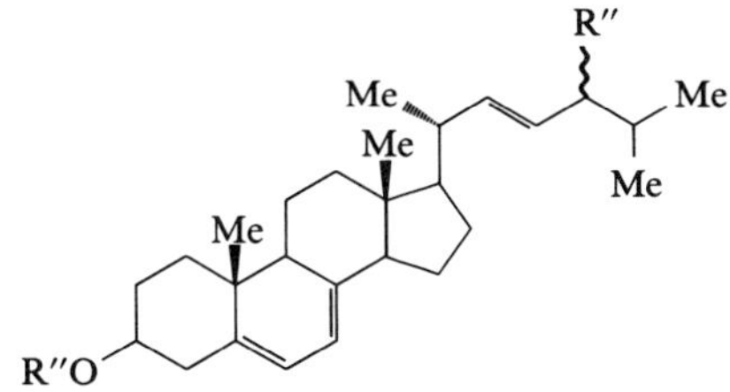	I4 Maleic hydrazide		342
	I4 Phthalic hydrazide		342
	I5 NPTAD	81–87	341, 343
{R′ = Me, Et; R″ = H, Bn, Ac, THP}			
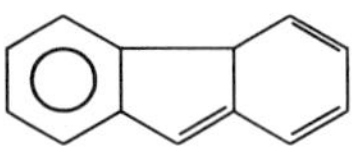	D5 NPM		344
	I5 NPTAD		344
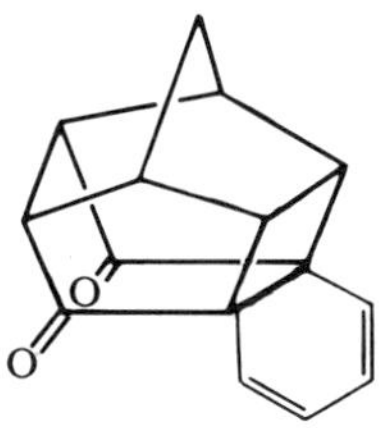	A1 MVK		345
	A7 DMM		345
	C7 1,4-Benzoquinones		345
	D3 MA		345
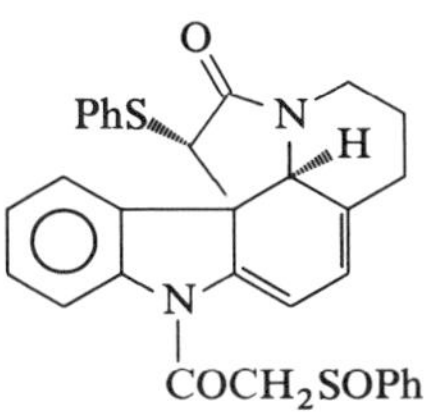	A1 Acryloyl chloride	65	346
	I5 NPTAD		346
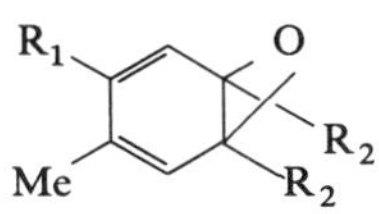	I5 NPTAD	66–86	365
{R_1 = CO_2Et, $PO(OMe)_2$; R_2 = Me, Ph}			
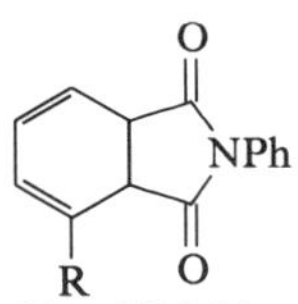 (R = H, Me)	D5 NPM		417

Seven-, Eight-, and Ten-Membered Rings

1,3-Cycloheptadienes 1,3-Cycloheptadienes react easily with *C*-nitroso compounds and singlet oxygen to give fundamental precursors for a range of natural products.[383,384]

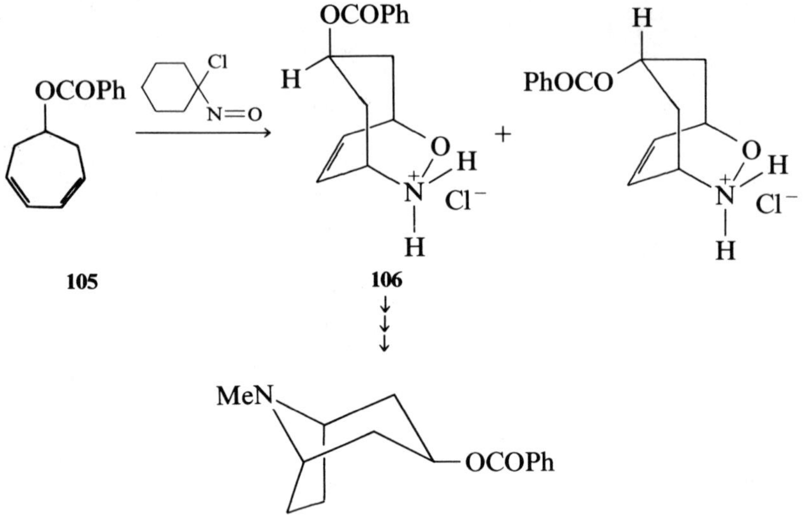

105 **106** **107**

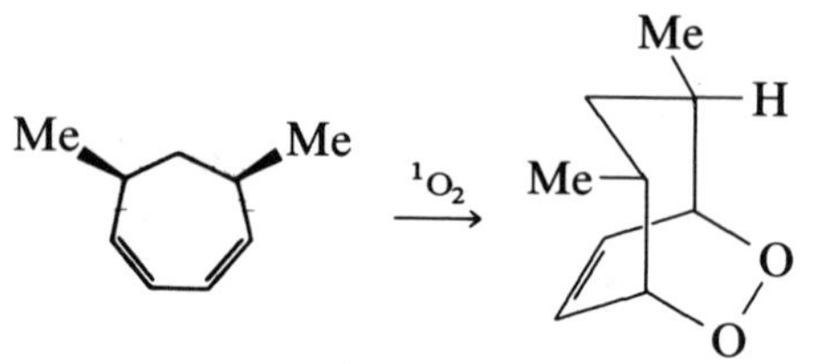

108 **109**

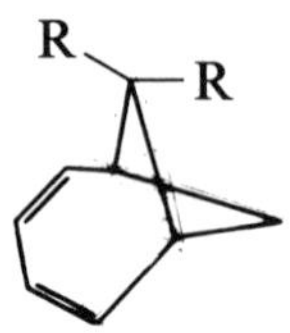

110 R = Me
111 R = H

The [4 + 2] nitrosocycloaddition of benzoate **105** gives 4 : 1 face selectivity in favor of **106**, which is then converted[384] into the tropane alkaloid tropacocaine (**107**). The photooxygenation of **108** prefers the opposite face and **109** is obtained as a single stereoisomer.

The presence of a small, bridged ring in the cycloheptadienic system affects the diene reactivity electronically and sterically.[385] The diene **110** is unreactive toward dienophiles such as NPTAD and TCNE but with benzyne gives [2 + 2] cycloaddition. The parent diene **111** behaves similarly except in the reaction with NPTAD, where a Diels–Alder adduct is obtained in 68% yield.[385]

3,5-Heptadienone (**112**) isomerizes to the 2,4-isomer **113**, and the two ketones have such a large difference in reactivity that it is possible to have selective Diels–Alder reactions. With NPM only the adduct from **113** is obtained, whereas singlet oxygen and NPTAD give chemoselectively 3,5-type Diels–Alder adducts.[390]

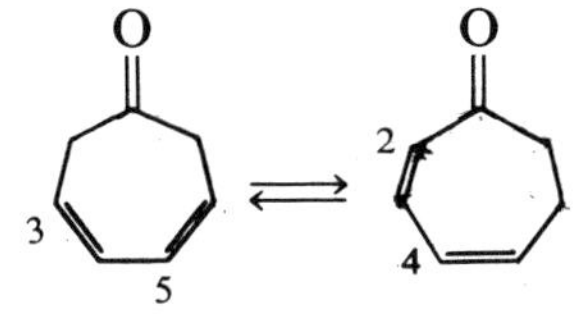

112 **113**

Cycloheptatrienes Cycloheptatriene (**114**) is an interesting molecule that can react with dienophile to give [4 + 2] and [6 + 2] cycloadditions and *ene* reactions.[398, 399, 453]

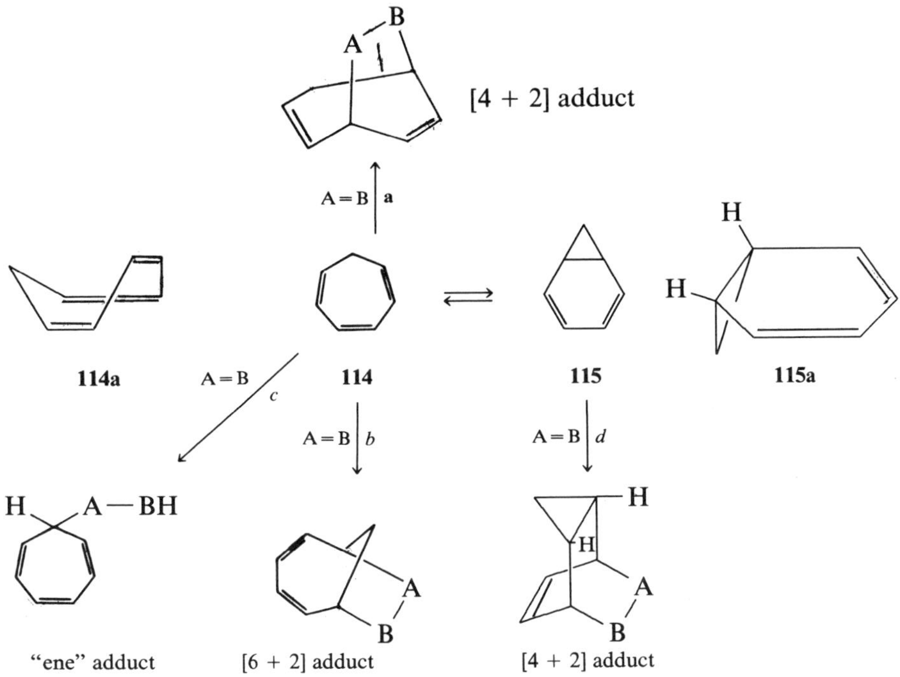

The Diels–Alder reaction proceeds almost exclusively through the valence isomer norcaradiene **115** and the [6 + 2] and ene reactions on the cyclohepta-triene form. The nature of dienophile and the reaction conditions strongly influence the route of the reaction. This behavior has been ascribed to the greater reactivity of the planar diene form **115a** than of the boat conformation **114a** of the triene one.

With very reactive dienophile (TCNE, NPTAD, MA) the Diels–Alder addition through the norcaradiene form **115** (*anti* addition with respect to cyclopropane ring) is preferred (route *d*); with less reactive dienophiles (acrylates, DMAD) the [6 + 2] and *ene* adducts are obtained (routes *c* and *b*).

The cycloheptatriene complexes with $Fe(CO)_3$ does not isomerize to the corresponding norcaradiene isomer, and with the powerful dienophile carbomethoxymaleic anhydride, undergoes a normal Diels–Alder reaction through the cycloheptatriene valence isomer **114** (route *a*). A similar behavior is observed in the reactions of benzofused analog with NPTAD and singlet oxygen.[386]

Tropylium ion (**116**) also exists as the sole cycloheptatriene valence isomer, and consequently, adducts **117** and **118** are obtained with high *endo* selectivity.

A vinyl group in the cycloheptatrienilic ring strongly influences the chemoselectivity of the reaction (Chap. 4).

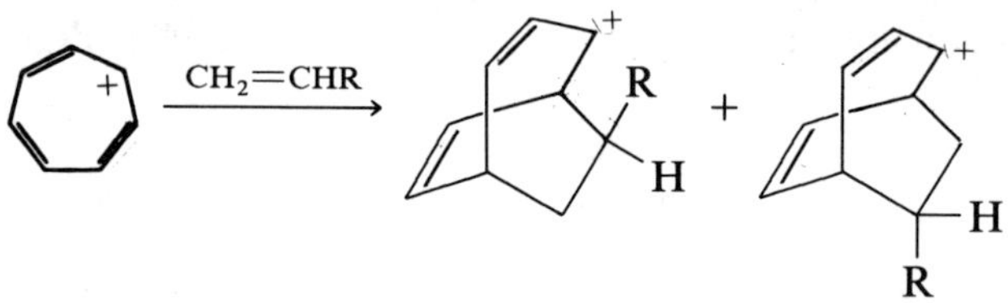

116 **117** exo **118** endo

Tropone (**119**: R = H) is a nonbenzenoid aromatic compound that undergoes various pericyclic reactions. With MA and norbornene, it gives Diels–Alder *endo* additions; with cyclopentadiene both [6 + 4] and [4 + 2] cycloadducts are obtained, whereas with isodicyclopentadiene a [6 + 4] bonding pathway predominates.[391,392a] The [4 + 2] cycloadditions of 2-substituted tropones **119** with ethene give two regioisomers **120** and **121**, and the regioselectivity of the reaction depends on the electronic nature of the substituents.

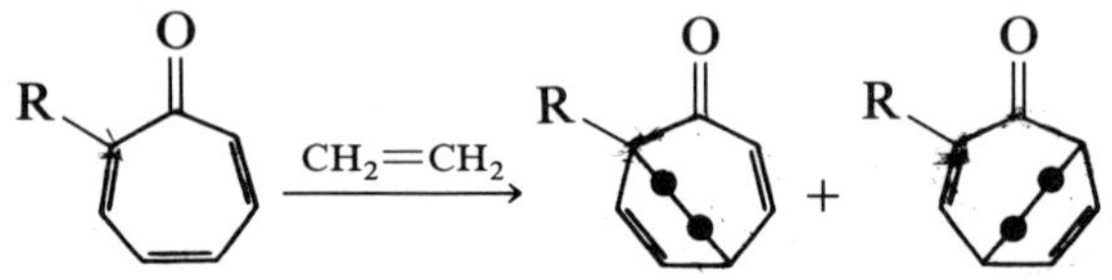

119 **120** **121**

(R = Cl, OMe, Ph)

Cyclooctatetraenes Cyclooctatetraenes also enter into Diels–Alder reactions presumably involving valence isomers.[382,405,406] The prevalent formation of *endo* adducts **124** and **127** in the cycloadditions of a 2 : 1 mixture of **122** and **125** with MA and acenaphtylene, respectively (*anti* addition with respect to cyclobutene ring), has been explained involving the presence of valence tautomers **123** and **126** in a multiple equilibrium.[406] When the cyclobutene-valence tautomer is energetically disfavored, the cycloaddition does not occur. Indeed, the hexa-annulated compound **128** reacts with NPTAD in 2 h, whereas the cyclopentane fused analog **129** does not give the cycloaddition because its valence isomers having a cyclobutene ring fused to a cyclopentane is too strained energetically.[405]

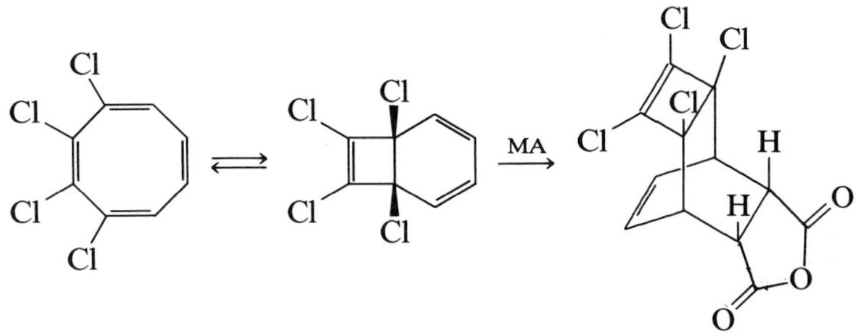

122 **123** **124**

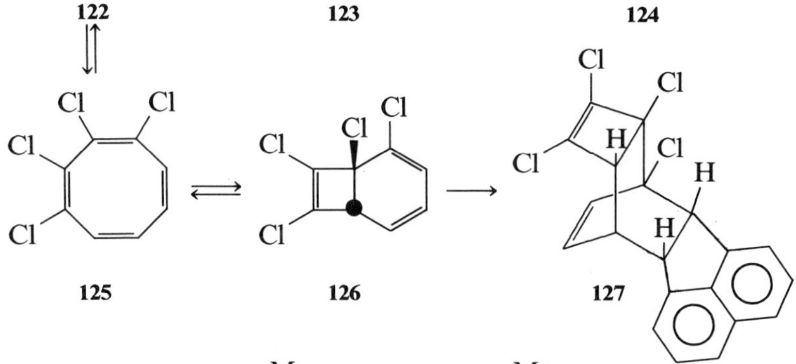

125 **126** **127**

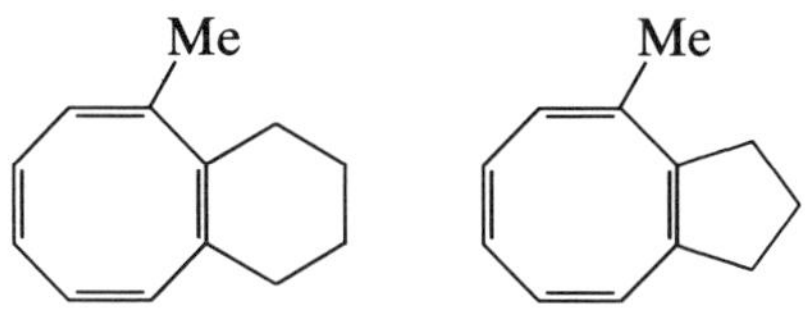

128 **129**

[10]Annulenes The cycloadditions of 1,6-methano[10]annulenes **130** with dienophiles (i.e., A=A) occur prevalently *anti-endo* to the bridge, and mono and bis adducts **131**, **132**, and **133** are obtained. Whereas the parent compound **130** (R = H) reacts with one equivalent of triazoline-3,5-diones (NPTAD and NMTAD), affording more bis adducts than mono adducts, the mono-substituted derivatives at C-11 react more slowly and only mono adducts are obtained.[363]

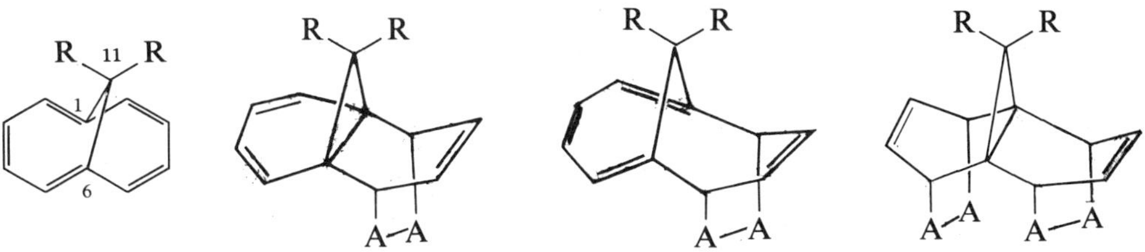

130 **131** **132** **133**

The monoadducts exist as norcaradienes **131**; however, an equilibrate norcaradiene **131**-heptafulvene **132** system has been observed in the cycloaddition of 11-methylene-1,6-methano[10]annulene with dicyanoacetylene.[360]

Tables of Diels–Alder Reactions See Tables 6.28 to 6.33; see also the general remarks about classifications, abbreviations, and symbols in the Introduction.

TABLE 6.28 Diels–Alder Reactions of 1,3-Cycloheptadienes

Diene			Dienophile	Yield (%)	Ref.
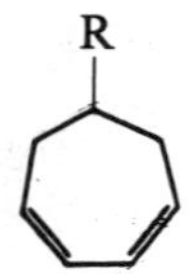	R = H	A10	(*E*)-1,2-Bis(phenylsulfonyl)ethene	77–87	48b, 382
		A12	TCNE (k)		63f
		I3	[Singlet oxygen]	100	383
		I4	2,3-Diazabicyclo[2.2.1]hept-2-ene	30	17
		J1	[Acylnitroso-compound]	85	384
		J1	Ethylnitroso-formate	71	384
	R = OCOPh	J1	1-Chloro-1-nitroso-cyclohexane	57	384
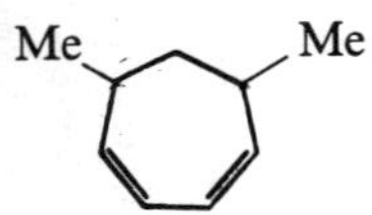		I3	[Singlet oxygen]	90	383
		E6	[Benzyne]	43	385
		I5	NPTAD	68	385
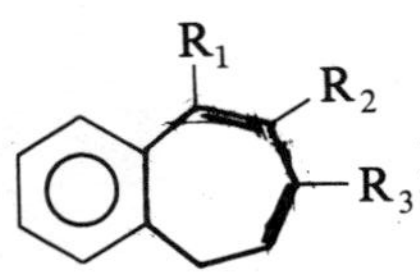	$R_1 = R_2 = R_3 = H$	I3	[Singlet oxygen]	39	386
		I5	NPTAD	55	386

	$R_1 = N(CH_2)_4$; $R_2 = R_3 = CO_2Me$	E3	DMAD	25	387
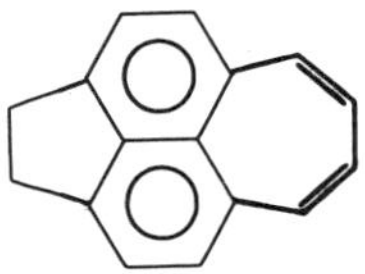		D3	MA		388
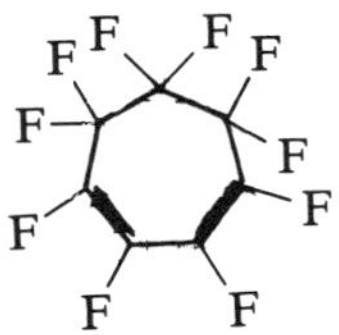		A9	Ethene		389
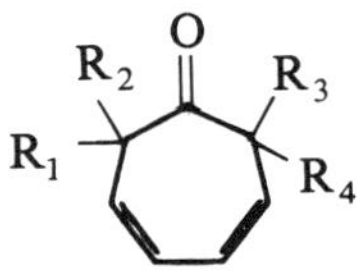	$R_1 = R_2 = R_3 = R_4 = H$	I5	NPTAD	78	390
	$R_1 = R_2 = R_3 = R_4 = Me$	I5	NPTAD	84	390
	$R_1 = Pr^i, R_2 = R_3 = R_4 = Me$	I5	NPTAD	93	390
	$R_1 = Bu^t, R_2 = R_3 = R_4 = Me$	I5	NPTAD	87	390

TABLE 6.29 Diels–Alder Reactions of 1,3,5-Cycloheptatrienes

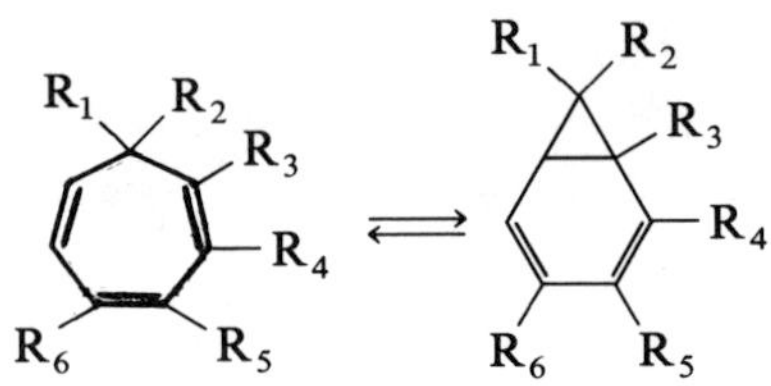

R_1	R_2	R_3	R_4	R_5	R_6		Dienophile	Yield (%)	Ref.
H	H	H	H	H	H	A10	1,2-Bis(phenylsulfonyl) ethene (E, Z)	92	48b, c
						C7	1,4-Benzoquinones		398
						D3	Carbomethoxymaleic anhydride		397
						E1	MP	41	399
						E4	Ethyne-bridged dimolybdenum complex		286, 296
						E5	Dicyanoacetylene (k)		325
						J1	Trifluoronitroso-methane	47	283
H	H	H	H	$-CH{=}CH_2$	H	A12	TCNE	75	325
						I5	NPTAD	95	325
Me	Me	H	H	Me	H	A8	ACN		401
						D3	MA		401
$C(N_2)R$ [R = $POPh_2$, $PO(OMe)_2$, PO(OMe)Ph, CO_2Et]	H	H	H	H	H	I5	NPTAD	70–90	422
$-O(CH_2)_2O-$		H	H	H	H	D3	MA	14	400
						I5	NPTAD	78	400
$-O(CH_2)_2O-$		H	H	$-(CH{=}CH)_2-$		I5	NPTAD		400

TABLE 6.30 Diels–Alder Reactions of Tropones and Benzotropones

Diene			Dienophile	Yield (%)	Ref.
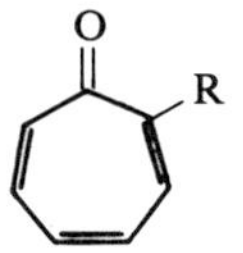	R = H	A8	Ethene	77	54a, 392b
		C1	Norbornene (k)		391
		D3	MA (k)		391, 396
	R = Me	A8	Ethene	92	392
	R = Ph	A8	Ethene	88	392
	R = OMe	A8	Ethene	100	392
		E1	MP		393
		E3	DMAD		393
	R = OCOPh	A8	Ethene	78	392
	R = Cl	A8	Ethene	100	392
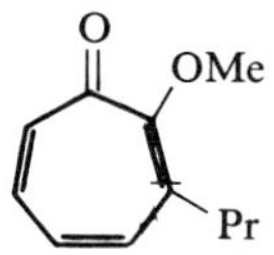		E3	DMAD	65	393
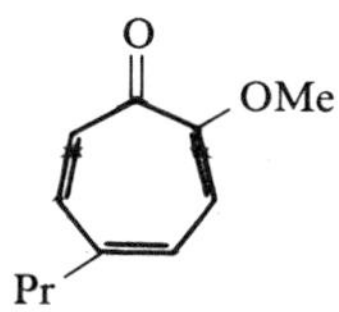		E3	DMAD	95	393
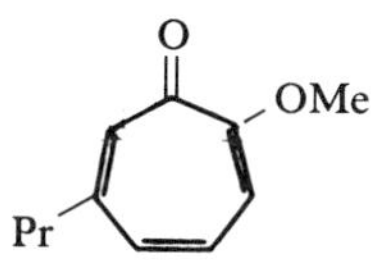		E3	DMAD	63	393
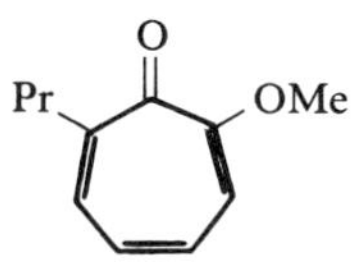		E3	DMAD	65	393
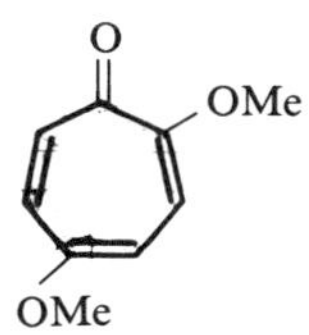		E3	DMAD	54	393
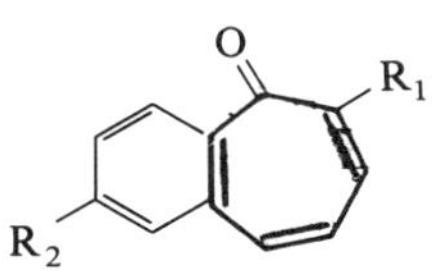	R_1 = H, R_2 = H	D3	MA	85	394
		D5	NPM	66	394
	R_1 = H, R_2 = OMe	D3	MA	72	394
		D5	NMM	81	394
	R_1 = Br, R_2 = H	D3	MA	40	394
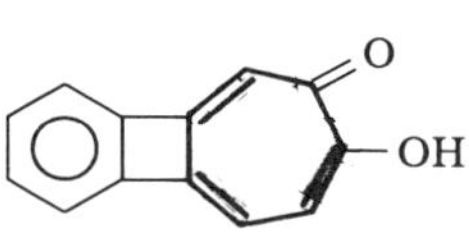		D3	MA	69	395

TABLE 6.31 Diels–Alder Reactions of Cyclooctatetraenes

Diene		Dienophile		Yield (%)	Ref.
		A10	(*E*)-1,2-Bis(phenylsulfonyl) ethene	85	48b, 382
		A10	Ethyne-bridged dimolybdenum complex	23	286, 296
		C1	Cyclooctatetraene		404
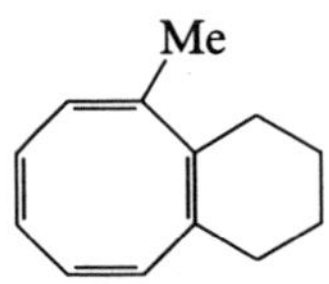		I5	NPTAD	15	405
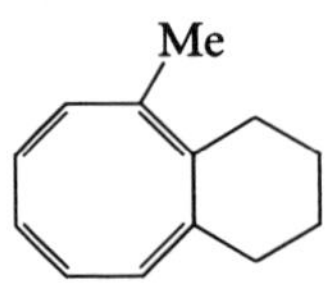		I5	NPTAD	4	405
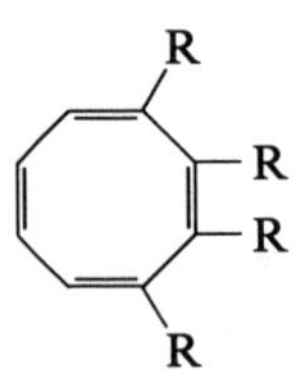	R = Me	I5	4-Endobornyl-1,2,4-triazoline-3,5-dione (*)	45	407
	R = Cl	C3	Acenaphthylene	60	406
		D3	MA	79	406

TABLE 6.32 Diels–Alder Reactions of Substituted 1,6-Methano[10]annulenes

Diene			Dienophile	Yield (%)	Ref.
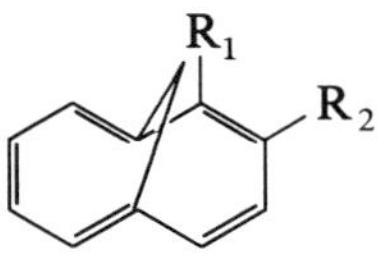					
R_1	**R_2**				
H	CN	I5	NMTAD	100	361
H	CO_2H	I5	NMTAD		364
H	CO_2Me	D3	MA	10	364
		D5	NPM	50	364
		I5	NMTAD	54	364
H	Br	I5	NMTAD	43	361
CN	H	I5	NMTAD	60	361
CO_2H	H	I5	NMTAD		364
CO_2ME	H	D3	MA	64	364
		D5	NMM		364
		I5	NMTAD		364
Br	H	I5	NMTAD	100	361
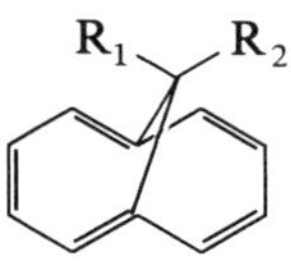					
R_1	**R_2**				
H	H	I5	NPTAD	50	361
H	Me	I5	NPTAD	70	361
H	CO_2Me	I5	NMTAD	51	361
H	Br	I5	NMTAD	75	361
Me	CN	I5	NPTAD	80	361
$=CH_2$		A10	(*Z*)-1,2-Dicyanoethene	27	362
		I5	NMTAD	17	361
		I5	NPTAD	50	362
CN	H	I5	NPTAD	100	363
CO_2H	H	I5	NPTAD	100	363
CO_2Me	CO_2Me	I5	NPTAD		361
OH	H	I5	NPTAD	100	363
OAc	H	I5	NPTAD	100	363
Br	H	I5	NPTAD	97	363
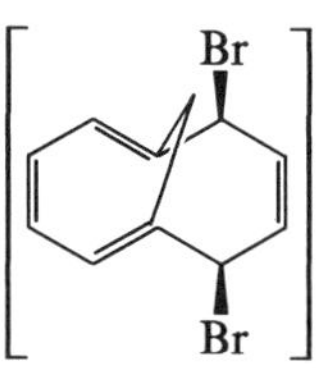		I5	NPTAD		415

TABLE 6.33 Diels–Alder Reactions of Miscellaneous Carbocyclic Dienes

Diene		Dienophile	Yield (%)	Ref.
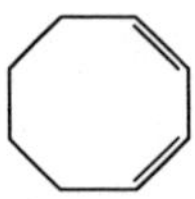	A7	DMM		403
	A8	Ethyne-bridged dimolybdenum complex	3	286, 402
	A12	1,1-Dicyano-2,2-bis(trifluoromethyl)ethene		109
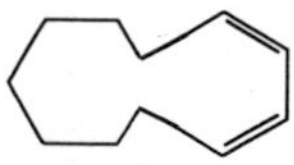	E5	Hexafluoro-2-butyne	76	408
	A10	(*Z*)-1,2-Diphenyl sulfonylethene	58	411
	A12	TCNE (k)		410
	E3	DMAD (k)		410
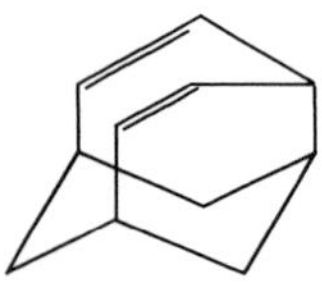	D3	MA	81	409
	E3	DMAD	22	409

REFERENCES FOR SECTION 6.1

1. a. H. Takemura, N. Komeshima, I. Takahashi, S.-I. Hashimoto, N. Ikota, K. Tomioka, and K. Koga, *Tetrahedron Lett.*, **28**, 5687 (1987).
   b. G. Bir and D. Kaufmann, *Tetrahedron Lett.*, **28**, 777 (1987).
   c. S.-I. Hashimoto, N. Komeshina, and K. Koga, *J. Chem. Soc. Chem. Commun.*, 437 (1979).
2. P. G. Gassman and D. A. Singleton, *Tetrahedron Lett.*, **28**, 5969 (1987).
3. K. Furuta, S. Hayashi, Y. Miwa, and H. Yamamoto, *Tetrahedron Lett.*, **28**, 5841 (1987).
4. R. M. Ortuno, R. Batllori, M. Ballesteros, M. Monsalvatje, J. Corbera, F. Sanchez-Ferrando, and J. Font, *Tetrahedron Lett.*, **28**, 3405 (1987).
5. S. Takano, A. Kurotaki, and K. Ogasawara, *Tetrahedron Lett.*, **28**, 3991 (1987).
6. K. H. Dotz, W. Kuhn, G. Muller, B. Huber, and H. G. Alt, *Angew. Chem. (Int. Ed. Engl.)*, **25**, 812 (1986).
7. N. Katagiri, T. Haneda, and C. Kaneko, *Chem. Pharm. Bull.*, **34**, 4875 (1986).
8. Y. Arai, S. Kuwayama, Y. Takeuchi, and T. Koizumi, *Synth. Commun.*, **16**, 233 (1986).
9. G. A. Krafft and P. M. Meinke, *J. Am. Chem. Soc.*, **108**, 1314 (1986).
10. a. S. G. Davies and J. C. Walker, *J. Chem. Soc. Chem. Commun.*, 609 (1986).
    b. J. W. Herndon, *J. Org. Chem.*, **51**, 2853 (1986).

11. H. Ogoshi, H. Mizushima, H. Toi, and Y. Aoyama, *J. Org. Chem.*, **51**, 2366 (1986).
12. O. De Lucchi, V. Lucchini, C. Marchioro, G. Valle, and G. Modena, *J. Org. Chem.*, **51**, 1457 (1986).
13. N. Ono, A. Kamimura, and A. Kaji, *Tetrahedron Lett.*, **27**, 15 (1986).
14. W. R. Dolbier, Jr., M. Seabury, D. Daly, and B. E. Smart, *J. Org. Chem.*, **51**, 974 (1986).
15. T. Cuvigny, C. Hervé du Penhoat, and M. Julia, *Tetrahedron*, **42**, 5321 (1986).
16. J. K. Whitesell, C.-L. Liu, C. M. Buchanan, H.-H. Chen, and M. A. Minton, *J. Org. Chem.*, **51**, 551 (1986).
17. S. F. Nelsen, S. C. Blackstock, and T. B. Frigo, *Tetrahedron*, **42**, 1769 (1986).
18. T. Katada, S. Eguchi, and T. Sasaki, *J. Org. Chem.*, **51**, 314 (1986).
19. F. Farina, M. C. Paredes, and V. Stefani, *Tetrahedron*, **42**, 4309 (1986).
20. H. Fisher, V. Gerbing, J. Riede, and R. Beun, *Angew. Chem.* (*Int. Ed. Engl*), **25**, 78, (1986).
21. P. A. Grieco, S. D. Larsen, and W. F. Fobare, *Tetrahedron Lett.*, **27**, 1975 (1986).
22. M. E. Jung and K. R. Buszek, *Tetrahedron Lett.*, **27**, 6165 (1986).
23. a. H.-J. Schneider and N. K. Sangwan, *J. Chem. Soc. Chem. Commun.*, 1787 (1986).
    b. K. Furuta, K. Iwanaga, and H. Yamamoto, *Tetrahedron Lett.*, **27**, 4507 (1986).
24. N. Ono, A. Kamimura, and A. Kaji, *J. Org. Chem.*, **51**, 2139 (1986).
25. a. K. Narasaka, M. Inque, and T. Yamada, *Chem. Lett.*, 1967 (1986).
    b. K. Narasaka, M. Inque, and B. Okada, *Chem. Lett.*, 1109 (1986).
26. a. *Jpn. Kokai Tokyo Koho*, JP 59, 164,731 [JP 84, 164,731] (1984); *Chem. Abstr.*, **102**, 61861c (1985).
    b. A. Inada, Y. Shimizu, Y. Ishii, and M. Ogawa, *Sekiyu Gakkai Shi*, **27**, 320 (1984).
27. E. Block, A. Wall, and J. Zubieta, *J. Am. Chem. Soc.*, **107**, 1783 (1985).
28. a. S. C. Cooper and P. G. Sammes, *J. Chem. Soc. Perkin Trans. 1*, 2407 (1984).
    b. A. P. Marchand, S. C. Surim, A. D. Earlywine, D. R. Powell, and D. Van der Helm, *J. Org. Chem.* **49**, 670 (1984).
    c. M. Oda and Y. Kanao, *Chem. Lett.*, 1547 (1981).
    d. T. J. Holmes, V. John, J. Vennerstrom, and K. E. Choi, *J. Org. Chem.*, **49**, 4736 (1984).
    e. G. Mehta, A. Srikrishna, A. V. Reddy, and S. M. Nair, *Tetrahedron*, **37**, 4543 (1981).
    f. I. Lee and E. S. Han, *Bull. Korean Chem. Soc.*, **4**, 41 (1983).
29. S. A. Harkin, O. Singh, and E. J. Thomas, *J. Chem. Soc. Perkin Trans. 1*, 1489 (1984).
30. S. M. Proust and D. D. Ridley, *Aust. J. Chem.*, **37**, 1677 (1984).
31. a. P. Laszlo and J. Lucchetti, *Tetrahedron Lett.*, **25**, 2147 (1984).
    b. R. Breslow, U. Maitra, and D. Rideout, *Tetrahedron Lett.*, **24**, 1901 (1983).
    c. R. J. Ardecky, F. A. J. Kerkesky, and M. P. Cava, *J. Org. Chem.*, **46**, 1483 (1981).
    d. F.-J. Gottschalk and P. Weyerstahl, *Chem. Ber.*, **113**, 555 (1980).

e. R. Breslow and U. Maitra, *Tetrahedron Lett.*, **25**, 1239 (1984).
f. J. Tamariz and P. Vogel, *Helv. Chim. Acta*, **64**, 188 (1981).

32. P. Camps, J. Castané, M. Feliz, and M. Figueredo, *Tetrahedron*, **40**, 5235 (1984).
33. X. Creary, P. A. Inocencio, T. L. Underiner, and R. Kostromin, *J. Org. Chem.*, **50**, 1932 (1985).
34. a. K. Hayakawa, N. Nishiyama, and K. Kanematsu, *J. Org. Chem.*, **50**, 512 (1985).
    b. G. Barbarella, M. Cinquini, and S. Colonna, *J. Chem. Soc. Perkin Trans. 1*, 1646 (1980).
    c. M. Hanack and F. W. Massa, *Tetrahedron Lett.*, **22**, 557 (1981).
35. T. Watabe and M. Oda, *Chem. Lett.*, 1791 (1984).
36. a. W. Oppolzer and C. Chapuis, *Tetrahedron Lett.*, **25**, 5383 (1984).
    b. T. Poll, J. O. Metter, and G. Helmchen, *Angew. Chem. (Int. Ed. Engl.)*, **24**, 112 (1985).
    c. W. Oppolzer, C. Chapuis, and M. J. Kelly, *Helv. Chim. Acta*, **66**, 2358 (1983).
    d. W. Oppolzer, C. Chapuis, G. M. Dao, D. Reichlin, and T. Godel, *Tetrahedron Lett.*, **23**, 4781 (1982).
    e. W. Oppolzer, M. Kurth, D. Reichlin, C. Chapuis, M. Mohnhaupt, and F. Moffatt, *Helv. Chim. Acta*, **64**, 2802 (1981).
    f. W. Choy, L. A. Reed III, and S. Masamune, *J. Org. Chem.*, **48**, 1137 (1983).
    g. D. P. Curran, B. H. Kim, H. P. Piyasena, R. J. Loncharich, and K. N. Houk, *J. Org. Chem.*, **52**, 2137 (1987).
    h. W. Oppolzer, C. Chapuis, and G. Bernardinelli, *Tetrahedron Lett.*, **25**, 5885 (1984).
    i. W. Oppolzer, M. J. Kelly, and J. Bernardinelli, *Tetrahedron Lett.*, **25**, 5889 (1984).
    j. J. Mulzer, M. Kappert, G. Huttner, and I. Jibril, *Tetrahedron Lett.*, **26**, 1631 (1985).
    k. T. Poll, G. Helmchen, and B. Bauer, *Tetrahedron Lett.*, **25**, 2191 (1984).
    l. W. Oppolzer, M. Kurth, D. Reichlin, and F. Moffatt, *Tetrahedron Lett.*, **22**, 2545 (1981).
    m. T. Poll, A. Sobczak, H. Hartmann, and G. Helmchen, *Tetrahedron Lett.*, **26**, 3095 (1985).
    n. R. J. Loncharich, T. R. Schwartz, and K. N. Houk, *J. Am. Chem. Soc.*, **109**, 14 (1987).
    o. A. A. Z. Samii, A. De Savignac, I. Rico, and A. Lattes, *Tetrahedron*, **41**, 3683 (1985).
37. W. Oppolzer, C. Chapuis, and G. Bernardinelli, *Helv. Chim. Acta*, **67**, 1397 (1984).
38. N. V. Vasil'ev, S. A. Khokhlova, A. F. Kolomiets, and G. A. Sokol'skii, *Zh. Org. Khim.*, **20**, 1983 (1984); E. E., 1810 (1984).
39. M. G. Voronkov, S. N. Adamovich, V. I. Rakhlin, R. G. Mirskov, and M. V. Sigalov, *Izv. Akad. Nauk SSSR, Ser. Khim.*, 2792 (1984).
40. S. D. Larsen and P. A. Grieco, *J. Am. Chem. Soc.*, **107**, 1768 (1985).
41. T. Molz, P. Konig, R. Goes, G. Gauglitz, and H. Meier *Chem. Ber.*, **117**, 833 (1984).
42. C. Maignan and R. A. Raphael, *Tetrahedron*, **39**, 3245 (1983).

43. D. Horton, T. Machinami, and Y. Takagi, *Carbohydr. Res.*, **121**, 135 (1983).
44. a. P. Bhaté and D. Horton, *Carbohydr. Res.*, **122**, 189 (1983).
    b. F. Shafizadeh, M. G. Essig, and D. D. Ward, *Carbohydr. Res.*, **114**, 71 (1983).
45. K. B. Wiberg, M. G. Matturro, P. J. Okarma, and M. E. Jason, *J. Am. Chem. Soc.*, **106**, 2194 (1984).
46. K. Uneyama, K. Takano, and S. Torii, *Bull. Chem. Soc. Jpn.*, **56**, 2867 (1983).
47. H. Tomioka, S. Miura, and Y. Izawa, *Tetrahedron Lett.*, **24**, 3353 (1983).
48. a. R. V. C. Carr, R. V. Williams, and L. A. Paquette, *J. Org. Chem.*, **48**, 4976 (1983).
    b. O. De Lucchi, V. Lucchini, L. Pasquato, and G. Modena, *J. Org. Chem.*, **49**, 596 (1984).
    c. O. De Lucchi and G. Modena, *J. Chem. Soc. Chem. Commun.*, 914, (1982).
49. R. S. Glass, K. Reineke, and M. Shanklin, *J. Org. Chem.*, **49**, 1527 (1984).
50. H. Link, K. Bernauer, and G. Englert, *Helv. Chim. Acta*, **65**, 2645 (1982).
51. I. Gupta and P. Yates, *J. Chem. Soc. Chem. Commun.*, 1227 (1982).
52. B. Gaede and T. M. Balthazor, *J. Org. Chem.*, **48**, 276 (1983).
53. a. J.-L. Gras, *J. Chem. Res.* (*S*), 300 (1982).
    b. W. R. Dolbier, Jr., and C. R. Burkholder, *Tetrahedron Lett.*, **21**, 785 (1980).
    c. W. Oppolzer and C. Chapuis, *Tetrahedron Lett.*, **24**, 4665 (1983).
54. a. H. Takeshita, T. Hatsui, I. Shimooda, and H. Mametsuka, *Bull. Chem. Soc. Jpn.*, **55**, 2291 (1982).
    b. H.-J. Liu and T. Ko Ngooi, *Synth. Commun.*, **12**, 715 (1982).
    c. E. C. Angell, F. Fringuelli, M. Huo, L. Minuti, A. Taticchi, and E. Wenkert, *J. Org. Chem.*, **53**, 4325 (1988).
55. M. M. Movsumzade, A. S. Kyazimov, N. V. Petrova, Z. A. Safarova, and G. Kh. Mamedova, *Dokl. Akad. Nauk Az. SSR*, **38**, 45 (1982).
56. P. Yates and S. P. Douglas, *Can. J. Chem.*, **60**, 2760 (1982).
57. M. G. Barlow, S. Culshaw, R. N. Haszeldine, and W. D. Morton, *J. Chem. Soc. Perkin Trans. 1*, 2105 (1982).
58. G. Maier and W. A. Jung, *Chem. Ber.*, **115**, 804 (1982).
59. a. H. Parlar and R. Baumann, *Angew. Chem.* (*Int. Ed. Engl.*), **20**, 1014 (1981).
    b. V. Branchadell, A. Oliva, and J. Bertran, *J. Mol. Struct.* (*Theochem.*), **21**, 85 (1985).
    c. R. V. Honeychuck, P. V. Bonnesen, J. Farahi, and W. H. Hersh, *J. Org. Chem.*, **52**, 5293 (1987).
    d. S. Danishefsky and M. Bednarski, *Tetrahedron Lett.*, **26**, 2507, (1985).
60. L. Nicolas, and M. Beugelmans-Verrier, *Tetrahedron*, **37**, 3847 (1981).
61. L. N. Kryukov, L. Yu. Kryukova, and A. F. Kolomiets, *Zh. Org. Khim.*, **17**, 2629 (1981).
62. M. E. Jung, K. Shishido, L. Light, and L. Davis, *Tetrahedron Lett.*, **22**, 4607 (1981).
63. a. H. K. Hall, Jr., P. Nogues, J. W. Rhoades, R. C. Sentman, and M. Detar, *J. Org. Chem.*, **47**, 1451 (1982).
    b. V. Macho, M. Kavala, M. Polievka, L. Komora, M. Matuskova, I. Repasova, and P. Suran, *Czech. CS*, 224,809; *Chem. Abstr.*, **103**, 104609s (1985).

c. G. B. Sergeev, V. S. Komarov, and A. V. Zvnov, *Zh. Obshch. Khim.*, **54**, 985 (1984).
d. R. Sustmann, M. Boehm, and J. Sauer, *Chem. Ber.*, **112**, 883 (1979).
e. G. Ahlgren, *Tetrahedron Lett.*, 915 (1979).
f. C. Rücker, D. Lang, J. Sauer, H. Friege, and R. Sustmann, *Chem. Ber.*, **113**, 1663 (1980).
g. R. L. Beddoes, J. M. Bruce, H. Finch, L. M. J. Heelam, I. D. Hunt, and O. S. Mills, *J. Chem. Soc. Perkin Trans.* **1**, 2670 (1981).

64. R. Bloch and J.-M. Denis, *Angew. Chem.* (*Int. Ed. Engl.*), **19**, 928 (1980).
65. D. Ranganathan, S. Ranganathan, and M. M. Mehrotra, *Tetrahedron*, **36**, 1869 (1980).
66. Y. Kobayashi, T. Yoshida, Y. Hanzawa, and I. Kumadak, *Tetrahedron Lett.*, **21**, 4601 (1980).
67. H. Horikawa, T. Nishitani, T. Iwasaki, Y. Mushika, I. Inoue, and M. Miyoshi, *Tetrahedron Lett.*, **21**, 4101 (1980).
68. O. Abou-Teim, R. B. Jansen, J. F. W. McOmie, and D. H. Perry, *J. Chem. Soc. Perkin Trans. 1*, 1841 (1980).
69. D. A.-J. Al-Sammerrai, J. T. Ralph, and D. E. West, *J. Heterocycl. Chem.*, **17**, 1705 (1980).
70. S. P. Briggs, D. I. Davies, R. F. Newton, and D. P. Reynolds, *J. Chem. Soc. Perkin Trans. 1*, 146 (1981).
71. J. Triepel and H. H. Otto, *Arch. Pharm.*, **315**, 295 (1982).
72. a. M. C. Carrenom, F. Farina, A. Galan, and J. L. Garcia Ruano, *J. Chem. Res.* (*S*), 297 (1979).
    b. F. Farina, A. Galan, and J. L. Garcia Ruano, *An. Quim.*, **74**, 954 (1978).
73. Y. Kobayashi, T. Nakano, R. Takahashi, and I. Kumadaki, *J. Chem. Soc. Perkin Trans. 1*, 2253 (1979).
74. R. N. Warrener, R. P. Johnson, C. W. Jefford, and D. A. Ralph, *Tetrahedron Lett.*, 2939 (1979).
75. K. Inoue, T. Iwasaki, and H. Horikawa, *Jpn. Kokai Tokyo Koho*, JP 79, 109,957 (1979); *Chem. Abstr.*, **92**, 41461u (1980).
76. a. A. Gorgues and A. Le Coq, *J. Chem. Soc. Chem. Commun.*, 767 (1979).
    b. A. Gorgues, A. Simon, A. Le Coq, and F. Corre, *Tetrahedron Lett.*, **22**, 625 (1981).
77. M. F. Ahern and G. W. Gokel, *J. Chem. Soc. Chem. Commun.*, 1019 (1979).
78. a. R. V. C. Carr and L. A. Paquette, *J. Am. Chem. Soc.*, **102**, 853 (1980).
    b. K. Laping and M. Hanack, *Tetrahedron Lett.*, 1309 (1979).
79. Y. Kobayashi, M. Honda, Y. Hanzawa, I. Kumadaki, and A. Ohsawa, *J. Chem. Soc. Perkin Trans. 1*, 1743 (1979).
80. K. Wald and H. Wamhoff, *Chem. Ber.*, **111**, 3519 (1978).
81. a. M. Chandler and R. J. Stoodley, *J. Chem. Soc. Perkin Trans. 1*, 1007 (1980).
    b. P. N. Preston, T. Winwick, and J. O. Morley, *J. Chem. Soc. Perkin Trans. 1*, 39 (1985).
82. W. Hoffmann and K. Von Fraumberg, *Ger. Offen.*, 2,719,976 (1978); *Chem. Abstr.*, **90**, 86872d (1979).

83. L. A. Badovskaya, O. B. Kravchenko, V. G. Kul'nevich, and I. A. Kuzovnikova, *Zh. Org. Khim.*, **15**, 50 (1979); *Chem. Abstr.*, **90**, 168012n (1979).
84. J. R. Lindsaysmith, R. O. C. Norman, and M. R. Stillings, *Tetrahedron*, **34**, 1381 (1978).
85. M. Hirama, Y. Koyama, Y. Shoji, and S. Ito, *Tetrahedron Lett.*, 2289 (1978).
86. M. A. Fox, R. Cardona, and N. J. Kiwiet, *J. Org. Chem.*, **52**, 1469 (1987).
87. R. J. Schmitt, J. C. Bottaro, R. Malhotra, and C. D. Bedford, *J. Org. Chem.*, **52**, 2294 (1987).
88. K. H. Doetz and W. Kuhn, *J. Organomet. Chem.*, **286**, C-23 (1985).
89. W. Adam and A. Berkessel, *Chem. Ber.*, **118**, 5018 (1985).
90. J. Bourguignon, G. Le Nard, and G. Queguiner, *Can. J. Chem.*, **63**, 2354 (1985).
91. L. A. Burke, *Theor. Chim. Acta*, **68**, 101 (1985).
92. D. E. Gaitanopoulos and J. Weinstock, *J. Heterocycl. Chem.*, **22**, 957 (1985).
93. O. De Lucchi, G. Licini, and L. Pasquato, *J. Chem. Soc. Chem. Commun.*, 418 (1985).
94. A. F. Benda, A. F. Elev, A. E. Kalinovskii, and G. A. Sokol'skii, *Zh. Org. Khim.*, **21**, 287 (1985).
95. a. B. Albert, C. Heller, R. Iden, G. Martin, H. D. Martin, B. Mayer, and A. Oftring, *Isr. J. Chem.*, **25**, 74 (1985).
    b. H. D. Martin, A. Oftring, R. Iden, E. Schwichtenberg, and H. J. Schivek, *Tetrahedron Lett.*, **23**, 841 (1982).
96. a. E. Wenkert and C. A. Broka, *Finn. Chem. Lett.*, 126 (1984).
    b. J. Nakayama, Y. Nakamura, and M. Hoshino, *Heterocycles*, **23**, 1119 (1985).
97. K. Baum and D. Tzeng, *J. Org. Chem.*, **50**, 2737 (1985).
98. M. G. Veliev, M. M. Guseinov, L. A. Yanovskaya, and K. Ya. Burstein, *Tetrahedron*, **41**, 749 (1985).
99. a. E. Block, and A. Wall, *Tetrahedron Lett.*, **26**, 1425 (1985).
    b. B. E. Smart and W. J. Middleton, *J. Am. Chem. Soc.*, **109**, 4982 (1987).
100. M. M. Latypova, V. V. Plemenkov, V. N. Kalinina, and I. G. Bolesov, *Zh. Org. Kim.*, **20**, 542 (1984); E. E., 489 (1984).
101. W. Donike, H. Ulrich, H. Thiemer, and J. Pieper, *Ger. Offen. DE*, 3,307,486 (1984); *Chem. Abstr.*, **101**, 210625n (1984).
102. R. Al-Hamdany and S. Salih, *J. Iraqi Chem. Soc.*, 61 (1982).
103. S. F. Nelsen, S. C. Blackstock, and T. B. Frigo, *J. Am. Chem. Soc.*, **106**, 3366 (1984).
104. C. Maignan, A. Guessous, and F. Rouessac, *Tetrahedron Lett.*, **25**, 1727 (1984).
105. A. Snow and H. Wohltjen, *Anal. Chem.*, **56**, 1411 (1984).
106. E. Baeuml and H. Mayer, *J. Org. Chem.*, **48**, 2600 (1983).
107. C. C. Hanstock and J. C. Tebby, *Phosphorus Sulfur*, **15**, 239 (1983).
108. T. R. Hoge, A. J. Caruso, and A. S. Magee, *J. Org. Chem.*, **47**, 4152 (1982).
109. W. J. Middleton and E. M. Bingham, *J. Fluorine Chem.*, **20**, 397 (1982).
110. P. D. Bartlett, G. L. Combs, A. X. T. Le, W. H. Watson, J. Galloy, and M. Kimura, *J. Am. Chem. Soc.*, **104**, 3131 (1982).

111. T. Liese, G. Splettstoesser, and A. De Meijere, *Angew. Chem. (Int. Ed. Engl.)*, **21**, 790 (1982).

112. L. A. Paquette and V. R. Williams, *Tetrahedron Lett.*, **22**, 4643 (1981).

113. Z. M. Ismail and H. M. R. Hoffmann, *J. Org. Chem.*, **46**, 3549 (1981).

114. T. K. Kostina, G. M. Baranov, and V. V. Perekalin, *Zh. Obshch. Khim.*, **51**, 712 (1981).

115. V. N. Nikonenko, L. I. Shipel'skaya, M. A. Trofimenko, L. G. Zhilina, and A. E. Pinsker, *Khim. Tekhno. (Kiev)*, 18 (1981).

116. M. G. Veliev, M. M. Guseinov, L. A. Yanovskaya, and R. F. Gakhramanov, *Izv. Akad. Nauk SSSR, Ser. Khim.*, 144 (1981).

117. G. F. Il'in, A. F. Kolomiets, and G. A. Sokol'skii, *Zh. Org. Khim.*, **17**, 1779 (1981).

118. J. M. Birchall, K. Burger, R. N. Naszeldine, and S. N. Nona, *J. Chem. Soc. Perkin Trans. 1*, 2080 (1981).

119. a. H. Vathke-Ernst and H. M. R. Hoffmann, *Chem. Ber.*, **114**, 1464 (1981).
b. H. M. R. Hoffmann and H. Vathke-Ernst, *Chem. Ber.*, **114**, 2208 (1981).

120. a. D. Ranganathan, C. B. Rao, S. Ranganathan, A. K. Mehrotra, and R. Iyengar, *J. Org. Chem.*, **45**, 1185 (1980).
b. T. Hayama, S. Tomoda, Y. Takeuchi, and Y. Nomura, *J. Org. Chem.*, **49**, 3235 (1984).

121. W. R. Roth, F. G. Klarner, and H. W. Lennartz, *Chem. Ber.*, **113**, 1818 (1980).

122. a. W. R. Dolbier and C. R. Burkholder, *J. Org. Chem.*, **49**, 2381 (1984).
b. W. R. Dolbier, C. R. Burkholder, G. E. Wicks, G. J. Palenik, and M. Gawron, *J. Am. Chem. Soc.*, **107**, 7183 (1985).

123. D. Hermeling, and H. J. Schäefer, *Angew. Chem. (Int. Ed. Engl.)*, **23**, 233 (1984).

124. D. A. Evans, K. T. Chapman, and J. Bisaha, *J. Am. Chem. Soc.*, **106**, 4261 (1984).

125. M. L. Durrant and E. J. Thomas, *J. Chem. Soc. Perkin Trans. 1*, 901 (1984).

126. H. Gnichtel, C. Gumprecht, and P. Luger, *Liebigs Ann. Chem.*, 1531 (1984).

127. W. D. Wulff and C. D. Yang, *J. Am. Chem. Soc.*, **106**, 7565 (1984).

128. R. A. Pabon, D. J. Bellville, and N. L. Bauld, *J. Am. Chem. Soc.*, **105**, 5158 (1983).

129. H. J. Liu, E. N. C. Browne, and P. R. Pednekar, *Can. J. Chem.*, **60**, 921 (1982).

130. W. D. Wulff and D. C. Yang, *J. Am. Chem. Soc.*, **105**, 6726 (1983).

131. D. Martina and F. Brion, *Tetrahedron Lett.*, **23**, 865 (1982).

132. a. L. Stella, and J. L. Boucher, *Tetrahedron Lett.*, **23**, 953 (1982).
b. J.-L. Boucher and L. Stella, *Tetrahedron*, **42**, 3871 (1986).
c. J. L. Boucher and L. Stella, *Tetrahedron*, **41**, 875 (1985).

133. P. G. Willard and S. E. de Laszlo, *J. Org. Chem.*, **50**, 3738 (1985).

134. B. K. Rammash, C. M. Gladstone, and J. L. Wong, *J. Org. Chem.*, **46**, 3036 (1981).

135. P. Chamberlain and A. E. Rooney, *Tetrahedron Lett.*, 383 (1979).

136. J. J. Gaiewski, *J. Am. Chem. Soc.*, **101**, 4393 (1979).

137. M. Baumann and W. Hoffmann, *Liebigs Ann. Chem.*, 743 (1979).
138. K. Hagakawa, M. Aso, and K. Kanematsu, *J. Org. Chem.*, **50**, 2036 (1985).
139. D. L. Boger and C. E. Brotherton, *Tetrahedron*, **42**, 2777 (1986).
140. J. F. Buzinkai, D. M. Hrubowchak, and F. X. Smith, *Tetrahedron Lett.*, **26**, 3195 (1985).
141. G. A. Tolstikov, V. P. Kazakov, E. E. Shults, R. G. Bulgakov, and R. G. Kantyukova, *Zh. Org. Khim.*, **20**, 313 (1984).
142. E. Kobayashi, S. Matsumura, and J. Furukawa, *Polym. Bull. (Berlin)*, **3**, 285 (1980).
143. K. Friedrich and M. Zamkanei, *Chem. Ber.*, **112**, 1867 (1979).
144. J. L. Ripoll and M. C. Lasne, *Tetrahedron Lett.*, 5201 (1978).
145. A. P. Davis and G. H. Whitham, *J. Chem. Soc. Chem. Commun.*, 639 (1980).
146. P. Dowd and W. Weber, *J. Org. Chem.*, **47**, 4777 (1982).
147. D. Martina and F. Brion, *Tetrahedron Lett.*, **23**, 865 (1982).
148. M. E. Jung and K. R. Buszek, *J. Org. Chem.*, **50**, 5440 (1985).
149. A. Krebs, E. Franken, and S. Mueller, *Tetrahedron Lett.*, **22**, 1675 (1981).
150. G. Maier, K. A. Schneider, M. K. Dieter, H. Irngartinger, and A. Lenz, *Angew. Chem. (Int. Ed. Engl.)*, **21**, 437 (1982).
151. P. T. Klemarczyk, J. M. Sanders, M. K. Vock, J. F. Vinals, F. L. Schmitt, and E. G. Granda, *U.S. Pat.*, 4,312,888 (1982); *Chem. Abstr.*, **96**, 161160z (1982).
152. G. Buchbauer and H. C. Rohner, *Liebigs Ann. Chem.*, 2093 (1981).
153. H. M. R. Hoffmann and O. Koch, *J. Org. Chem.*, **51**, 2939 (1986).
154. J. P. Nallet, C. Arnaud, and J. Huet, *Tetrahedron Lett.*, 2583 (1979).
155. G. Grundke and H. M. R. Hoffmann, *J. Org. Chem.*, **46**, 5428 (1981).
156. V. I. Koshutin, S. V. Ovcharova, and V. A. Zeverev, *Zh. Obshch. Khim.*, **48**, 2500 (1978).
157. L. M. Tolbert, J. C. Gregory, and C. P. Brock, *J. Org. Chem.*, **50**, 548 (1985).
158. R. Mercier, J. Douglade, J. Amaudrut, J. Sala-Pala, and J. E. Guerchais, *J. Organomet. Chem.*, **244**, 145 (1983).
159. R. Kh. Geivandovm, *Zh. Org. Chim.*, **19**, 878 (1983).
160. R. Kaiser and D. Lamparsky, *Helv. Chim. Acta*, **67**, 1884 (1984).
161. U. H. Brinker and I. Fleischhauer, *Tetrahedron*, **37**, 4495 (1981).
162. P. T. Klemarczyk, J. M. Sanders, M. H. Vock, J. F. Vinals, F. L. Schmitt, and E. J. Granda, *U.S. Pat.*, 4,284,824 (1981); *Chem. Abstr.*, **95**, 20343d (1981).
163. D. G. Ibbott, N. C. Payne, and A. Shaver, *Inorg. Chem.*, **20**, 2193 (1981).
164. P. Yates and I. Gupta, *J. Chem. Soc. Chem. Commun*, 449 (1981).
165. a. I. Fleming and R. V. Williams, *J. Chem. Soc. Perkin Trans. 1*, 684 (1981).
     b. W. R. Roush and T. E. D'Ambra, *J. Org. Chem.*, **46**, 5045 (1981).
166. I. Fleming and J. P. Michael, *J. Chem. Soc. Perkin Trans. 1*, 1549 (1981).
167. G. I. Dmitrienko, M. E. Savard, R. W. Friesen, and M. S. Thandi, *Tetrahedron Lett.*, **26**, 1691 (1985).
168. J. W. Wilts and S. Z. Ahmed, *J. Org. Chem.*, **44**, 4000 (1979).

169. E. L. Clennan and M. M. E. Mehrsheikh, *J. Am. Chem. Soc.*, **106**, 7112 (1984).
170. D. J. Burnell, H. B. Goodbrand, S. M. Kaiser, and Z. Valenta, *Can. J. Chem.*, **62**, 2398 (1984).
171. F. K. Brown, K. N. Houk, D. J. Burnell, and Z. Valenta, *J. Org. Chem.*, **52**, 3050 (1987).
172. P. von Rague Schleyer, P. Grubmüller, W. F. Maier, O. Vostrowsky, L. Skatteboel, and K. H. Holm, *Tetrahedron Lett.*, **21**, 921 (1980).
173. N. J. Lewis, D. B. Knigh, and W. J. Collins, *J. Med. Chem.*, **22**, 1505 (1979).
174. V. D. Kiselev, I. M. Shakirov, and A. I. Konovalov, *Zh. Org. Khim.*, **21**, 665 (1985).
175. K. Hirao, A. Ando, T. Hamada, and O. Yonemitsu, *J. Chem. Soc. Chem. Commun.*, 300 (1984).
176. N. M. Guseinov, F. F. Muganlinskii, T. A. Umudov, and A. S. Kyazimov, *Dokl. Akad. Nauk. Az. SSR*, **38**, 43, (1982).
177. D. J. Burnell and Z. Valenta, *J. Chem. Soc. Chem. Commun.*, 1247 (1985).
178. a. M. M. Guseinov, T. G. Kyazimova, R. S. Babaev, and E. G. Aslanova, *Dokl. Akad. Nauk Az. SSR*, **36**, 51, (1980).
   b. F. F. Muganlinskii and M. M. Guseinov, *Azerb. Khim. Zh.*, (3), 52 (1982).
179. M. E. Jung and C. D. Radcliffe, *Tetrahedron Lett.*, **21**, 4397 (1980).
180. a. M. S. Salakhov, N. F. Musaeva, S. N. Suleimanov, and M. M. Guseinov, *Nauch Tr. Azerb. In-Ta Nefti Khim.*, 76 (1979); *Chem. Abstr.*, **92**, 93601j (1980).
   b. V. A. Palm and M. F. Musajeva, *Org. React. (Tartu)*, **23**, 96 (1986).
181. a. M. S. Salakhov, N. F. Musaeva, and S. N. Suleimanov, *Org. React. (Tartu)*, **16**, 54 (1979).
   b. N. F. Musaeva, M. S. Salakhov, and S. N. Suleimanov, *Org. React. (Tartu)*, **16**, 65 (1979).
   c. M. S. Salakhov, E. M. Treivus, R. S. Salakhova, M. M. Guseinov, and V. S. Umaeva, *Vopr. Stereokhim.*, **6**, 36 (1977); *Chem. Abstr.*, **90**, 203373e (1979).
   d. M. S. Salakhov, N. F. Musaeva, S. A. Mamedov, and S. N. Suleimanov, *Org. React. (Tartu)*, **15**, 501 (1978).
   e. M. S. Salakhov, N. F. Musaeva, S. N. Suleimanov, A. A. Bairamov, and A. A. Aleskerov, *Org. React. (Tartu)*, **15**, 429 (1978).
   f. M. S. Salakhov, N. F. Musaeva, A. K. Karaeva, and S. N. Suleimanov, *Org. React. (Tartu)*, **15**, 297 (1978).
182. E. D. Weiler and J. J. Brennan, *J. Heterocycl. Chem.*, **15**, 1299 (1978).
183. a. N. F. Musaeva, M. F. Salakhov, R. S. Salakhova, V. S. Umaeva, and T. A. Kopylova, *Org. React. (Tartu)*, **16**, 349 (1979).
   b. N. F. Musaeva, M. S. Salakhov, S. N. Suleimanov, R. S. Salakhova, and A. A. Gasanova, *Org. React. (Tartu)*, **16**, 386 (1979).
   c. M. S. Salakhov, N. F. Musaeva, S. A. Mamedov, T. A. Kopylova, and A. A. Gasanova, *Org. React. (Tartu)*, **16**, 402 (1979).
184. *Jpn. Kokai Tokyo Koho*, JP 58, 222,041 [JP 83, 222,041] (1983); *Chem. Abstr.*, **100**, 209239n (1984).
185. F. F. Muganlinskii, T. A. Umudov, M. M. Guseind, H. Ali-Ammar, and A. S. Kyazimov, *Azerb. Khim. Zh.*, 51 (1980).

186. P. H. Mazzocchi, B. Stahly, J. Dodd, N. G. Rondan, L. N. Domelsmith, M. D. Rozeboom, P. Caramella, and K. N. Houk, *J. Am. Chem. Soc.*, **102**, 6482 (1980).

187. M. Papadopoulos and G. Jenner, *Tetrahedron Lett.*, **23**, 1889 (1982).

188. G. Jenner, M. Papadopoulos, and W. J. Le Noble, *Nouv. J. Chim.*, **7**, 687 (1983).

189. M. G. Voronkov, S. N. Adamovich, V. I. Rakhlin, R. G. Mirskov, and M. V. Sigalov, *Izv. Akad. Nauk SSSR, Ser. Khim.*, 2792 (1984).

190. A. Dondoni, M. Fogagnolo, A. Mastellari, P. Pedrini, and F. Ugozzoli, *Tetrahedron Lett.*, **27**, 3915 (1986).

191. a. A. M. Mustafaev, S. N. Novikov, M. M. Guseinov, and N. R. Adigezalov, *Org. React. (Tartu)*, **23**, 3 (1986).
 b. A. M. Mustafaev, N. R. Adigezalov, M. M. Guseinov, and Ch. A. Chalabiev, *Azerb. Khim. Zh.*, 36 (1979).
 c. A. M. Mustafaev and N. R. Adigezalov, *Monomery Polim.*, 69 (1983); *Chem. Abstr.*, **101**, 230292e (1984).
 d. A. M. Mustafaev, N. R. Adigezalov, E. A. Berdnikov, G. G. Butenko, and A. I. Konovalov, *Zh. Org. Khim.*, **17**, 88 (1981).

192. A. M. Mustafaev, N. A. Alekperov, M. M. Guseinov, A. B. Imamaliev, and S. I. Mekhtiev, *Zh. Org. Khim.*, **21**, 1691 (1985).

193. P. H. Daniels, J. L. Wong, J. L. Atwood, L. G. Canada, and R. D. Rogers, *J. Org. Chem.*, **45**, 435 (1980).

194. a. U. O'Connor and W. Rosen, *Tetrahedron Lett.*, 601 (1979).
 b. U. O'Connor and W. Rosen, *J. Org. Chem.*, **45**, 1824 (1980).

195. E. M. Movsumzade, S. M. Movsumzade, and G. M. Mamedov, *Azerb. Khim. Zh.*, 49 (1980).

196. R. G. Hall and S. Trippett, *Tetrahedron Lett.*, **23**, 2603 (1982).

197. D. A. Kidwell and K. Blemann, *Anal. Chem.*, **54**, 2462 (1982).

198. T. A. Umudov, F. F. Muganlinskii, M. M. Guseinov, M. A. Y. Tabit, and E. K. Mamedova, *Azerb. Khim. Zh.*, **6**, 38 (1983).

199. a. M. S. Salakhov, S. A. Ismailov, and N. A. Alekperov, *Azerb. Khim. Zh.*, 42 (1978).
 b. M. S. Salakhov and S. A. Ismailov, *Zh. Org. Khim.*, **20**, 98 (1984).

200. R. N. Warener, R. A. Russell, and G. J. Collinm, *Tetrahedron Lett.*, 4447 (1978).

201. H. W. Thompson, J. K. Wong, R. A. Lalancette, J. A. Boyko, and A. M. Robertiello, *J. Org. Chem.*, **50**, 2115 (1985).

202. M. G. Veliev, A. A. Ragimov, D. D. Guseinov, and M. R. Mirzoeva, *Azerb. Khim. Zh.*, 44 (1980).

203. a. M. G. Veliev, M. M. Guseinov, E. S. Mamedov, and R. F. Gakhramanov, *Synthesis*, 337 (1984).
 b. M. G. Veliev, M. M. Guseinov, L. A. Yanovskaya, E. Mamedov, and A. A. Bairamov, *Izv. Akad. Nauk SSSR, Ser. Khim.*, 1358 (1980).
 c. M. M. Guseinov, M. G. Veliev, and E. S. Mamedov, *Monomery Polim.*, 21 (1983); *Chem. Abstr.*, **101**, 130285m (1984).

204. a. T. G. Kyazimova, M. M. Guseinov, and R. S. Babaev, *Monomery Polim.*, 13 (1983); *Chem. Abstr.*, **101**, 130284k (1984).
 b. T. G. Kyazimova, *Azerb. Khim. Zh.*, (6), 81 (1979).

205. I. M. Akhmediv, Z. M. Zeinalova, and F. E. Dzhavadov, *Azerb. Khim. Zh.*, 50 (1978).
206. M. S. Salakhov, E. M. Treivus, V. S. Umaeva, I. B. Katkova, R. I. Islamova, M. M. Guseinov, and R. M. Salakhova, *Neth. Appl.*, 79, 08,453 (1979); *Chem. Abstr.*, **95**, 204941d (1981).
207. T. A. Umudov, F. F. Muganlinskii, M. M. Guseinov, and H. Ali-Ammar, *Azerb. Khim. Zh.*, 75 (1981).
208. M. T. Reetz and G. Ncumeier, *Chem. Ber.*, **112**, 2209 (1979).
209. F. Gavina, A. M. Costero, P. Gil, B. Palazov, and S. V. Luis, *J. Am. Chem. Soc.*, **103**, 1797 (1981).
210. P. G. Baraldi, A. Barco, S. Benetti, G. P. Pollini, E. Polo, and D. Simoni, *J. Chem. Soc. Chem. Commun.*, 1040 (1984).
211. F. A. Amer, A. Khalaf, and Y. M. Darwish, *Indian J. Chem., Sect. B*, **17B**, 575 (1979).
212. S. Greenfield and K. Mackenzie, *J. Chem. Soc. Perkin Trans. 2*, 1651 (1986).
213. L. Capuano, T. Triesch, V. Schramm, and W. Hiller, *Chem. Ber.*, **117**, 2785 (1984).
214. V. V. Plemenkov, and V. P. Kostin, *Zh. Org. Khim.*, **15**, 1217 (1979).
215. T. L. Chan, N. Z. Huang, and F. Sondheimer, *Tetrahedron*, **39**, 427 (1983).
216. A. Oku, H. Hasegawa, H. Shimazu, J. Nishimura, and T. Harada, *J. Org. Chem.*, **46**, 4152 (1981).
217. N. Almirante, M. L. Gelmi, D. Pocar, M. Valsecchi, M. Ballabi, and B. Gioia, *Tetrahedron*, **42**, 51 (1986).
218. J. Nakayama, A. Sakai, and M. Hoshino, *J. Org. Chem.*, **49**, 5084 (1984).
219. J. W. Barton, M. K. Sheperd, and R. J. Willis, *J. Chem. Soc. Perkin Trans. 1*, 967 (1986).
220. R. N. Warrener, R. Y. S. Tanand, and R. A. Russel, *Tetrahedron Lett.*, 2943 (1979).
221. O. Tsuge, T. Ohnishi, and H. Watanabe, *Heterocycles*, **16**, 2085 (1981).
222. W. J. Leigh, *Can. J. Chem.*, **63**, 2736 (1985).
223. K. Gollnick and H. Hartmann, *Tetrahedron Lett.*, **23**, 2651 (1982).
224. M. Yasuda, K. Harano, and K. Kanematsu, *Tetrahedron Lett.*, **21**, 627 (1980).
225. M. Yasuda, K. Harano, and K. Kanematsu, *J. Am. Chem. Soc.*, **103**, 3120 (1981).
226. M. Mori, A. Hayamizu, and K. Kanematsu, *J. Chem. Soc. Perkin Trans. 1*, 1259 (1981).
227. F. E. Arnold, *U.S. Pat.*, 811,346 (1977); *Chem. Abstr.*, **90**, 103653c (1979).
228. Y. Yamashita and M. Masumura, *Heterocycles*, **14**, 29 (1980).
229. K. Harano, M. Yasuda, and K. Kanematsu, *J. Org. Chem.*, **47**, 3736 (1982).
230. G. I. Fray, G. R. Geen, K. Mackenzie, and D. L. Williams-Smith, *Tetrahedron*, **35**, 1173 (1979).
231. V. V. Plemenkov, Kh. Z. Giniyatov, Ya. Ya. Villen, N. V. Villen, L. S. Surmina, and I. G. Bolesov, *Dokl. Akad. Nauk SSSR*, **254**, 895 (1980).
232. S. Mondal and A. J. Bhattacharya, *Curr. Sci.*, **53**, 676 (1984).
233. M. V. Basilevskii, N. N. Weinberg, and V. M. Zhulin, *Croat. Chem. Acta*, **57**, 1423 (1985).

234. Ya. D. Samuilov, R. L. Nurullina, and A. I. Konovalov, *Zh. Org. Skhim.*, **20**, 2492 (1984).
235. S. Mondal, T. K. Bandyopadhyay, and A. J. Bhattacharya, *Indian J. Chem., Sect. B*, **22B**, 225 (1983).
236. S. Mondal, T. K. Bandyopadhyay, and A. J. Bhattacharya, *Indian J. Chem., Sect. B*, **22B**, 448 (1983).
237. T. K. Bandyopadhyay and A. J. Bhattacharya, *Indian J. Chem., Sect. B*, **21B**, 91 (1982).
238. S. Eguchi, K. Ishiura, T. Noda, and T. Sasaki, *J. Org. Chem.*, **52**, 496 (1987).
239. F. Gavina, A. M. Costero, A. M. Gonzalez, and S. V. Luis, *J. Org. Chem.*, **52**, 2997 (1987).
240. U. Szeimies-Seebach, A. Schoeffer, R. Roemer, and G. Szeimies, *Chem. Ber.*, **114**, 1767 (1981).
241. B. Uebersax and M. Neuenschwander, *Chimia*, **35**, 400 (1981).
242. a. K. Sakai, T. Kobori, and T. Fujisawa, *Tetrahedron Lett.*, **22**, 115 (1981).
b. T. Fujisawa, K. Sakai, and T. Kobori, *Jpn. Kokai Tokyo Koho*, JP 79, 90,165 (1979), *Chem. Abstr.*, **92**, 6152x (1980).
243. K. Kanematsu, S. Morita, S. Fukushima, and E. Osawa, *J. Am. Chem. Soc.*, **103**, 5211 (1981).
244. A. Oku, Y. Nozaki, H. Hasegawa, J. Nishimura, and T. Harada, *J. Org. Chem.*, **48**, 4374 (1983).
245. D. B. Knigh, R. W. Hall, and D. G. Cleary, *J. Heterocycl. Chem.*, **18**, 1649 (1981).
246. a. L. A. Paquette, R. V. C. Carr, P. Charumilind, and J. F. Blount, *J. Org. Chem.*, **45**, 4922 (1980).
b. L. A. Paquette, K. E. Green, and L. Y. Hsu, *J. Org. Chem.*, **49**, 3650 (1984).
247. a. L. A. Paquette and H. Kunzer, *J. Am. Chem. Soc.*, **108**, 7431 (1986).
b. M. C. Bohm, R. V. C. Carr, R. Gleiter, and L. A. Paquette, *J. Am. Chem. Soc.*, **102**, 7218 (1980).
c. W. H. Watson, J. Galloy, P. D. Bartlett, and A. M. Roof, *J. Am. Chem. Soc.*, **103**, 2022 (1981).
248. L. A. Paquette, R. V. C. Carr, M. C. Bohm, and R. Gleiter, *J. Am. Chem. Soc.*, **102**, 1186 (1980).
249. L. A. Paquette, P. C. Hayes, P. Charumilind, M. C. Bohm, R. Gleiter, and J. F. Blount, *J. Am. Chem. Soc.*, **105**, 3148 (1983).
250. a. L. A. Paquette, J. W. Fischer, A. R. Browne, and C. W. Doecke, *J. Am. Chem. Soc.*, **107**, 686 (1985).
b. L. A. Paquette, A. L. Browne, C. W. Doecke, and R. V. Williams, *J. Am. Chem. Soc.*, **105**, 4113 (1983).
251. P. Charumilind and L. A. Paquette, *J. Am. Chem. Soc.*, **106**, 8225 (1984).
252. L. A. Paquette, P. Charumilind, M. C. Bohm, R. Gleiter, L. S. Bass, and J. Clardy, *J. Am. Chem. Soc.*, **105**, 3136 (1983).
253. L. A. Paquette, V. R. Williams, R. V. C. Carr, P. Charumilind, and J. F. Blount, *J. Org. Chem.*, **47**, 4566 (1982).
254. L. A. Paquette, M. Gugelehuk, and Y.-L. Hsu, *J. Org. Chem.*, **51**, 3864 (1986).
255. P. D. Bartlett and C. Wu, *J. Org. Chem.*, **49**, 1880 (1984).

256. L. A. Paquette, P. Charumilind, and J. C. Gallucci, *J. Am. Chem. Soc.*, **105**, 7364 (1983).
257. L. A. Paquette, T. M. Kravetz, M. C. Bohm, and R. Gleiter, *J. Org. Chem.*, **48**, 1250 (1983).
258. J. C. Gallucci, T. M. Kravetz, K. E. Green, and L. A. Paquette, *J. Am. Chem. Soc.*, **107**, 6592 (1985).
259. L. A. Paquette, K. E. Green, R. Gleiter, W. Schafer, and J. C. Gallucci, *J. Am. Chem. Soc.*, **106**, 8232 (1984).
260. L. A. Paquette, T. M. Kravetz, and L. Y. Hsu, *J. Am. Chem. Soc.*, **107**, 6598 (1985).
261. J. Ipaktschi, J. Herber, H.-O. Kalinowski, and R. Boese, *Tetrahedron Lett.*, **28**, 3467 (1987).
262. L. A. Paquette, M. Gugelchuk, and M. L. McLaughlin, *J. Org. Chem.*, **52**, 4732 (1987).
263. L. A. M. Turkenburg, P. M. L. Blok, W. H. de Wolf, and F. Bickelhaupt, *Tetrahedron Lett.*, **22**, 3317 (1981).
264. W. E. Noland, L. L. Landucci, and V. Kameswaran, *J. Org. Chem.*, **45**, 3456 (1980).
265. J. R. Bull and R. I. Thomson, *J. Chem. Soc. Chem. Commun.*, 451 (1986).
266. D. Schomburg, M. Thielmann, and E. Winterfeldt, *Tetrahedron Lett.*, **27**, 5833 (1986).
267. P. G. Gassmann, D. A. Singleton, J. J. Wilwerding, and S. P. Chavan, *J. Am. Chem. Soc.*, **109**, 2182 (1987).
268. A. Gorgues and A. Le Coq, *Tetrahedron Lett.*, 4829 (1979).
269. D. W. Reynolds, K. T. Lorenz, H. S. Chion, D. J. Bellville, R. A. Pabon, and N. L. Bauld, *J. Am. Chem. Soc.*, **109**, 4960 (1987).
270. G. Weber, K. Menke, and H. Hopf, *Chem. Ber.*, **113**, 531 (1980).
271. D. N. Gupta, P. Hodge, and N. Khan, *J. Chem. Soc. Perkin Trans. 1*, 689 (1981).
272. D. Stephan, A. Gorgues, and A. Le Coq, *Tetrahedron Lett.*, **27**, 4295 (1981).
273. J. Jurczak and M. Tkacz, *J. Org. Chem.*, **44**, 3347 (1979).
274. G. Huybrechts, H. Poppelsdorf, L. Maesschalck, and B. Van Mele, *Int. J. Chem. Kinet.*, **16**, 93 (1984).
275. a. R. S. Garigipati, R. Cordova, M. Parvez, and S. M. Weinreb, *Tetrahedron*, **42**, 2979 (1986).
    b. S. W. Remiszewski, J. Yang, and S. M. Weinreb, *Tetrahedron Lett.*, **27**, 1853 (1986).
276. G. C. Calhoun and G. B. Schuster, *Tetrahedron Lett.*, **27**, 911 (1986).
277. a. P. G. Gassman and D. A. Singleton, *J. Am. Chem. Soc.*, **106**, 7993 (1984).
    b. F.-G. Klarner, B. M. J. Dogan, O. Ermer, W. E. Doering, and M. P. Cohen, *Angew. Chem. (Int. Ed. Engl.)*, **25**, 108 (1986).
    c. D. J. Bellville, D. D. Wirth, and N. L. Bauld, *J. Am. Chem. Soc.*, **103**, 718 (1981).
    d. N. L. Bauld and D. J. Bellville, *U.S. Pat.*, 4,503,195; *Chem. Abstr.*, **103**, 22206u (1985).
    e. G. C. Calhoun and G. B. Schuster, *J. Am. Chem. Soc.*, **106**, 6870 (1984).

f. P. Laszlo and J. Lucchetti, *Tetrahedron Lett.*, **25**, 1567 (1984).
g. N. L. Bauld, D. J. Bellville, S. A. Gardner, Y. Migron, and G. Cogswell, *Tetrahedron Lett.*, **23**, 825 (1982).
h. J. Mattay, J. Gersdorf, and J. Mertes, *J. Chem. Soc. Chem. Commun.*, 1088 (1985).

278. H. Natsugari, R. R. Whittle, and S. M. Weinreb, *J. Am. Chem. Soc.*, **106**, 7867 (1984).
279. M. Sabuni, G. Kresze, and H. Braun, *Tetrahedron Lett.*, **25**, 5377 (1984).
280. H. Felber, G. Kresze, H. Braun, and A. Vasella, *Tetrahedron Lett.*, **25**, 5381 (1984).
281. A. S. Kyazimov, M. M. Movsumzade, Z. A. Safarova, and N. K. Pashaeva, *Azerb. Khim. Zh.*, (4), 66 (1983).
282. Y. K. Rao and M. Nagarajan, *Synthesis*, 757 (1984).
283. M. G. Barlow, R. N. Haszeldine, and K. W. Murray, *J. Chem. Soc. Perkin Trans. 1*, 1960 (1980).
284. D. A.-J. Al-Sammerrai, J. T. Ralph, and D. E. West, *J. Heterocycl. Chem.*, **17**, 1705 (1980).
285. M. Shen and A. G. Schultz, *Tetrahedron Lett.*, **22**, 3347 (1981).
286. S. A. R. Knox, R. F. D. Stansfield, F. G. A. Stone, M. J. Winter, and P. Woodward, *J. Chem. Soc. Chem. Commun.*, 934 (1979).
287. G. E. Keck and S. A. Fleming, *Tetrahedron Lett.*, 4763 (1978).
288. a. G. O. Torosyan, S. L. Paravyan, K. Ts. Tagmazyan, and A. T. Babayan, *Arm. Khim. Zh.*, **32**, 177 (1979); *Chem. Abstr.*, **92**, 58589q (1980).
b. R. H. Hill, M. G. Newton, N. S. Pantaleo, and K. M. Collins, *J. Org. Chem.*, **45**, 1593 (1980).
289. I. U. Khand, P. L. Pauson, and M. J. A. Habib, *J. Chem. Res. (S)*, 346 (1978).
290. J. Jurczak and M. Tkacz, *Synthesis*, 42 (1979).
291. a. J. Mlcoch and E. Steckhan, *Tetrahedron Lett.*, **28**, 1081 (1987).
b. J. Mlcoch and E. Steckhan, *Angew. Chem. (Int. Ed. Engl.)*, **24**, 412 (1985).
292. a. G. Kresze, E. Kysela, and W. Dittel, *Liebigs Ann. Chem.*, 210 (1981).
b. G. Kresze, and E. Kysela, *Liebigs Ann. Chem.*, 202 (1981).
293. E. Kesler, *J. Heterocycl. Chem.*, **17**, 1113 (1980).
294. a. P. H. Harland and P. Hodge, *Synthesis*, 419 (1983).
b. D. N. B. Mallen and J. M. Smith, *Biomed. Mass Spectrom.*, **9**, 125 (1982).
295. Grelan Pharmaceutical Co., *Jpn. Kokai Tokyo Koho*, JP 82, 128,641 (1982); *Chem. Abstr.*, **97**, 197984f (1982).
296. S. A. R. Knox, R. F. D. Stansfield, F. G. A. Stone, M. J. Winter, and P. Woodward, *J. Chem. Soc. Dalton Trans.*, 67 (1982).
297. a. K. Tolkiehn and K. Krohn, *Chem. Ber.*, **113**, 1575 (1980).
b. K. Krohn and E. Broser, *J. Org. Chem.*, **49**, 3766 (1984).
c. T. R. Kelly, L. Ananthasubramanian, K. Borah, J. W. Gillard, R. N. Goerner, P. F. King, J. M. Lyding, W.-G. Tsang, and J. Vaya, *Tetrahedron*, **40**, 4569 (1984).
d. T. R. Kelly, A. Whiting, and N. S. Chandrakumar, *J. Am. Chem. Soc.*, **108**, 3510 (1986).

e. K. Krohn, *Tetrahedron Lett.*, **21**, 3557 (1980).
f. T. R. Kelly and M. Montury, *Tetrahedron Lett.*, 4311 (1978).
g. G. A. Kraus and S. H. Woo, *J. Org. Chem.*, **52**, 4841 (1987).

298. K. T. Potts, D. Bhattacharjee, and E. B. Walsh, *J. Chem. Soc. Chem. Commun.*, 114 (1984).
299. J. P. Gouesnard, *Tetrahedron*, **34**, 2089 (1978).
300. W. B. T. Cruse, I. Fleming, P. T. Gallagher, and O. Kennari, *J. Chem. Res. (S)*, 372 (1979).
301. P. A. Harland and P. Hodge, *Synthesis*, 223 (1982).
302. A. V. Rama Rao and D. R. Reddy, *Synth. Commun.*, **16**, 97 (1986).
303. D. J. Bellville and N. L. Bauld, *J. Am. Chem. Soc.*, **104**, 2665 (1982).
304. a. A. Korgesaar, T. Tensing, and M. Nechaeva, *Org. React. (Tartu)*, **21**, 462 (1984).
b. A. Korgesaar, N. Johanson, and I. Annus, *Org. React. (Tartu)*, **21**, 456 (1984).
c. A. Korgesaar, V. Siilats, and J. Reiljan, *Org. React. (Tartu)*, **16**, 534 (1979).
305. J. E. Baldwin, P. D. Bailey, G. Gallacher, K. A. Singleton, and P. M. Wallace, *J. Chem. Soc. Chem. Commun.*, 1049 (1983).
306. R. A. Pabon, D. J. Bellville, and N. L. Bauld, *J. Am. Chem. Soc.*, **106**, 2730 (1984).
307. M. Kakushima, *Can. J. Chem.*, **57**, 2564 (1979).
308. K. Jankowski, *J. Org. Chem.*, **47**, 3649 (1982).
309. A. J. Birch and S. F. Dyke, *Aust. J. Chem.*, **31**, 1625 (1978).
310. a. M. Ihara, Y. Ishida, K. Fukumoto, and T. Kametani, *Chem. Pharm. Bull.*, **33**, 4102 (1985).
b. D. L. Boger and M. Patel, *J. Org. Chem.*, **49**, 4098 (1984).
c. D. L. Boger, M. Patel, and F. Takusagawa, *J. Org. Chem.*, **50**, 1911 (1985).
311. T. W. Hart, R. N. Haszeldine, and A. E. Tipping, *J. Chem. Soc. Perkin Trans. 1*, 1544 (1980).
312. A. G. Davies, R. S. Hay-Motherwell, J. C. Evans, and C. C. Rowland, *J. Chem. Soc. Chem. Commun.*, 1513 (1986).
313. T. Kurihara, M. Kodana, J. Sasaki, and S. Ito, *Koen Yoshishu-Tennen Yuki Kagobutsu Toronkai*, **22**, 228 (1979); *Chem. Abstr.*, **93**, 26569n (1980).
314. H. Upadek and K. Bruns, *Eur. Pat. Appl.*, 53,717 (1980); *Chem. Abstr.*, **97**, 198414g (1982).
315. M. Noguchi, S. Kakimoto, and S. Kajigaeshi, *Chem. Lett.*, 151 (1985).
316. K. B. Becker, *Synthesis*, 238 (1980).
317. H. M. Sirat, E. J. Thomas, and N. D. Tyrrell, *J. Chem. Soc. Chem. Commun.*, 36 (1979).
318. G. Kresze, M. M. Weib, and W. Dittel, *Liebigs Ann. Chem.*, 203 (1984).
319. G. Kresze, W. Dittel, and H. Melzer, *Liebigs Ann. Chem.*, 224 (1981).
320. I. Erden, *Synth. Commun.*, **14**, 989 (1984).
321. J. C. Berridge, A. Gilbert, and G. N. Taylor, *J. Chem. Soc. Perkin Trans. 1*, 2174 (1980).

322. A. Bertsch, W. Grimme, and G. Reinhardt, *Angew. Chem.* (*Int. Ed. Engl.*), **25**, 377 (1986).

323. M. E. Jung and C. A. McCombs, *J. Am. Chem. Soc.*, **100**, 5207 (1978).

324. K. B. White and W. Reusch, *Tetrahedron*, **34**, 2439 (1978).

325. M. Aga, K. Okoda, and M. Oda, *Tetrahedron Lett.*, **27**, 5653 (1986).

326. Toray Ind. Inc., *Jpn. Kokai Tokyo Koho*, JP 82, 80,391 (1982); *Chem. Abstr.*, **97**, 163242x (1982).

327. T. Hamada, H. Iijima, T. Yamamoto, N. Numao, K.-I. Hirao, and O. Yonemitsu, *J. Chem. Soc. Chem. Commun.*, 696 (1980).

328. A. Murai, S. Sato, and T. Masamune, *Bull. Chem. Soc. Jpn.*, **57**, 2276 (1984).

329. A. Murai, S. Sato, and T. Masamune, *Chem. Lett.*, 429 (1981).

330. K. Mach, F. Turecek, H. Atropiusova, and V. Hanus, *Organometallics*, **5**, 1215 (1986).

331. D. Kaufmann and A. De Meijere, *Chem. Ber.*, **116**, 1897 (1983).

332. K. Hayakawa, I. Fujii, and K. Kanematsu, *J. Org. Chem.*, **48**, 166 (1983).

333. P. R. Crabbendam, L. Maat, and H. C. Beyerman, *Recl. J. R. Neth. Chem. Soc.*, **100**, 293 (1981).

334. P. R. Crabbendan, T. S. Lie, J. T. M. Linders, and L. Maat, *Recl. Trav. Chim. Pays-Bas*, **103**, 296 (1984).

335. A. C. Ghosh, D. E. Portlock, H. C. Dalzen, C. Malmberg, P. Herlihy, R. K. Razdan, W. L. Duax, and G. D. Smith, *J. Org. Chem.*, **48**, 4137 (1983).

336. A. Singh, S. Archer, K. Hoogsten, and J. Hirshfield, *J. Org. Chem.*, **48**, 173 (1983).

337. G. A. Tolstikov, E. E. Shultz, and L. V. Spirikhin, *Tetrahedron*, **42**, 591 (1986).

338. L. Maat, J. A. Peters, and M. A. Prazeres, *Recl. Trav. Chim. Pays-Bas*, **104**, 205 (1985).

339. J. D. Friedrich, *J. Org. Chem.*, **52**, 2442 (1987).

340. E. E. Van Tamelen and S. R. Zawacky, *Tetrahedron Lett.*, **26**, 2833 (1985).

341. R. I. Yakhimovich, N. F. Fursaeva, and V. E. Pashinnik, *Khim. Prir. Soedin.*, 580 (1980); *Chem. Abstr.*, **94**, 15985q (1981).

342. N. A. Bogoslovskii, G. E. Litvinova, I. A. Titova, G. I. Samokhvalov, V. G. Mairanovskii, V. M. Gurevich, and T. M. Filippova, *Zh. Org. Khim.*, **17**, 1909 (1981).

343. M. Tada and A. Oikawa, *J. Chem. Soc. Perkin Trans. 1*, 1858 (1979).

344. D. Neuhaus and C. W. Rees, *J. Chem. Soc. Chem. Commun.*, 318 (1983).

345. J. M. Coxon, M. J. O'Connell, and P. J. Steel, *J. Org. Chem.*, **52**, 4726 (1987).

346. P. Magnus, T. Gallagher, P. Brown, and J. C. Huffman, *J. Am. Chem. Soc.*, **106**, 2105 (1984).

347. M. E. Jung, C. A. McCombs, Y. Takeda, and Y. G. Pan, *J. Am. Chem. Soc.*, **103**, 6677 (1981).

348. I. Landheer and D. Ginsburg, *Tetrahedron*, **37**, 133 (1981).

349. P. Ashkenazi, R. Gleiter, W. Von Philipsborn, P. Bigler, and D. Ginsburg, *Tetrahedron*, **37**, 127 (1981).

350. M. Peled and D. Ginsburg, *Tetrahedron*, **37**, 151 (1981).
351. J. Kalo and D. Ginsburg, *Tetrahedron*, **34**, 2155 (1978).
352. M. Kaftory, M. Peled, and D. Ginsburg, *Helv. Chim. Acta*, **62**, 1326 (1979).
353. J. Kalo, D. Ginsburg, and J. J. Bloomfield, *Tetrahedron*, **34**, 2153 (1978).
354. P. Ashkenazi, J. Kao, A. Ruttimann, and D. Ginsburg, *Tetrahedron*, **34**, 2161 (1978).
355. M. Peled and D. Ginsburg, *Tetrahedron*, **37**, 161 (1981).
356. P. Ashkenazi, J. Oliker, and D. Ginsburg, *Tetrahedron*, **34**, 2171 (1978).
357. P. Ashkenazi, M. Kaftory, W. Grimme, K. Heger, E. Vogel, and D. Ginsburg, *Bull. Soc. Chim. Belg.*, **88**, 841 (1979).
358. P. Ashkenazi, M. Kaftory, T. Sayrac, G. Maier, and D. Ginsburg, *Helv. Chim. Acta*, **66**, 2709 (1983).
359. P. Ashkenazi, M. Kaftory, and D. Ginsburg, *Helv. Chim. Acta*, **66**, 2712 (1983).
360. T. Tsuji and S. Nishida, *Tetrahedron Lett.*, **24**, 3361 (1983).
361. P. Ashkenazi, E. Vogel, and D. Ginsburg, *Helv. Chim. Acta*, **66**, 615 (1983).
362. F.-G. Klarner, B. M. J. Dogan, R. Weider, D. Ginsburg, and E. Vogel, *Angew. Chem. (Int. Ed. Engl.)*, **25**, 346 (1986).
363. P. Ashkenazi, E. Vogel, and D. Ginsburg, *Tetrahedron*, **34**, 2167 (1978).
364. P. Ashkenazi, M. Peled, E. Vogel, and D. Ginsburg, *Tetrahedron*, **35**, 1321 (1979).
365. K. L. Hoffmann, G. Maas, and M. Regitz, *J. Org. Chem.*, **52**, 3851 (1987).
366. A. G. Schultz, J. P. Dittami, F. P. Lavieri, C. Salowey, P. Sundararaman, and M. B. Szymula, *J. Org. Chem.*, **49**, 4429 (1984).
367. H. Auksi and P. Yates, *Can. J. Chem.*, **59**, 2510 (1981).
368. a. M. Bertrand, P. Teisseire, and G. Pelerin, *Tetrahedron Lett.*, **21**, 2055 (1980).
b. M. Bertrand, P. Teisseire, and G. Pelerin, *Tetrahedron Lett.*, **21**, 2051 (1980).
369. M. Matsumoto and K. Kuroda, *Angew. Chem. (Int. Ed. Engl.)*, **21**, 382 (1982).
370. Y. He Hsu, L. C. Kuo, and C. C. Liao, *J. Chin. Chem. Soc. (Taipei)*, **31**, 199 (1984).
371. C. C. Liao and H. S. Lin, *J. Chin. Chem. Soc. (Taipei)*, **30**, 69 (1983).
372. H. D. Becker, B. W. Skelton, and A. H. White, *Aust. J. Chem.*, **36**, 1361 (1983).
373. L.-C. Kuo and C.-C. Liao, *J. Chin. Chem. Soc. (Taipei)*, **31**, 263 (1984).
374. C. C. Liao and H. S. Lin, *Proc. Natl. Sci. Connc. Repub. China, Part B*, **7**, 413 (1983); *Chem. Abstr.*, **101**, 130615u (1984).
375. R. Al-Hamdany and S. Salih, *J. Iraqi Chem. Soc.*, 53 (1982).
376. V. Horak, F. Y. Foster, R. de Levie, J. W. Jones, and P. Svoronos, *Tetrahedron Lett.*, **22**, 3577 (1981).
377. C.-C. Liao, H.-S. Lin, and J.-T. Lin, *J. Chin. Chem. Soc. (Taipei)*, **27**, 87 (1980).
378. V. V. Plemenkov, M. M. Latypova, I. G. Bolesov, Ya. Ya. Villen, and N. V. Villen, *Dokl. Akad. Nauk SSSR*, **272**, 1395 (1983).
379. J. L. Pyle, A. A. Shaffer, and J. S. Cantrell, *J. Org. Chem.*, **46**, 115 (1981).
380. T. Sasaki, K. Hayakawa, T. Manabe, and S. Nishida, *J. Am. Chem. Soc.*, **103**, 565 (1981).
381. J. L. Pyle, R. L. Lunsford, and J. S. Cantrell, *J. Org. Chem.*, **44**, 2391 (1979).

382. O. De Lucchi and G. Modena, *Tetrahedron Lett.*, **24** 1653 (1983).

383. A. J. Pearson and H. S. Bausal, *Tetrahedron Lett.*, **27**, 283 (1986).

384. H. Iida, Y. Watanabe, and C. Kibayashi, *J. Org. Chem.*, **50**, 1818 (1985).

385. T.-K. Yin, J. G. Lee, and W. T. Borden, *J. Org. Chem.*, **50**, 531 (1985).

386. B. Atasoy and M. Balci, *Tetrahedron*, **42**, 1461 (1986).

387. W. E. Noland and V. Kameswaran, *J. Org. Chem.*, **46**, 1940 (1981).

388. Y. Sugihara, K. Hashimoto, M. Yoshikawa, and I. Murata, *Tetrahedron Lett.*, **23**, 1925 (1982).

389. A. E. M. Khalil, R. Stephens, and J. C. Tatlow, *J. Fluorine Chem.*, **22**, 43 (1983).

390. W. Adam, I. Erden, and O. Cox, *J. Org. Chem.*, **44**, 861 (1979).

391. H. Takeshita, S. Sugiyama, and T. Hatsui, *Bull. Chem. Soc. Jpn.*, **58**, 2490 (1985).

392. a. L. A. Paquette, S. J. Hathaway, and P. F. T. Schirch, *J. Org. Chem.*, **50**, 4199 (1985).
b. T. Uyehara and Y. Kitahara, *Bull. Chem. Soc. Jpn.*, **52**, 3355 (1979).

393. T. Uyehara, S. Migakoshi, and Y. Kitahara, *Bull. Chem. Soc. Jpn.*, **52**, 2023 (1979).

394. A. Hassner, D. Middlemiss, J. Muray-Rust, and P. Murray-Rust, *Tetrahedron*, **38**, 2539 (1982).

395. M. Sato, S. Ebine, K. Nishijima, and J. Tsunetsugu, *Bull. Chem. Soc. Jpn.*, **54**, 766 (1981).

396. H. Takeshita, S. Sugiyama, and T. Hatsui, *Chem. Lett.*, 1855 (1984).

397. Z. Goldschmidt, S. Antebi, H. E. Gottlieb, D. Cohen, U. Shmuely, and Z. Stein, *J. Organomet. Chem.*, **282**, 369 (1985).

398. A. Mori, H. Mametsuka, and H. Takeshita, *Bull. Chem. Soc. Jpn.*, **58**, 2072 (1985).

399. M. Papadopoulos and G. Jenner, *Tetrahedron Lett.*, **26**, 3335 (1985).

400. I. Heise and J. Leitich, *Chem. Ber.*, **118**, 332 (1985).

401. E. N. Manukov, V. A. Chuiko, and E. S. Mardilovich, *Vesti Akad. Navuk B. SSR, Ser. Khim. Navuk*, 115 (1985); *Chem. Abstr.*, **103**, 53726t (1985).

402. S. A. R. Knox, R. F. D. Stansfield, F. G. A. Stone, M. J. Winter, and P. Woodward, *J. Chem. Soc. Dalton Trans.*, 167 (1982).

403. T. G. Lenz and L. S. Hegedus, *Sun Proc. Int. Sol. Energy Soc. Silver Jubilee Congr.*, **1**, 154 (1979); *Chem. Abstr.*, **93**, 153127s (1980).

404. A. P. Humphries and S. A. R. Knox, *J. Chem. Soc. Dalton Trans.*, 1514 (1978).

405. L. A. Paquette, T. Z. Wang, and C. E. Cottrell, *J. Am. Chem. Soc.*, **109**, 3730 (1987).

406. R. J. Atkins and G. I. Fray, *Tetrahedron*, **35**, 1177 (1979).

407. J. M. Gardlik and L. A. Paquette, *Tetrahedron Lett.*, 3597 (1979).

408. P. G. Gassman, S. R. Korn, T. F. Bailey, T. H. Johnson, J. Finer, and J. Clardy, *Tetrahedron Lett.*, 3401 (1979).

409. R. Yamaguchi, M. Ban, and M. Kawanisi, *J. Chem. Soc. Chem. Commun.*, 826 (1984).

410. G. Jenner and M. Papadopoulos, *Tetrahedron Lett.*, **23**, 4333 (1982).

411. O. De Lucchi and G. Modena, *Phosphorus Sulfur*, **14**, 229 (1983).

412. P. Yates and S. P. Douglas, *Can. J. Chem.*, **60**, 2760 (1982).
413. J. Fink, W. Rosch, U.-J. Vogelbacher, and M. Regitz, *Angew. Chem.* (*Int. Ed. Engl.*), **25**, 280 (1986).
414. Toray Industries Inc. *Jpn. Kokai Tokyo Koho*, JP 58, 83,688 and JP 58 83,689; *Chem. Abstr.*, **99**, 158222c (1983).
415. T. Schooll, J. Lex, and E. Vogel, *Angew. Chem.* (*Int. Ed. Engl*), **21**, 920 (1982).
416. R. Kh. Geivandov and E. I. Kovshev, *Zh. Org. Khim.*, **17**, 543 (1981).
417. M. P. Shusherina, T. L. Nesterova, and O. V. Polyakova, *Zh. Org. Khim.*, **16**, 1285 (1980).
418. M. M. Latypova, V. V. Plemenkov, V. B. Tuzova, Kh. Z. Giniyatov, and I. G. Bolesov, *Zh. Org. Khim.*, **18**, 1650 (1982).
419. V. V. Plemenkov and L. V. Katerinich, *Zh. Org. Khim.*, **18**, 835 (1982).
420. D. W. Jones, A. Pomfret, and R. L. Wife, *J. Chem. Soc. Perkin Trans. 1*, 459 (1983).
421. T. Kurihara, K. Nasu, S. Haginaga, and K. Mihara, *Chem. Pharm. Bull.*, **32**, 4410 (1984).
422. W. Bethauser, M. Regitz, and W. Theis, *Tetrahedron Lett.*, **22**, 2535 (1981).
423. G. Maier and K. A. Schneider, *Angew. Chem.* (*Int. Ed. Engl.*), **19**, 1022 (1980).
424. A. I. Konavalov, *Russ. Chem. Rev.*, **52**, 1064 (1983).
425. J. Sauer and H. Wiest, *Angew. Chem.*, **74**, 353 (1962).
426. A. M. Mustafaer, N. R. Adigezalov, V. D. Kiselev, A. I. Konovalov, and M. M. Guseinov, *Zh. Org. Khim.*, **16**, 2549 (1980).
427. A. S. Onischenko, "Diene Synthesis," Israel Program of Scientific Translation, Jerusalem (1964).
428. J. G. Martin and R. K. Hill, *Chem. Rev.*, **61**, 537 (1961).
429. Y. Kobuke, T. Fueno, and J. Furukawa, *J. Am. Chem. Soc.*, **92**, 6548 (1970).
430. E. C. Angell, F. Fringuelli, L. Minuti, F. Pizzo, A. Taticchi, and E. Wenkert, *J. Org. Chem.*, **51**, 5177 (1986).
431. J. Sauer and R. Sustmann, *Angew. Chem.* (*Int. Ed. Engl.*), **19**, 779 (1980).
432. V. Mark, *J. Org. Chem.*, **39**, 3179, 3181 (1974).
433. W. H. Watson, Ed., "Stereochemistry and Reactivity of Systems Containing π Electrons," *Methods of Stereochemical Analysis*, Verlag Chemie, Weinheim, West Germany (1983).
434. K. L. Williamson, Y. F. Li Hsu, R. Lacko, and C. He Youn, *J. Am. Chem. Soc.*, **91**, 6129 (1969).
435. K. L. Williamson and Y. F. Li Hsu, *J. Am. Chem. Soc.*, **92**, 7385 (1970).
436. N. T. Anh, *Tetrahedron*, **29**, 3227 (1973).
437. G. Gleiter and L. A. Paquette, *Acc. Chem. Res.*, **16**, 328 (1983).
438. L. A. Paquette, H. Künzer, K. E. Green, O. De Lucchi, G. Licini, L. Pasquato, and G. Valle, *J. Am. Chem. Soc.*, **108**, 3453 (1986).
439. J. P. Hagenburch, P. Vogel, A. A. Pinkerton, and D. Schwarzenbach, *Helv. Chim. Acta*, **64**, 1818 (1981).
440. J. M. Mellor and C. F. Webb, *J. Chem. Soc. Perkin Trans. 2*, 26 (1974).
441. J. C. Jagt and A. M. Van Leusen, *J. Org. Chem.* **39**, 564 (1974).

442. T. Fukuyama and Y. M. Yung, *Tetrahedron Lett.*, **22**, 3759 (1981).
443. A. K. Saksena, *Tetrahedron Lett.*, **21**, 133 (1980).
444. W. Oppolzer, *Angew. Chem. (Int. Ed. Engl.)*, **23**, 876 (1984).
445. K. J. Stone and R. D. Little, *J. Org. Chem.*, **49**, 1849 (1984).
446. D. Craig, J. J. Shipman, J. Keihl, F. Widmer, R. Fowler, and A. Hawthorne, *J. Am. Chem. Soc.*, **76**, 4573 (1954).
447. V. D. Kiselev, I. M. Shakirov, and A. I. Konovalov, *Zh. Org. Khim.*, **22**, 311 (1986).
448. G. M. Rubbottom and D. S. Krueger, *Tetrahedron Lett.*, 611 (1977).
449. M. E. Jung and Y. G. Pan, *Tetrahedron Lett.*, **21**, 3127 (1980).
450. J. E. Baldwin, P. D. Bailey, G. Gallacher, M. Otsuka, K. A. Singleton, P. M. Wallace, K. Prout, and W. Wolf, *Tetrahedron*, **40**, 3695 (1984).
451. Yu. A. Arbuzov and T. A. Pisha, *Dokl. Acad. Nauk SSSR*, **116**, 71 (1957).
452. S. L. Mattes and S. Farid, *Org. Photochem.*, **6**, 233 (1983).
453. G. Jenner and M. Papadopoulos, *J. Org. Chem.*, **51**, 585 (1986).

6.2. AROMATIC HYDROCARBONS

Reactivity, Selectivity, Synthesis Aromatic systems such as benzene and naphthalene are relatively unreactive when normal dienophiles are used, but with the highly reactive dicyanoacetylene they give [4 + 2] cycloadducts.[68] The Diels–Alder "adduct" benzene–naphthalene **1** is a known compound that decomposes into its components at 0°C but cannot be obtained by cycloaddition reaction of the two aromatic hydrocarbons.[69]

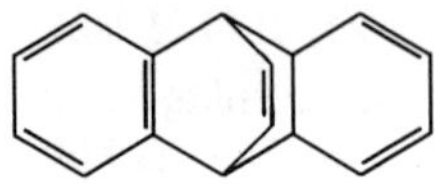

1

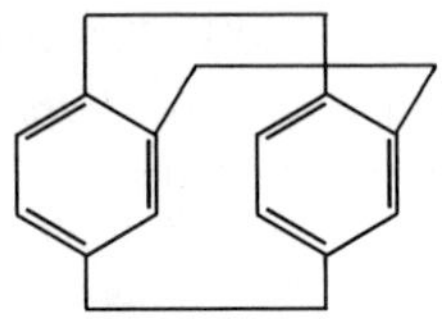

2

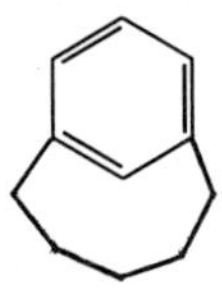

3

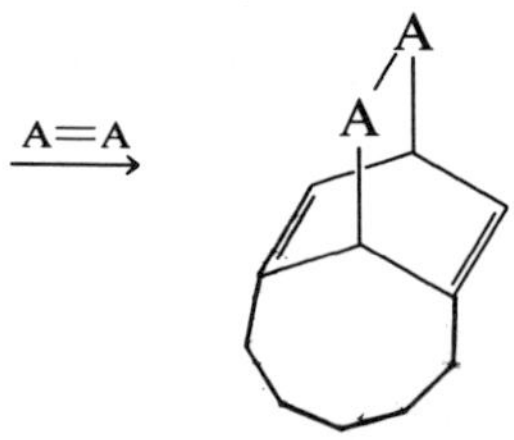

4

Highly strained bridged benzenes such as cyclophanes **2** and **3** show a markedly reduced aromaticity and react easily with various dienophiles, such as NPTAD, TCNE, MA, and DMAD. The metacyclophane **3** is more reactive than **2**.[2,3]

The relative importance of electronic and steric effects of the Diels–Alder reactions of aromatic systems can be evaluated from the cycloadditions of 1,4-dimethylnaphthalene and $[n](1,4)$naphthalenophanes **5** ($n = 8, 9, 10, 14$) with dicyanoacetylene. The 1,4-dimethylnaphthalene gives, as the prevalent adduct, that derived from the addition on the more electron-rich substituted ring, while all the bridged naphthalenes **5** afford the adduct **6** as the unique or most prevalent compound.[8]

5

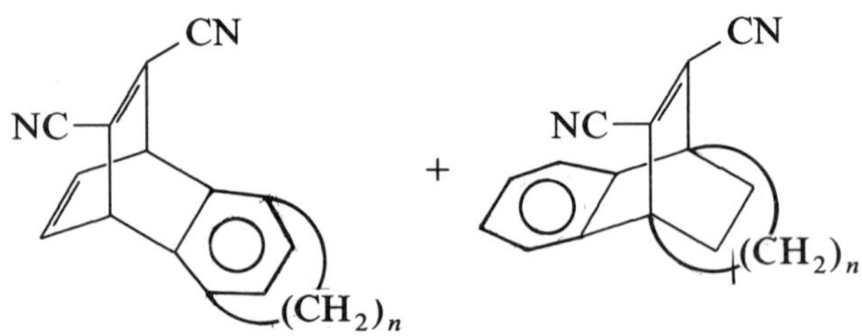

6 **7**

Anthracene and its derivatives give Diels–Alder reactions at diene sites C-9 and C-10, in agreement with the lowest benzenoid character of the central ring, and the cycloadditions occur more easily than those with benzene and naphthalene. The cycloaddition of anthracene and MA is complete at 20°C in the absence of a catalyst, after 200 days, and after about 90 seconds in the presence of $AlCl_3$ or $GaCl_3$.[43,70]

The Lewis catalyst can change the transition state from symmetric to unsymmetric. This has been hypothesized observing that the Diels–Alder reactions of anthracene with dienophiles containing one or two chiral moieties [i.e., methyl(-)bornylfumarate and di(-)bornylfumarate] exhibit cooperativity in asymmetric induction only in the absence of Lewis acids.[12] The reaction rate of substituted anthracenes is increased by alkyl substituents and reduced by the halogen atoms at C-9. The rate of addition of TCNE to 9-substituted anthracenes at 25°C follows the order Me > Et > Bu^t > Pr^i > Me_3Si > H > Br.[46] The 9,10-dimethylanthracene is about 3600 and 400 times more reactive than the unsubstituted compound and the cyclopentadiene, respectively.[46] Bulky substituents (Ph_3Si, Ph_3Sn) at C-9 prevent the reaction.[46]

In Scheme 6.5 the regioselectivity of the cycloadditions of some 9-substituted anthracenes with acrylic acid is reported. The *ortho* regioisomer is generally preferred.[55] Mixtures of regioisomers are also obtained in the condensation of 1,8-disubstituted anthracenes (Scheme 6.6) with benzynes derived from 6-substituted anthranilic acids. The ratio of tripticene regioisomers depends on the nature of the substituents in the anthracene and benzene units.[58]

A systematic study[41] of the Diels–Alder reactivity of a large number of polycyclic aromatic hydrocarbons shows that the rate constants of the cycloaddition with MA increase on linear annulation and decrease with angular annulation. Some examples are reported in Table 6.34. The reactions are generally regioselective; the dots in the formulas of Table 6.34 indicate the most reactive diene sites. Linear-type hydrocarbons give monoadducts almost

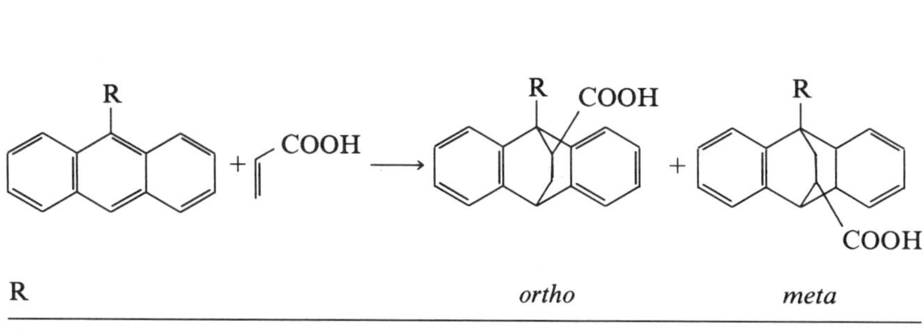

R	*ortho*	*meta*
CN	65%	35%
COOH	44%	56%
CHO	100%	0%
NO_2	19%	81%
OMe	100%	0%

Scheme 6.5

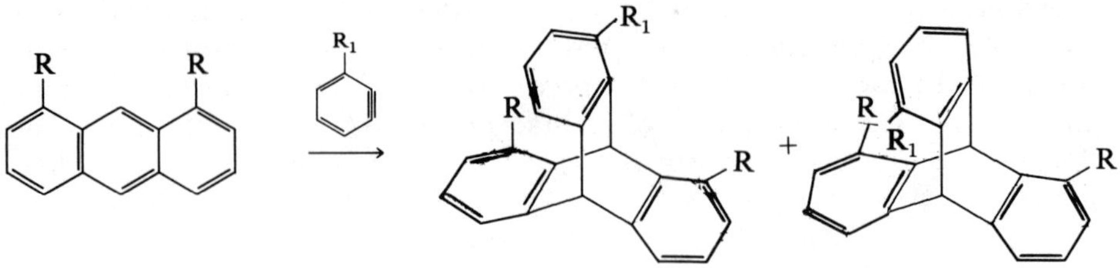

R	R_1		
Cl	Me	75%	25%
CO_2Me	Me	69%	31%
CO_2Me	CO_2Me	24%	76%
CN	CO_2Me	1%	99%

Scheme 6.6

TABLE 6.34 Relative Reaction Constants of Acene-Type Hydrocarbons with MA in 1,2,4-Trichloro-benzene at 91.5°C

Acene	k_{rel}	Acene	k_{rel}
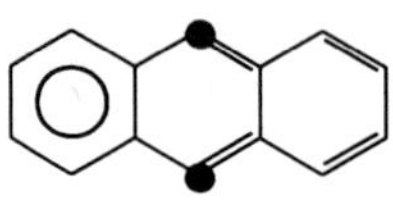	1	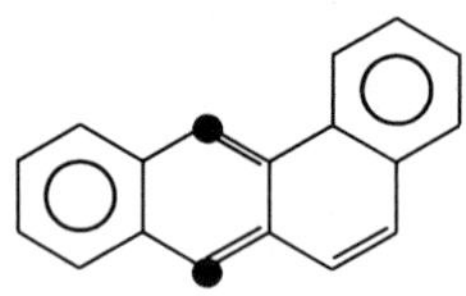	6×10^{-3}
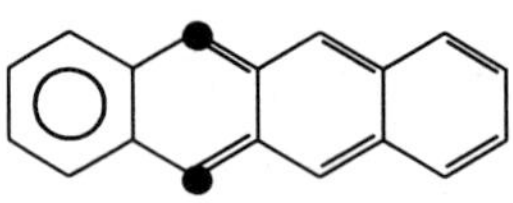	20.3	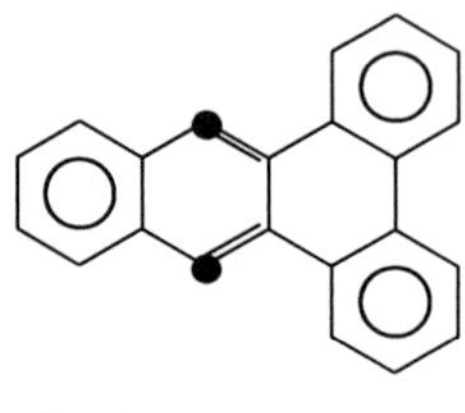	3×10^{-3}
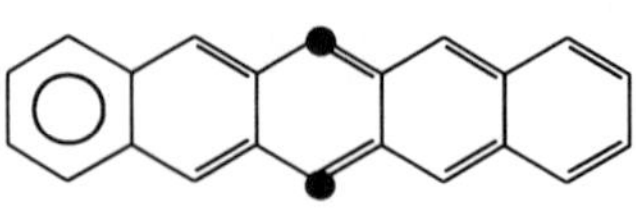	722.5	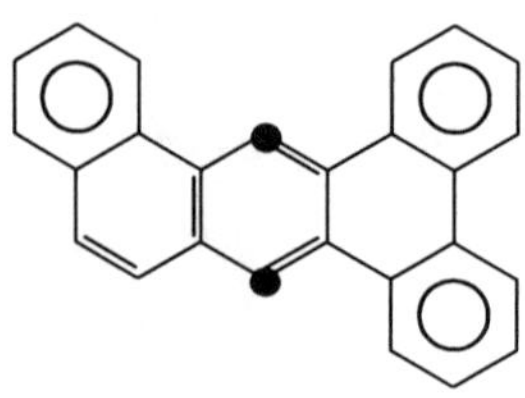	1.5×10^{-3}
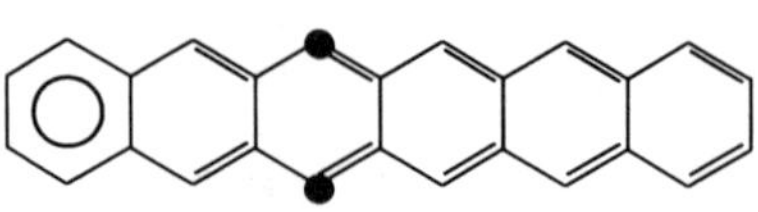	1447	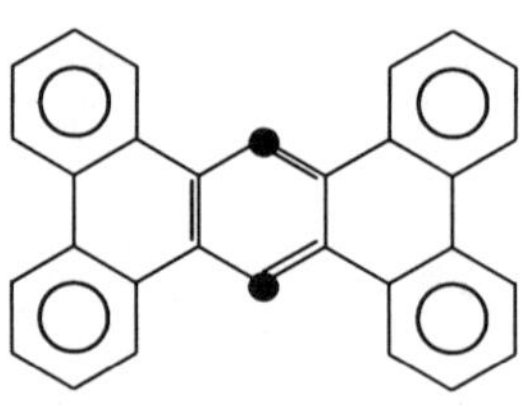	0.8×10^{-3}

exclusively, while angular annulated-type hydrocarbons can add two or three molecules of MA. The results were interpreted by using several theoretical models. The simplest is Clar's sextet theory,[71] which qualitatively describes the experimental data without theoretical calculations. According to Clar, the increasing reactivity going from anthracene to hexacene is due to a gradual loss of benzenoid character of the aromatic hydrocarbon while the decreasing reactivity with the angular annulation is the consequence of the generation of a new π-sextet when an angular benzo rest is added. The sextet concept also accounts for the regioselectivity; the adduct with the largest possible number of sextets is preferentially formed. The adduct **8** of the reaction of 1,2-benzotetracene with the general dienophile A=A is favored over **9** because it has three sextets. Similarly, one can predict that the most reactive positions of linearly fused thioaromatic compounds **10**–**12** are those indicated by dots in the structural formulas.[67]

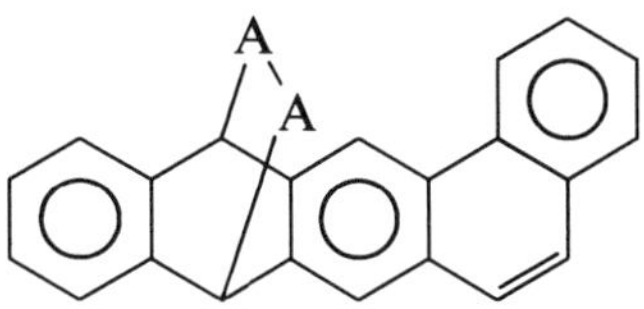

8

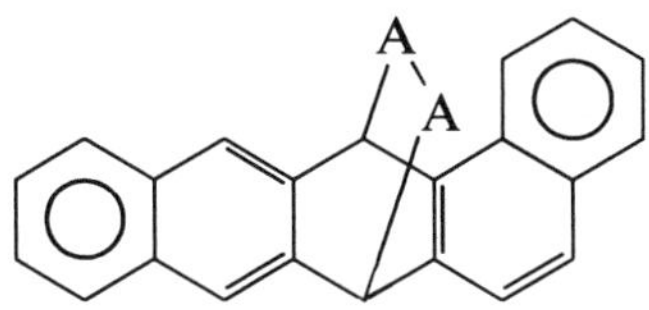

9

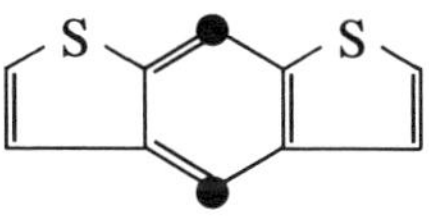

10

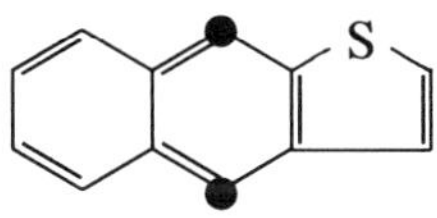

11

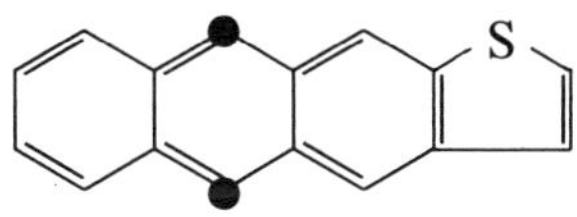

12

Tables of Diels–Alder Reactions See Tables 6.35 to 6.41; see also the general remarks about classifications, abbreviations, and symbols in the Introduction.

TABLE 6.35 Diels–Alder Reactions of Benzenes and Cyclophanes

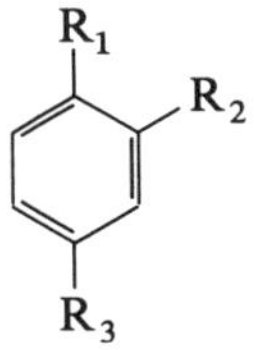

Diene		Dienophile	Yield (%)	Ref.
$R_1 = R_2 = R_3 = H$	E6	[Tetrachlorobenzyne]	75–86	1
$R_1 = R_3 = Me, R_2 = OMe$	E6	[Tetrachlorobenzyne]	95	1
$R_1 = OMe, R_2 = R_3 = H$	E6	[Tetrachlorobenzyne]	100	1
$R_1 = OMe, R_2 = Me\ R_3 = H$	E6	[Tetrachlorobenzyne]	95	1
$R_1 = OMe, R_2 = R_3 = Me$	E6	[Tetrachlorobenzyne]	90	1

TABLE 6.35 (*Continued*)

Diene		Dienophile	Yield (%)	Ref.
	A7	DMM	77	2
	D3	MA	100	2
	E3	DMAD	100	2
	E5	Hexafluoro-2-butyne	100	2
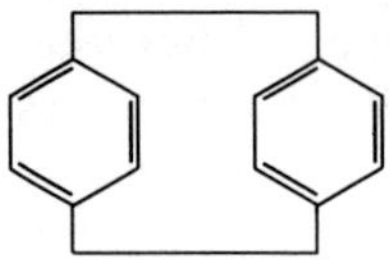	E5	Dicyanoacetylene	71	3
	I5	NPTAD	99	3
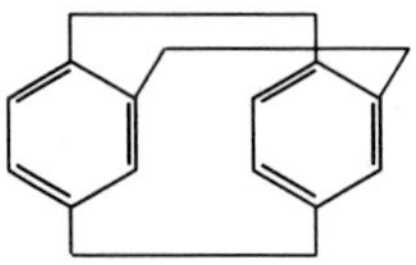	A12	TCNE	59	3
	D3	MA	53	3
	E3	DMAD	61	3
	I5	NPTAD	79	3
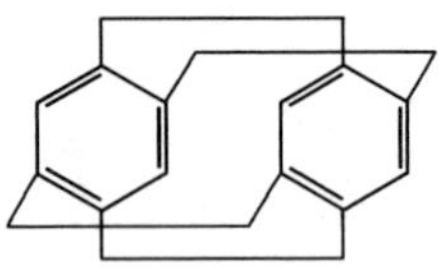	I5	NPTAD	100	3

TABLE 6.36 Diels–Alder Reactions of Naphthalenes

Diene		Dienophile	Yield (%)	Ref.
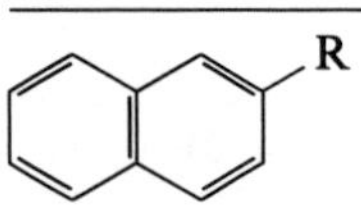				
R = H	D3	MA		4a
	I5	NMTAD		4b
R = Me	D3	MA	15	5
R = OH	A10	(*Z*)-1,2-Bis(phenylsulfonyl)-ethene	80	6
	D3	MA		7
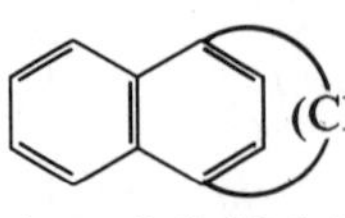 $(CH_2)_n$ ($n = 8, 9, 10, 14$)	E5	Dicyanoacetylene	19–61	8

TABLE 6.37 Diels–Alder Reactions of Anthracene

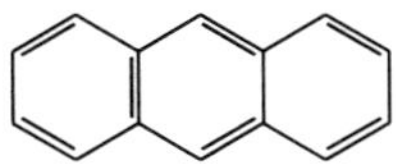

Dienophile		Yield (%)	Ref.
A1	MVK	85	13
A1	Pantolactone acrylate (*)		27
A3	Crotonic acid		7
A3	(*E*)-6,6-Dimethoxy-3-hexen-2-one	77	22
A7	DMF		12
A7	Methyl(-)bornyl fumarate (*)	75	12
A7	Di(-)bornyl fumarate (*)	95	12
A7	Di(-)menthyl fumarate (*)		12
A7	Maleic acid	86	15
A7	Fumaric acid	94	7, 15
A7	Methyl(-)menthyl fumarate (*)	92	12, 23
A8	Phenylvinylsulfone	96	32, 35
A8	Vinyltriphenylphosphonium bromide	25	44
A10	1,2-Bis(phenylsulfonyl)ethene (*E* and *Z*)	46–89	6
A10	1-(Phenylsulfonyl)-2-(trimethylsilyl)-ethene (*E* and *Z*)	35–98	35
A12	TCNE (k)		10, 19, 26, 42, 46
A12	Tetranitroethene	99	18
C1	3-Methoxy-*trans*-cycloheptene		24
C1	[Bicyclo[3.1.0]hex-(1,5)-ene]		28
C2	7-Methylenenorbornene	72	29
C2	7-Methylenenorbornadiene	41	29
C2	[1,7-Dehydroquadricyclane]	26	47
C2	[Tricyclo[4.1.0.0^{2,7}]hept-(1(7)-ene]	31	48
C4	Cyclobutendione	35	14
C7	BQ		16
C8	1,2,3,6-Tetrahydrobenzocyclobutene-3,6-dione	83	31
C9	Naphthoquinone sulfoxide	65	33
D3	MA (k)		11, (37), 40, 41
D3	Carbomethoxymaleic anhydride	55	34
D3	Maleic anhydrides		43
D4	1-(Ethoxycarbonyl)-1,2,3,6-tetrahydropyridine		17
D5	*N*-Arylmaleimides		16, 19, 40
D5	NPM		19
D5	1,5-Bis(3,5-dioxo-Δ′-1,2,4-triazolin-4-ylmethyl)-2,4-dimethylbenzene	68	39
D9	4-Isothiazolin-3-one-1,1-dioxide	41	21
D10	Benzo-1,4-dithiadiene-disulfone	94	20
E3	But-2-ynedial	43	30

TABLE 6.37 (*Continued*)

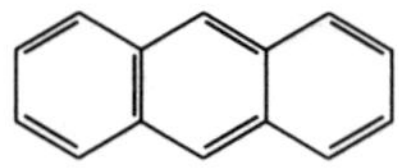

Dienophile		Yield (%)	Ref.
E6	[3,4,5,6-Dicyclopentabenzyne]	6	9
E6	[4,4′-Methylenebis(benzyne)]	46	36
E6	[Five-membered hetaryne]	20	45
E6	4,5-Dehydrotropone		51
I4	[Phthalazine-1,4-diones]	4–89	25
I6	Digermene	67	38

TABLE 6.38 Diels–Alder Reactions of Monosubstituted Anthracenes

Diene	Dienophile		Yield (%)	Ref.
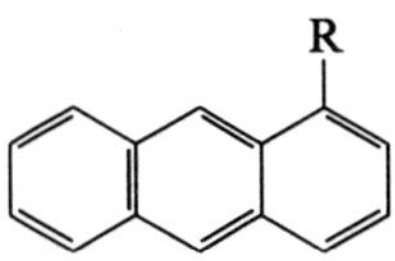				
R				
NAc_2	C7	BQ		49
F	A12	TCNE (k)		50
Cl	A12	TCNE (k)		50
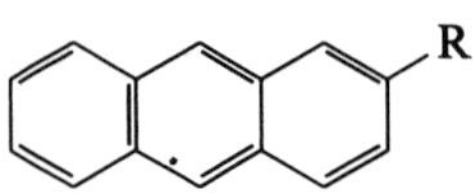				
R				
Me	A12	TCNE (k)		42, 50
	E6	4,5-Dehydrotropone		51
	I4	[Phthalazine-1,4-dione]	40	25a
F	A12	TCNE (k)		50
Cl	A12	TCNE (k)		50
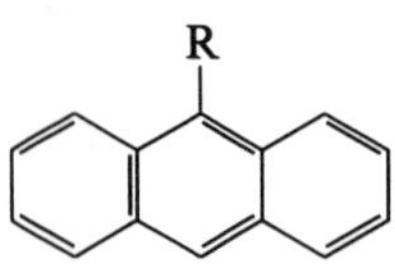				
R				
D	C2	[Tricyclo[4.1.0.0^{2,7}]hept-1(7)-ene]	11	48
Me	A8	Vinyltriphenylphosphonium bromide	30	44

TABLE 6.38 (*Continued*)

Diene		Dienophile	Yield (%)	Ref.
	A12	TCNE (k)		42, 46, 50
	C2	[Tricyclo[4.1.0.0^{2,7}]hept-1(7)-ene]	44	48
	D3	Maleic anhydrides (k)		43
	I4	[Phthalazine-1,4-diones]	35	25
Et	A12	TCNE (k)		46
Pr	A12	TCNE (k)		46
Pr^i	A12	TCNE (k)		46
$CH_2CH{=}CH_2$	E6	[Arynes]	15–80	52
$C{\equiv}CPh$	E5	9-Phenylethynylanthracene		54
CN	A1	Acrylic acid	85	55
	C2	[Tricyclo[4.1.0.0^{2,7}]hept-1(7)-ene]	1	48
CHO	A1	Acrylic acid	95	55
	A8	ACN	85	55
CO_2H	A1	Acrylic acid	81	55
	A8	ACN	91	55
α-Py	A1	Acrylamide	47	53
NO_2	A1	Acrylic acid	94	55
OMe	C2	[Tricyclo[4.1.0.0^{2,7}]hept-1(7)-ene]	34	48
	C2	[1,7-Dehydroquadricyclane]	38	47
	D3	MA		40
F	A12	TCNE (k)		50
Cl	A12	TCNE (k)		50
	C2	[Tricyclo[4.1.0.0^{2,7}]hept-1(7)-ene]	14	48
	D3	Maleic anhydrides (k)		43
Br	A12	TCNE (k)		42, 46, 50
	C2	[Tricyclo[4.1.0.0.^{2,7}]hept-1(7)-ene]	5	48
I	A12	TCNE (k)		50
$SiMe_3$	A12	TCNE (k)		46
SBu^t	E6	[Arynes]	38–49	56
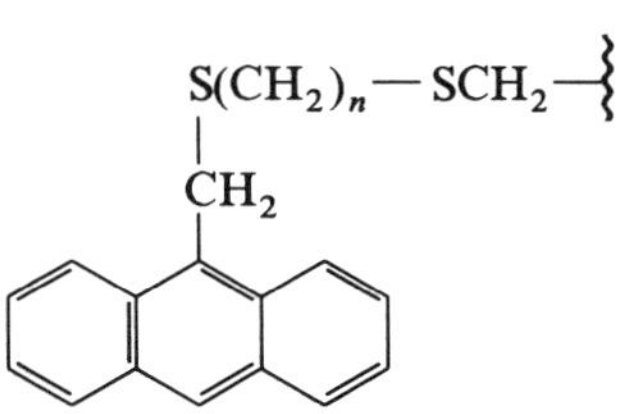 (n = 8, 12)	D5	1,5-Bis(3,5-dioxo-Δ′-1,2,4-triazolin-4-ylmethyl)-2,4-dimethylbenzene	38–43	39

TABLE 6.39 Diels–Alder Reactions of Disubstituted Anthracenes

Diene		Dienophile	Yield (%)	Re[f.]
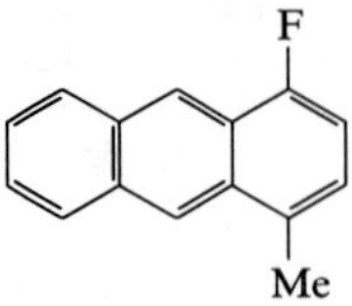	C7	2,3,5,6-Tetrafluorocyclohexa-2,5-diene-1,4-dione		5
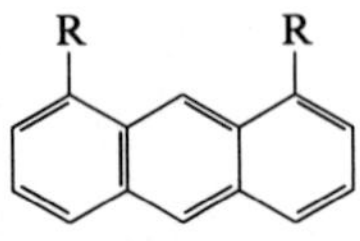 (R = CN, CO_2H, Cl)	E6	[Arynes]	20–74	5
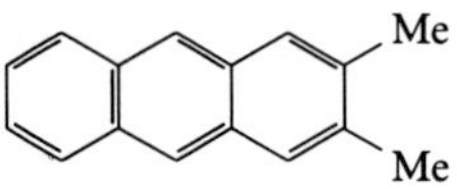	E6	4,5-Dehydrotropone		5
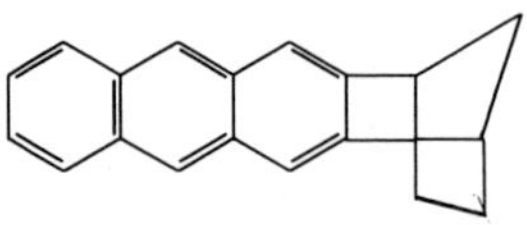	D3	MA	96	5
	E3	DMAD	90	5
	I5	NMTAD	95	5
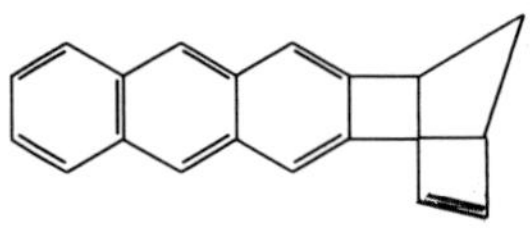	E3	DMAD	79	[illegible]
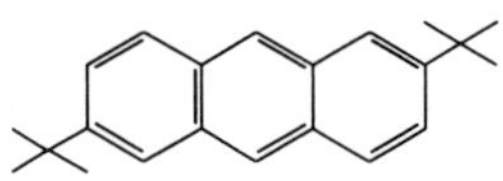	A12	TCNE (k)		[illegible]
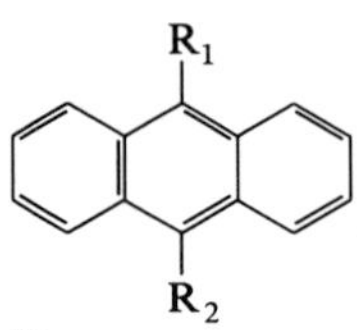				
R_1 **R_2**				
Me Me	A1	MAC		[illegible]
	A8	Vinyl phosphonate		[illegible]
	A8	Vinyltriphenylphos-phoniumbromide	40	[illegible]
	A12	TCNE (k)		[illegible]
	D3	MA		40, [illegible]
	D3	Maleic anhydrides		43, [illegible]
	D5	Maleimide		[illegible]
	E6	[Five-membered hetaryne]	52	[illegible]
	I4	[Phthalazine-1,4-dione]	68	2[illegible]
	J1	[Nitroso methyl ketone]	90	[illegible]
Me Cl	I4	[Phthalazine-1,4-dione]	30	[illegible]
Me Br	A12	TCNE (k)		[illegible]
OMe OMe	C2	[Tricyclo[4.1.0.0^{2,7}]hept-1(7)-ene]	5	[illegible]
Br Br	C2	Tricyclo[4.1.0.0^{2,5}]deca-3,7,9-trienes		[illegible]

TABLE 6.40 Diels–Alder Reactions of Tri- and Tetrasubstituted Anthracenes

Diene			Dienophile	Yield (%)	Ref.
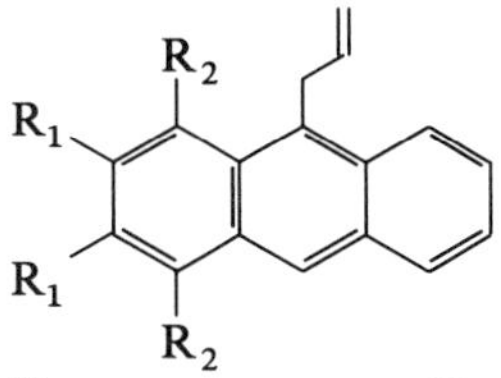					
R_1	**R_2**				
H	Me	E6	[Benzyne]	65	52
Cl	H	E6	[Benzyne]	65	52
Cl	Cl	E6	[Chlorobenzyne]	10	52
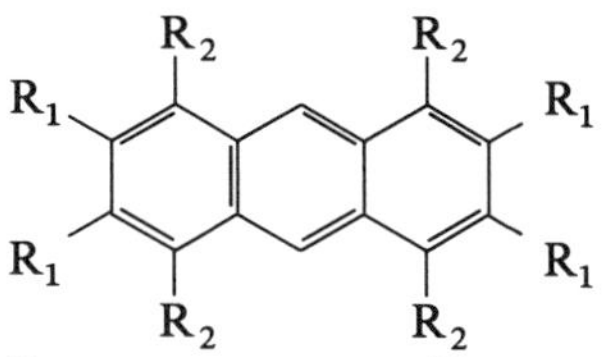					
R_1	**R_2**				
H	OAc	C7	BQ		64
Me	H	E6	4,5-Dehydrotropone		51

TABLE 6.41 Diels–Alder Reactions of Miscellaneous Polynuclear Aromatics

Diene		Dienophile	Yield (%)	Ref.
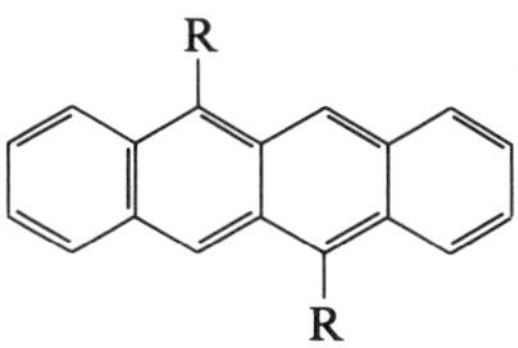 (R = H, Cl)	C7	BQ (k)		19a
	D3	MA (k)		19a, 41
	D5	NPM (k)		19a
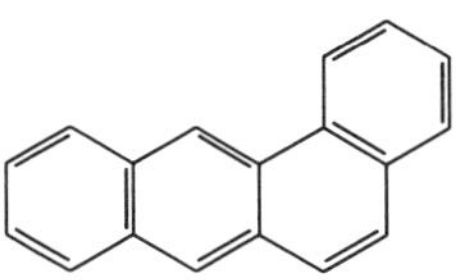	A12	TCNE (k)		42
	D3	MA (k)		41
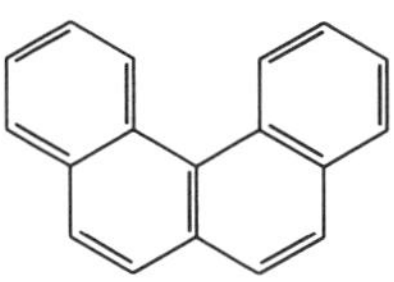	D3	MA (k)		41

TABLE 6.41 (*Continued*)

Diene		Dienophile	Yield (%)	Ref.
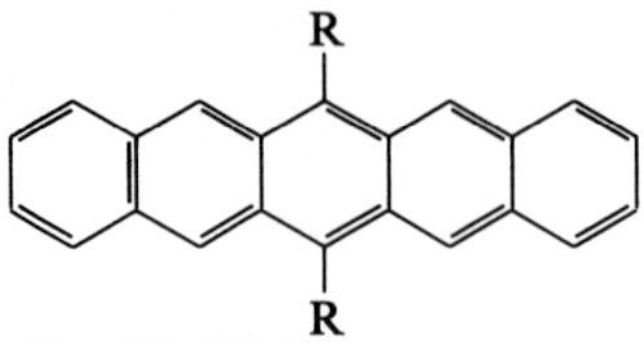 (R = H, Cl)	C7	BQ (k)		19a
	D3	MA (k)		19a, 41
	D5	NPM (k)		19a
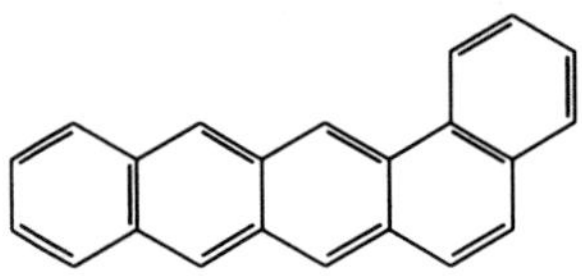	D3	MA (k)		41
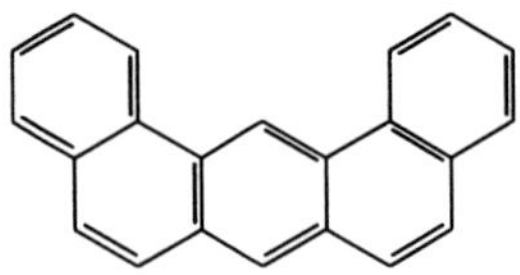	D3	MA (k)		41
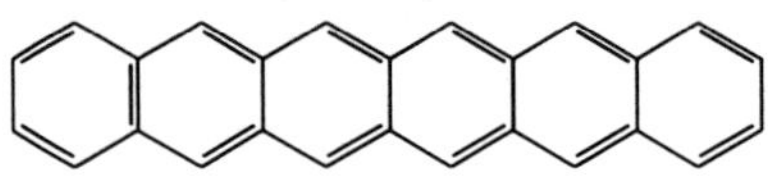	D3	MA (k)		41
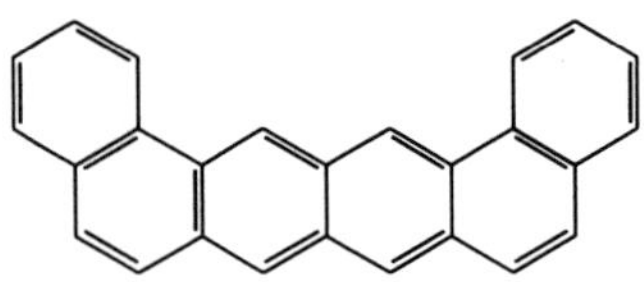	D3	MA (k)		41
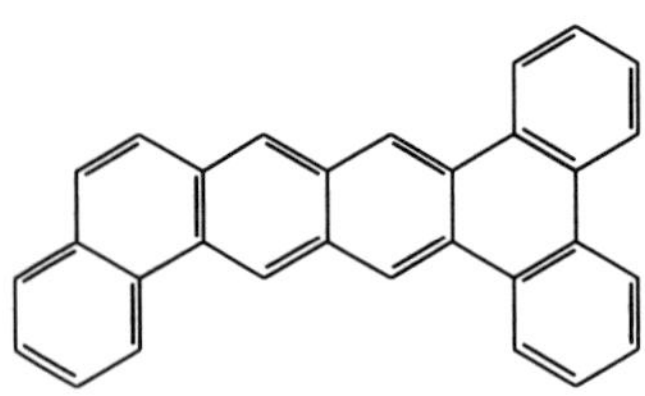	D3	MA (k)		41
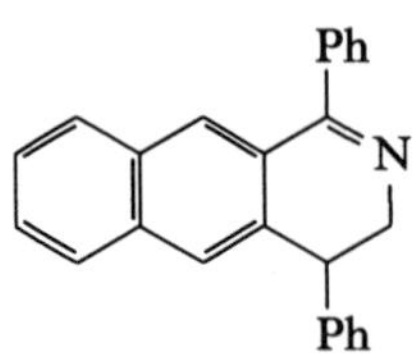	A8	ACN	75	65
	D3	MA	40	65

TABLE 6.41 (*Continued*)

Diene		Dienophile	Yield (%)	Ref.
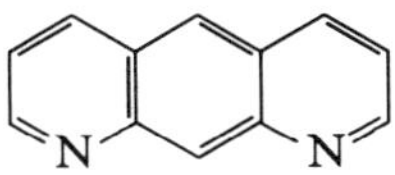	A10	(*E*)-1,2-Dichloroethene		66
	D3	MA	49	66
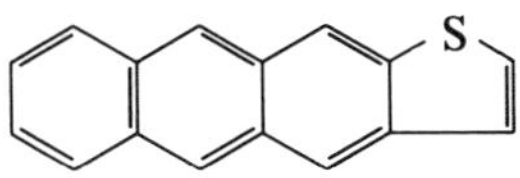	D3	MA	75	67
	E3	DMAD	47	67
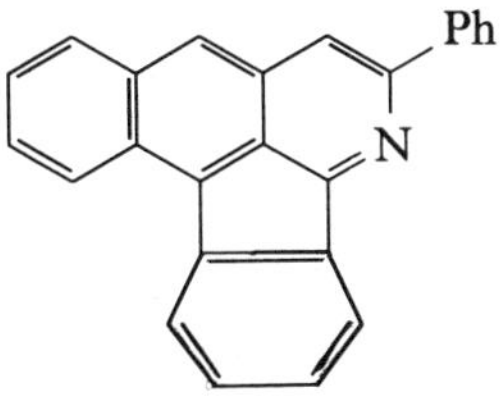	A8	ACN		65
	D3	MA	44	65

References for Section 6.2

1. N. J. Hales, H. Heaney, J. H. Hollinshead, and P. Singh, *Org. Synth.*, **59**, 71 (1979).
2. L. A. M. Turkenburg, P. M. L. Blok, W. H. De Wolf, and F. Bickelhaupt, *Angew. Chem.* (*Int. Ed. Engl.*), **21**, 298 (1982).
3. A. F. Murad, J. Kleinschroth, and H. Hopf, *Angew. Chem.* (*Int. Ed. Engl.*), **19**, 389 (1980).
4. a. V. S. Abramov, A. A. Zharov, and V. M. Zhulin, *Dokl. Akad. Nauk SSSR*, **258**, 1127 (1981).
   b. D. P. Kjell and R. S. Sheridan, *J. Photochem.*, **28**, 205 (1985).
5. S. M. Verma and N. Singh, *Indian J. Chem., Sect. B*, **19B**, 368 (1980).
6. O. De Lucchi, V. Lucchini, L. Pasquato, and G. Modena, *J. Org. Chem.*, **49**, 596 (1984).
7. A. K. Singh, M. Verma, and S. M. Verma, *Indian J. Chem., Sect. B*, **23B**, 631 (1984).
8. K. B. Wiberg and M. J. O'Donnel, *J. Am. Chem. Soc.*, **101**, 6660 (1979).
9. R. P. Thummel and T. Fuchigami, *J. Org. Chem.*, **46**, 5217 (1981).
10. a. V. D. Kiselev, G. V. Mavrin, and A. I. Konovalov, *Zh. Org. Khim.*, **16**, 1435 (1980).
    b. Ya. D. Samuilov, A. I. Movchan, S. E. Soloveva, and A. I. Konovalov, *Zh. Org. Khim.*, **20**, 2494 (1984).
11. K. R. Bansal, *Indian J. Chem. Educ.*, **6**, 36 (1979).
12. L. M. Tolbert and M. B. Ali, *J. Am. Chem. Soc.*, **106**, 3806 (1984).
13. W. E. Hahn, A. Sokolowska, W. Szalecki, and M. Siekierska, *Pol. J. Chem.*, **54**, 349 (1980).

14. a. H. D. Martin, A. Oftring, R. Iden, E. Schwichtenberg, and H.-J. Schiwek, *Tetrahedron Lett.*, **23**, 841 (1982).
    b. R. Albert, C. Heller, R. Iden, G. Martin, H.-D. Martin, B. Mayer, and A. Oftring, *Isr. J. Chem.*, **25**, 74 (1985).
15. J. Horyna, *Czech. CS*, 198,678 (1982); *Chem. Abstr.*, **99**, 38250p (1983).
16. a. V. D. Kiselev, D. G. Khuzyasheva, and A. I. Konovalov, *Zh. Org. Khim.*, **19**, 1268 (1983).
    b. V. D. Kiselev, I. M. Shakirov, and A. I. Konovalov, *Zh. Org. Khim.*, **20**, 1454 (1984).
17. P. Rajagopalan, *J. Med. Chem.*, **27**, 946 (1984).
18. K. Baum and D. Tzeng, *J. Org. Chem.*, **50**, 2736 (1985).
19. a. Ya. D. Samuilov, V. G. Urgadov, L. F. Urgadova, and A. I. Konovalov, *Zh. Org. Khim.*, **21**, 1249 (1985).
    b. V. D. Kiselev, D. G. Khuzyasheva, O. A. Zabotina, E. A. Veisman, and A. I. Konovalov, *Zh. Org. Khim.*, **19**, 1617 (1983).
20. E. Wenkert and C. A. Broka, *Finn. Chem. Lett.*, 126 (1984).
21. E. D. Weiler and J. J. Brennan, *J. Heterocycl. Chem.*, **15**, 1299 (1978).
22. P. Yates and S. P. Douglas, *Can. J. Chem.*, **60**, 2760 (1982).
23. G. Helmchen and R. Schmierer, *Angew. Chem. (Int. Ed. Engl.)*, **20**, 205 (1981).
24. H. Jendralla, *Angew. Chem. (Int. Ed. Engl.)*, **19**, 1032 (1980).
25. a. K. J. Gould, N. P. Hacker, J. F. W. McOmie, and D. H. Perry, *J. Chem. Soc. Perkin Trans. 1*, 1834 (1980).
    b. O. Abou-Teim, R. B. Jansen, J. F. W. McOmie, and D. H. Perry, *J. Chem. Soc. Perkin Trans. 1*, 1841 (1980).
26. M. Lofti and R. M. G. Roberts, *Tetrahedron*, **35**, 2137 (1979).
27. G. Helmchen, K. Ihrig, and H. Schindler, *Tetrahedron Lett.*, **28**, 183 (1987).
28. K. B. Wiberg and G. Bonneville, *Tetrahedron Lett.*, **23**, 5385 (1982).
29. D. N. Butler and I. Gupta, *J. Chem. Soc. Chem. Commun.*, 596 (1980).
30. A. Gorgues and A. Le Coq, *J. Chem. Soc. Chem. Commun.*, 767 (1979).
31. M. Oda and Y. Kanao, *Chem. Lett.*, 1547 (1981).
32. R. V. C. Carr and L. A. Paquette, *J. Am. Chem. Soc.*, **102**, 853 (1980).
33. G. A. Kraus and S. H. Woo, *J. Org. Chem.*, **51**, 114 (1986).
34. H. K. Hall, Jr., P. Nogues, J. W. Rhoades, R. C. Sentman, and M. Detar, *J. Org. Chem.*, **47**, 1451 (1982).
35. R. V. C. Carr, R. V. Williams, and L. A. Paquette, *J. Org. Chem.*, **48**, 4976 (1983).
36. J. Nakajama, A. Sakai, and M. Hoshino, *J. Org. Chem.*, **49**, 5084 (1984).
37. C. Chen, *J. Chin. Chem. Soc. (Taipei)*, **30**, 21 (1983).
38. Y. Nakadaira, H. Tobita, and H. Sakurai, *New Front. Organomet. Inorg. Chem., Proc. China–Jpn.–USA, Trilateral Seminar*, **2** (1982); *Chem. Abstr.*, **102**, 185188h (1985).
39. K. Wald and H. Wamhoff, *Chem. Ber.*, **111**, 3519 (1978).
40. V. D. Kiselev, G. V. Mavrin, and A. I. Konovalov, *Zh. Org. Khim.*, **18**, 2505 (1982); E.E., 2210 (1982).

41. a. D. Biermann and W. Schmidt, *J. Am. Chem. Soc.*, **102**, 3163 (1980).
b. D. Biermann and W. Schmidt, *J. Am. Chem. Soc.*, **102**, 3173 (1980).
42. S. Fukuzumi and J. K. Kochi, *Tetrahedron*, **38**, 1035 (1982).
43. V. D. Kiselev, I. M. Shakirov, and A. I. Konovalov, *Zh. Org. Khim.*, **21**, 1215 (1985); E.E., 1105 (1985).
44. C. C. Haustock and J. C. Tebby, *Phosphorus Sulfur*, **15**, 239 (1983).
45. M. G. Reinecke and H. H. Ballard, *Tetrahedron Lett.*, 4981 (1979).
46. M. Lofti and R. M. G. Roberts, *Tetrahedron*, **35**, 2131 (1979).
47. O. Baumgartel, J. Harnisch, G. Szeimies, M. Van Meerssche, G. Germain, and J.-P. Declercq, *Chem. Ber.*, **116**, 2205 (1983).
48. U. Szeimies-Seebach, A. Schoffer, R. Romer, and G. Szeimies, *Chem. Ber.*, **114**, 1767 (1981).
49. S. Mahanti, H. Maurya, and S. M. Verma, *J. Indian Chem. Soc.*, **61**, 1034 (1984).
50. R. M. G. Roberts and F. Yavari, *Tetrahedron*, **37**, 2657 (1981).
51. T. Nakazawa, N. Abe, K. Kubo, and I. Murata, *Tetrahedron Lett.*, 4995 (1979).
52. S. Hatakeyama, T. Mitsuhashi, and M. Oki, *Bull. Chem. Soc. Jpn.*, **53**, 731 (1980).
53. A. W. Czarnik, *Tetrahedron Lett.*, **25**, 4875 (1984).
54. H. D. Becker and K. Andersson, *J. Photochem.*, **26**, 75 (1984).
55. P. V. Alston, R. M. Ottenbrite, and J. Newby, *J. Org. Chem.*, **44**, 4939 (1979).
56. N. Nakamura, *Chem. Lett.*, 1795 (1983).
57. D. A. Burgess, M. F. Finn, D. Jordan, I. D. Rae, and B. D. Walters, *Aust. J. Chem.*, **37**, 1769 (1984).
58. M. E. Rogers and B. A. Averill, *J. Org. Chem.*, **51**, 3308 (1986).
59. P. C. Hayes and L. A. Paquette, *J. Org. Chem.*, **48**, 1257 (1983).
60. J. Fleischhauer, A. N. Asaad, W. Schleker, and H.-D. Scharf, *Liebigs Ann. Chem.*, 306 (1981).
61. V. D. Kiselev, O. G. Khuzyasheva, I. M. Shakirov, and A. I. Konovalov, *Zh. Org. Khim.*, **19**, 2064 (1983).
62. G. E. Keck, *Tetrahedron Lett.*, 4767 (1978).
63. I. G. Dinulescu, I. I. Pogany, A. Dimitriu, E. Georgescu, and M. Avram, *Rev. Roum. Chim.*, **25**, 969 (1980); *Chem. Abstr.*, **94**, 139300g (1981).
64. H. Iwamura, T. Oosumi, and F. Toda, *Koen Yoshishu-Hibenzenkei Hokozoku Kagaku Toronkai* [oyobi] *Kozo Yuki Kagaku Toronkai*, **12**, 241 (1979); *Chem. Abstr.*, **92**, 198151r (1980).
65. L. V. Gorshkova, V. P. Zvolinskii, N. S. Prostakov, A. V. Varlamov, M. Torres, and V. F. Zakharov, *Khim. Geterotsikl. Soedin.*, 1645 (1980).
66. H. Quast and N. Schön, *Liebigs Ann. Chem.*, 877 (1984).
67. W. A. Lindley and D. W. M. MacDowell, *J. Org. Chem.*, **48**, 4419 (1983).
68. E. Ciganek, *Tetrahedron Lett.*, 3321 (1967).
69. W. Grimme and H. G. Köser, *Angew. Chem.* (*Int. Ed. Engl.*), **19**, 307 (1980).
70. P. Yates and P. Eaton, *J. Am. Chem. Soc.*, **82**, 4436 (1960).
71. E. Clar, "The Aromatic Sextet", Wiley, New York (1972).

6.3 HETEROCYCLIC DIENES

Since Diels and Alder[160] found that furan gives 1 : 1 adducts easily with MA at room temperature, the [4 + 2] cycloadditions of heterocyclic compounds possessing a conjugated diene system have been investigated extensively.[223] The course and facility with which the reaction proceeds depends on the nature and number of heteroatoms, as well as the aromatic character of the compound and the reactivity of the dienophile. Substitution–addition reactions are sometimes pathways accompanying the cycloaddition process. The adducts cannot always be isolated because they rearrange easily, isomerize, and give retro Diels–Alder reactions.

Five-Membered Rings

Heterocyclodienes with One Heteroatom Furan and its 2-methyl- and 2,5-dimethyl derivatives (**1**: $R_1 = R_2 = H$; $R_1 = H$; $R_2 = Me$; $R_1 = R_2 = Me$) react easily with MA. The kinetically controlled product is the *endo* **3**, which isomerizes to the *exo* **2** via a retro Diels–Alder reaction under the influence of the solvent and temperature.[275,276] The carbomethoxymaleic anhydride reacts more rapidly than MA; with furan, the *exo* adduct is the prevalent isomer at room temperature.[10] It is interesting to observe that both MA and its carbomethoxy derivative give *endo* cycloadditions with cyclopentadiene.[10] Dimethylmaleic anhydride, on the contrary, does not react with furan even at pressures up to 40 kbar. A synthetic equivalent is the 2,5-dihydrothiophene-3,4-dicarboxylic anhydride, which under 15 kbar pressure, affords a mixture of **5** (85%) and **6** (15%). The major adduct is then converted into the natural product cantharidin **7**.[277]

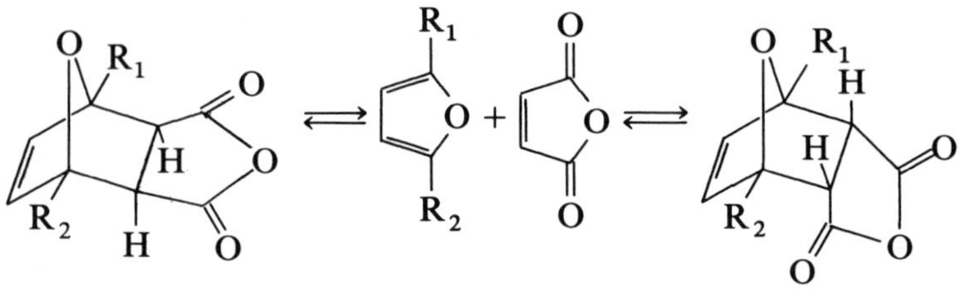

2 **1** **3**

The 7-oxabicyclo[2.2.1]heptane system **8** ($R = CO_2H, CO_2Et, CO_2Me, CN$) is a key compound in the synthesis of many natural products (prostaglandines, nucleosides, antibiotics). Its synthesis by Diels–Alder reaction between furan and acrylates requires long reaction times and occurs with low yields.[36] Exceptions are cycloadditions with acryloyl chloride and nitroethene.[36] Attempts to accelerate the reactions by Lewis acids such as $FeCl_3$, $SnCl_4$, $ZnCl_2$, and $AlCl_3$ or by increasing the reaction temperature have given polymers, retro Diels–Alder reactions, and substitution products. The reaction can be forced and the yields increased by applying high pressure[15d,16b] or by using catalysts such as K-10 bentonite clay dropped with Fe^{III},[40] cupric fluoborate–hydroquinone,[36] and tetrakis(acetonitrile) copper(I) fluoborate.[36]

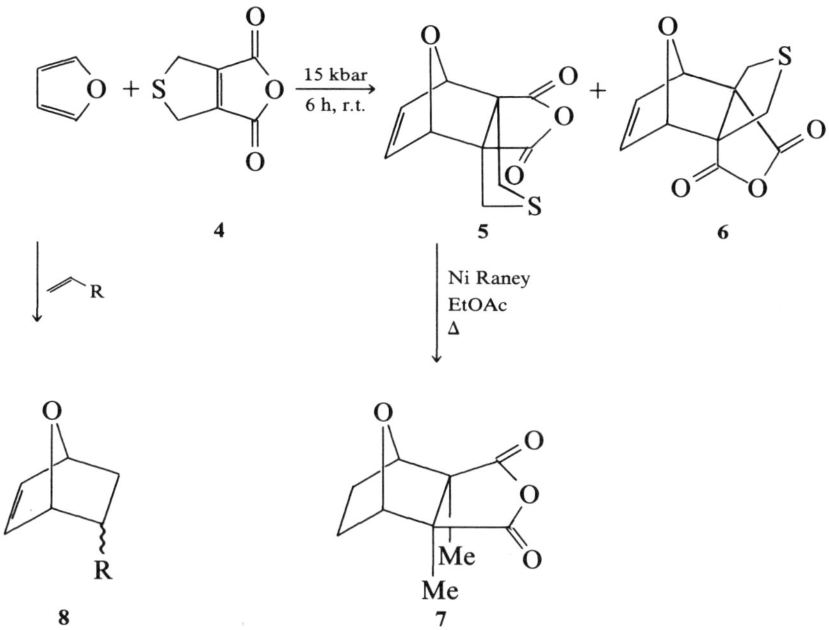

Furan and its derivatives react easily with singlet oxygen to produce endoperoxides (ozonides), which can be converted into functionalized organic compounds (i.e., **9** and **10**).[278] A kinetic study[11a] has pointed out that symmetrical and unsymmetrical 2,5-disubstituted furans exhibit different selectivity toward singlet oxygen, and it has been hypothesized that the geometry of the exciplex and its transition state for collapse to the endoperoxide are different for the two classes of compounds.

The facial selectivity of the Diels–Alder reactions of furans generally resembles that of cyclopentadienes, while differences are observed in the *endo-exo* diastereoselectivity. Thus the cycloaddition reaction of (norborn-2-eno)[*c*]furan **11** (X = O) with MA is completely *exo-anti* (with respect to the ethano bridge) diastereoselective, and the isolated adduct **12** (X = O) is the product of thermodynamic control.[77] The cyclopentadiene analogous to **11** (X = CH_2) gives a 2 : 1 mixture of **12** and **13**.

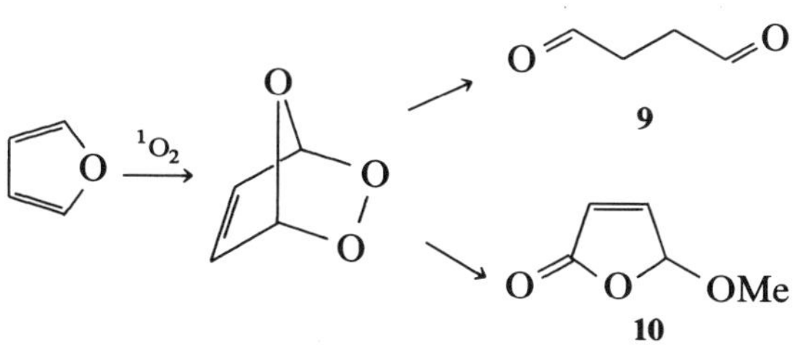

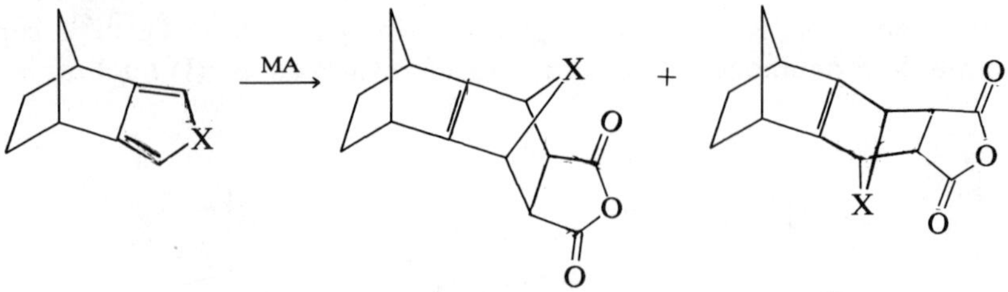

11 12 13

Unlike furan, thiophene is a very poor reactive diene and it reacts with MA under very high pressure to give the exo adduct **14**.[84] Thiophene derivatives that are able to react with MA are thiophene-1,1-dioxide and 2,5-dimethoxythiophene. The Diels–Alder reactions of thiophenes are sometimes followed by loss of the heteroatom. Thiophene traps benzyne, affording naphthalene[85] after sulfur extrusion; the highly reactive tetrachlorothiophene-*S,S*-dioxide undergoes cycloadditions, followed by loss of sulfur dioxide, with a variety of olefinic compounds and with azulene.[86–88]

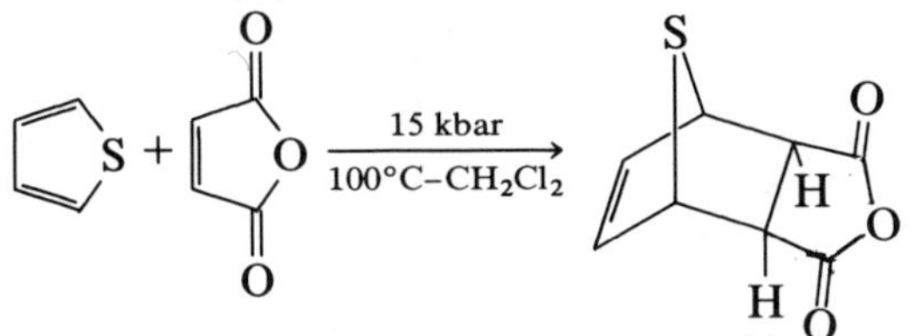

14

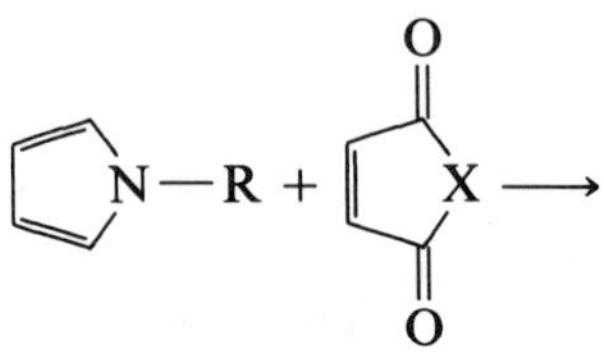

15

R = Bn, COPh, COMe
X = O, NMe, NPh

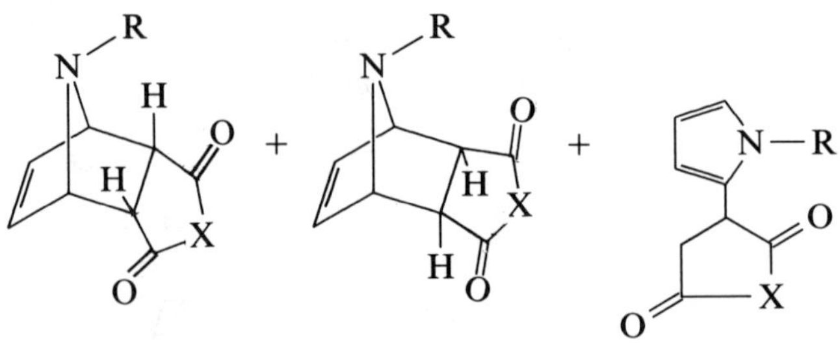

16 17 18

Pyrroles behave similarly to thiophenes. *N*-alkyl and *N*-acylpyrroles react with highly reactive dienophiles such as arynes or oxyallyl cations. The cycloaddition reactions can be accelerated by $AlCl_3$ or by high pressure. In addition to Diels–Alder adducts, 2-substituted pyrroles can also be obtained (i.e., **15** → **16**, **17**, **18**).[4,5] Thus the cycloaddition of *N*-acylpyrroles with MA and NPM under high pressure gives high yields and *endo* and *exo* adducts **16** and **17** uncontaminated by side products, while *N*-methyl and *N*-benzyl derivatives react with NPM and MA, affording only α-substituted pyrroles **18**. Cycloadducts **16** and **17** are precursors of compounds of pharmacological interest.

The pentachloropyrrole (**19**) is in equilibrium with the isomers **20** and **21**, and spectroscopic data indicate that the diene exists almost exclusively in the 1-azadiene form **20**. However, Diels–Alder reactions with styrene and vinyl acetate do not afford the expected adduct **22** (R = Ph, OAc) but rather give the 2-azaisomer **23** (R = Ph, OAc).[103,104] It has been argued that the 2-azadiene form **21** is present in a very small amount and must be the most kinetically active. The predominant 1-azadiene **20** has not been trapped, even with high reactive dienophiles such as TCNE, NPTAD, and benzyne.[104]

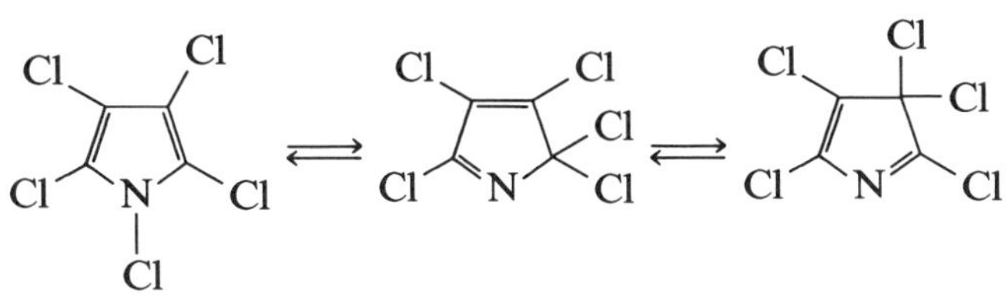

19 **20** **21**

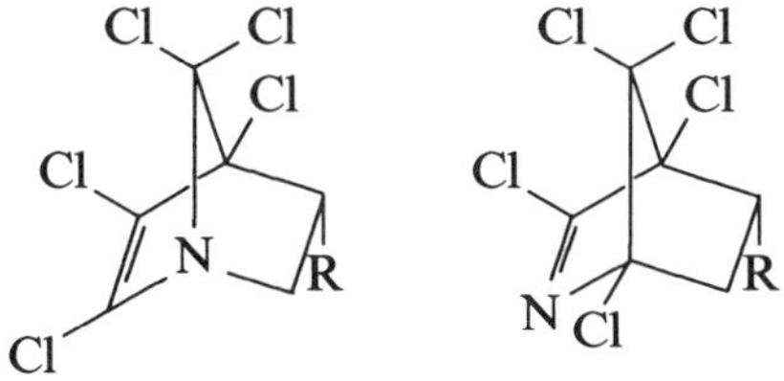

22 **23**

Diels–Alder reactions of phosphole oxides afford adducts that lose the phosphorus bridge under the reaction conditions (i.e., **24** → **26**). Under high pressure and at room temperature the formation of 7-phosphanorbornadienes **25** is observed.[96]

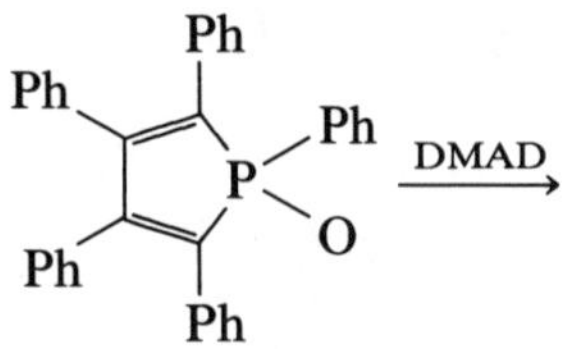

24

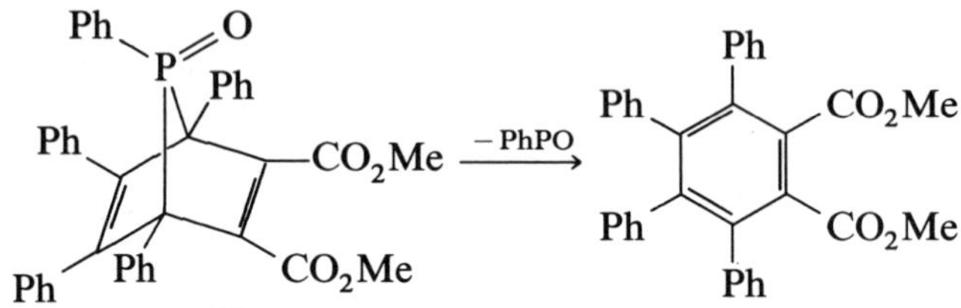

25 26

The reactivities of 1,1-dimethylsilole (**27**: X = Si) and 1,1-dimethylgermole (**27**: X = Ge) resemble that of furan, and *endo* adducts **28** are isolated by reaction with MA at room temperature.[92,101]

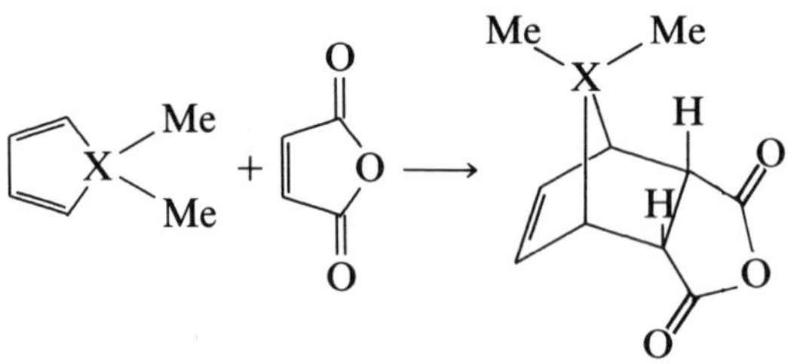

27 28

Heterocyclodienes with More Than One Heteroatom Among the dienic five membered heterocycles with two heteroatoms, the oxazoles are the most frequently used in the Diels–Alder reactions because they allow the preparation of compounds of synthetic interest. The course of the cycloadditions depends on the structure of oxazoles, the dienophile, and the reaction conditions. Ethylenic dienophiles generally afford unstable Diels–Alder adducts that rearrange to pyridines (Scheme 6.7).[9,279]

A methoxy group at C-5 in the oxazole ring increases the reactivity of the compounds, and the presence of a *p*-nitrophenyl group increases the stability of cycloadducts. The oxazole **29**, with NPTAD and diethylazodicarboxylate, gives the adducts **30** and **31**, respectively.[218a]

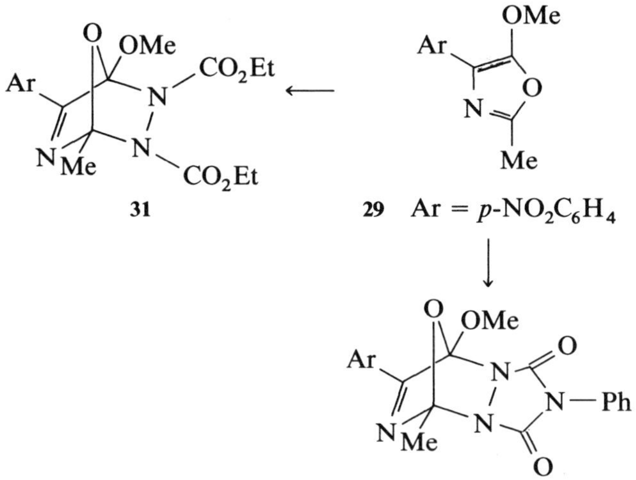

31 **29** Ar = $p\text{-}NO_2C_6H_4$ **30**

The reaction of oxazole with acetylenic dienophiles affords furans, which are originated by a retro Diels–Alder reaction of the initial adduct with loss of nitrile. A second Diels–Alder reaction can then occur on the furanic ring.

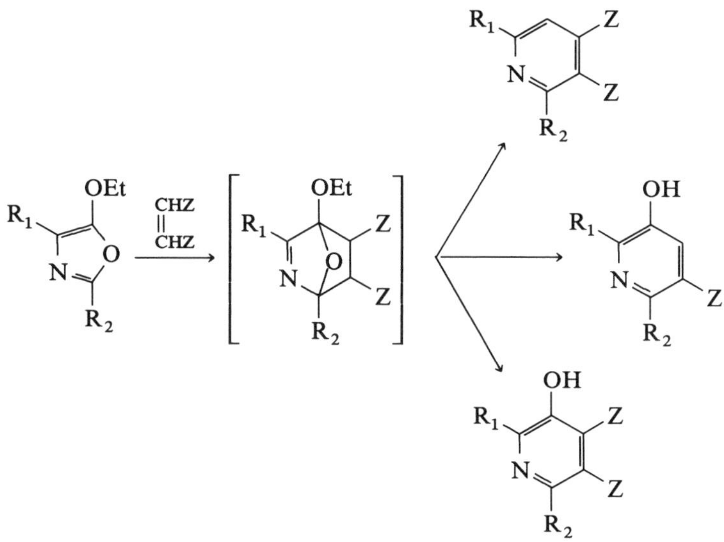

R_1 = H, Me; R_2 = H, $(CH_2)_3$—CO_2Et

Z,Z = Cl, COOH, —C(=O)—O—C(=O)—

Scheme 6.7

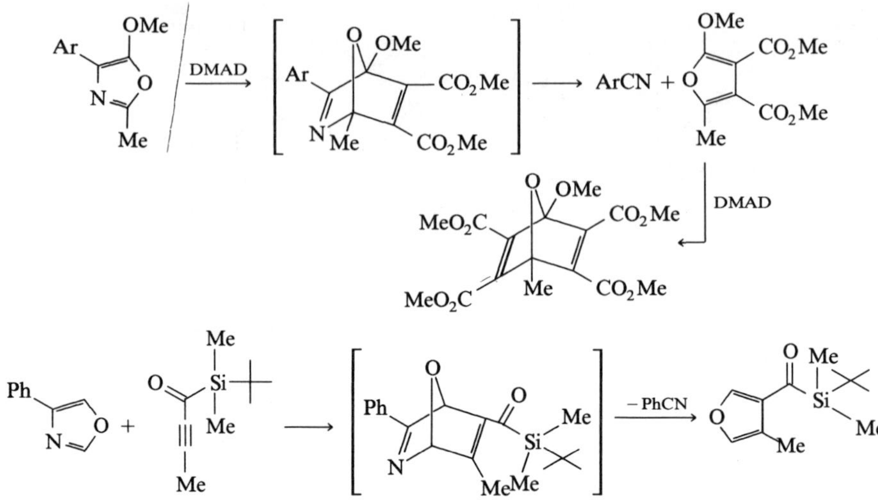

Scheme 6.8

Examples are reported in Scheme 6.8.[215b,218a] The primary adducts generally are not able to be isolated. A methoxy group at C-5 increases both the reactivity and the regioselectivity of the cycloaddition.

Diels–Alder reactions of other five-membered heterocycles having more than one heteroatom are rare. Scheme 6.9 reports examples of isopyrazoles[211] and imidazoles[280] which participate in inverse electron demand Diels–Alder reactions with electron-rich dienophiles and reactions of 1,2-thiaphosphole-2-sulfides generated in situ.[212]

Tables of Diels–Alder Reactions See Tables 6.42 to 6.50; see also the general remarks about classifications, abbreviations, and symbols in the Introduction.

Benzofused Five-Membered Rings

Reactivity, Selectivity, Synthesis Benzo[*c*]furans (isobenzofurans) are very reactive but generally unstable dienes which are prepared in situ and trapped. Exceptions are the 1-phenyl- and 1,3-diphenylisobenzofuran **32** and the 5,6-methylenedioxy derivatives **33**.[142,281]

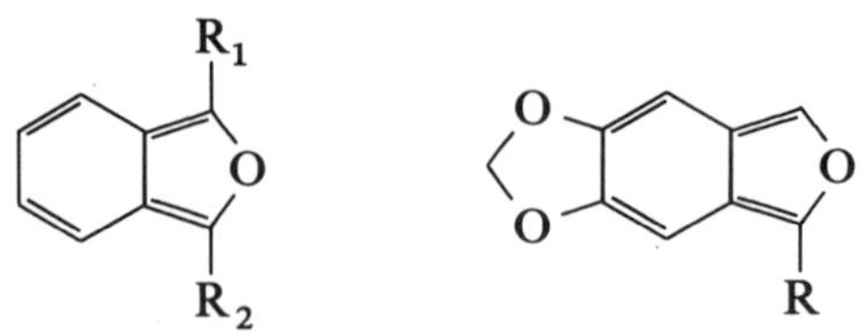

32 $R_1 = H, R_2 = Ph$; $R_1 = R_2 = Ph$ 33 $R = CN, CO_2Me$

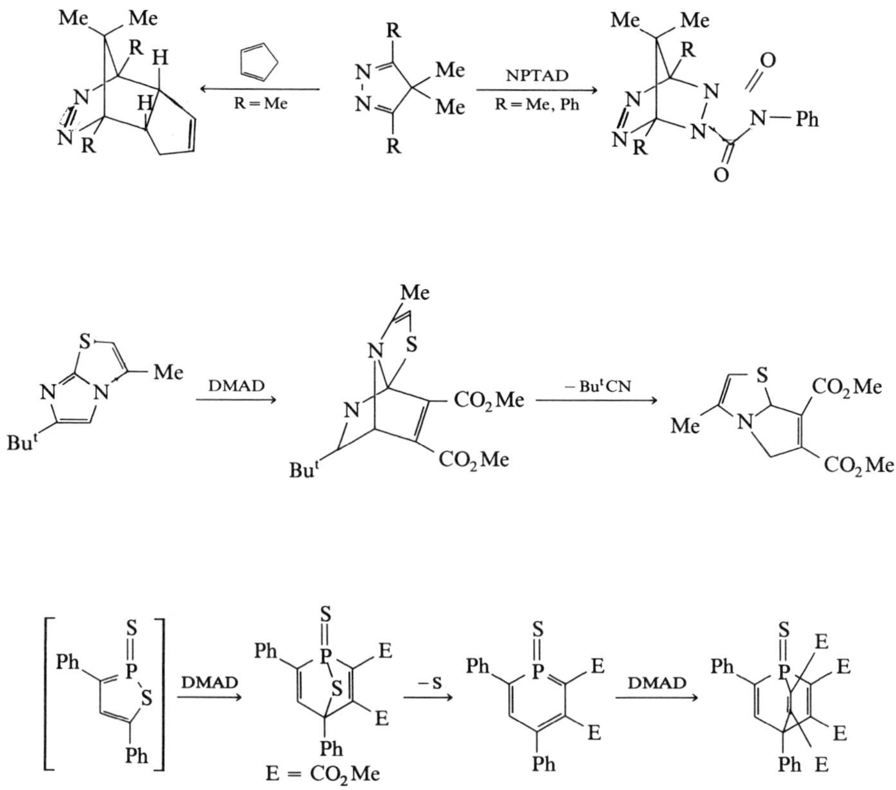

Scheme 6.9

The isobenzofuran **34** (R = H) is about 10^6 times more reactive than 1,3-butadiene in the cycloaddition reaction with MA.[283] A study of substituent effect of the cycloaddition of 1- and 1,3-disubstituted isobenzofurans shows that the 1,3-diphenyl derivative **32** ($R_1 = R_2$ = Ph) is 12 times less reactive than the unsubstituted **34** (R = H) and 1-ethoxy isobenzofuran **34** (R = OEt) is the most reactive diene prepared in solution to date.[283]

The bimolecular rate reaction constants of **32** ($R_1 = R_2$ = Ph) with dimethyl-, methyl(-)bornyl-, and di(-)bornylfumarate are 4.1×10^{-2}, 2.22×10^{-2} and 1.7×10^{-2}, respectively. Lewis acid catalysts enhance the rate by at least two orders of magnitude and alter the diastereoisomeric ratios. The uncatalyzed reactions show cooperativity in asymmetric induction, and this has been used as a probe to investigate the geometry of the transition state of the cycloaddition reaction.[130]

TABLE 6.42 Diels–Alder Reactions of Pyrroles

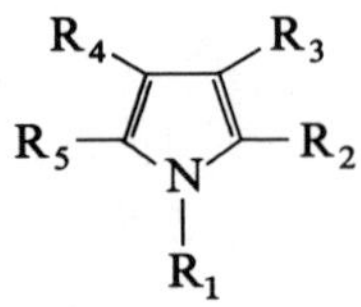

R_1	R_2	R_3	R_4	R_5		Dienophile	Yield (%)	Ref.
H	H	H	H	H	C2	Hexakis(trifluoro-methyl)benzvalene	12	1
Me	H	H	H	H	E6	[Difluorobenzyne]	44	2
Bu^t	H	H	H	H	E6	[Naphthyne]	78	3
Ac	H	H	H	H	D3	MA	35	5
					D5	Maleimides	77–84	5
					E3	DMAD	83	4, 5
COPh	H	H	H	H	D3	MA	20–25	5
					D5	NMM	77	5
					E3	DMAD	40	5
CO_2Et	H	H	H	H	D3	MA	26	5
					D5	NPM	46	5
					E3	DMAD	35	5
CO_2Bn	H	H	H	H	D5	NPM	90	5
COC_6H_4-pCl	H	H	H	H	D5	NPM	90	5
Me	$-(CH_2)_2-$	CH_2CMe_2	H	H	E6	[Fluorobenzyne]	12	2
Me	Me	H	H	Me	E6	[Benzynes]	22–23	2
NH_2	Me	H	H	Me	E3	DMAD		6
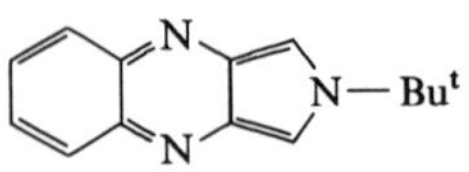					D5	NMM	87–91	114
					E3	DMAD	94	114
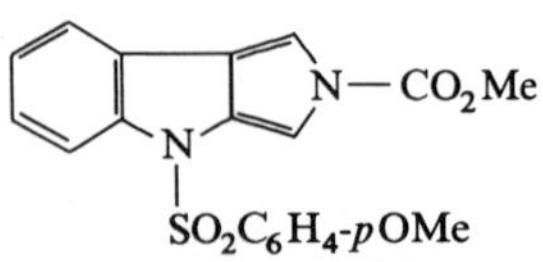					D5	NPM	76	7
					E6	[Benzyne]	75	7
Me	Me	Me	Me	Me	E6	[Fluorochloro-benzynes]	20	2
NH_2	Me	Me	Me	Me	E3	DMAD		6
$NHCO_2Me$	Me	CO_2Et	Me	Et	E3	DMAD		8
$NHCO_2Me$	Me	CO_2Et	Et	Me	E3	DMAD		8
$NHCO_2Me$	Me	Ph	$-(CH_2)_3-$		E3	DMAD		8
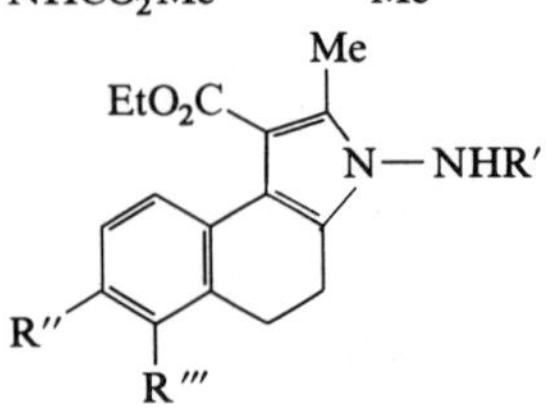		R′ = H, CO_2Me; R″ = H, OMe; R‴ = H, Me			E1	MP	70	9a
					E3	DMAD	50–90	8

TABLE 6.43 Diels–Alder Reactions of Furan

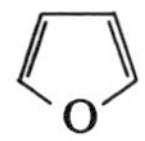

Dienophile		Yield (%)	Ref.
A1	Acrylates	17–95	12
A1	MAC	2–76	13, 36
A1	Acryloyl chloride	76	13
A1	3-(2-Pyridylsulfinyl) acrylates (*)	69–80	30
A1	Acrylic acid	48	36
A1	AC	24–65	40
A1	MVK	48–81	40
A4	Maleamic acid		26
A7	Maleates	55–96	12, 13b
A7	Fumaroyl chloride	100	17
A8	Phenylvinyl sulfonate	84	42
A8	ACN	48–100	13b, 31, 36
A9	α-Chloroacrylonitrile	33–60	13b, 15
A10	(*E*)-1,2-Bis(phenylsulfonyl)ethene	95	23
A10	1,1,1-Trichloro-3-nitro-2-propene	86	41
A13	Allenic ketones	> 90	20
A13	1,3-Dicarboalkoxyallenes	90–97	28
A13	Phenylsulfonylpropadiene	61	29
A13	Fluoroallene	60	39, 55
A13	Bis(trimethylsilyl)allene dicarboxylate	88	46
B1	2,2-Difluoromethylenecyclopropane	38	139
C1	[Dewar thiophene]		14
C1	Cyclopropenes	100	22
C1	[3-Methoxy-*trans*-cycloheptene]	95	37
C1	2-Cyano-3-methyl-5-thiabicyclo[2.1.0]pent-2-ene	100	38
C2	Hexakis(trifluoromethyl)benzvalene	86	1
C2	Hexafluoro-3-oxatricyclo[3.2.0.0^{2,4}]-hept 6-ene	42	32
C2	[1,7-Dehydroquadricyclane]	61	44
C2	[Tricyclo[4.1.0.0^{2,7}]hept-1(7)ene]	47	47
C4	[Bicyclo[3.3.1]non-1-en-3-ones]	53–86	34
C7	1,4-Benzoquinones	72–85	16
D3	MA		10
D3	Anydrides	12–41	19
D3	Dihydrothiophene anhydride	100	25
D4	2-Ethoxycarbonylhexafluoro-2-azabicyclo[3.2.0] hepta-3,6-diene	95	27

TABLE 6.43 (*Continued*)

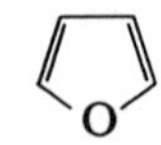

Dienophile		Yield (%)	Ref.
D6	Thiirene sulfoxide peroxide	86	21
E2	Ethyl-β-phenylsulfonyl propiolate	73–75	18
E2	Ethyl(diethoxyphosphinyl) propinoate		24
E3	DMAD	84	43
E6	[Naphthyne]	72	3
E6	[5,6-Didehydrodi-benzo[*a*, *e*]cyclooctene]	51	33, 45
E6	[1,3-Naphthodiynes]	40–45	35
E6	5,6,11,12-Tetradehydrodi-benzo[*a*, *e*]cyclooctene		45
I3	[Singlet oxygen] (k)		11

TABLE 6.44 Diels–Alder Reactions of Monosubstituted Furans

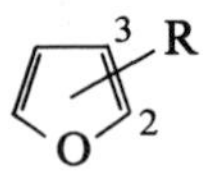

R		Dienophile	Yield (%)	Ref.
		2-Substituted		
Me	A1	MAC	100	49
	A1	AC	100	49
	A7	DMM	100	49
	A7	DMF	95	49
	A8	Phenylvinyl sulfonate	85	42
	A8	ACN	100	49
	A9	α-Chloroacrylonitrile	62	15c, d
	A9	α-Acetoxyacrylonitrile	85	15d
	A10	1,1,1-Trichloro-3-nitro-2-propene	83	41
	C2	[Tricyclo[4.1.0.0^{2,7}]hept-1(7)ene]	19	47
	C2	Hexakis(trifluoro-methyl)benzvalene	92	1
	D3	MA (k)		10d
	E6	[Fluorobenzyne]	55	48
	I3	[Singlet oxygen]		11

TABLE 6.44 (*Continued*)

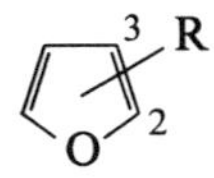

R		Dienophile	Yield (%)	Ref.
		2-Substituted		
Et	E6	[Fluorobenzyne]		48
$CH{=}CH_2$	A7	DMM	5	3
pR′—C_6H_4 (R′ = H, Me, CN, CO_2Et, Cl, Br, NO_2)	I3	[Singlet oxygen] (k)		11a
CN	I3	[Singlet oxygen] (k)		11a
CH_2NHR' (R′ = Bn, $PhCH_2CH_2$, Ar)	D3	MA	58–99	50
$CH{=}N{-}NMe_2$	A10	Fumaronitrile	13	54
	C9	NQ	77	54
	D3	MA	94	54
	D5	NEM	90	54
CHO	E3	DMAD		54
	I3	[Singlet oxygen] (k)		11a
CH_2OH	D3	MA		51
	A13	Bis(trimethylsilyl)allene-dicarboxylate	55	46
CH_2OBn	A13	Bis(trimethylsilyl)allene-dicarboxylate	52	46
$CH(OCH_2)_2$	D3	MA	30–48	52
$C(OAc){=}CH_2$	A7	DMM	50	53
OMe	C9	1,4-Naphthoquinones sulfoxide	65	60
	D3	MA		59
	I3	[Singlet oxygen] (k)		11a
OAc	D3	Chloromethylmaleic anhydride	76	61
$OC_6H_4{-}pR'$ (R′ = H, Me)	D3	MA	82	62
	D5	NPM	89	62
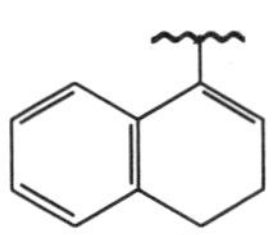	A10	Fumaronitrile		57
	E3	DMAD		57
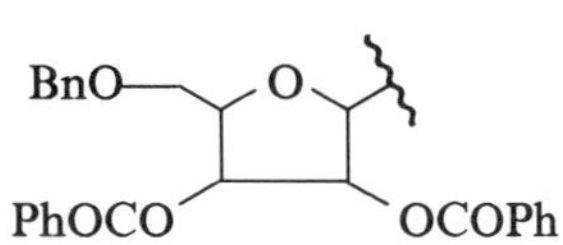	E3	DMAD	94	58
	D5	Maleimide	94	58

TABLE 6.44 (*Continued*)

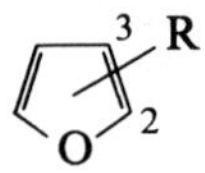

R		Dienophile	Yield (%)	Ref.
		2-Substituted		
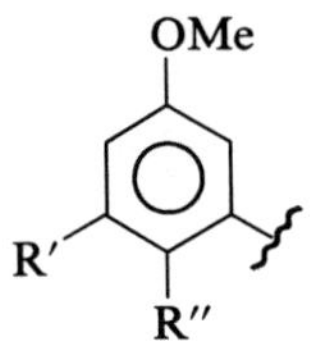 (R′, R″ = H, OMe)	D3	MA	88–95	56
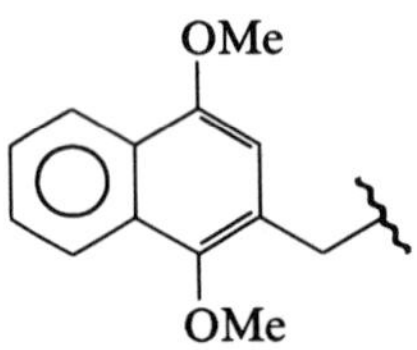	D3	MA	70	56
		3-Substituted		
NHCOPh	D3	MA	76	63
$NHCO_2Bu^t$	A3	Methyl-3-nitroacrylate	10	64
	E3	DMAD	24	64
Br	I3	[Singlet oxygen] (k)		11a

TABLE 6.45 Diels–Alder Reactions of Disubstituted Furans

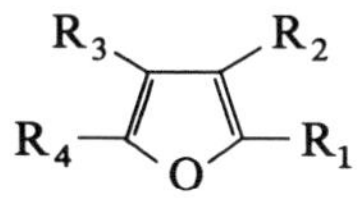

R_1	R_2	R_3	R_4	Dienophile	Yield (%)	Ref.
			2, 3			
[OMe	OMe	H	H]	E2 Ethylphenyl propiolate	23	65
[OMe	Br	H	H]	D3 MA	97	65
				E2 Ethylphenyl propiolate	10	65
			2, 4			
$OSiMe_3$	H	OMe	H	D3 MA	80	66
			2, 5			
Me	H	H	Me	A1 MVK	19–65	40
				A1 AC	25–42	40
				A9 α-Chloroacrylonitrile	58	15c
				C2 [Bicyclo[3.1.0]hex-(1, 5)ene]		69
				C2 [Tricyclo[4.1.0.0^{2,7}]-hept-1(7)ene]	53	47
				D3 MA (k)		10a, d
				E3 DMAD	15	43
				E3 Acetylendicarbonyl-dichloride	49	67
				E6 [1,3-Naphthodiynes]	22–47	35
				E6 [4,4′-Methylenebis-(benzyne)]	61	68
				I3 [Singlet oxygen]		11
Me	H	H	CN	I3 [Singlet oxygen] (k)		11a
Me	H	H	CHO	I3 [Singlet oxygen] (k)		11a
Me	H	H	$CH{=}NNMe_2$	D3 MA	74	54
				D5 NEM	65	54
Bu^t	H	H	Bu^t	I3 [Singlet oxygen] (k)		11a
Ph	H	H	Ph	I3 [Singlet oxygen] (k)		11a
Ar	H	H	Ar	I3 [Singlet oxygen] (k)		11a
(Ar = p-$R'C_6H_4$; R′ = Cl, Br, Me)						
CHO	H	H	CHO	I3 [Singlet oxygen] (k)		11a
CO_2R'	H	H	CO_2R'	I3 [Singlet oxygen] (k)		11a
(R′ = Me, Et)						
OMe	H	H	OMe	C4 Cyclobutenedione		70
$SiMe_3$	H	H	$SiMe_3$	E3 DMAD	100	71
Br	H	H	CO_2Et	I3 [Singlet oxygen] (k)		11a
Br	H	H	Br	I3 [Singlet oxygen] (k)		11a

TABLE 6.45 (*Continued*)

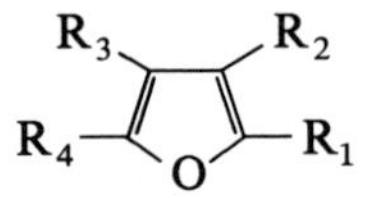

R_1	R_2	R_3	R_4		Dienophile	Yield (%)	Ref.
				3, 4			
H	Me	Me	H	A7	Maleamic acid		26
H	CHO	$(CH_2)_6CO_2Me$	H	D3	MA	64	72
				D5	NMM	79	72
H	OMe	OMe	H	C4	Cyclobutenedione	20	76
				C7	1,4-Benzoquinones	86–89	74
				D3	Dichloromaleic-anhydride	67	75
				E2	Methylphenyl-propiolate	71	65
H	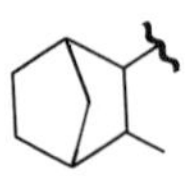		H	D3	MA		77
H	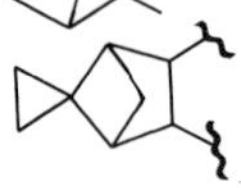		H	D5	NPM	53	78
				E3	DMAD	15	78
				E6	[Benzyne]	26	78

TABLE 6.46 Diels–Alder Reactions of Tri- and Tetrasubstituted Furans

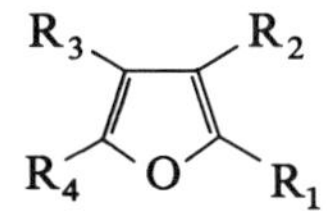

R_1	R_2	R_3	R_4		Dienophile	Yield (%)	Ref.
					Trisubstituted		
H	Me	CN	NH_2	A1	MVK		79
$OSiMe_3$	H	Me	$OSiMe_3$	E3	DMAD	73–100	71
$OSiMe_3$	H	Ph	$OSiMe_3$	A1	Ethyl acrylate	97	71
					Tetrasubstituted		
Me	Me	Me	Me	C1	Tris(trifluoromethyl)trifluoroacetylcyclopropene		22d
				C2	Hexakis(trifluoromethyl)benzvalene	68	1
Me	Me	CN	NH_2	A1	Vinyl ketones	40–89	79
				A1	MVK	78–95	81
				A1	MAC	49–63	81
Ph	Ph	CN	NH_2	A1	MVK	56	79
				A1	MAC	47	81
$-(CH_2)_4-$		$-(CH_2)_4-$		A1	Acrylic acid		80
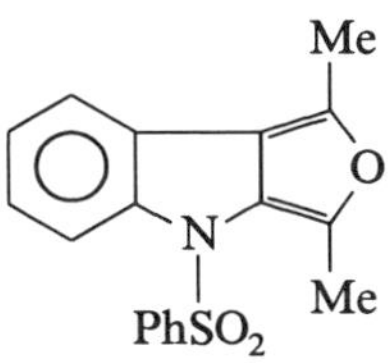				D5	NPM	100	83
				E3	DMAD	100	83
				E6	[Benzyne]	99	83
				E7	[3,4-Pyridyne]	38	83
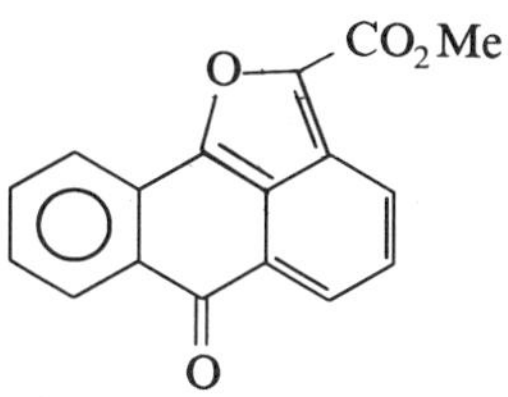				A8	ACN	95	82

TABLE 6.47 Diels–Alder Reactions of Miscellaneous Thiophenes

Diene			Dienophile	Yield (%)	Ref.
		D3	MA	8–47	84
		E6	[Benzyne]	33	85
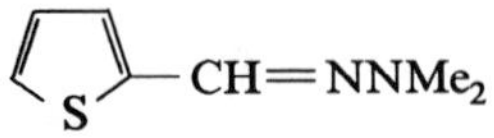		D3	MA	6	54
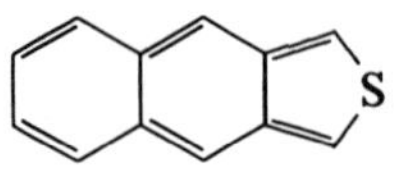		D5	NPM	7	90
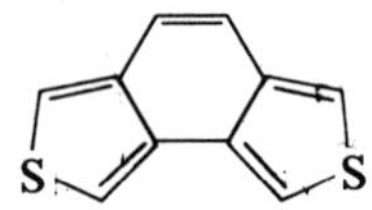		A12	TCNE	67	91
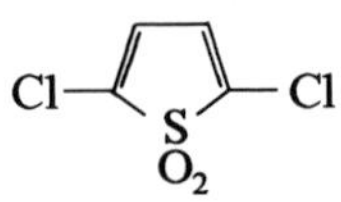		D1	2,5-Dimethylfuran	77	86
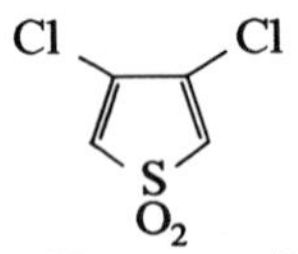		D4	1,4-Dihydronicotinates	58–97	220
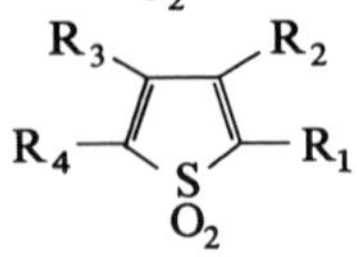	R_1–R_4 = Cl	A1	MAC	91	88
		A8	Ethene	89	88
		A8	ACN	83	88
		C1	Cycloalkenes	27–92	88
		C3	Azulene	30	87
		C7	BQ	74	88
		C9	NQ	68	88
		D1	Furans	62–93	86
		D3	MA	61	88
		D4	*N*-Methylpyrrole	44	88
		D5	NMM	90	88
		D6	Thiophene	61	88
		E4	Phenylacetylene	86	88
	R_1–R_4 = Br	D1	2,5-Dimethylfuran	90	86
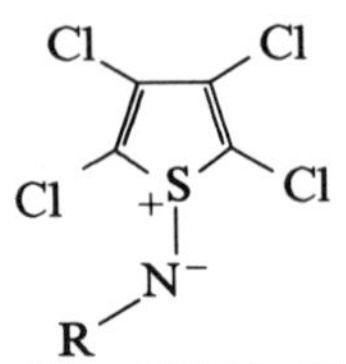 (R = CO_2Et, CO_2Ph, *p*-$MeC_6H_4SO_2$)		C1	Cyclohexene	95	89
		C2	Norbornene	93	
		C3	Indene	69	89
		C3	Acenaphtylene	100	89
		D1	Benzofuran	50	89
		D6	Thiophene	17	89

TABLE 6.48 Diels–Alder Reactions of Oxazoles

Diene			Dienophile	Yield (%)	Ref.
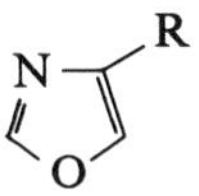	R = Me	D1	3-Methylsulfonyl-2,5-dihydrofuran	75	214
	R = Ph	E5	Acetylenes	37–95	215
		E5	1-(*t*-Butyldimethylsilyl)-2-butyn-1-one	37	273
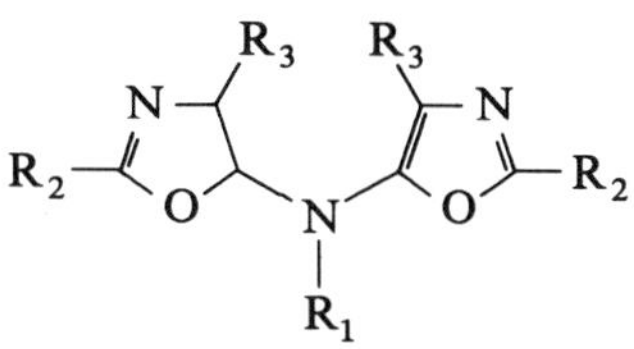 (R_1 = Me, Pr; R_2 = H, Me; R_3 = H, Et)		D5	Maleimide	100	216
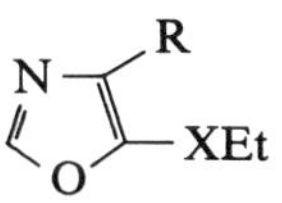 (R = H, Me; X = O, S)		A1	MVK	8–91	9b, 217
		A1	Ethyl acrylate	91	217
		A1	Acrylic acid	85	217
		A8	ACN	85	217
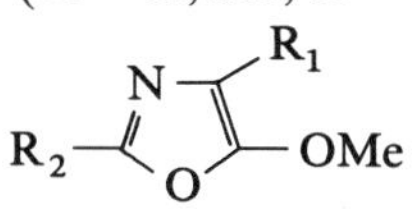	R_1 = *p*-$NO_2C_6H_4$, R_2 = Me	D5	NMM	86–90	218
		D5	NPM	60–100	218
		E3	DMAD		218
		I5	NPTAD	69	218
	R_1 = *p*-$NO_2C_6H_4$, R_2 = Et	D5	Maleimides	74–99	218a
	R_1 = Et, Pr^i, Bu^i; R_2 = Me, Bu^i	D6	2-Thiolen-4-one-*S,S*-dioxide	35–60	219

TABLE 6.49 Diels–Alder Reactions of Miscellaneous Five-Membered Heterocycles Having One Heteroatom

Diene			Dienophile	Yield (%)	Ref.
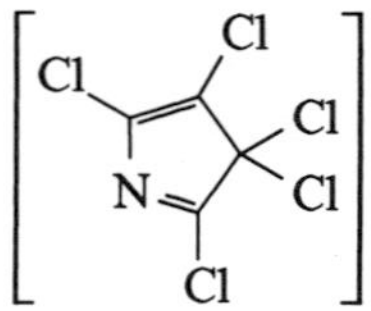		A8	Styrene		103
		A8	Vinyl acetate	83	104
		A8	Vinyl chloride		105
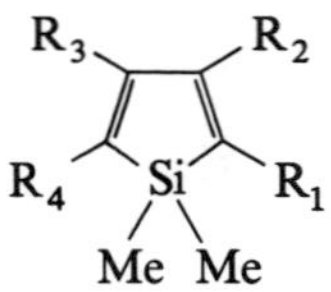	R_1–R_4 = H	E5	Acetylenes		93
		D3	MA	62	92
		D13	1,1-Dimethylsilole	58	92
	$R_2 = R_3 = H, R_1 = R_4 = Ph$	E3	DMAD		93
		E6	[Arynes]	50–58	93
	R_1–R_4 = Ph	E6	[Arynes]	34–75	93
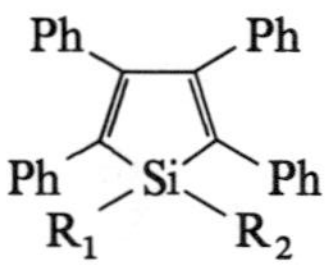	$R_1 = Me, R_2 = Cl$	E5	Acetylenes	37–80	93
	$R_1 = Ph, R_2 = Me$	E5	Acetylenes	76–90	93
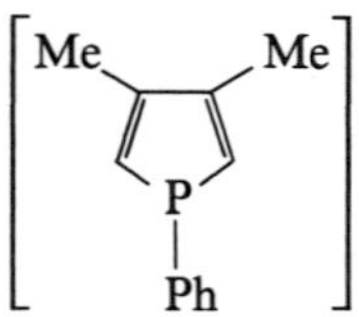		A8	*cis*-Phospholevinyl-phosphine Pd-complex		94

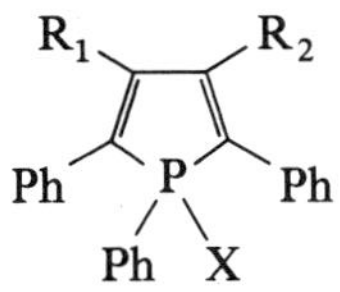	$R_1 = R_2 = H$, $X = O$	C1	Cyclopentadiene	40	95
	$X = S$	C1	Cyclopentadiene	40	95
	$X = Se$	C1	Cyclopentadiene	35	95
	$R_1 = R_2 = Ph$, $X = O$	E3	DMAD	16	96
		E3	Diterbutylacetylene-dicarboxylate	10	96
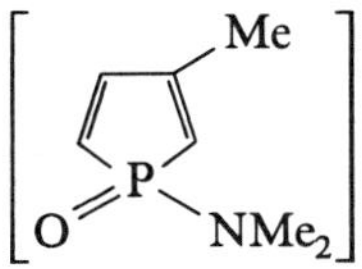		D5	NPM		97
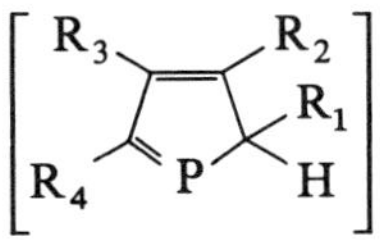	R_1–$R_4 = H$	D12	2*H*-Phosphole		98, 99
	$R_1 = R_4 = H$, $R_2 = R_3 = Me$	D12	3,4-Dimethyl-2*H*-phosphole	60	98, 99
	$R_1 = R_4 = Ph$, $R_2 = R_2 = H$	D12	2,5-Dipheynl-2-*H*-phosphole	50	98
		E5	Diphenylacetylene	28	100
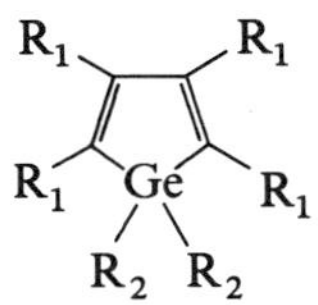	$R_1 = H$, $R_2 = Me$	D3	MA		101
	$R_1 = Ph$, $R_2 = Me$	E6	[Arynes]	65–80	102
	$R_1 = R_2 = Ph$	E6	[3,4,6-Trimethylbenzyne]	70	102

TABLE 6.50 Diels–Alder Reactions of Miscellaneous Five-Membered Heterocycles Having Two Heteroatoms

Diene		Dienophile	Yield (%)	Ref.
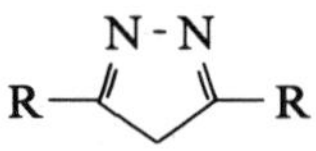 (R = H, Me)	C1	Cyclopentadiene	62–71	211
	C2	Norbornene	15–51	211
	C2	Norbornadiene	24–30	211
	C3	Benzonorbornadiene	54–66	211
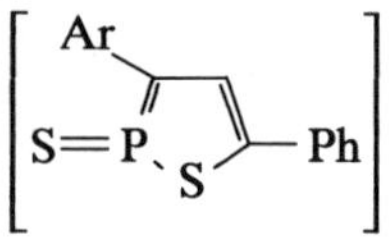 (Ar = Ph, pMeC$_6$H$_4$)	E3	DMAD	59–72	212
	E4	Phenylacetylene	28–38	212
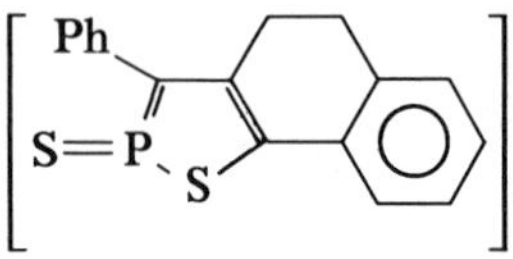	E3	DMAD	74	212
	E4	Phenylacetylene	86	212
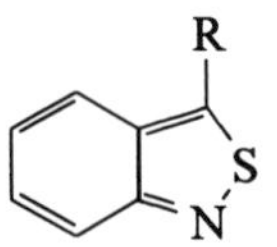 (R = H, Me)	E1	MP	2	213
	E3	DMAD	5–23	213
	E3	Diethylacetylene dicarboxylate	5–23	213

Isobenzofurans are generally more reactive than the analogous furans. When heated the adducts give a retro Diels–Alder reaction or can loss the oxygen bridge by action of hydracids, giving naphthalenes or dihydronaphthalenes[118] (i.e., **35** → **36**). The *ortho-endo* selectivity of the cycloadditions is predominant (i.e., **36**, **37**, **39**: Z = O).[109,118,139,142]

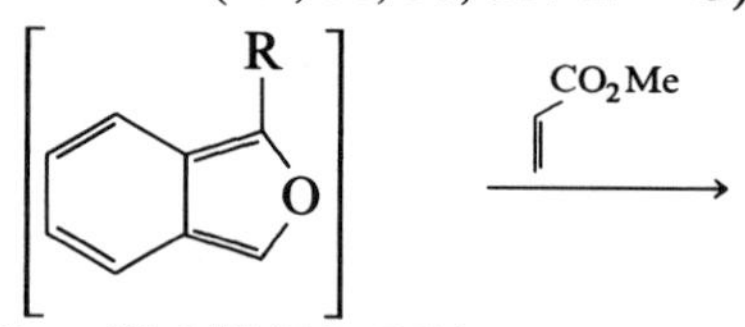

34 R = H, NR′R″, OR′

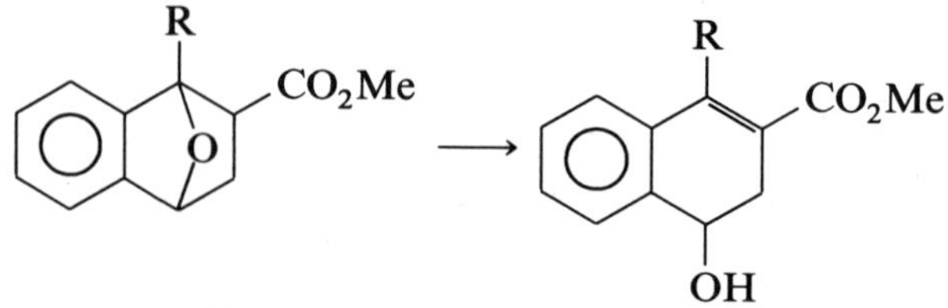

35 **36**

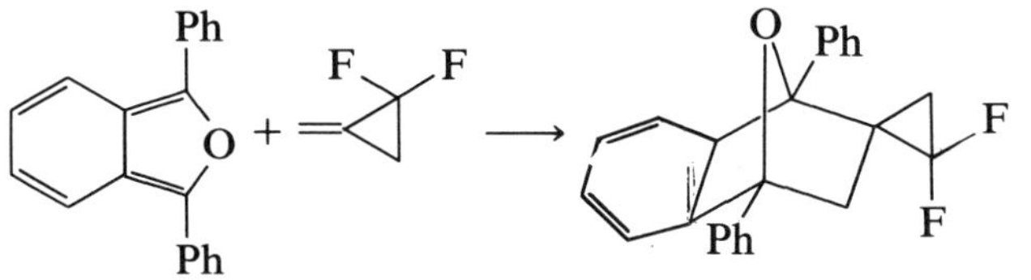

37

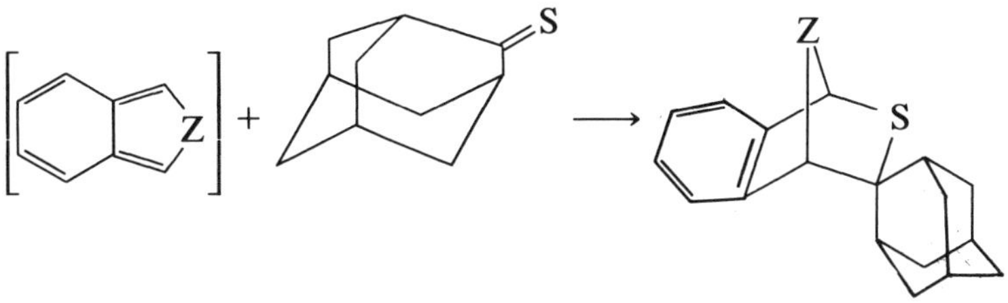

38 39

The benzo[*f*]- and benzo[*e*]isobenzofurans have three formal diene moieties but the 4π system of the furan ring is greatly the most reactive in the [4 + 2] cycloadditions. The benzo[*f*]- (**40**) is generated in situ while the benzo[*e*]- (**41**) is a crystalline compound which can be isolated; with MA, both give a mixture of *endo* **42** and *exo* **43** adducts.[118,147,148]

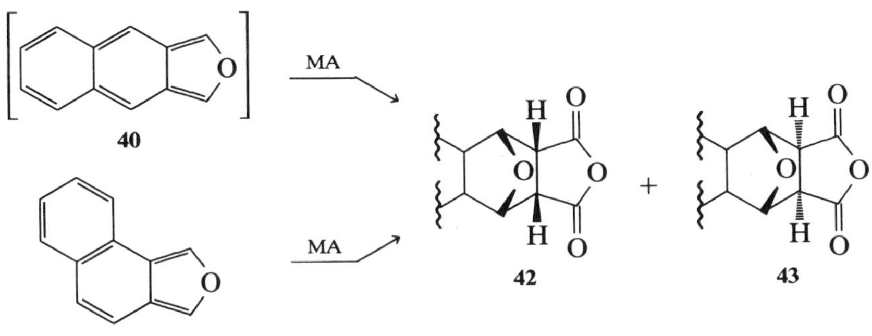

40 41 42 43

The isobenzothiophenes behave similarly to oxygen analogs (i.e., **44** → **45**), but they are less reactive. As a consequence of the loss of reactivity of the ring containing the heteroatom, the NPM reacts with benzo[*f*]isobenzothiophen both at the diene moiety of the thiophene ring and at that of the central ring of the molecule.[90]

Indole and its derivatives react to give prevalently *endo* monoadducts (i.e., **39**: Z = NCO_2Bu^t and **47**)[109] but sometimes bisadducts have also been isolated (i.e., **46**). Fluoroindoles have been used to prepare carcinogenic polyfluorobenz[*a*]anthracenes (i.e., **48** → **49**).[113]

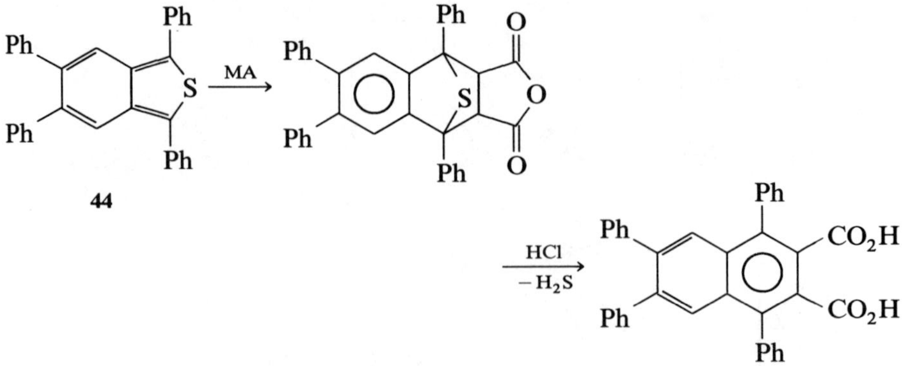

44

45

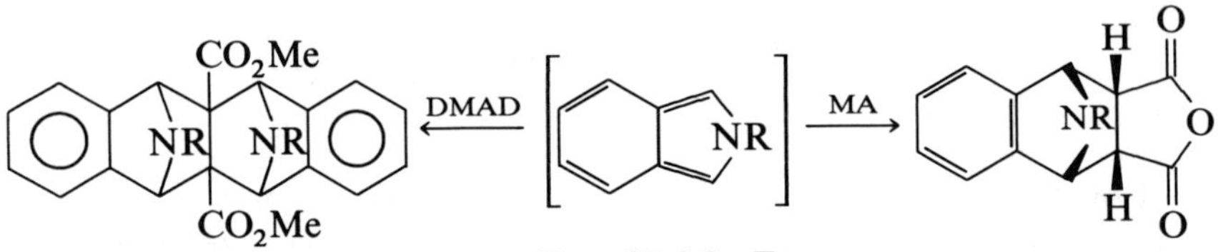

46 R = H, Me, Et 47

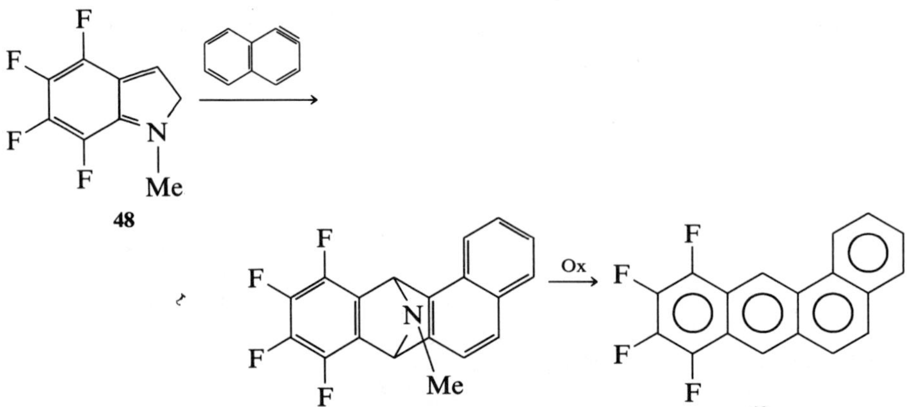

48

49

Tables of Diels–Alder Reactions See Tables 6.51 to 6.53; see also the general remarks about classifications, abbreviations, and symbols in the Introduction.

TABLE 6.51 Diels–Alder Reactions of Isoindoles

Diene		Dienophile	Yield (%)	Ref.
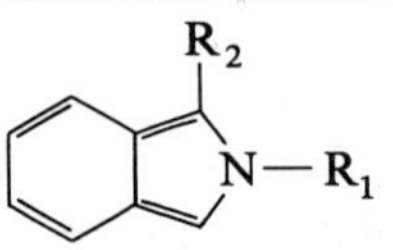	R_1 = Me, R_2 = H	C2 [Bicyclobutenes]	25	108
		D4 9-Alkyl-1,4-dihydro-naphthalen-1,4-imines	25	106
		D5 Maleimides		107

TABLE 6.51 (*Continued*)

Diene			Dienophile	Yield (%)	Ref.
		E6	[2,3-Naphthalyne]	52	3
		E6	[1,2-Naphthalyne]	26	113
	R_1 = Ph, R_2 = SEt	D5	NPM	90	184
	R_1 = CO_2Bu^t, R_2 = H	F2	Adamanthanethione	96	109
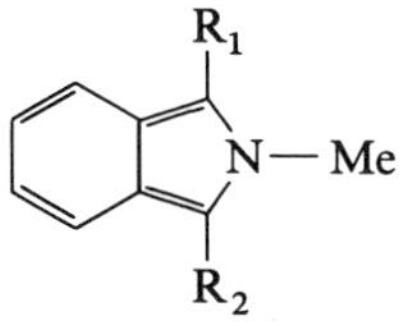	R_1 = Me, R_2 = H	C2	[Bicyclobutenes]	61	108
	R_1 = R_2 = Me	C2	[Bicyclobutenes]	35–61	108
		C2	[1,7-Dehydroquadricyclane]	63	44
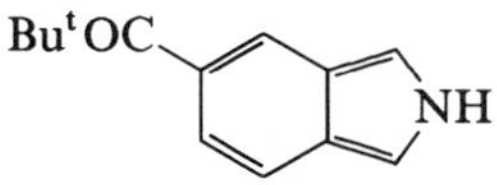		D5	*N*-Alkylmaleimides	95	110
		D5	*N*-Arylmaleimides	95	110
		E3	DMAD		110
		I1	Dimethylazodicarboxylate		110
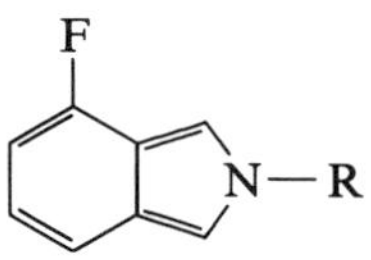	R = Me	D4	9-Alkyl-1,4-dihydronaphthalen-1,4-imines	6	106
	R = Bn	D4	9-Alkyl-1,4-dihydronaphthalen-1,4-imines		106
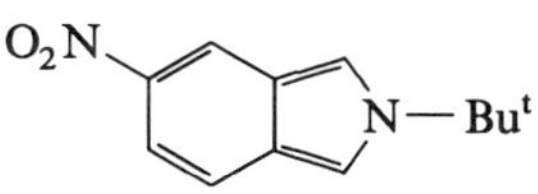		D5	NMM		111
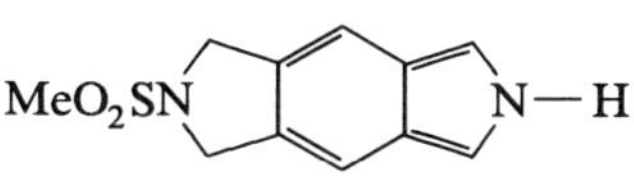		D5	Maleimides		112
		E3	DMAD		112
		I1	Dimethylazodicarboxylate		112
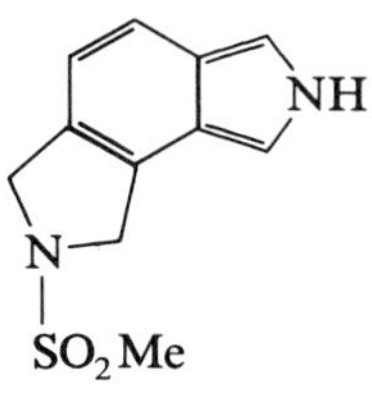		D5	Maleimides		112
		E3	DMAD		112
		I1	Dimethylazodicarboxylate		112
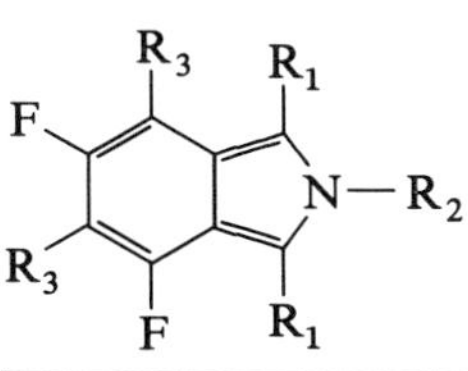	R_1 = R_2 = H, R_3 = F	E6	[1,2-Naphthalyne]	61	113
	R_1 = R_2 = H, R_3 = Cl	E6	[1,2-Naphthalyne]	57	113
	R_2 = H, R_1 = Me, R_3 = F	E6	[1,2-Naphthalyne]	91	113
	R_2 = H, R_1 = Me, R_3 = Cl	E6	[1,2-Naphthalyne]	39	113
	R_1 = H, R_2 = Me, R_3 = F	E6	[2,3-Naphthalyne]	68	3

TABLE 6.52 Diels–Alder Reactions of Isobenzofurans

Diene			Dienophile	Yield (%)	Ref.
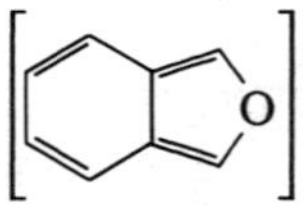		A7	DMF	70	115
		C1	Cyclohexene	5	147
		C2	Norbornene	44–60	117, 147
		C7	BQ	25	115
		C9	NQ	70	115
		D3	MA	54	115
		D3	Dimethylmaleic anhydride	100	116
		E3	DMAD	30–80	115, 147
		F2	Adamanthamethione	79	109
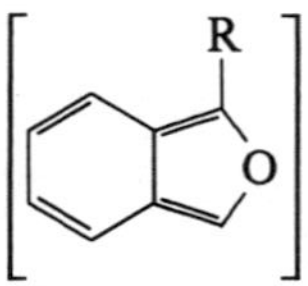	$R = CH_2C_6H_4$-pR′	A1	MAC		120
	(R′ = H, OMe)	A1	MAC	52	118
	$R = N(Pr^i)_2$	A1	*N*-Methyl-*N*-phenylacrylamide	56	118
		A7	DMM	75	118
		A8	Phenylvinylsulfone	39	118
		C4	2-Cyclohexenone	44	118
		E3	DMAD	49	118
	R = OEt	A7	Diethyl maleate	66	119
		D3	MA	56	119
		E2	Ethylphenyl propiolate	62	140
		E3	DMAD	76	140
	R = OMe	A9	α-Chloroacrylonitrile	73	117
		C2	Norbornene	19	117
		D3	MA	75	117
		E3	DMAD	42	117

	R_1	R_2	R_3	R_4				
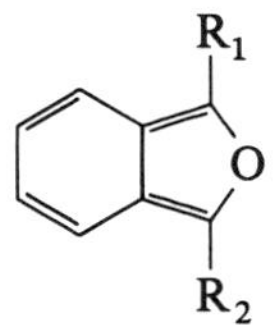	$R_1 = R_2 = Me$				A3	Methyl cinnamate	73	121
					A3	*o*-Methyl methylcinnamate	73	121
	$R_1 = R_2 = Bu^t$				I3	[Singlet oxygen]	96	122
	$R_1 = R_2 = SO_2Me$				I5	DNTAD	> 90	123
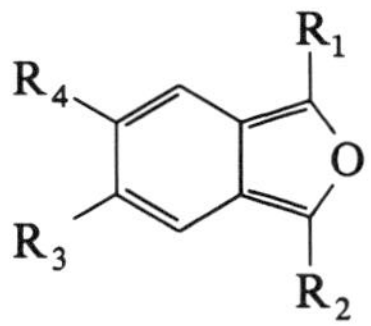	H	H	$-(CH{=}CH)_2-$		C2	Norbornene	57	147
					D2	Butenolide	67	147
					D3	MA	40	147
					E3	DMAD	27	147
	[H	H	OMe	OMe]	A1	MVK	55	141
	[H	H	OMe	OMe]	A1	MVK	78	143b
					C9	NQ	86	143b
					D2	2-Butenolide	72	143b
	[H	H	$-OCH_2O-$]		E3	DMAD	96	143b
	[H	Me	OMe	OMe]	E3	DMAD	96	143b
	[H	C_6H_{11}	$-OCH_2O-$]		A1	MAC	90	142
					E3	DMAD	70	143
	[H	Ar	OMe	OMe]	E3	DMAD	65	143
	[H	Ar	$-OCH_2O-$]		E3	DMAD	62–65	143
	[H	CN	$-OCH_2O-$]		A1	MVK	95	142
					A1	MAC	95	142
					A8	ACN	95	142
	[H	CO_2Me	$-OCH_2O-$]		A1	MVK	95	142
					A8	ACN	90	142
	[H	OEt	OMe	OMe]	E3	DMAD	71	140
	[Me	Ar	OMe	OMe]	E3	DMAD	72	143b
	Ph	Ph	$-(CH{=}CH)_2-$		A10	2-Butene (*E* and *Z*)		146
					D5	Trimethylmaleimide	56	146

TABLE 6.52 (*Continued*)

Diene					Dienophile		Yield (%)	Ref.
	Ph	Ph	CHO	CHO	D5	NPM		144
	$OSiMe_2Bu^t$	Ar	OMe	OMe	A7	DMF	> 70	145
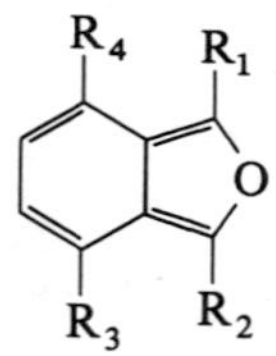	[H	Bn	Me	Me]	A1	MAC	38	120
	[H	*p*-OMeBn	Me	Me]	A1	MAC	66	120
	[$OSiMe_2Bu^t$	H	Ar	H]	A7	DMF	78–87	145
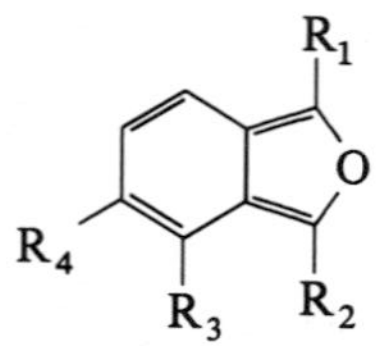	H	H	—$(CH{=}CH)_2$—		A9	α-Acetoxyacrylonitrile	18–87	148
					D2	2-Butenolide	60	148
					D3	MA	64	148
					E3	DMAD	71	148
	[H	H	OMe	OMe]	E3	DMAD	34	143b
	[H	Ar	OMe	OMe]	E3	DMAD	62	143b
	[H	OMe	—$(CH{=}CH)_2$—]		D3	MA	20	148
					E3	DMAD	40	148
	[H	OEt	—$(CH{=}CH)_2$—]		D3	MA	20	148
					E3	DMAD	47	148
	[OEt	H	—$(CH{=}CH)_2$—]		A9	α-Acetoxyacrylonitrile	78	148
					D3	MA	56	148
					E3	DMAD	93	148
	$OSiMe_2Bu^t$	H	Ph	Me	A7	DMF	80	145
	[$SiMe_3$	$OSiMe_3$	—$(CH{=}CH)_2$—]		D3	MA		149
					E6	[Arynes]	8–66	149

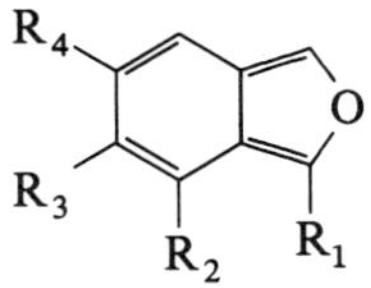	[Ar	OMe	OMe	OMe]	E3	DMAD	73	143b
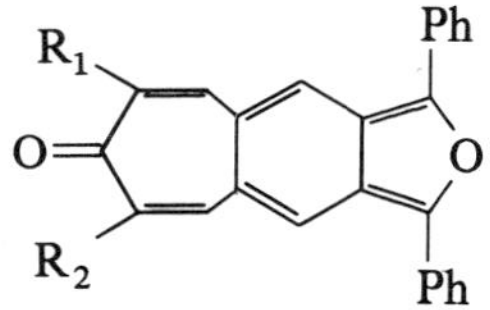 (R$_1$, R$_2$ = H, Ph)					D5	NPM		144
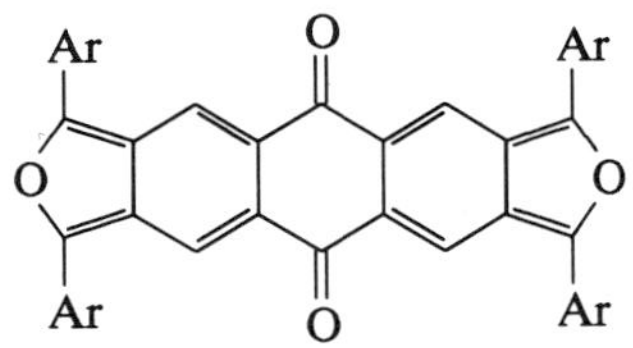					D3	MA	70	150
					D5	NMM	90	150
					D5	*N*-4-*t*-Butylphenyl) maleimide	49	150
					D5	1,4-*N*, *N*′-Dimaleimidobenzene	52	150
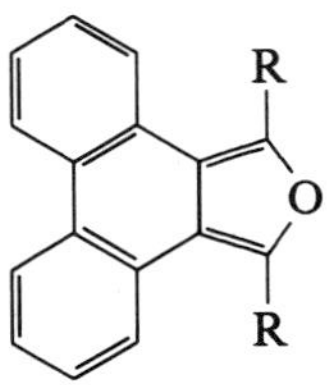	R = Ph				C3	7 Isopropylidenebenzo-norbornadiene	82	131
	R = H				C1	2,3-Diphenylcyclopropene		125b

TABLE 6.53 Diels–Alder Reactions of 1,3-Diphenylisobenzofuran

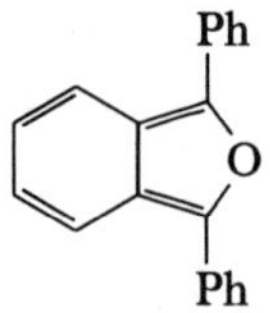

	Dienophile	Yield (%)	Ref.
A1	MAC (k)		137
A4	(*E*)-6,6-Dimethoxy-3-hexen-2-one	74	135
A7	Methyl(-)bornyl fumarate (*) (k)		130
A7	Di(-)bornyl fumarate (*) (k)		130
A7	DMF (k)		130
A8	Phenylvinyl selenoxide (k)		125a
A8	Phenylvinyl selenide (k)	64	125a
A8	Phenylvinyl sulfide (k)		125a
A9	Vinyl phosphonate (k)		137
A9	3-[(Trimethylsilyl)oxy]-3-buten-2-one	54	269
B1	2,2-Difluoromethylenecyclopropane	89	139
C1	3,3-Diphenylcyclopropene		125a
C1	2-Cyano-3-methyl-5-thiabicyclo[2.1.0]pent-2-ene	100	38
C1	Cyclopentene		129
C1	3,4-Dimethylenecyclobutene	93	132
C1	Cycloheptatriene	37	132
C1	9-Oxabicyclo[6.1.0]non-*trans*-4-ene	57	133
C1	1-Methyl-1-phenylcyclopropene		138
C2	[Tricyclo[4.1.0.0^{2,7}]hept-1(7)-ene]	42	47
C2	[Bicyclo[3.1.0]hex-1(5)-ene]	80	69
C2	[Bicyclo[4.1.0]hept-1(6)-ene]	71	69
C2	7-Isopropylidenebenzonorbornadiene	95	131
C2	[Bicyclo[4.4.1]undec-1(11)-ene]	28	136
C2	[Carbenes]		274
C3	2,3-Dehydro-1,2-dihydro-1,1-dimethylnaphthalene	20	124
C4	Cyclopent-4-ene-1,3-dione	25	128
D1	2,5-Dihydrofuran		129
D4	*p*-Nitrobenzyl-7-oxo-1-azabicyclo[3.2.0]hept-2-ene-2-carboxylate	56	268
D6	3-Chloro-2*H*-thiene-1,1-dioxide	35	126
E2	Ethyl(diethoxyphosphinyl)propinoate		24
E6	Dehydrodibenzo[*a,e*]cyclooctanes	30–57	45
E6	[5,6,9,10-Tetradehydrobenzocyclooctene]	6	127
E6	[13,14-Didehydrotribenzo[*a,c,e*]cyclooctene]	26	127
G4	2-Phenyl-1-azaspiro[2.2]pent-1-ene	76	134

Six-Membered Rings

Heterocyclodienes with One Heteroatom The most representative compounds belonging to this group of dienes are the heteroaromatics and their 1,2-dihydroderivatives, together with 2-pyrones and 2-pyridones. The six-membered heteroaromatic dienes are of great interest in both organic synthesis and the theoretical field. They allow one to relate the reactivity, the regio- and stereoselectivity of the Diels–Alder reaction to the type of heteroatoms, the aromaticity of the compounds, and the nature and position of substituents present in the ring.

The heteroatoms of pyridine, stibabenzene, and bismabenzene belong to the VA group of the periodic table; however, the three heterocycles show very different behavior toward the dienophiles. Only pyridines substituted with electron-donating groups give Diels–Alder adducts which readily loss the heteroatom (**50** → **51**).[284] On the contrary, stibabenzenes **53** (X = Sb) and bismabenzenes **53** (X = Bi) readily give [4 + 2] cycloadditions with alkynes (**53** → **54**) and head-to-head Diels–Alder dimers **52**.[207]

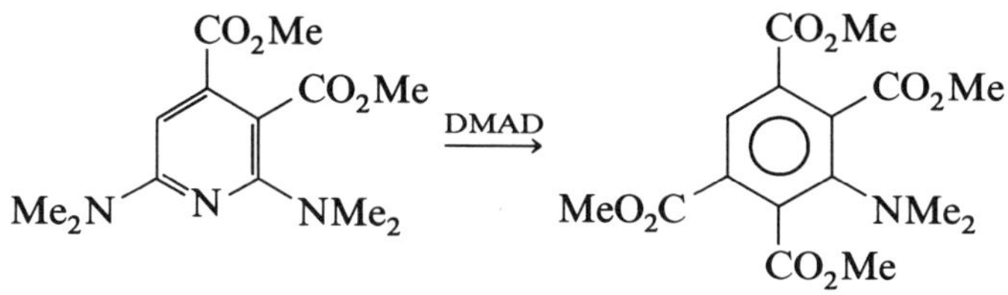

50 **51**

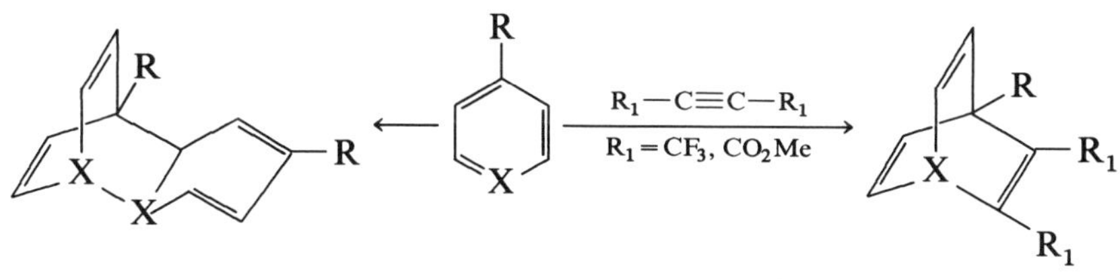

52 **53** **54**

X = Sb, Bi
R = H, Me, Bu^t

The 1,2-dihydropyridines are reactive toward dienophiles, but the reactions are strongly affected by the substituents present on the heterocyclic ring and by the nature of dienophile. 1,2-Dihydropyridines, which do not have electron-withdrawing groups (i.e., **57**) behave as enamines rather than as dienes in reactions with MAC, DMAD, and MVK. At −10°C the cycloaddition of **57** and MAC gives the [2 + 2] cycloadduct **58**, which at 80°C is converted into the thermodynamically stable adducts **55** and **56** (**55**/**56** = 3.2 : 1).[153]

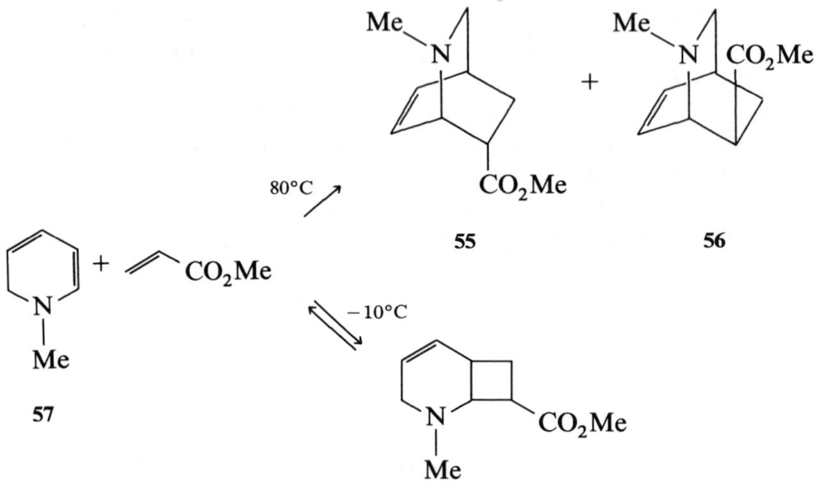

55 **56** **57** **58**

Similarly, the azocine derivative **60** is the result of an electrocyclic ring opening of the initial [2 + 2] cycloadduct **59**.

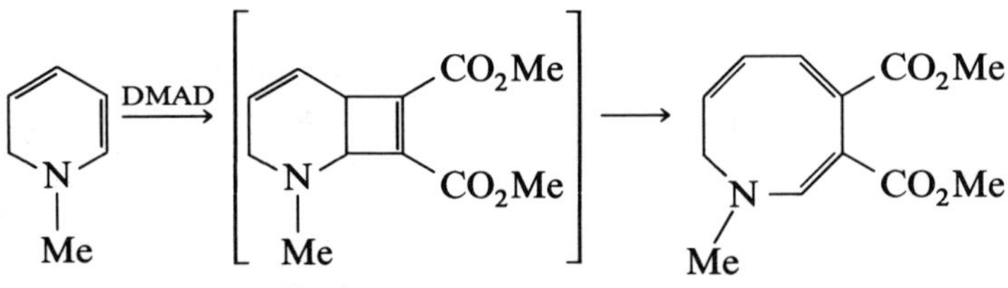

57 **59** **60**

The enamine reactivity of 1,2-dihydropyridines is suppressed by the presence of electron-withdrawing groups on the nitrogen atom. An example is the Diels–Alder reaction of *N*-acyl- and *N*-carboxymethyl-1,2-dihydropyridines **61** with nitrosocarbonyl compounds[155] (Scheme 6.10). Two regioisomer *endo* adducts are obtained (**62** and **63**) and the regioselectivity of the reaction depends on the electronic effects of the substituents present in the diene and dienophile.[155] The reaction has been used to synthesize racemic diamino sugars.[156d] The cycloaddition of 1-carbethoxy-1,2-dihydro-3-methylpyridine with MAC is, on the contrary, regioselective but not diastereoselective. Only *meta-endo* and *meta-exo* adducts are obtained in the ratio 6 : 4.[162]

By replacing the nitrogen atom of 1,2-dihydropyridines with a boron atom, the effect of the substituents is very different. Boracyclohexadienes **64** having + M-type substituents at the boron atom behave as dienes, and with hexafluorobutyne give adducts **65** that can be isolated.[152]

The 2-pyrones react as either dienes or dienophiles, depending on their reaction partners. With simple substituted butadienes the methylcoumalate **67** (R_1 = H, R_2 = CO_2Me) behaves as dienophile involving the 5,6-double bond (**67** → **68**), but with high-electron-rich alkenes (i.e., acetal ketene) and dienes (i.e., 3,4-dimethoxyfuran) it reacts as diene (**67** → **66**).[73] This last reaction is probably not concerted, but rather proceeds through a zwitterion intermediate.[73] 3-Sulfonyl derivatives **67** (R_1 = Ts, R_2 = H) undergo inverse electron demand Diels–Alder cycloadditions with enantiomeric alkylvinyl ethers with good diastereoselectivity.[190] Chloropyrones (i.e., **69**) give with benzvalene (**70**) selectively cycloadducts (i.e., **71**) which on heating eliminate CO to form chlorocyclohexadienes such as **72**.[193]

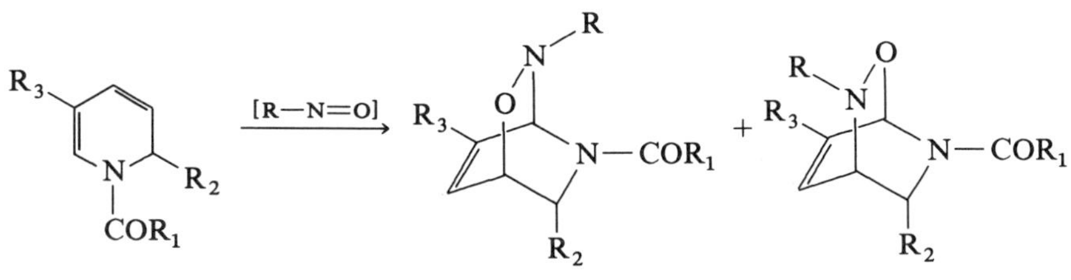

61 **62** **63**

R_1	R_2	R_3	R	Adducts
OMe	Bu	H	COPh	**62** ≫ **63**
OMe	H	H	CO_2Ph	**62** ⩾ **63**
Me	Ph	H	CO_2Bn	**62** ≃ **63**
OMe	Bu	CO_2Me	Ph	**62** ≪ **63**

Scheme 6.10

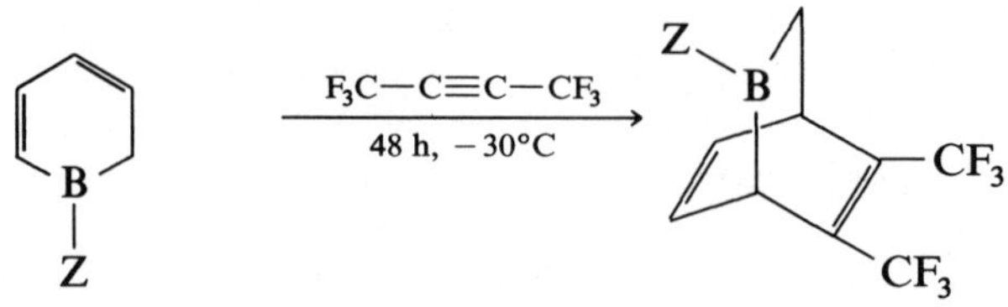

64 Z = OMe, NH_2 65

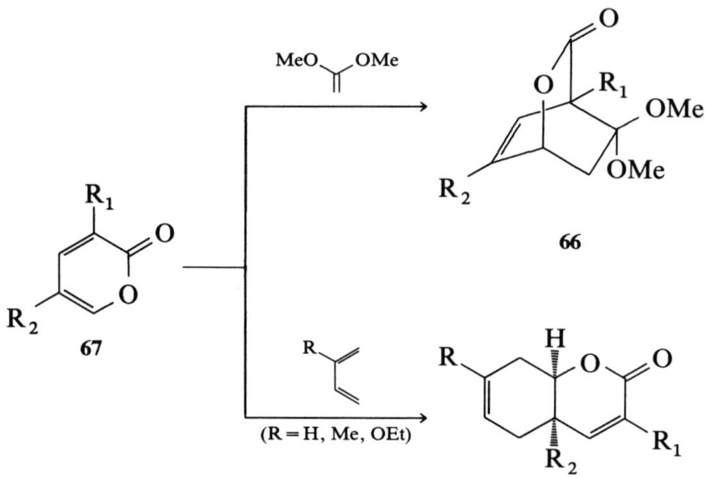

66 67 68

2-Pyridones have higher aromatic character than 2-pyrones and therefore give [4 + 2] cycloadditions with highly reactive dienophiles or when electron-withdrawing substituents are present in the molecule. *N*-unsubstituted-2-pyridones exist as a mixture of tautomers (i.e., **73**, **74**) and they react with benzyne to give mixtures of Diels–Alder adducts (i.e., **75**) and Michael-type products (i.e., **76** and **77**).[178b] The reactivity and chemoselectivity depend on the substituents present in the diene, and the most favored reaction is the formation of 2-phenoxypyridine.[178b] *N*-methyl-2-pyridones do not exist as tautomers, and only Diels–Alder adducts are observed.[178a] Sulfonyl-2-pyridones **78** are sufficiently electron deficient to react with electron-rich dienophile in inverse electron demand Diels–Alder reactions.[176] The yields are generally low and some improvement can be achieved by using high pressure. The adducts of the cycloadditions of 1-benzyl-2-pyridone with MAC and MA have been used as starting materials in the synthesis of epi-ibogamine and desethylibogamine, respectively.[172]

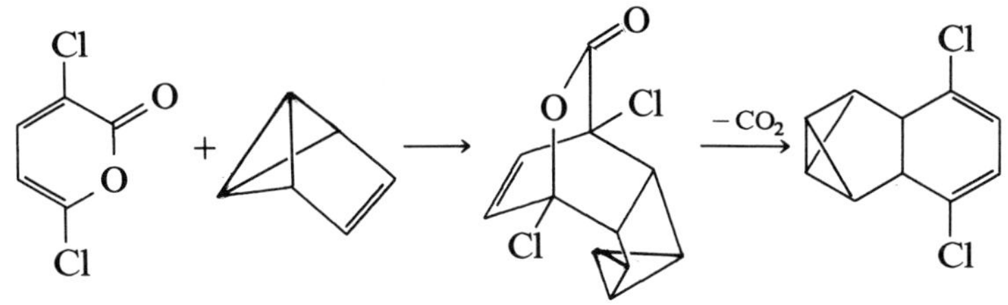

69 70 71 72

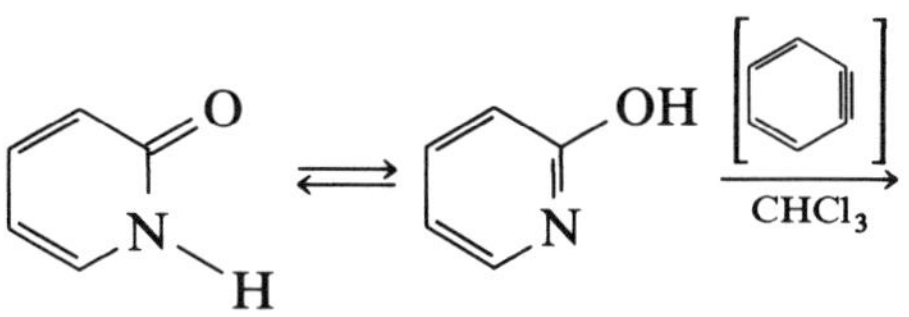

73 74

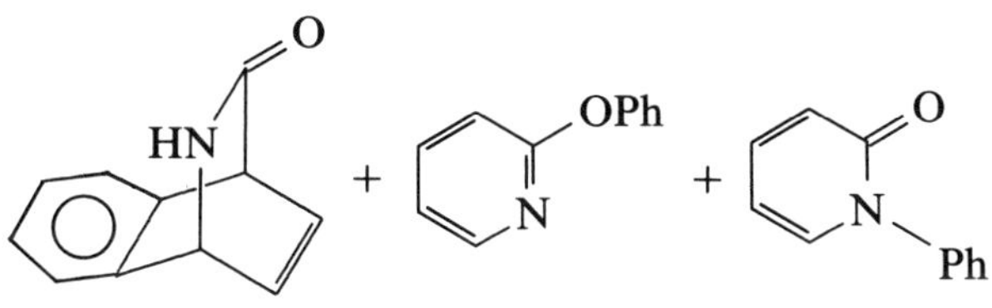

75 76 77

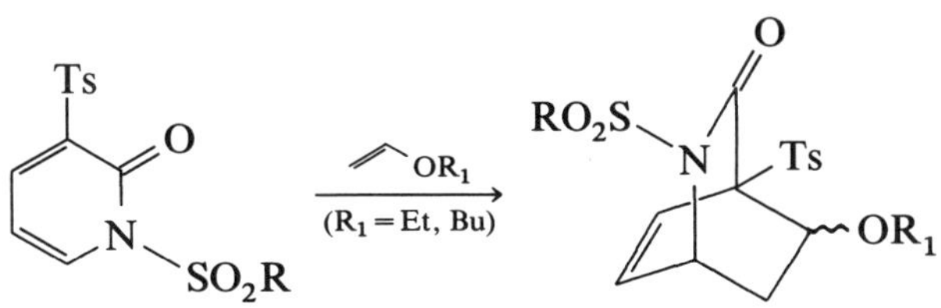

78

Heterocyclodienes with More Than One Heteroatom Silapyrans, disilabenzenes, and azines are the dienes of this group, which in the past and in recent times have stimulated the greatest interest on theoretical and/or synthetic grounds.

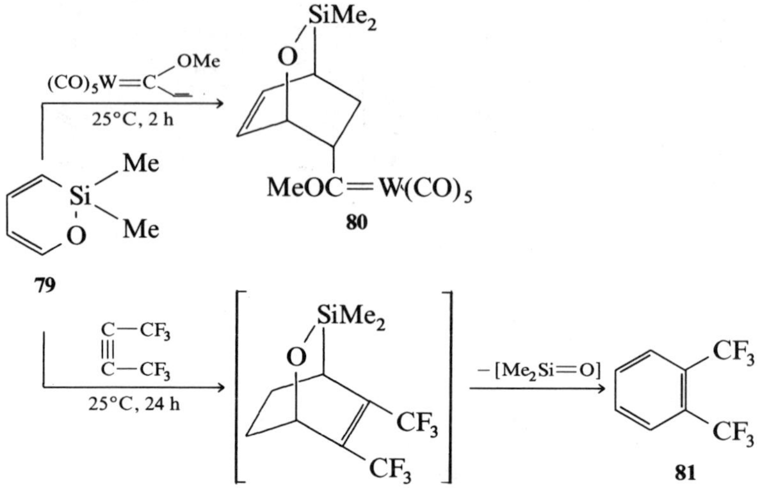

The 2-silapyran (or 1,2-oxasilin) **79** with perfluoro-2-butyne gives an adduct which, at room temperature, loses the silanone bridge forming the *o*-bis(trifluoromethyl)benzene **81**.[168] Adducts that are detectable and can be isolated have been obtained in the cycloadditions with Fischer carbene complexes[226] (**79** → **80**) and MA.[168]

Hexamethyl-1,4-disilabenzene (**82**), generated in situ, reacts with alkynes to give stable disilabarralenes **83**.

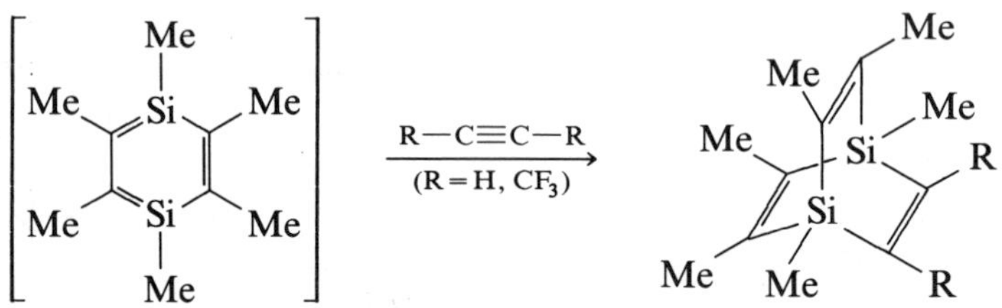

82 **83**

Di-, tri- and tetraazines are dienes of great interest in the synthesis of natural heterocyclic compounds, and therefore their Diels–Alder reaction has been investigated extensively. The chemical behavior in [4 + 2] cycloadditions is summarized in Schemes 6.11 to 6.17.

Most of the azines have electron-withdrawing substituents and therefore participate in inverse Diels–Alder reactions with electron-rich dienophiles. The presence of strong electron-donating substituents can allow the azine to react with normal Diels–Alder reaction. Carbon–carbon and/or carbon–nitrogen (see Schemes 6.11 to 6.17) are the atoms involved in the

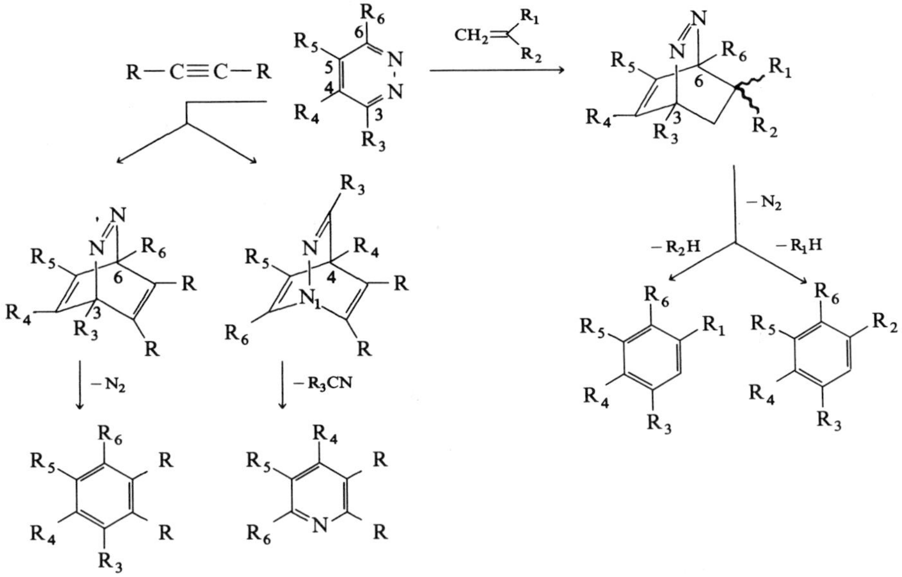

Scheme 6.11 1,2-Diazines (Pyridazines). (From Ref. 290.)

cycloadditions. The regio- and stereoselectivity of the reaction is controlled by the electronic nature of substituents of diene and dienophile. The cycloadditions are often exothermic and accompanied by the evolution of nitrogen and/or the extrusion of nitriles, hydracids, alcohols, amines, and silylols. The six-membered ring formed at the end of the process is of the carbocyclic or heterocyclic type.

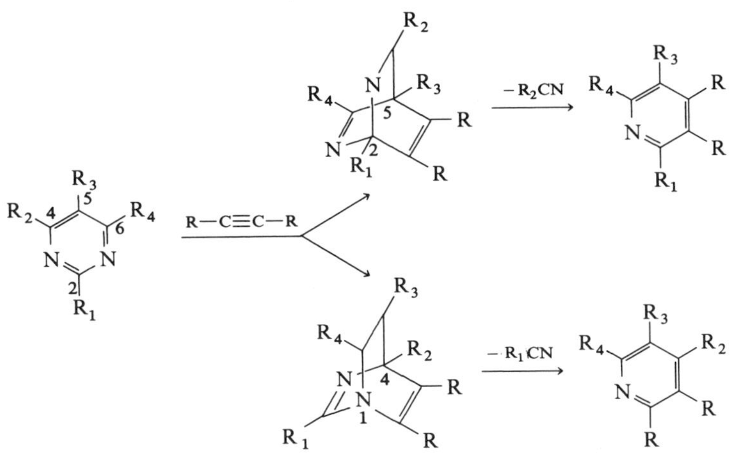

Scheme 6.12 1,3-Diazines (Pyrimidines). (From Refs. 235b, 291, and 292.)

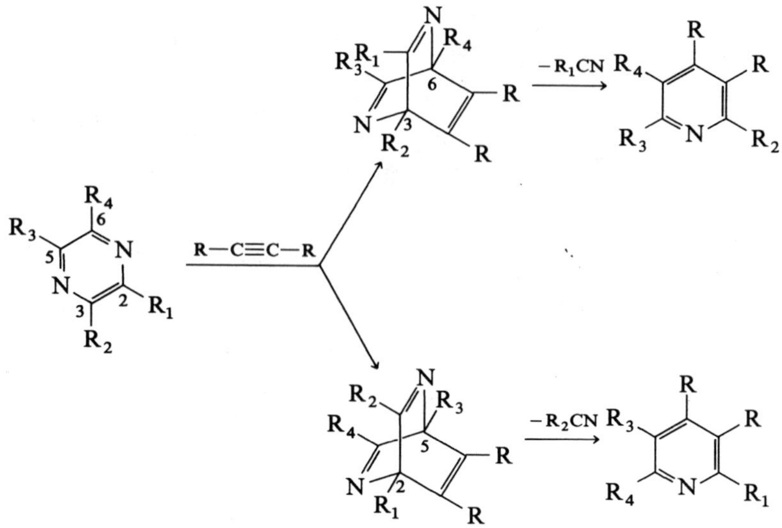

Synthesis of Pyridones

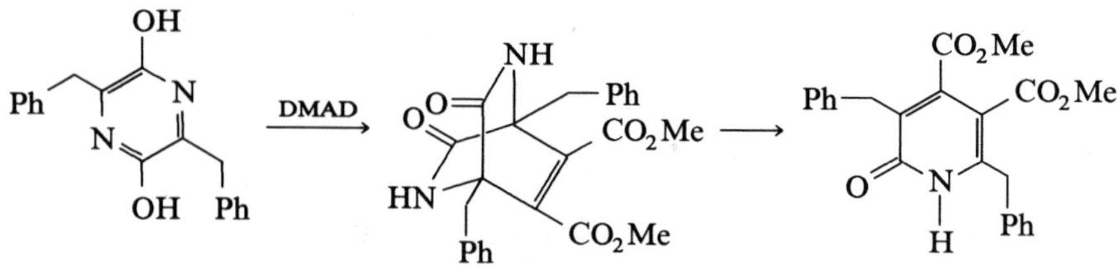

Scheme 6.13 1,4-Diazines (Pyridazines). (From Ref. 237.)

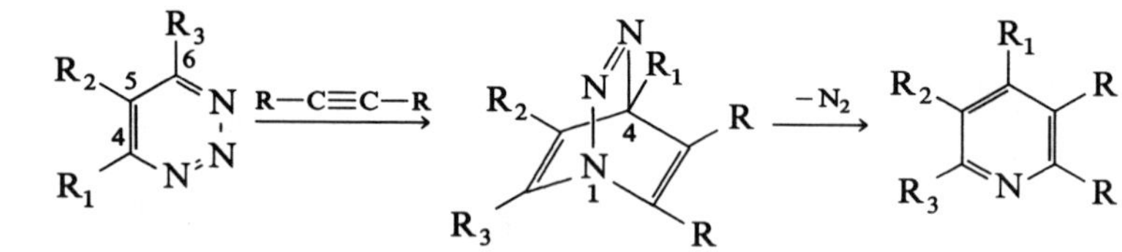

Synthesis of Fabianine[286, 287]

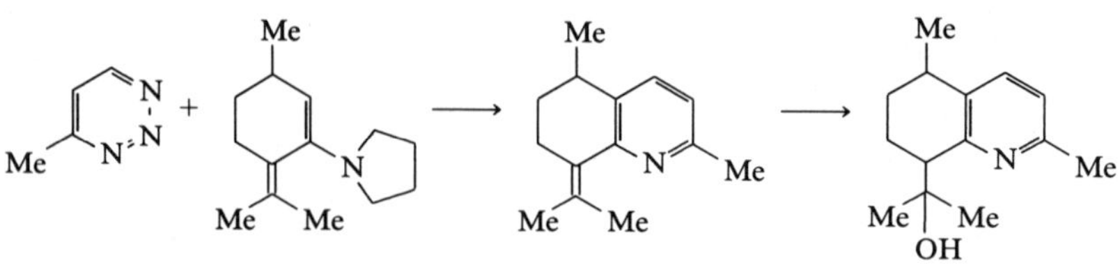

Scheme 6.14 1,2,3-Triazines. (From Refs. 238, 286, and 287.)

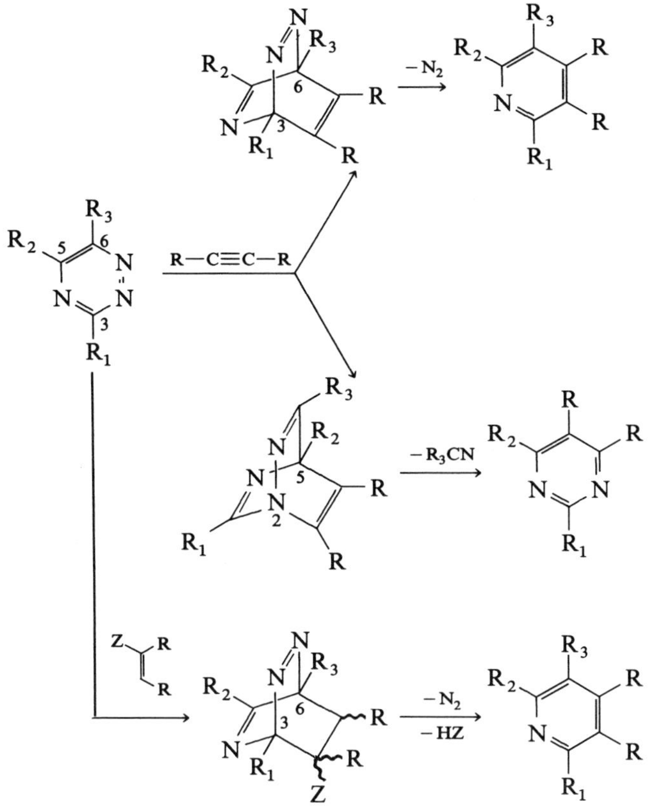

Synthesis of Lavendamicin[282, 285]

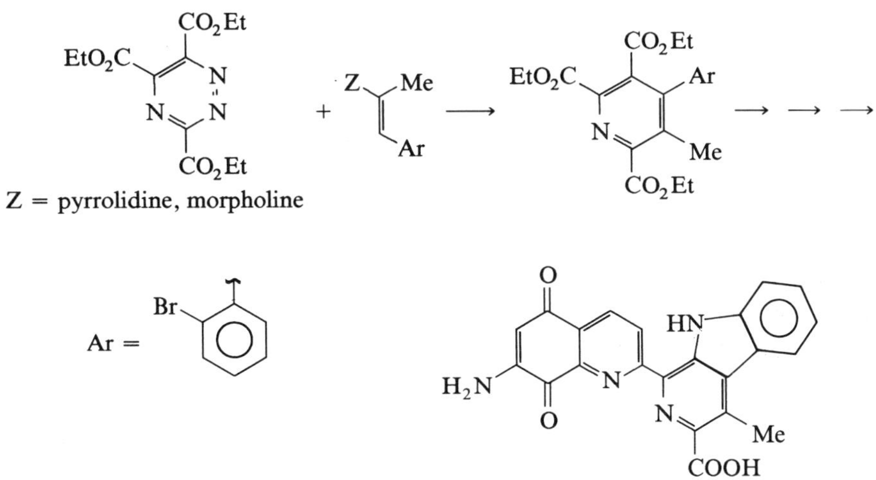

Scheme 6.15 1,2,4-Triazines. (From Refs. 247 and 285.)

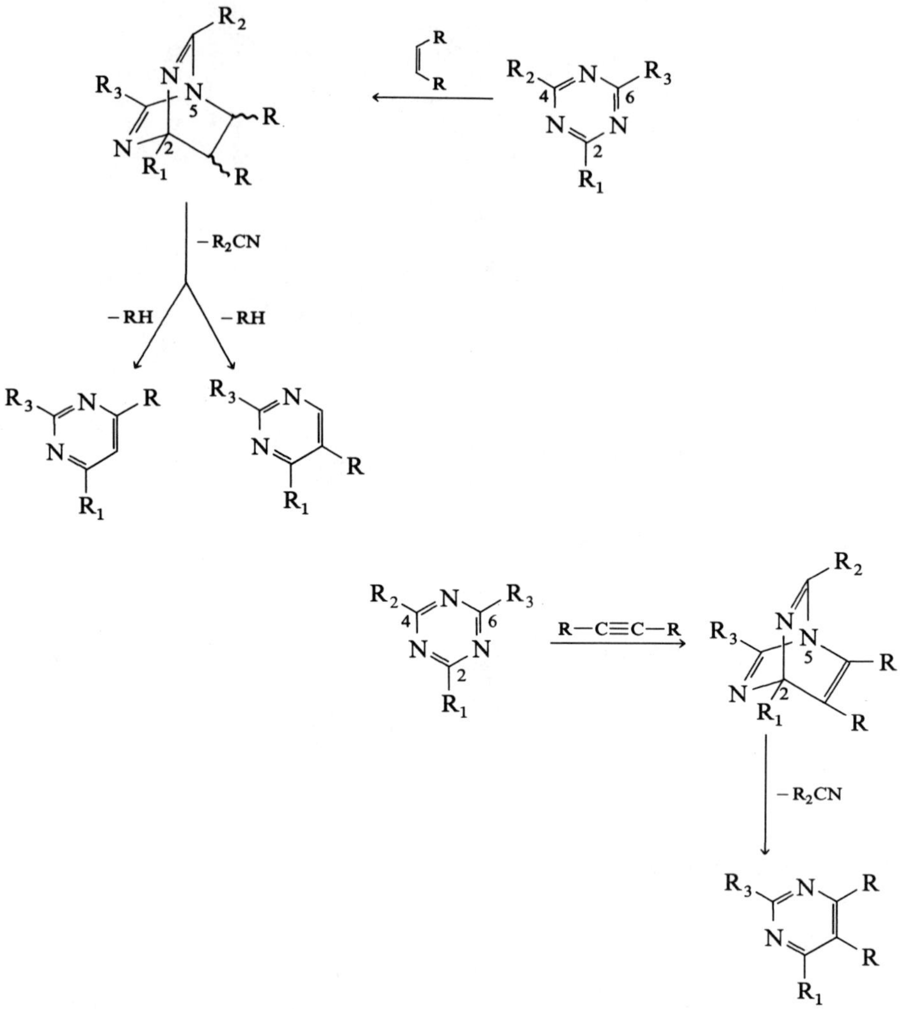

Scheme 6.16 1,3,5-Triazines. (From Ref. 293.)

The azines have been used as dienes in the synthesis of streptonigrin,[239] lavendamicin,[282,285] octamethylporphirin,[259,288] prodigiosin,[289] fabianine,[286,287] PDE-I,[254b] PDE-II,[254] and pyridazine-substituted pyrimidines nucleosides.[243] Some of these syntheses are illustrated in Schemes 6.14 to 6.17.

Tables of Diels–Alder Reactions See Tables 6.54 to 6.62; see also the general remarks about classifications, abbreviations, and symbols in the Introduction.

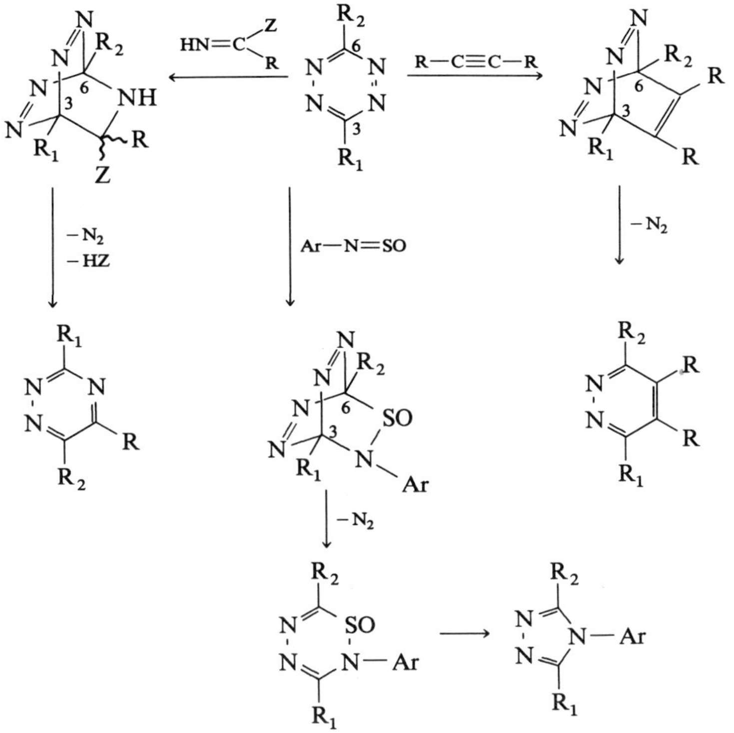

Synthesis of PDE-II[254]

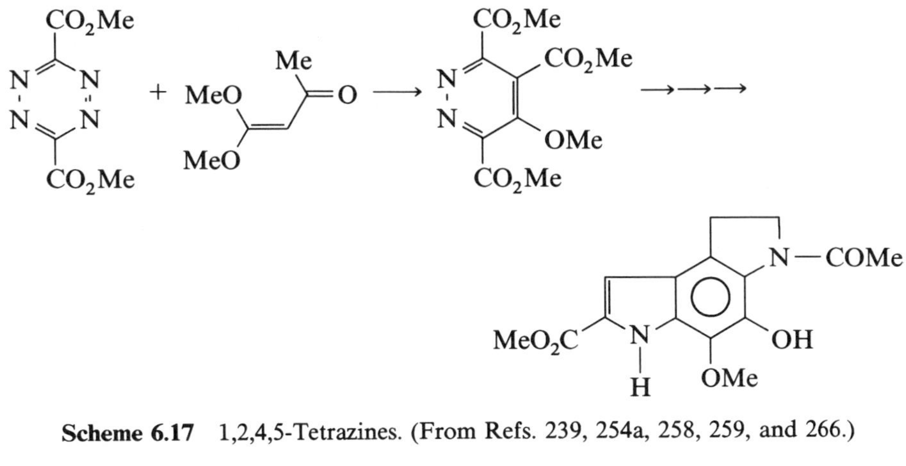

Scheme 6.17 1,2,4,5-Tetrazines. (From Refs. 239, 254a, 258, 259, and 266.)

TABLE 6.54 Diels–Alder Reactions of *N*-Substituted 1,2-Dihydropyridines

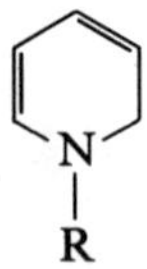

R		Dienophile	Yield (%)	Ref.
Me	A1	MAC	25	153
	A4	Ethyl-α-(*N*-methylindol-2-yl) acrylate	37	153
	A4	Methyl methacrylate	34	153
CO_2Me	A1	AC	70	154
	J1	[Nitrosocarbonylbenzene]	84	155
	J1	Nitrosoformates	25–41	155
	J1	Nitrosobenzene	67–86	156
CO_2Et	A4	Alkyl-2-(1-benzensulfonyl-indol-2-yl)acrylates	42–67	157, 225
$CO_2CH_2CCl_3$	A4	Alkyl-2-(1-benzensulfonyl-indol-2-yl)acrylates	31	158
CO_2Bu^t	A4	Alkyl-2-(1-benzensulfonyl-indol-2-yl)acrylates	65	158
CO_2Bn	A4	Alkyl-2-(1-benzensulfonyl-indol-2-yl)acrylates	73	158
	J1	Nitrosobenzene	75	156d

TABLE 6.55 Diels–Alder Reactions of 1,2-Dihydropyridines and Dihydroisoquinolines

Diene		Dienophile	Yield (%)	Ref.
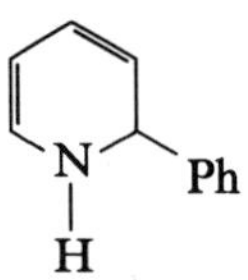	J2	*N*-Sulfinylcar-bomethoxyamine	81	159
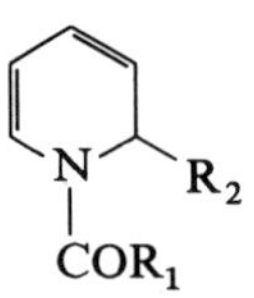	D3	MA	73–90	159, 161
	D5	Maleimides	50–90	161
	J1	Nitrosobenzene	42–83	156c, d
	J1	[Nitrosocarbonyl-benzene]	55–70	155
(R_1 = Me, OMe; R_2 = Me, Bu, Ph)	J1	Nitroso formates	24–44	155

TABLE 6.55 (*Continued*)

Diene		Dienophile	Yield (%)	Ref.
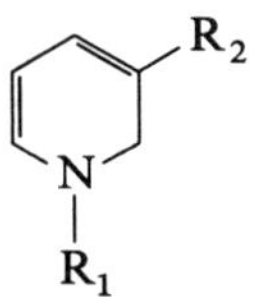 (R_1 = CO_2Et, CO_2Bn, Bn; R_2 = Me, Et, OMe, $SiMe_3$, CN)	A1	MAC	37–85	162
	A1	Ethyl acrylate		163
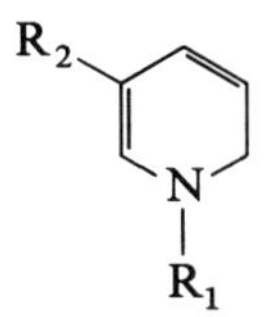 (R_1 = Me, Ph, Bn, CO_2Me; R_2 = Et, CN, CO_2NH_2, CO_2Me)	A1	Ethyl acrylate		163
	A9	Dimethylmethylene-malonate	68	165
	E3	DMAD		164
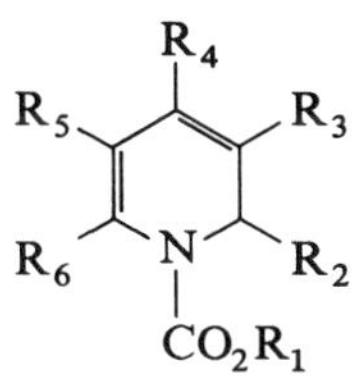 (R_1 = Et; R_2 = Me, Bu^n; R_3 = H, Me, Et; R_4 = H, Me; R_5 = H, Me, CO_2Me; R_6 = H, Me, Bn)	D5	NPM	25–94	166
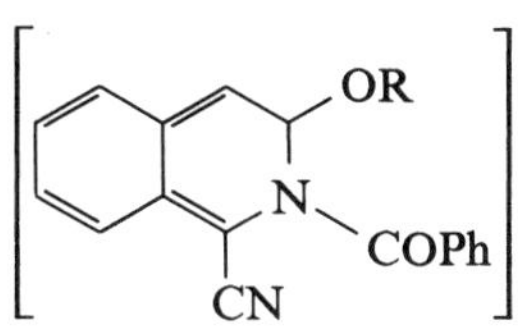 (R = Me, Et)	D5	NPM		167

TABLE 6.56 Diels–Alder Reactions of 2-Pyridones, 2-Thiopyridones, and 3-Isoquinolones

Diene			Dienophile		Yield (%)	Ref.
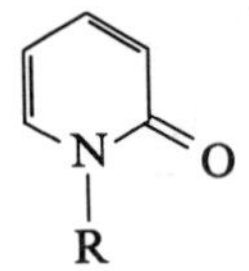	R = Me		A1	MAC	29	169
			A1	MVK	16	170
			A8	ACN	22	169
			E3	DMAD	12–16	171
			E6	[Benzyne] (k)		178b
	R = Bn		A1	MAC	17	172
	R = CH = CHPh		D3	MA	92	173
			E3	DMAD	22	173
			E6	[Benzyne]	25	175
			E6	[4,5-Dimethoxy-benzyne]	2	175
	R = CH = CHMe		A1	MVK	43	173
			E3	DMAD	21	173
	R = 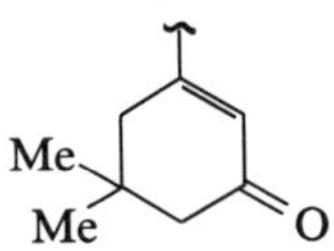		A1	MVK	41	173
			E3	DMAD	21	173, 174
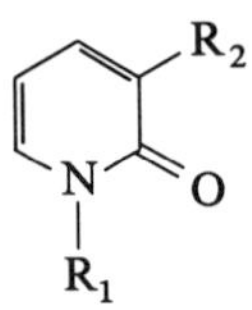	**R_1**	**R_2**				
	H	Me	E6	[Benzyne]	4	178
	Me	Me	E3	DMAD	0–40	171, 179
			E6	[Benzyne] (k)		178b
	Me	Me	E3	DMAD	0–40	171, 179
			E6	[Benzyne] (k)		178b
	Et	Me	D5	NPM		181
	Pr	Me	D5	NPM		181
	Bn	OBn	E6	[Benzyne]	10	175
			E6	[4,5-Dimethoxybenzyne]	9	175
	SO_2Ar	Ts	A8	Vinyl ethers	0–78	176
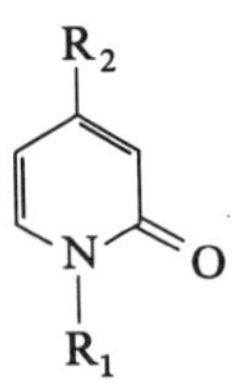	**R_1**	**R_2**				
	H	Me	E6	[Benzyne]	12	178
	Me	Me	E6	[Benzyne] (k)	10	178
	Me	Ph	E3	DMAD	0–22	171, 179
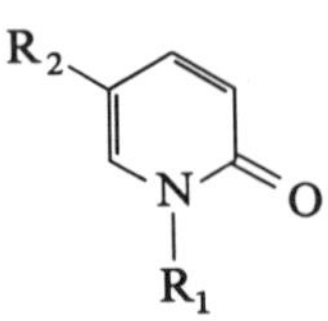 (R_1 = H, Me; R_2 = Me)			E6	[Benzyne] (k)	17	178

TABLE 6.56 (*Continued*)

Diene		Dienophile	Yield (%)	
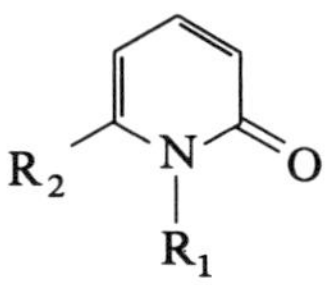	E3	DMAD	0–20	171, 179
	E6	[Benzyne] (k)	1–20	178
(R_1 = H, Me; R_2 = Me, OMe, Cl)				
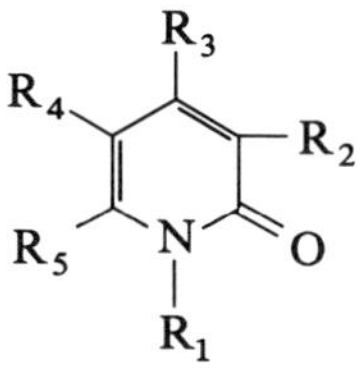	D5	NPM	18–55	177
	E3	DMAD	0–70	171, 179
	E6	[Benzyne] (k)	1–30	178
R_1 = H, Me, OMe; R_2 = Me, CN; R_3 = Me, Ph; R_4 = Me, Pr, Ph, Br; R_5 = Me; R_4, R_5 = —$(CH_2)_3$—				
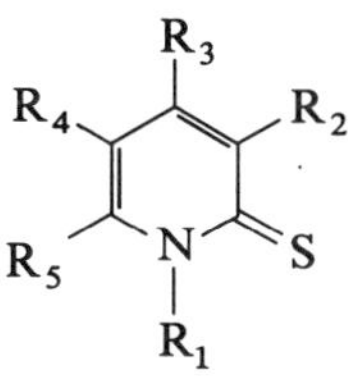	D5	NPM	30–63	181–183
R_1 = H, Me, Et, Pr; R_2 = H, Me; R_3 = H; R_4 = H, Me, Pr; R_5 = H, Me				
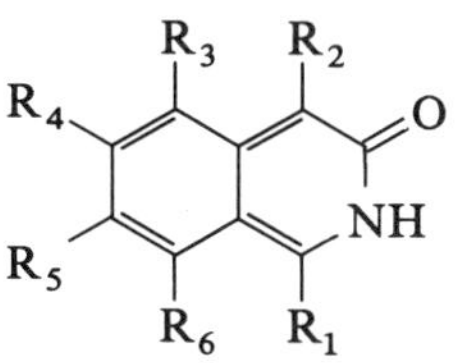	D3	MA	37–72	180
	D5	NPM	23–97	180
	D5	*N*-Carbamoyl-maleimide	85	180b
R_1 = Ph, Py, *p*-ClC_6H_4; R_2 = H, Me, Pr^i, Bn, $CH_2C_6H_4N$; R_3 = H, Me; R_4 = H, OMe; R_5 = H, Cl, OMe; R_6 = H; $R_3, R_4 = R_4, R_5 = R_5, R_6$ = —$(CH{=}CH)_2$—				

TABLE 6.57 Diels–Alder Reactions of 2-Pyrones and Related Compounds

Diene			Dienophile	Yield (%)	Ref.
			2-Pyrone		
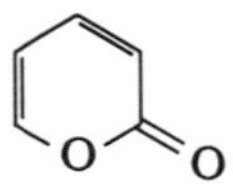		C7	BQ		186
		D5	NPM	85	187
		E5	Diphosphorilacetylenes	60–70	185
			Monosubstituted 2-Pyrones		
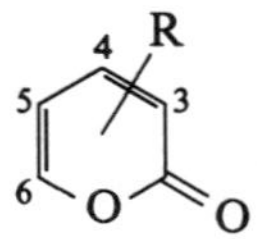	3R = OH	C9	1,4-Naphthoquinones	14–52	188
	3R = OMe	C9	5,8-Quinoxalinedione	80	189
		C9	2,1,3-Benzothiadiazole-4,7-dione	60	189
	3R = Ts	A8	Vinyl ethers (*)	75–95	190
	3R = Br	D5	NPM	85	187
	5R = CO_2H	A14	1,3-Butadiene	100	192
	5R = CO_2Me	D1	Furans	38–52	73
		D5	NPM	95	187
		E4	Ethynyltrichlorosilane	90	191
	6R = Me	D5	NPM	99	187
	6R = Ph	D5	NPM	97	187
	6R = pBr—C_6H_4	D5	NPM	95	187
			Disubstituted 2-Pyrones		
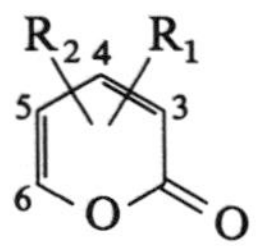	$3R_1$ = Cl, $5R_2$ = Cl	C2	Benzvalene	56	193
	$4R_1$ = Me, $6R_2$ = Me	D5	NPM	95	187
	$4R_1$ = CO_2Me, $6R_2$ = CF_3	A8	Vinyl acetate	71	194
		C1	Cyclopentene	92	194
		C1	Cyclooctene	87	194

		E5	Acetylenes	39–91	194
	$4R_1$ = OMe, $6R_2$ = OMe	D3	MA	63	195
		D5	NPM	34	195
	$4R_1$ = $OSiMe_3$, $6R_2$ = $OSiMe_3$	E3	DMAD	48	195
	$4R_1$ = OAc, $6R_2$ = OMe	E3	DMAD	73	195
	$5R_1$, $6R_2$ = —$Ar(CH_2)_2$—	A9	1,1-Dimethoxyethene	75	203
			Trisubstituted 2-Pyrones		
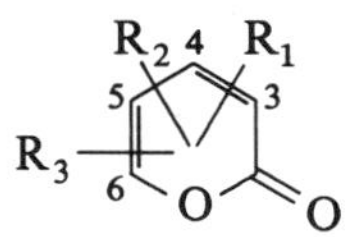	$3R_1$ = Cl, $4R_2$ = Cl, $5R_3$ = Cl	C2	Benzvalene		193
	$3R_1$ = CO_2Me, $5R_2$,	A8	*N*-Vinyl-2-pyrrolidinone	98	196, 204
	$6R_3$ = —$(CH_2)_4$—	A8	Ethyl vinyl ether	51	204
		A9	1,1-Dimethoxyethene	50–86	196, 201
		A11	1,1,2-Trimethoxyethene	57	196
		D2	Vinylene carbonate	83	196, 204
		E5	Acetylenes	43–57	204
	$3R_1$ = CO_2Me; $5R_2$, $6R_3$ = —$Ar(CH_2)_2$—	A9	1,1-Dimethoxyethene	73–78	204
	$3R_1$ = CO_2Me; $5R_2$, $6R_2$ = —$C_5H_8CH_2$—	A9	1,1-Dimethoxyethene	90	204
	$4R_1$ = Me, $5R_2$ = CO_2Et, $6R_3$ = Me	A11	Ethenemorpholines	45–55	197
		D3	MA		199
		E5	But-2,3-yn-1,4-diolo		198
	$4R_1$ = Me, $5R_2$ = CO_2Me, $6R_3$ = Me	C1	Cycloalkenylmorpholines	41–55	197
	$4R_1$ = Me, $5R_2$ = Ac, $6R_3$ = Me	E1	MP	84	200
		E3	DMAD	93	200

TABLE 6.57 (*Continued*)

Diene		Dienophile	Yield (%)	Ref.
		Trisubstituted 2-Pyrones		
$4R_1$ = Me; $5R_2$, $6R_3$ = —$COCH_2CMe_2O$—	E1	MP	100	202
	E3	DMAD	76	202
	E4	Trimethylsilylacetylene	100	202
	E5	Diphenylacetylene	78	202
		Tetrasubstituted 2-Pyrones		
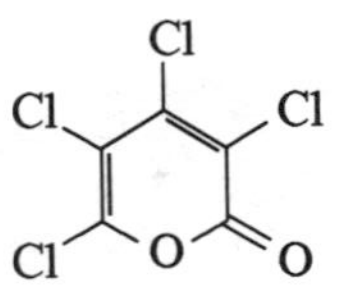	C2	Benzvalene	82	193
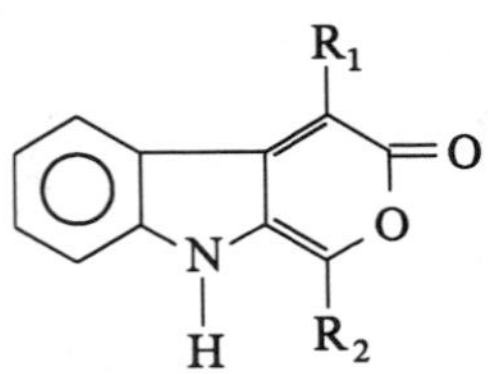 (R_1 = H, Me; R_2 = Me)	E3	DMAD	67–81	205
	E3	Dibenzoylacetylene	52–76	205
	E6	[Benzyne]	38–44	205
	E7	[3,4-Didehydropyridine]	20	205b

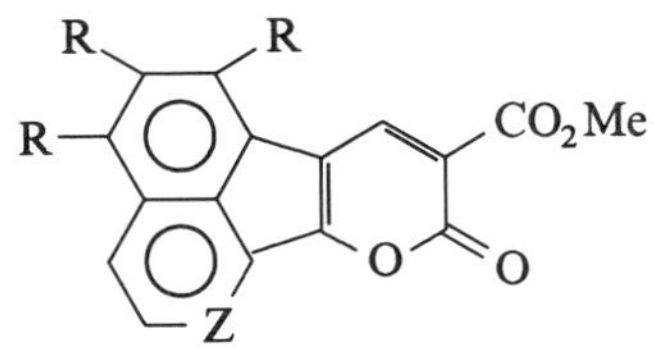 (Z = CH, N; R = H, OMe)	A1 A9	1,1,2-Trimethoxyethene 1,1-Dimethoxyethene	98 83	206 204, 206
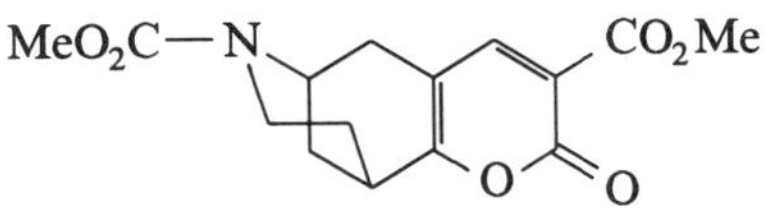	A9	1,1-Dimethoxyethene	80	204
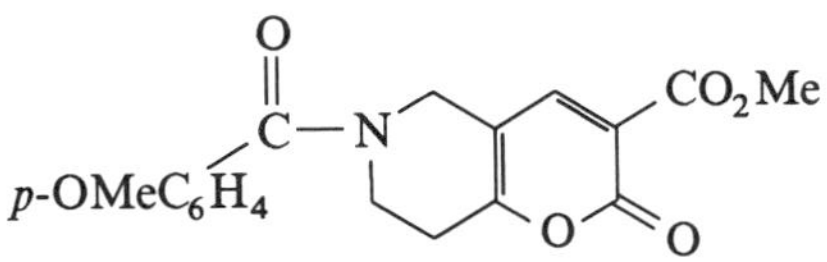	A9	1,1-Dimethoxyethene	91	204
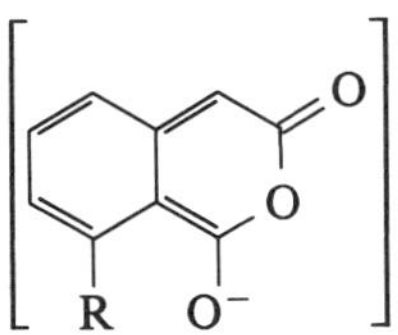 (R = H, OMe)	C9 E1 E3 E3	NQ MP DMAD Diethylacetylene dicarboxylate	68–88 49 79–94 96	221 221 221 221

TABLE 6.58 Diels–Alder Reactions of Diazines and Triazines

Diene	R_1	R_2		Dienophile	Yield (%)	Ref.
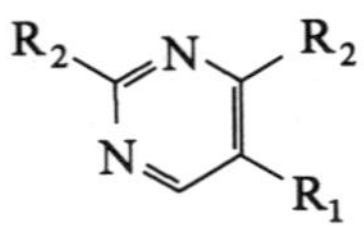	H	CN	E5	1-Diethylamino-1-propyne	97	236
			E5	Diethylaminophenylacetylene	20	236
	Me	CN	E5	1-Diethylamino-1-propyne	36	236
	NO_2	H	A11	1,1-Dimorpholinoethene		235b
			A11	Ketene-*N*,*N*-acetals	52–57	235a
			A11	Ketene-*O*,*O*-acetals	17–80	235a
			C1	Pyrrolidinoenamine	60	235a
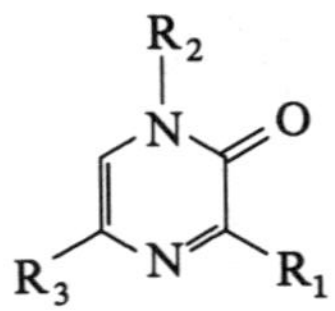						
R_1 = CN, OMe, Cl;			E1	Ethyl propynoate	69–95	237
R_2 = Me, Ph;			E3	DMAD	59–92	237
R_3 = H, Cl			E4	Phenylacetylene	36–99	237
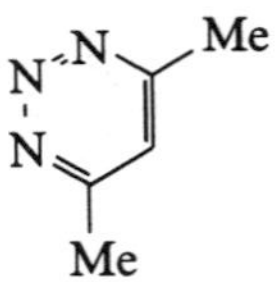			A8	Phenylvinylsulfone	35	238
			E4	Phenylacetylene	38	238
			E5	1-Diethylamino-1-propyne	50	238

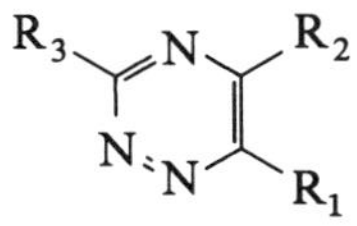

R_1	R_2	R_3				
H	H	H	A11	Pyrrolidinoenamines	22–78	241
H	H	CO_2Et	A11	Pyrrolidinoenamines	0–49	239, 241b
			C3	Benzocyclopropene	33	240
—$(CH_2)_3$—		Me	C1	4-(1-Cyclohexen-1-yl) morpholine	5	245
—$(CH_2)_3$—		Ph	C1	1-(1-Cyclopenten-1-yl) pyrrolidine	88	245
—$(CH_2)_3$—		CO_2Me	C3	1-(6-Methoxy-3,4-dihydro-naphthyl) pyrrolidines	21–35	245
Ph	Ph	Me	F7	Benzamidine	66	264
Ph	Ph	α-Py	F7	Benzamidine	63	264
α-Py	Ph	Ph	F7	Benzamidine	66	264
CO_2Me	CO_2Me	CN	D1	Oxepin	16	244
CO_2Et	H	H	A11	[Pyrrolidinoenamine]	19	241b
CO_2Et	CO_2Et	H	C3	Benzocyclopropene	41	240
CO_2Et	CO_2Et	CN	C3	Benzocyclopropene	70	240
CO_2Et	CO_2Et	CO_2Et	A11	Pyrrolidinoenamine	0–79	239
			C3	Benzocyclopropene	67	240
			D4	[Lactam acetals]	89–98	253
			E4	5-Ethynyl-2′-deoxyuridine	30	243
F	F	F	C1	Cyclopentene	55	246
			C1	Cyclooctene	13	246
			C2	Bicyclo[2.2.1]hept-2-ene	52	246
			E5	Bis(trimethylstannyl)acetylene	11	246
SBn	Me	H	C1	1-Pyrrolidino-1-cyclopentene	95	242
SBn	Ph	H	C1	1-Pyrrolidino-1-cyclopentene		242
SOBn	Me	H	C1	1-Pyrrolidino-1-cyclopentene	20–88	242
SOBn	Pr^i	H	C1	1-Pyrrolidino-1-cyclopentene		242
SOBn	Ph	H	C1	1-Pyrrolidino-1-cyclopentene		242
SO_2Bn	Me	H	C1	1-Pyrrolidino-1-cyclopentene	66	242
SO_2Bn	Ph	H	C1	1-Pyrrolidino-1-cyclopentene		242

TABLE 6.58 (*Continued*)

Diene		Dienophile	Yield (%)	Ref.
Cl Cl Cl	A8	Ethene	1	247, 250
	A10	(*Z*)-But-2-ene	75	247
	C1	Cyclopentene	77	247, 250
	C1	(*Z*)-Cyclooctene	80	247, 250
	C2	Bicyclo[2.2.1]hept-2-ene	95	247, 250
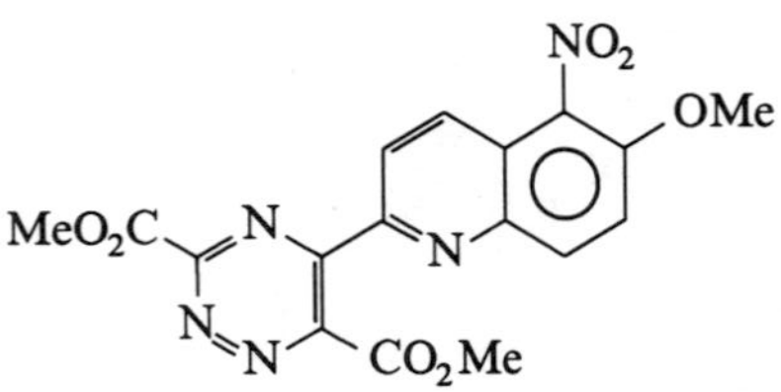	A11	Enamines	65–68	239

TABLE 6.59 Diels–Alder Reactions of Tetrazines

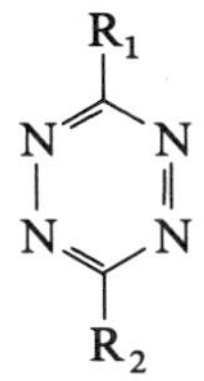

R_1	R_2		Dienophile	Yield (%)	Ref.
H	Ar	A10	(*Z*)-1-Nitro-2-dimethylaminoethene	32–46	266
		A10	(*Z*)-1-Cyano-2-morpholinoethene	36–60	266
Ph	Me	F6	Benzamidine	77	264
Ph	α-Py	F6	Benzamidine	71	264
Ph	Ph	D4	[Lactamacetals]	90–98	253
		D4	Δ1-Pyrrolinketimines	59–78	257
		D4	3*H*-Indolketimine	23	257
		E6	Cyclooctine	84	262
		F6	Amidines	22–89	263, 264
α-Py	Me	F6	Benzamidine	35	264
α-Py	α-Py	A10	(*Z*)-1-Nitro-2-dimethylaminoethene	34	266
		A10	(*Z*)-1-Cyano-2-morpholinoethene	60	266
		C2	Octahydrodimethanonaphthalene (k)		265
		F6	Benzamidine	38	264
γ-Py	γ-Py	A10	(*Z*)-1-Nitro-2-dimethylaminoethene	32	266
		A10	(*Z*)-1-Cyano-2-morpholinoethene	56	266
CO_2Me	CO_2Me	A5	4,4-Dimethoxybut-3-en-2-one	71	254
		A9	1,1-Dimethoxyethene	65	259
		A10	Crotonaldehyde-*N*,*N*-dimethylhydrazone	75	261
		A10	Cinammaldehyde-*N*,*N*-dimethylhydrazone	95	261
		A11	2-[(Triethylsilyl)oxy]-2-butene	87	259
		C1	Enamines	69–85	259
		C3	1-(6-Methoxy-3,4-dihydro-2-naphthyl)pyrrolidine	88	245
		D1	Oxepin	34	244

TABLE 6.59 (*Continued*)

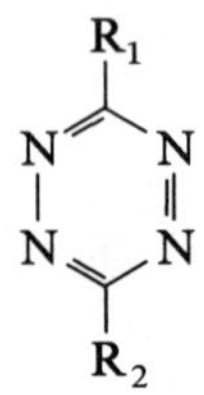

R_1	R_2		Dienophile	Yield (%)	Ref.
		D1	3,4,6-Tri-*O*-acetyl-1,5-anhydro-2-deoxy-D-arabino-hex-1-enitol (*)	76	251
		D1	2-Methoxyfuran		256
		D4	Δ1-Pyrrolinketimine	46–52	257
		D4	3*H*-Indolketimine	75	257
		D4	Indoles	30–80	258
		D8	5-Ethynyl-2′-deoxyuridine	71	243
		D8	*N*-Methylimidazole		256
		E2	4-(Trimethylsilyl)-3-butyn-2-one	56	252
		E2	1-(Trimethylsilyl)-1-pentyn-3-one	49	252
		E4	Benzyloxyacetylene	82	259
		E5	2-Butyne	Traces	259
		E5	Silylalkynes	25–69	252
		F1	Thioformamides		248
		F1	[Thiobenzaldehyde]		248
		F7	Amidines	27–82	239, 255
		G4	Oxazolines	30–95	249
		G4	Pyrazoline	90	249
		H1	*N*-Cyanoamines		260
CF_3	CF_3	F1	Thioformamides		248
		F1	[Thiobenzaldehyde]		248
		G4	Oxazolines	35–45	249
		G4	Thiazoline	46	249
α-Th	α-Th	D4	3*H*-Indolketimine	26	257

TABLE 6.60 Diels–Alder Reactions of Miscellaneous Six-Membered Heterocyclic Dienes Having One Heteroatom

Diene		Dienophile	Yield (%)	Ref.
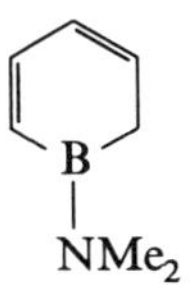	E6	Hexafluoroacetylene		152
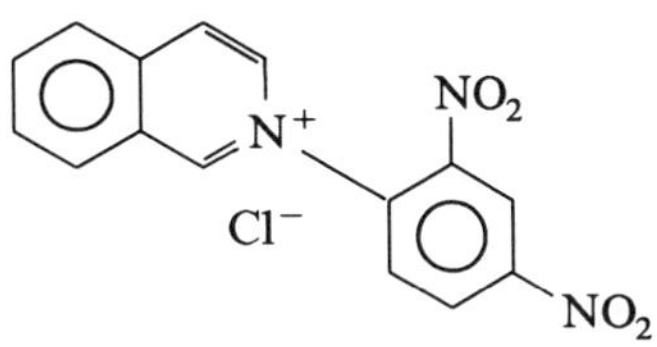	A9	Vinyl ethers	45–95	267
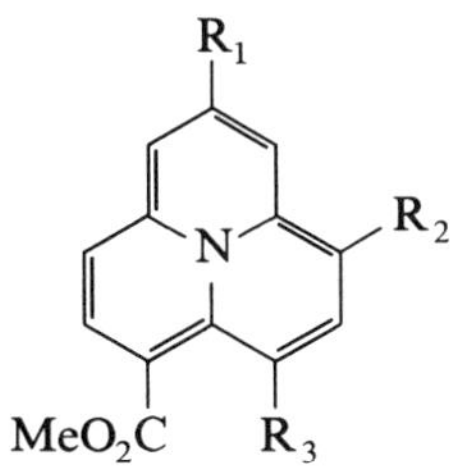 (R_1 = H, Me; R_2 = R_3 = CN, CO_2Me)	E1	MP		151
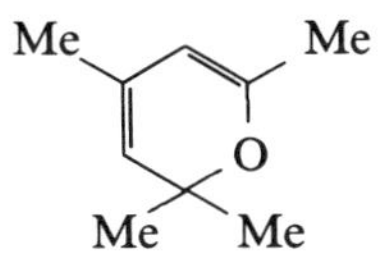	E3	DMAD		168
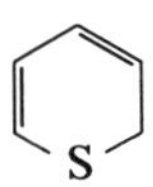	A1	MAC	32	272
	D3	MA	19	272
	D5	NPM	53	272
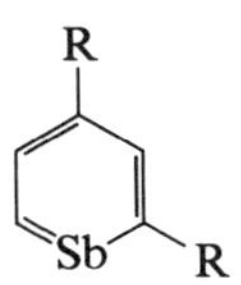 (R = H, Me)	D14	Stibabenzene		207
	E6	Hexafluorobutyne		207
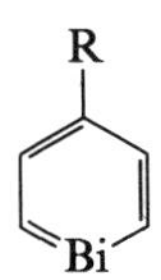 (R = H, Me)	D15	Bismabenzene		207
	E6	Hexafluorobutyne		207

TABLE 6.61 Diels–Alder Reactions of Miscellaneous Six-Membered Heterocyclic Dienes Having Two Heteroatoms

Diene		Dienophile	Yield (%)	Ref.
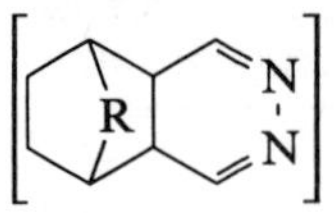 ($R = CH_2, (CH_2)_2, CH{=}CH$)	C1	Cyclopentadiene	53–66	224
	C2	Norbornene	39–60	224
	C2	Norbornadiene	7–28	224
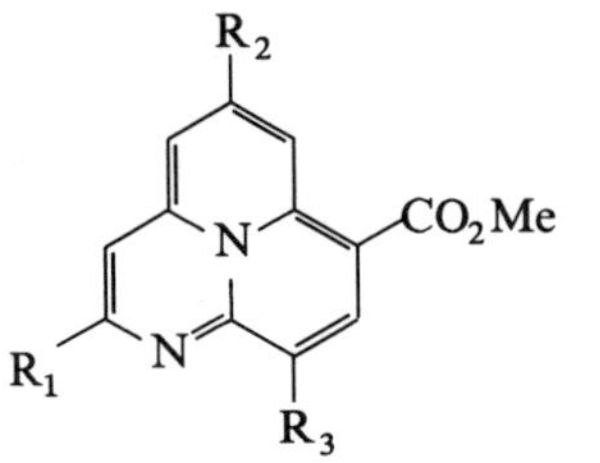 (R_1 = H, Me; R_2 = H, Me; R_3 = CN, CO_2Et)	D5	NPM	80	151
	E1	MP	40–60	151
	E3	DMAD	80	151
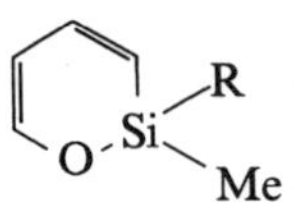 R = Me	A1	MAC		226
	A8	Methoxyvinyltungsten complex		226
	D3	MA	80	168
	E5	Perfluoro-2-butyne	72	168
R = $SiMe_3$	D3	MA	33	168
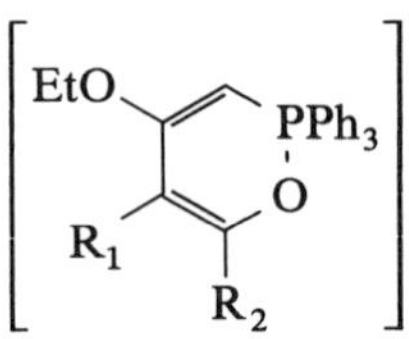 (R_1 = H, Me, Ph; R_2 = Pr^i, OEt, Ph, NHMe, NHPh, p-$NO_2C_6H_4$)	D5	NMM	30–95	228
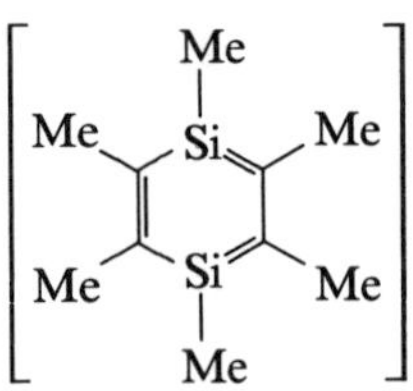	E4	Acetylene		227
	E5	Perfluoro-2-butyne	88	227
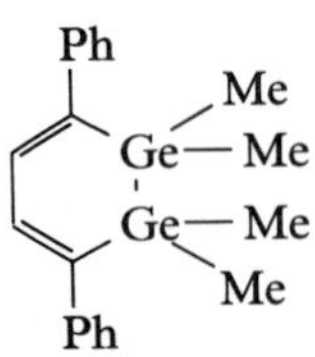	E6	[Benzyne]	46	229

TABLE 6.62 Diels–Alder Reactions of Miscellaneous Heterocyclic Dienes

Diene		Dienophile	Yield (%)	Ref.
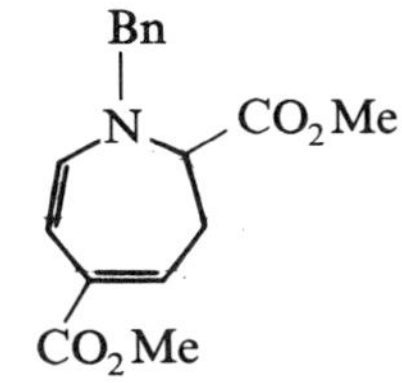	E3	DMAD	35–100	208
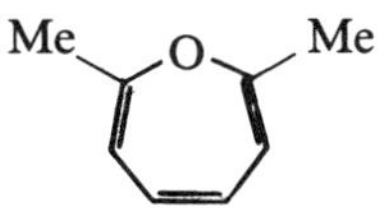	I1	Dimethylazo-dicarboxylate	60	210
	I1	Bis(trichloroethyl)-azodicarboxylate	92	210
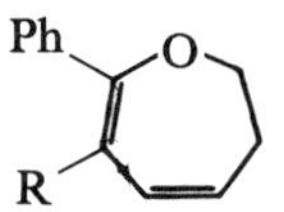 (R = CO_2Me, Ar)	I5	NPTAD		209
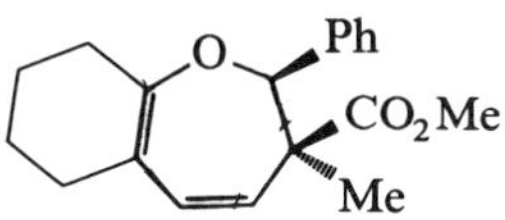	I5	NPTAD	100	222
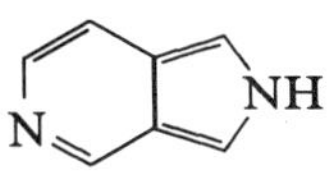	D5	NPM	96	270
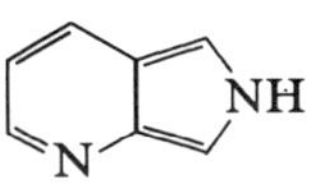	D5	NPM	64	270
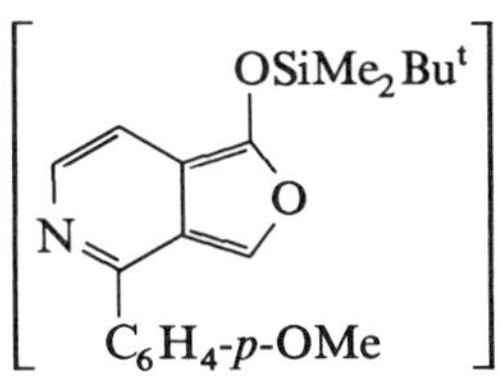	A7	DMF	38	145
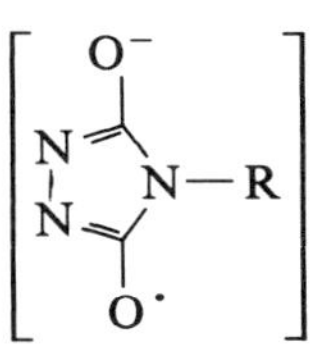 (R = Me, Ph, Bn)	I5	1,2,4-Triazolinediones	35–100	230

TABLE 6.62 (*Continued*)

Diene		Dienophile	Yield (%)	Ref.
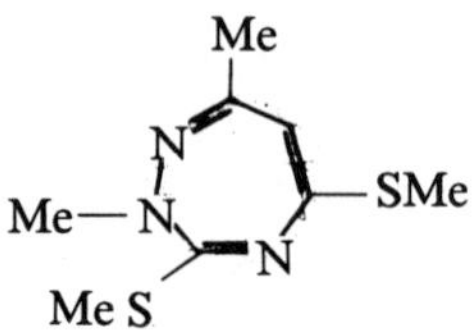	D3	MA	> 60	234
	E3	DMAD	> 60	234
	I5	NPTAD	85	234
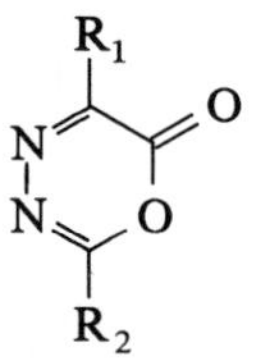	A10	(*E*)-1-Propenyl-pyrrolidine	49	232
	A11	2-Methylpropenyl-pyrrolidine	55	231
	C2	Benzvalene	75	233
(R_1 = Me, Ph; R_2 = Ph, Bn)	D1	2,3-Dihydrofuran	42–54	232
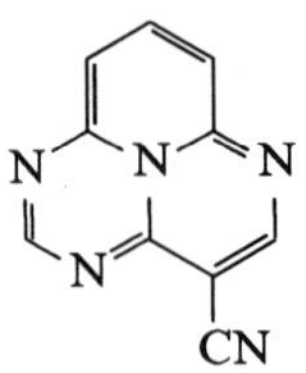	E3	DMAD	20	151
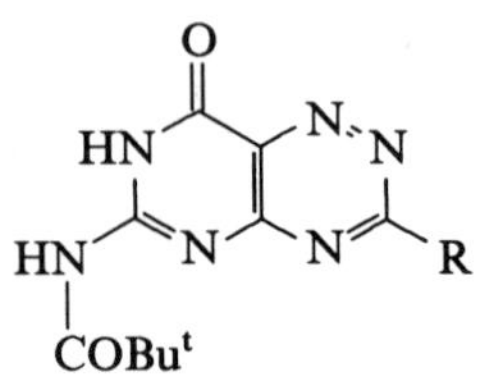 (R = p-MeC_6H_4, SMe)	C1	1-Morpholino-cycloalkenes	3–57	271
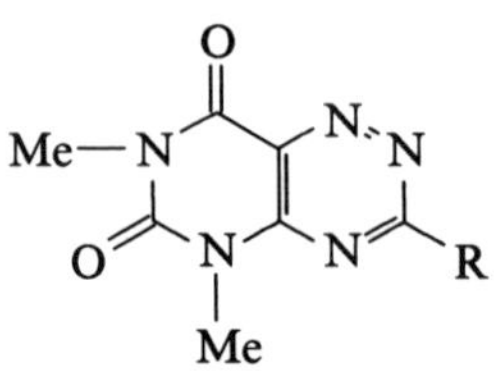 (R = H, SMe, p-MeC_6H_4)	C1	1-Morpholino-cycloalkenes	12–74	271

REFERENCES FOR SECTION 6.3

1. Y. Kobayashi, M. Honda, Y. Hanzawa, I. Kumadaki, and A. Ohsawa, *J. Chem. Soc. Perkin Trans. 1*, 1743 (1979).
2. G. W. Gribble, M. P. Sibi, S. Kumar, and W. J. Kelly, *Synthesis*, 502 (1983).
3. C. S. Le Houllier and G. W. Gribble, *J. Org. Chem.*, **48**, 2364 (1983).
4. N. Kotsuki, Y. Mori, H. Nishizawa, M. Ochi, and K. Matsuoka, *Heterocycles*, **19**, 1915 (1982).
5. M. G. B. Drew, A. V. George, N. S. Isaacs, and H. S. Rzepa, *J. Chem. Soc. Perkin Trans. 1*, 1277 (1985).
6. A. G. Schultz, M. Shen, and R. Ravichandran, *Tetrahedron Lett.*, **22**, 1767 (1981).
7. C.-K. Sha, K.-S. Chuang, and J.-J. Young, *J. Chem. Soc. Chem. Commun.*, 1552 (1984).
8. A. G. Schultz and M. Shen, *Tetrahedron Lett.*, 2969 (1979).
9. a. A. G. Schultz and M. Shen, *Tetrahedron Lett.*, **22**, 1775 (1981).
   b. B. A. Johnsen and K. Undheim, *Acta Chem. Scand., Ser. B*, **B 37**, 907 (1983).
10. a. Y. A. Ragab and G. B. Butler, *J. Polym. Sci. Polym. Chem. Ed.*, **19**, 1175 (1981); *Chem. Abstr.*, **95**, 7941c (1981).
    b. D. P. Dolata and R. Bergman, *Tetrahedron Lett.*, **28**, 707 (1987).
    c. H. K. Hall, Jr., P. Nogues, J. W. Rhoades, R. C. Sentman, and M. Detar, *J. Org. Chem.*, **47**, 1451 (1982).
    d. M. J. S. Dewar and A. B. Pierini, *J. Am. Chem. Soc.*, **106**, 203 (1984).
    e. B. G. Sparks and B. E. Poling, *AIChE J.*, **29**, 534 (1983); *Chem. Abstr.*, **99**, 143117y (1983).
    f. K. J. Daumn, *Energy Res. Abstr.*, **9**(18) (1984), Abstr. 35908; *Chem. Abstr.*, **102**, 23888j (1985).
    g. B. E. Poling, *Energy Res. Abstr.*, **9**(16) (1984), Abstr. 30974; *Chem. Abstr.*, **102**, 9698r (1985).
11. a. E. L. Clennan and M. E. Mehrsheikh-Mohammadi, *J. Am. Chem. Soc.*, **106**, 7112 (1984).
    b. E. Fiedler, *Ber. Max-Planck Inst. Stromungsforsch.*, (1983); *Chem. Abstr.*, **100**, 102561j (1984).
12. H. Kotsuki, H. Nishizawa, M. Ochi, and K. Matsuoka, *Bull. Chem. Soc. Jpn.*, **55**, 496 (1982).
14. a. H. Kotsuki, K. Asao, and H. Ohnishi, *Bull. Chem. Soc. Jpn.*, **57**, 3339 (1984).
    b. F. Brion, *Tetrahedron Lett.*, **23**, 5299 (1982).
    c. P. Deshong and N. E. Lowmaster, *Synth. Commun.*, **13**, 537 (1983).
14. W. A. Rendall, M. Torres, and O. P. Strausz, *J. Am. Chem. Soc.*, **107**, 723 (1985).
15. a. J. Moursounidis and D. Wege, *Aust. J. Chem.*, **36**, 2473 (1983).
    b. E. Vieira and P. Vogel, *Helv. Chim. Acta*, **65**, 1700 (1982).
    c. P. F. Schuda and J. M. Bennett, *Tetrahedron Lett.*, **23**, 5525 (1982).
    d. H. Kotsuki and H. Nishizawa, *Heterocycles*, **16**, 1287 (1981).
16. a. M. Oda and Y. Kanao, *Chem. Lett.*, 1547 (1981).
    b. J. Jurczak, T. Kożluk, S. Filipek, and C. H. Eugster, *Helv. Chim. Acta*, **66**, 222 (1983).
17. C. L. D. Jennings-White, A. B. Holmes, and P. R. Raithby, *J. Chem. Soc. Chem. Commun.*, 542 (1979).

18. M. Shen and A. G. Schultz, *Tetrahedron Lett.*, **22**, 3347 (1981).
19. G. Maier and W. A. Jung, *Tetrahedron Lett.*, **21**, 3875 (1980).
20. M. Bertrand, J.-L. Gras, and B. S. Galledou, *Tetrahedron Lett.*, 2873 (1978).
21. W. Ando, Y. Hanyu, T. Takata, T. Sakurai, and K. Kobayashi, *Tetrahedron Lett.*, **25**, 1483 (1984).
22. a. W. A. Rendall, M. Torres, and O. P. Strausz, *J. Org. Chem.*, **50**, 3034 (1985).
    b. Y. Apeloig, D. Arad, M. Kapon, and M. Wallerstein, *Tetrahedron Lett.*, **28**, 5917 (1987).
    c. R. Bloch and J.-M. Denis, *Angew. Chem. (Int. Ed. Engl.)*, **19**, 928 (1980).
    d. Y. Kobayashi, T. Yoshida, Y. Hanzawa, and I. Kumadaki, *Tetrahedron Lett.*, **21**, 4601 (1980).
23. O. De Lucchi, V. Lucchini, L. Pasquato, and G. Modena, *J. Org. Chem.*, **49**, 596 (1984).
24. R. G. Hall and S. Trippett, *Tetrahedron Lett.*, **23**, 2603 (1982).
25. W. G. Dauben, J. M. Gerdes, and D. B. Smith, *J. Org. Chem.*, **50**, 2576 (1985).
26. A. Vila Casas, *Span. ES*, 517,063 (1983); *Chem. Abstr.*, **103**, 141822b (1985).
27. M. G. Barlow, S. Culshaw, R. N. Haszeldine, and W. D. Morton, *J. Chem. Soc. Perkin Trans. 1*, 2105 (1982).
28. A. P. Kozikowski and A. Ames, *J. Am. Chem. Soc.*, **103**, 3923 (1981).
29. A. J. Guildford and R. W. Turner, *J. Chem. Soc. Chem. Commun.*, 466 (1983).
30. H. Takayama, A. Iyobe, and T. Koizumi, *J. Chem. Soc. Chem. Commun.*, 771 (1986).
31. C. C. Wah, A. H. Fawcett, J. N. Mulemwa, and C. E. Tan, *Polymer*, **26**, 1268 (1985).
32. M. G. Barlow, R. Haszeldine, and C. J. Peck, *J. Chem. Soc. Chem. Commun.*, 158 (1980).
33. N. Z. Huang, *Chem. Ind. (London)*, 364 (1981).
34. a. H. O. House, M. B. DeTar, R. F. Sieloff, and D. Van Derveer, *J. Org. Chem.*, **45**, 3545 (1980).
    b. H. O. House, M. B. DeTar, and D. Van Derveer, *J. Org. Chem.*, **44**, 3793 (1979).
35. G. W. Gribble, R. B. Perni, and K. D. Onan, *J. Org. Chem.*, **50**, 2934 (1985).
36. J. A. Moore and E. M. Partain III, *J. Org. Chem.*, **48**, 1105 (1983).
37. H. Jendralla, *Angew. Chem. (Int. Ed. Engl.)*, **19**, 1032 (1980).
38. J. A. Baltrop, A. C. Day, and E. Irving, *J. Chem. Soc. Chem. Commun.*, 966 (1979).
39. W. R. Dolbier and C. R. Burkholder, *J. Org. Chem.*, **49**, 2381 (1984).
40. P. Laszlo and J. Lucchetti, *Tetrahedron Lett.*, **25**, 4387 (1984).
41. T. M. Balthazor, B. Gaede, D. E. Korte, and H.-S. Shieh, *J. Org. Chem.*, **49**, 4547 (1984).
42. L. L. Klein and T. M. Deeb, *Tetrahedron Lett.*, **26**, 3935 (1985).
43. J. Jurczak, S. Belmak, T. Kozluk, S. Pikul, and P. Salanski, *Bull. Pol. Acad. Sci. Chem.*, **32**, 135 (1984).
44. O. Baumgartel, J. Harnisch, G. Szeimies, M. Van Meerssche, G. Germain, and J.-P. Declercq, *Chem. Ber.*, **116**, 2205 (1983).

45. H. N. C. Wong and F. Sondheimer, *Tetrahedron*, **37**, Suppl. 1, 99 (1981).
46. T. Suzuki, S. Kagaya, A. Tomino, K. Unno, and T. Kametani, *J. Chem. Soc. Perkin Trans. 1*, 2801 (1980).
47. U. Szeimies-Seebach, A. Schoffer, R. Romer, and G. Szeimies, *Chem. Ber.*, **114**, 1767 (1981).
48. G. W. Gribble and J. W. Kelly, *Tetrahedron Lett.*, **22**, 2475 (1981).
49. J. Rimmelin, G. Jenner, and P. Rimmelin, *Bull. Soc. Chim. Fr.*, 461 (1978).
50. S. Takano, Y. Oshima, F. Ito, and K. Ogasawara, *Yakugaku Zasshi*, **100**, 1194 (1980); *Chem. Abstr.*, **95**, 24871x (1981).
51. A. Pelter and B. Singaram, *J. Chem. Soc. Perkin Trans. 1*, 1383 (1983).
52. a. S. Takano, F. Ito, and K. Ogasawara, *Heterocycles*, **14**, 453 (1980).
    b. T. Suzuki, A. Tomino, K. Unno, and T. Kametani, *Heterocycles*, **14**, 439 (1980).
53. H. Kotsuki, A. Kondo, H. Nishizawa, M. Ochi, and K. Matsuoka, *J. Org. Chem.*, **46**, 5454 (1981).
54. K. T. Potts and E. B. Walsh, *J. Org. Chem.*, **49**, 4099 (1984).
55. W. R. Dolbier, Jr., and C. R. Burkholder, *Tetrahedron Lett.*, **21**, 785 (1980).
56. G. A. Kraus and M. D. Hagen, *J. Org. Chem.*, **48**, 3265 (1983).
57. L. M. Gomes and J. Cabares, *C. R. Acad. Sci. Ser. C*, **290**, 29 (1980); *Chem. Abstr.*, **93**, 26178j (1980).
58. I. Maeba, F. Usami, and H. Furukawa, *J. Org. Chem.*, **49**, 1534 (1984).
59. A. V. R. Rao, B. M. Chanda, and A. V. Wadekar, *Indian J. Chem., Sect. B*, **20 B**, 248 (1981).
60. G. A. Kraus and S. H. Woo, *J. Org. Chem.*, **51**, 114 (1986).
61. D. Gravel, R. Deziel, F. Brisse, and L. Hechler, *Can. J. Chem.*, **59**, 2997 (1981).
62. a. N. P. Soloveva, Yu. N. Sheinker, A. F. Oleinik, and E. V. Adamskaya, *Khim. Geterotsikl. Soedin.*, 613 (1984).
    b. A. F. Aleinik, E. V. Adamskaya, K. Yu. Novitskii, N. P. Soloveva, N. I. Fadeeva, I. N. Degtyareva, N. B. Lapaeva, and G. N. Pershin, *Khim.-Farm. Zh.*, **14**, 49 (1980).
63. J. N. Bridson, *Can. J. Chem.*, **57**, 314 (1979).
64. M. M. Campbell, A. D. Kaye, and M. Sainsbury, *Tetrahedron*, **38**, 2783 (1982).
65. E. McDonald, A. Suksamrarn, and R. D. Wilye, *J. Chem. Soc. Perkin Trans. 1*, 1893 (1979).
66. A. Pelter, R. Al-Bayati, and W. Lewis, *Tetrahedron Lett.*, **23**, 353 (1982).
67. G. Maier and W. A. Jung, *Chem. Ber.*, **115**, 804 (1982).
68. J. Nakayama, A. Sakai, and M. Hoshino, *J. Org. Chem.*, **49**, 5084 (1984).
69. K. B. Wiberg and G. Bonneville, *Tetrahedron Lett.*, **23**, 5385 (1982).
70. H.-D. Martin, A. Oftring, R. Iden, E. Schwichtenberg, and H.-J. Schiwek, *Tetrahedron Lett.*, **23**, 841 (1982).
71. P. Brownbridge and T.-H. Chan, *Tetrahedron Lett.*, **21**, 3423 (1980).
72. M. F. Ansell, M. P. L. Caton, and P. C. North, *Tetrahedron Lett.*, **23**, 2811 (1982).
73. N. E. Jung, L. J. Street, and Y. Usui, *J. Am. Chem. Soc.*, **108**, 6810 (1986).
74. J. Jurczak, A. L. Kawczynski, and T. Kozluk, *J. Org. Chem.*, **50**, 1106 (1985).

75. K. Matsumoto, Y. Ikemi, S. Hashimoto, H. S. Lee, and Y. Okamoto, *J. Org. Chem.*, **51**, 3729 (1986).
76. B. Albert, C. Heller, R. Iden, G. Martin, H.-D. Martin, B. Mayer, and A. Oftring, *Isr. J. Chem.*, **25**, 74 (1985).
77. M. Avenati, J.-P. Hagenbuch, C. Mahaim, and P. Vogel, *Tetrahedron Lett.*, **21**, 3167 (1980).
78. L. A. Paquette, K. E. Green, R. Gleiter, W. Schafer, and J. C. Gallucci, *J. Am. Chem. Soc.*, **106**, 8232 (1984).
79. A. A. Fatmi, N. A. Vaidya, W. B. Iturrian, and C. De Witt Blanton, Jr., *J. Med. Chem.*, **27**, 772 (1984).
80. A.-G. Bayer and A.-G. Degussa, *Belg. BE*, 893,636 (1982); *Chem. Abstr.*, **99**, 22295w (1983).
81. W. J. Nixon, Jr., J. T. Garland, and C. De Witt Blanton, Jr., *Synthesis*, 56 (1980).
82. M. V. Gorelik and R. A. Alimova, *Zh. Org. Khim.*, **20**, 818 (1984).
83. G. W. Gribble, M. G. Saulnier, M. P. Sibi, and J. A. Obaza-Nutaitis, *J. Org. Chem.*, **49**, 4518 (1984).
84. H. Kotsuki, H. Nishizawa, S. Kitagawa, M. Ochi, N. Yamasaki, K. Matsuoka, and T. Tokoroyama, *Bull. Chem. Soc. Jpn.*, **52**, 544 (1979).
85. D. Del Mazza and M. G. Reinecke, *Heterocycles*, **14**, 647 (1980).
86. M. S. Raasch, *J. Org. Chem.*, **45**, 867 (1980).
87. Y. N. Gupta and K. N. Houk, *Tetrahedron Lett.*, **26**, 2607 (1985).
88. M. S. Raasch, *J. Org. Chem.*, **45**, 856 (1980).
89. O. Meth-Cohn and G. van Vuuren, *J. Chem. Soc. Chem. Commun.*, 1144 (1984).
90. J. Bornstein and R. P. Hardy, *J. Chem. Soc. Chem. Commun.*, 612 (1980).
91. L.-T. Wilson Lin and H. Hart, *J. Org. Chem.*, **49**, 1027 (1984).
92. A. Laporterie, J. Dubac, P. Mazerolles, and H. Iloughmane, *J. Organomet. Chem.*, **216**, 321 (1981).
93. B. Mayer and W. P. Neumann, *Tetrahedron Lett.*, **21**, 4887 (1980).
94. M. S. Holt, J. H. Nelson, P. Savignac, and N. W. Alcock, *J. Am. Chem. Soc.*, **107**, 6396 (1985).
95. H. Tomioka, S. Miura, and Y. Izawa, *Tetrahedron Lett.*, **24**, 3353 (1983).
96. K. Matsumoto, S. Hashimoto, and S. Otani, *Heterocycles*, **22**, 2713 (1984).
97. L. D. Quin, J. Szewczyk, K. M. Szewczyk, and A. T. McPhail, *J. Org. Chem.*, **51**, 3341 (1986).
98. G. De Lauzon, C. Charrier, H. Bonnard, and F. Mathey, *Tetrahedron Lett.*, **23**, 511 (1982).
99. C. Charrier, H. Bonnard, G. De Lauzon, and F. Mathey, *J. Am. Chem. Soc.*, **105**, 6871 (1983).
100. G. De Lauzon, C. Charrier, H. Bonnard, F. Mathey, J. Fisher, and A. Mitschler, *J. Chem. Soc. Chem. Commun.*, 1272 (1982).
101. A. Laporterie, G. Manuel, J. Dubac, P. Mazerolles, and H. Iloughmane, *J. Organomet. Chem.*, **210**, C-33 (1981).
102. W. P. Neumann and M. Schriewer, *Tetrahedron Lett.*, **21**, 3273 (1980).
103. P. H. Daniels, J. L. Wong, J. L. Atwood, L. G. Canada, and P. D. Rogers, *J. Org. Chem.*, **45**, 435 (1980).

104. M. E. Jung and J. J. Shapiro, *J. Am. Chem. Soc.*, **102**, 7862 (1980).
105. P. Marsh and D. E. Wiliams, *Acta Crystallogr.*, *Sect. B*, **B 35**, 2241 (1979).
106. G. W. Gribble and C. S. Le Houllier, *Tetrahedron Lett.*, **22**, 903 (1981).
107. V. A. Kovtunenko, Z. V. Voitenko, A. K. Tyltin, A. V. Turov, and F. S. Babichev, *Ukr. Khim. Zh.*, **49**, 1287 (1983); *Chem. Abstr.*, **100**, 173943w (1984).
108. H.-G. Zoch, A.-D. Schlüter, and G. Szeimies, *Tetrahedron Lett.*, **22**, 3835 (1981).
109. T. Katada, S. Eguchi, and T. Sasaki, *J. Org. Chem.*, **51**, 314 (1986).
110. R. Kreher, N. Kohl, and G. Use, *Angew. Chem.* (*Int. Ed. Engl.*), **21**, 621 (1982).
111. G. Use and R. Kreher, *Chem. -Ztg.*, **106**, 143 (1982).
112. R. Kreher and D. J. Herd, *Chem. -Ztg.*, **106**, 305 (1982).
113. G. W. Gribble, C. S. Le Houllier, M. P. Sibi, and R. W. Allen, *J. Org. Chem.*, **50**, 1611 (1985).
114. R. Kreher and G. Use, *Tetrahedron Lett.*, 4671 (1978).
115. J. G. Smith and P. W. Dibble, *J. Org. Chem.*, **48**, 5361 (1983).
116. J.P. McCormick and T. Shinmyozu, *J. Org. Chem.*, **47**, 4011 (1982).
117. M. A. Makhlouf and B. Rickborn, *J. Org. Chem.*, **46**, 2734 (1981).
118. C.-W. Chen and P. Beak, *J. Org. Chem.*, **51**, 3325 (1986).
119. L. Contreras and D. B. MacLean, *Can. J. Chem.*, **58**, 2580 (1980).
120. J. G. Smith, S. S. Welankiwar, N. G. Chu, E. H. Lai, and S. J. Sondheimer, *J. Org. Chem.*, **46**, 4658 (1981).
121. L. A. Levy and S. Kumar V. P., *Tetrahedron Lett.*, **24**, 1221 (1983).
122. I. Saito, A. Nakata, and T. Matsuuza, *Tetrahedron Lett.*, **22**, 1697 (1981).
123. D. W. Jones, *J. Chem. Soc. Chem. Commun.*, 766, (1982).
124. B. Miller and X. Shi, *J. Am. Chem. Soc.*, **109**, 578 (1987).
125. a. H. J. Reich, W. W. Willis, Jr., and P. D. Clark, *J. Org. Chem.*, **46**, 2775 (1981).
     b. M. M. Latypova, V. V. Plemenkov, V. B. Tuzova, Kh. Z. Giniyatov, and I. G. Bolesov, *Zh. Org. Khim.*, **18**, 1650 (1982).
126. T. C. Sedergran, M. Yokoyama, and D. Dittmer, *J. Org. Chem.*, **49**, 2408 (1984).
127. T.-L. Chan, N. Z. Huang, and F. Sondheimer, *Tetrahedron*, **39**, 427 (1983).
128. D. W. Jones, A. Pomfret, and R. L. Wife, *J. Chem. Soc. Perkin Trans. 1*, 459 (1983).
129. V. V. Plemenkov and L. V. Katerinich, *Zh. Org. Khim.*, **18**, 835 (1982).
130. L. M. Tolbert and B. M. Ali, *J. Am. Chem. Soc.*, **106**, 3806 (1984).
131. T. Sasaki, K. Hayakawa, T. Manabe, and S. Nishida, *J. Am. Chem. Soc.*, **103**, 565 (1981).
132. G. Kaupp, H.-W. Grüter, and E. Teufel, *Chem. Ber.*, **116**, 618 (1983).
133. P. F. Newton and G. H. Whitham, *J. Chem. Soc. Perkin Trans. 1*, 3067 (1979).
134. O. Tsuge, T. Ohnishi, and H. Watanabe, *Heterocycles*, **16**, 2085 (1981).
135. P. Yates, and S. P. Douglas, *Can. J. Chem.*, **60**, 2760 (1982).
136. K. B. Becker and J. L. Chappuis, *Helv. Chim. Acta*, **62**, 34 (1979).
137. V. D. Kiselev, D. G. Khuzyasheva, I. M. Shakirov, and A. I. Konovalov, *Zh. Org. Khim.*, **19**, 2064 (1983).
138. V. V. Plemenkov, Kh. Z. Giniyatov, Ya. Ya. Villem, N. V. Villem, L. S. Surmina, and I. G. Bolesov, *Dokl. Akad. Nauk SSSR*, **254**, 895 (1980).

139. W. R. Dolbier, Jr., M. Seabury, D. Daly, and B. E. Smart, *J. Org. Chem.*, **51**, 974 (1986).
140. L. Contreras and D. B. MacLean, *Can. J. Chem.*, **58**, 2573 (1980).
141. B. A. Keay and R. Rodrigo, *Can. J. Chem.*, **63**, 735 (1985).
142. R. Rodrigo and S. M. Knabe, *J. Org. Chem.*, **51**, 3973 (1986).
143. a. H. P. Plaumann, J. G. Smith, and R. Rodrigo, *J. Chem. Soc. Chem. Commun.*, 354 (1980).
   b. B. A. Keay, H. P. Plaumann, D. Rajapaksa, and R. Rodrigo, *Can. J. Chem.*, **61**, 1987 (1983).
144. D. Villessot and Y. Lepage, *J. Chem. Res. (S)*, 302 (1979).
145. M. Iwao, H. Inone, and T. Kuraishi, *Chem. Lett.*, 1263 (1984).
146. D. W. Jones, A. Pomfret, and R. L. Wife, *J. Chem. Soc. Chem. Commun.*, 463 (1980).
147. B. Mir-Mohamed-Sadeghy and B. Rickborn, *J. Org. Chem.*, **48**, 2237 (1983).
148. J. J. Cornejo, S. Ghodsi, R. D. Johnson, R. Woodling, and B. Rickborn, *J. Org. Chem.*, **48**, 3869 (1983).
149. D. J. Pollart and B. Rickborn, *J. Org. Chem.*, **51**, 3155 (1986).
150. T. Chiba, P. W. Kenny, and L. L. Miller, *J. Org. Chem.*, **52**, 4327 (1987).
151. H. Gotou, K. Kurata, Y. Tominaga, and Y. Matsuda, *J. Org. Chem.*, **50**, 4028 (1985).
152. G. Maier, J. Henkelmann, and H. P. Reisenauer, *Angew. Chem. (Int. Ed. Engl.)*, **24**, 1065 (1985).
153. B. Weinstein, L. C. Chin, and F. W. Fowler, *J. Org. Chem.*, **45**, 1657 (1980).
154. P. A. Wender, J. M. Schaus, and D. C. Torney, *Tetrahedron Lett.*, 2485 (1979).
155. S. K. Dubey and E. E. Knaus, *J. Org. Chem.*, **50**, 2080 (1985).
156. a. A. V. Pavlov and V. B. Mochalin, *Zh. Org. Khim.*, **19**, 234 (1983).
   b. J. Streith, G. Augelmann, H. Fritz, and H. Strub, *Tetrahedron Lett.*, **23**, 1909 (1982).
   c. E. E. Knaus, K. Avasthi, and C. S. Gian, *Can. J. Chem.*, **58**, 2447 (1980).
   d. G. Augelmann, J. Streith, and H. Fritz, *Helv. Chim. Acta*, **68**, 95 (1985).
157. R. J. Sundberg and J. D. Bloom, *J. Org. Chem.*, **45**, 3382 (1980).
158. R. J. Sundberg and J. D. Bloom, *J. Org. Chem.*, **46**, 4836 (1981).
159. A. V. Pavlov and V. B. Mochalin, *Zh. Org. Khim.*, **19**, 234 (1983).
160. O. Diels and K. Alder, *Chem. Ber.*, **62**, 544 (1929); *Liebigs Ann. Chem.*, **490**, 243 (1931).
161. G. R. Krow, J. T. Carey, D. E. Zacharias, and E. E. Knaus, *J. Org. Chem.*, **47**, 1989 (1982).
162. R. J. Sundberg, G. Hamilton, and C. Trindle, *J. Org. Chem.*, **51**, 3672 (1986).
163. R. A. Joshi, N. K. Ponkshe, and T. Ravindranathan, *Indian J. Chem. Sect. B*, **23 B**, 263 (1984).
164. R. M. Acheson and G. Paglietti, *J. Chem. Soc. Perkin Trans. 1*, 591 (1979).
165. S. Raucher and R. F. Lawrence, *Tetrahedron Lett.*, **24**, 2927 (1983).
166. G. R. Krow, K. C. Cannon, J. T. Carey, Y. B. Lee, S. W. Szczepanski, and H. G. Ramjit, *J. Heterocycl. Chem.*, **22**, 131 (1985).

167. G. W. Kirby, H. McGuigan, S. L. Tan, and B. C. Uff, *J. Chem. Res.* (*S*), 273 (1985).

168. G. Hussmann, W. D. Wulff, and T. J. Barton, *J. Am. Chem. Soc.*, **105**, 1263 (1983).

169. H. Tomisawa, H. Hongo, H. Kato, T. Naraki, and R. Fujita, *Chem. Pharm. Bull.*, **27**, 670 (1979).

170. G. G. Arsenault, K. Jankowski, and E. Luce, *Nouv. J. Chim.*, **5**, 79 (1981).

171. K. Matsumoto, Y. Ikemi-Kono, T. Uchida, and R. M. Achenson, *J. Chem. Soc. Chem. Commun.*, 1091 (1979).

172. H. Tomisawa, H. Hongo, H. Kato, and R. Fujita, *Heterocycles*, **16**, 1947 (1981).

173. P. S. Mariano, P. L. Huesmann, R. L. Beamer, and D. Dunaway-Mariano, *Tetrahedron*, **34**, 2617 (1978).

174. K. Matsumoto, Y. Ikemi, and S. Nakamura, *Heterocycles*, **19**, 499 (1982).

175. F. W. Muellner and L. Bauer, *J. Heterocycl. Chem.*, **20**, 1581 (1983).

176. G. H. Posner and C. Switzer, *J. Org. Chem.*, **52**, 1642 (1987).

177. V. S. Pilipenko and N. P. Shusherina, *Zh. Org. Khim.*, **17**, 2122 (1981); E.E., 1891 (1982).

178. a. M. Kuzuya, A. Noguchi, S. Kamiya, and T. Okuda, *Chem. Pharm. Bull.*, **33**, 2313 (1985).
   b. M. Kuzuya, A. Noguchi, E. Mano, and T. Okuda, *Bull. Chem. Soc. Jpn.*, **58**, 1149 (1985).

179. G. P. Gisby, S. E. Royall, and P. G. Sammes, *J. Chem. Soc. Perkin Trans. 1*, 169 (1982).

180. a. L. Hazai, Gy. Deak, G. Toth, A. Schnitta, A. Szollosy, and J. Tamas, *Acta Chim. Hung.*, **113**, 237 (1983).
   b. L. Hazai, A. Schnitta, Gy. Deak, G. Toth, and A. Szollosy, *Acta Chim. Hung.*, **117**, 99 (1984).

181. N. P. Shusherina, V. S. Pilipenko, O. K. Kireeva, B. I. Geller, and A. U. Stepanyants, *Zh. Org. Khim.*, **16**, 2390 (1980).

182. V. S. Pilipenko, *Viniti*, 3782, 172 (1979); *Chem. Abstr.*, **94**, 139657d (1981).

183. N. P. Shusherina, V. S. Pilipenko and L. V. Betaneli, *Zh. Org. Khim.*, **15**, 1277 (1979).

184. V. A. Kovtunenko, T. T. Kobrenko, Z. V. Voitenko, A. K. Tyltin, and F. S. Babichev, *Khim. Geterotsikl. Soedin.*, 1200 (1984).

185. E. P. Kyba, S. P. Rines, P. W. Owen, and S. S. P. Chou, *Tetrahedron Lett.*, **22**, 1875 (1981).

186. T. Shimo, K. Somekawa, and S. Kumamoto, *Nippon Kagaku Kaishi*, (12), 1927 (1982); *Chem. Abstr.*, **98**, 125817b (1983).

187. T. L. Nesterova, N. P. Shusherina, and E. V. Shulishov, *Zh. Org. Khim.*, **21**, 427 (1985).

188. P. Cano, A. Echavarren, P. Prados, and F. Farina, *J. Org. Chem.*, **48**, 5373 (1983).

189. J. D. Warren, V. J. Lee, and R. B. Angier, *J. Heterocycl. Chem.*, **16**, 1617 (1979).

190. G. H. Posner and D. G. Wettlaufer, *Tetrahedron Lett.*, **27**, 667 (1986).

191. K. Tamao, J. Yoshida, M. Akita, Y. Sugihara, T. Iwahara, and M. Kumada, *Bull. Chem. Soc. Jpn.*, **55**, 255 (1982).

192. T. Imagawa, T. Nakagawa, M. Kawanisi, and K. Sisido, *Bull. Chem. Soc. Jpn.*, **52**, 1506 (1979).
193. M. Christl and S. Freund, *Chem. Ber.*, **118**, 979 (1985).
194. P. Martin, J. Streith, G. Rihs, T. Winkler, and D. Bellus, *Tetrahedron Lett.*, **26**, 3947 (1985).
195. A. P. Kozikowski and R. Schmiesing, *Tetrahedron Lett.*, 4241 (1978).
196. D. L. Boger and M. D. Mullican, *Tetrahedron Lett.*, **24**, 4939 (1983).
197. H. L. Gingrich, D. M. Roush, and W. A. Van Saun, *J. Org Chem.*, **48**, 4869 (1983).
198. K. D. Showa and M. K. Ishikawa, *Jpn. Kokai Tokyo Koho, JP* 80, 85,544 (1980); *Chem. Abstr.*, **94**, 46980p (1981).
199. S. M. Verma and H. Maurya, *Indian J. Chem. Sect. B*, **24 B**, 447 (1985).
200. M. E. Jung and R. W. Brown, *Tetrahedron Lett.*, **22**, 3355 (1981).
201. D. L. Boger and M. D. Mullican, *Tetrahedron Lett.*, **23**, 4551 (1982).
202. S. D. Burke, J. O. Saunders, and C. W. Murtiashaw, *J. Org. Chem.*, **46**, 2425 (1981).
203. D. L. Boger and M. D. Mullican, *Tetrahedron Lett.*, **23**, 4555 (1982).
204. D. L. Boger and M. D. Mullican, *J. Org. Chem.*, **49**, 4033 (1984).
205. a. C. J. Moody, *J. Chem. Soc. Chem. Commun.*, 925 (1984).
   b. C. May and C. J. Moody, *J. Chem. Soc. Chem. Commun.*, 926 (1984).
206. D. L. Boger and C. E. Brotherton, *J. Org. Chem.*, **49**, 4050 (1984).
207. A. J. Ashe, T. R. Diephouse, and M. Y. El-Sheikh, *J. Am. Chem. Soc.*, **104**, 5693 (1982).
208. W. Eberbach and J. C. Carré, *Tetrahedron Lett.*, **21**, 1145 (1980).
209. W. Eberbach, G. Koenig, and V. Trostmann, *Tetrahedron Lett.*, 4649 (1979).
210. W. H. Rastetter and T. J. Richard, *Tetrahedron Lett.*, 2999 (1978).
211. K. Beck, A. Hohn, S. Hunig, and F. Prokschy, *Chem. Ber.*, **117**, 517 (1984).
212. H. Tanaka and S. Motoki, *Bull. Chem. Soc. Jpn.*, **59**, 2047 (1986).
213. M. R. Bryce, R. M. Achenson, and A. J. Rees, *Heterocycles*, **20**, 489 (1983).
214. W. Böll and H. König, *Liebigs Ann. Chem.*, 1657 (1979).
215. a. D. Liotta, M. Saindane, and W. Ott, *Tetrahedron Lett.*, **24**, 2473 (1983).
   b. H. J. Reich, J. M. Kelly, R. E. Olson, and R. C. Holtan, *Tetrahedron*, **39**, 949 (1983).
216. M. A. Aitzhanova and G. Ya. Kondrat'eva, *Izv. Akad. Nauk SSSR, Ser. Khim.*, (10), 2340 (1982).
217. B. A. Johnsen and K. Undheim, *Acta Chem. Scand., Ser. B*, **B 37**, 127 (1983).
218. a. T. Ibata, S. Nakano, H. Nakawa, J. Toyoda, and Y. Isogami, *Bull. Chem. Soc. Jpn.*, **59**, 433 (1986).
   b. T. Ibata, H. Nakawa, Y. Isogami, and K. Matsumoto, *Bull. Chem. Soc. Jpn.*, **59**, 3197 (1986).
219. G. A. Tolstikov and É. É. Shul'ts, *Zh. Org. Khim.*, **20**, 2230 (1984); E.E., 2032 (1984).
220. K. Akiba, A. Ohtani, and Y. Yamamoto, *J. Org. Chem.*, **51**, 5328 (1986).

221. Y. Tamura, M. Sasho, K. Nakagawa, T. Tsugoshi, and Y. Kita, *J. Org. Chem.*, **49**, 473 (1984).

222. W. Eberbach, E. Hädicke, and U. Trostmann, *Tetrahedron Lett.*, **22**, 4953 (1981).

223. A. R. Katritzky and N. Dennis, *Chem. Rev.*, **89**, 827 (1989).

224. S. Hunig and F. Prokschy, *Chem. Ber.*, **117**, 534 (1984).

225. R. J. Sundberg and J. D. Bloom, *Tetrahedron Lett.*, 5157 (1978).

226. W. D. Wulff and D. C. Yang, *J. Am. Chem. Soc.*, **105**, 6726 (1983).

227. J. D. Rich and R. West, *J. Am. Chem. Soc.*, **104**, 6884 (1982).

228. H.-J. Bestmann and K. Roth, *Angew. Chem. (Int. Ed. Engl.)*, **20**, 575 (1981).

229. N. Yasuhiro, T. Hiromi, and S. Hideki, *New Front. Organomet. Inorg. Chem., Proc. China–Jpn.–USA, Trilateral Seminar*, **2** (1982); *Chem. Abstr.*, **102**, 185188h (1985).

230. D. W. Borhani and F. D. Greene, *J. Org. Chem.*, **51**, 1563 (1986).

231. A. Padwa and P. Eisenbarth, *Tetrahedron Lett.*, **25**, 5489 (1984).

232. A. Padwa and P. Eisenbarth, *Tetrahedron*, **41**, 283 (1985).

233. M. Christl, U. Lanzendörfer, and F. Silke, *Angew. Chem. (Int. Ed. Engl)*, **20**, 674 (1981).

234. A. Hasnaoni, M. El Messaoudi, and J.-P. Lavergne, *J. Heterocycl. Chem.*, **22**, 25 (1985).

235. a. V. N. Charushin and H. C. van der Plas, *Tetrahedron Lett.*, **23**, 3965 (1982).
b. H. C. van der Plas, A. T. M. Marcelis, D. M. W. van der Ham, and J. W. Verhoeven, *J. Org. Chem.*, **51**, 4070 (1986).

236. J. C. Martin, *J. Heterocycl. Chem.*, **17**, 1111 (1980).

237. M. Tutonda, D. Vanderzande, J. Vekemans, S. Toppet, and G. Hoornaert, *Tetrahedron Lett.*, **27**, 2509 (1986).

238. T. Itoh, A. Ohsawa, M. Okada, T. Kaihoh, and H. Igeta, *Chem. Pharm. Bull.*, **33**, 3050 (1985).

239. D. L. Boger and J. S. Panek, *J. Am. Chem. Soc.*, **107**, 5745 (1985).

240. M. L. Maddox, J. C. Martin, and J. M. Muchowski, *Tetrahedron Lett.*, **21**, 7 (1980).

241. a. D. L. Boger and J. S. Panek, *J. Org. Chem.*, **46**, 2179 (1981).
b. D. L. Boger, J. S. Panek, and M. M. Meier, *J. Org. Chem.*, **47**, 895 (1982).

242. E. C. Taylor and J. E. Macor, *Tetrahedron Lett.*, **26**, 2415 (1985).

243. L. Maggiora and M. P. Mertes, *J. Org. Chem.*, **51**, 950 (1986).

244. R. Dhar, W. Hühnermann, T. Kämpchen, W. Overheu, and G. Seitz, *Chem. Ber.*, **116**, 97 (1983).

245. H. Neunhoeffer and H.-J. Metz, *Liebigs Ann. Chem.*, 1476 (1983).

246. M. G. Barlow, R. N. Haszeldine, and D. J. Simpkin, *J. Chem. Soc. Chem. Commun.*, 658 (1979).

247. M. G. Barlow, R. N. Haszeldine, and D. J. Simpkin, *J. Chem. Soc. Perkin Trans. 1*, 1245 (1982).

248. G. Seitz, R. Mohr, W. Overheu, R. Allmann, and M. Nagel, *Angew. Chem. (Int. Ed. Engl)*, **23**, 890 (1984).

249. G. Seitz, R. Dhar, R. Mohr, and W. Overheu, *Arch. Pharm.*, **317**, 237 (1984).
250. M. G. Barlow, R. N. Haszeldine, and D. J. Simpkin, *J. Chem. Soc. Chem. Commun.*, 658 (1979).
251. H. Gnichtel and C. Gumprecht, *Liebigs Ann. Chem.*, 628 (1985).
252. L. Birkofer and E. Hansel, *Chem. Ber.*, **114**, 3154 (1981).
253. H. Mohrle and H. Dwuletzki, *Chem. Ber.*, **119**, 3600 (1986).
254. a. D. L. Boger and R. S. Coleman, *J. Org. Chem.*, **51**, 3250 (1986).
b. D. L. Boger and R. S. Coleman, *J. Am. Chem. Soc.*, **109**, 2717 (1987).
255. D. L. Boger and J. S. Panek, *Tetrahedron Lett.*, **24**, 4511 (1983).
256. G. Seitz and T. Kämpchen, *Arch. Pharm.*, **311**, 728 (1978).
257. U. Berger, G. Dannhardt, and R. Obergrusberger, *Arch. Pharm.*, **315**, 428 (1982).
258. S. C. Benson, C. A. Palabrica, and J. K. Snyder, *J. Org. Chem.*, **52**, 4610 (1987).
259. D. L. Boger, R. S. Coleman, J. S. Panek, and D. Yohannes, *J. Org. Chem.*, **49**, 4405 (1984).
260. G. Seitz and W. Overheu, *Chem. -Ztg.*, **103**, 230 (1979).
261. G. Seitz and W. Overheu, *Arch. Pharm.*, **314**, 376 (1981).
262. T. Molz, P. Konig, R. Goes, G. Gauglitz, and H. Meier, *Chem. Ber.*, **117**, 833 (1984).
263. H. P. Figeys, A. Mathy, and A. Dralants, *Synth. Commun.*, **11**, 655 (1981).
264. H. P. Figeys and A. Mathy, *Tetrahedron Lett.*, **22**, 1393 (1981).
265. M. N. Paddon-Row, K. H. Patney, and R. N. Warrener, *J. Org. Chem.*, **44**, 3908 (1979).
266. A. T. M. Marcelis and H. C. Van der Plas, *Heterocycles*, **23**, 683 (1985).
267. R. B. Gupta and R. W. Franck, *J. Am. Chem. Soc.*, **109**, 5393 (1987).
268. J. H. Bateson, R. Southgate, J. W. Tyler, and S. C. M. Fell, *J. Chem. Soc. Perkin Trans. 1*, 973 (1986).
269. R. J. Ardecky, F. A. J. Kerdesky, and M. P. Cava, *J. Org. Chem.*, **46**, 1483 (1981).
270. C.-Y. Tsai and C.-K. Sha, *Tetrahedron Lett.*, **28**, 1419 (1987).
271. E. C. Taylor, K. F. McDaniel, and J. C. Warner, *Tetrahedron Lett.*, **28**, 1977 (1987).
272. R. H. Fleming and B. M. Murray, *J. Org. Chem.*, **44**, 2280 (1979).
273. H. J. Reich, J. M. Kelly, R. E. Olson, and R. C. Holtan, *Tetrahedron*, **39**, 949 (1983).
274. P. Warner and S.-C. Chang, *Tetrahedron Lett.*, 4141 (1979).
275. W. E. Doering, M. F. Neumann, E. Hasselmann, and R. Kaye, *J. Am. Chem. Soc.*, **94**, 3833 (1972).
276. M. W. Lee and W. C. Herndon, *J. Org. Chem.*, **43**, 518 (1978).
277. W. Dauben, C. R. Kessel, and K. H. Takemura, *J. Am. Chem. Soc.*, **102**, 6893 (1980).
278. J. Hamer, Ed., "1,4-Cycloadditions Reactions: The Diels–Alder Reaction in Heterocyclic Syntheses", Academic Press, New York (1967).
279. R. Breslow, A. W. Czarnik, M. Lauer, R. Leppkes, J. Winkler, and S. Zimmermann, *J. Am. Chem. Soc.*, **108**, 1969 (1986).

280. N. Abe, T. Nishiwaki, and N. Komoto, *Bull. Chem. Soc. Jpn.*, **53**, 3308 (1980); *Chem. Lett.*, 223 (1980).

281. D. Tobia and B. Rickborn, *J. Org. Chem.*, **51**, 3849 (1986).

282. D. L. Boger, S. R. Duff, J. S. Panek, and M. Yasuda, *J. Org. Chem.*, **50**, 5790 (1985).

283. D. Tobia and B. Rickborn, *J. Org. Chem.*, **52**, 2611 (1987).

284. R. Gompper and U. Heinemann, *Angew. Chem.* (*Int. Ed. Engl.*), **19**, 216, 217 (1980).

285. D. L. Boger, S. R. Duff, J. S. Panek, and M. Yasuda, *J. Org. Chem.*, **50**, 5782 (1985).

286. T. Sugita, J. Koyama, K. Tagahara, and Y. Suzuta, *Heterocycles*, **24**, 29 (1986).

287. T. Sugita, J. Koyama, K. Tagahara, and Y. Suzuta, *Heterocycles*, **23**, 2789 (1985).

288. D. L. Boger, *Tetrahedron*, **39**, 2869 (1983).

289. D. L. Boger and M. Patel, *J. Org. Chem.*, **52**, 3934 (1987).

290. H. Neunhoeffer and G. Werner, *Liebigs Ann. Chem.*, 437 (1973).

291. A. T. M. Marcelis and H. C. van der Plas, *J. Org. Chem.*, **51**, 67 (1986).

292. D. A. de Bie, B. Geurtsen, and H. C. van der Plas, *J. Org. Chem.*, **51**, 71 (1986).

293. D. L. Boger, J. Schumacher, M. D. Mullican, M. Patel, and J. S. Panek, *J. Org. Chem.*, **47**, 2673 (1982).

INDEX

T = Table